a Glance

D0774369

BARTS & THE LONDON QMSMD	
CL⠀⠀⠀⠀	QT17 DES 2003
CIRC ⠀PE	OWL
SUPPLIER	ML - 20/7/06
READING LIST	
OLD ⠀⠀ECK	

QM Library

23 1286181 7

👑 Barts and The London
Queen Mary's School of Medicine and Dentistry

WHITECHAPEL LIBRARY, TURNER STREET, LOND

020 7882

ONE WEEK

are to be

of Physiology

5th edition, completely revised
and expanded

Agamemnon Despopoulos, M.D.
Professor
Formerly: Ciba Geigy
Basel

Stefan Silbernagl, M.D.
Professor
Head of Department
Institute of Physiology
University of Wuerzburg
Wuerzburg, Germany

186 color plates by
Ruediger Gay and
Astried Rothenburger

Thieme
Stuttgart · New York

Library of Congress Cataloging-in-Publication Data

is available from the publisher

1st German edition 1979	1st Czech edition 1984
2nd German edition 1983	2nd Czech edition 1994
3rd German edition 1988	
4th German edition 1991	1st French edition 1985
5th German edition 2001	2nd French edition 1992
	3rd French edition 2001
1st English edition 1981	
2nd English edition 1984	1st Turkish edition 1986
3rd English edition 1986	2nd Turkish edition 1997
4th English edition 1991	1st Greek edition 1989
1st Dutch edition 1981	1st Chinese edition 1991
2nd Dutch edition 2001	
	1st Polish edition 1994
1st Italian edition 1981	
2nd Italian edition 2001	1st Hungarian edition 1994
	2nd Hungarian edition 1996
1st Japanese edition 1982	
2nd Japanese edition 1992	1st Indonesion edition 2000
1st Spanish edition 1982	
2nd Spanish edition 1985	
3rd Spanish edition 1994	
4th Spanish edition 2001	

This book is an authorized translation of the 5th German edition published and copyrighted 2001 by Georg Thieme Verlag, Stuttgart, Germany.
Title of the German edition:
Taschenatlas der Physiologie

Translated by Suzyon O'Neal Wandrey, Berlin, Germany

Illustrated by Atelier Gay + Rothenburger, Sternenfels, Germany

© 1981, 2003 Georg Thieme Verlag
Rüdigerstraße 14, D-70469 Stuttgart, Germany
http://www.thieme.de
Thieme New York, 333 Seventh Avenue,
New York, N.Y. 10001, U.S.A.
http://www.thieme.com

Cover design: Cyclus, Stuttgart
Typesetting by: Druckhaus Götz GmbH,
    Ludwigsburg, Germany
Printed in Germany by: Appl Druck
    GmbH & Co. KG, Wemding, Germany

IV

ISBN 3-13-545005-8 (GTV)
ISBN 1-58890-061-4 (TNY)   1 2 3 4 5

Important Note: Medicine is an ever-changing science undergoing continual development. Research and clinical experience are continually expanding our knowledge, in particular our knowledge of proper treatment and drug therapy. Insofar as this book mentions any dosage or application, readers may rest assured that the authors, editors, and publishers have made every effort to ensure that such references are in accordance with **the state of knowledge at the time of production of the book**.

Nevertheless, this does not involve, imply, or express any guarantee or responsibility on the part of the publishers in respect to any dosage instructions and forms of applications stated in the book. **Every user is requested to examine carefully** the manufacturers' leaflets accompanying each drug and to check, if necessary in consultation with a physician or specialist, whether the dosage schedules mentioned therein or the contraindications stated by the manufacturers differ from the statements made in the present book. Such examination is particularly important with drugs that are either rarely used or have been newly released on the market. Every dosage schedule or every form of application used is entirely at the user's own risk and responsibility. The authors and publishers request every user to report to the publishers any discrepancies or inaccuracies noticed.

Some of the product names, patents, and registered designs referred to in this book are in fact registered trademarks or proprietary names even though specific reference to this fact is not always made in the text. Therefore, the appearance of a name without designation as proprietary is not to be construed as a representation by the publisher that it is in the public domain.

This book, including all parts thereof, is legally protected by copyright. Any use, exploitation, or commercialization outside the narrow limits set by copyright legislation, without the publisher's consent, is illegal and liable to prosecution. This applies in particular to photostat reproduction, copying, mimeographing or duplication of any kind, translating, preparation of microfilms, and electronic data processing and storage.

QIVI LIBRARY
(MILE END)

Preface to the Fifth Edition

The base of knowledge in many sectors of physiology has grown considerably in magnitude and in depth since the last edition of this book was published. Many advances, especially the rapid progress in sequencing the human genome and its gene products, have brought completely new insight into cell function and communication. This made it necessary to edit and, in some cases, enlarge many parts of the book, especially the chapter on the fundamentals of cell physiology and the sections on neurotransmission, mechanisms of intracellular signal transmission, immune defense, and the processing of sensory stimuli. A list of physiological reference values and important formulas were added to the appendix for quick reference. The extensive index now also serves as a key to abbreviations used in the text.

Some of the comments explaining the connections between pathophysiological principles and clinical dysfunctions had to be slightly truncated and set in smaller print. However, this base of knowledge has also grown considerably for the reasons mentioned above. To make allowances for this, a similarly designed book, the *Color Atlas of Pathophysiology* (S. Silbernagl and F. Lang, Thieme), has now been introduced to supplement the well-established *Color Atlas of Physiology*.

I am very grateful for the many helpful comments from attentive readers (including my son Jakob) and for the welcome feedback from my peers, especially Prof. H. Antoni, Freiburg, Prof. C. von Campenhausen, Mainz, Dr. M. Fischer, Mainz, Prof. K.H. Plattig, Erlangen, and Dr. C. Walther, Marburg, and from my colleagues and staff at the Institute in Würzburg. It was again a great pleasure to work with Rüdiger Gay and Astried Rothenburger, to whom I am deeply indebted for revising practically all the illustrations in the book and for designing a number of new color plates. Their extraordinary enthusiasm and professionalism played a decisive role in the materialization of this new edition. To them I extend my sincere thanks. I would also like to thank Suzyon O'Neal Wandrey for her outstanding translation. I greatly appreciate her capable and careful work. I am also indebted to the publishing staff, especially Marianne Mauch, an extremely competent and motivated editor, and Gert Krüger for invaluable production assistance. I would also like to thank Katharina Völker for her ever observant and conscientious assistance in preparing the index.

I hope that the 5th Edition of the *Color Atlas of Physiology* will prove to be a valuable tool for helping students better understand physiological correlates, and that it will be a valuable reference for practicing physicians and scientists, to help them recall previously learned information and gain new insights in physiology.

Würzburg, December 2002
*Stefan Silbernagl**

---

* e-mail: stefan.silbernagl@mail.uni-wuerzburg.de

Preface to the First Edition

In the modern world, visual pathways have outdistanced other avenues for informational input. This book takes advantage of the economy of visual representation to indicate the simultaneity and multiplicity of physiological phenomena. Although some subjects lend themselves more readily than others to this treatment, inclusive rather than selective coverage of the key elements of physiology has been attempted.

Clearly, this book of little more than 300 pages, only half of which are textual, cannot be considered as a primary source for the serious student of physiology. Nevertheless, it does contain most of the basic principles and facts taught in a medical school introductory course. Each unit of text and illustration can serve initially as an overview for introduction to the subject and subsequently as a concise review of the material. The contents are as current as the publishing art permits and include both classical information for the beginning students as well as recent details and trends for the advanced student.

A book of this nature is inevitably derivative, but many of the representations are new and, we hope, innovative. A number of people have contributed directly and indirectly to the completion of this volume, but none more than *Sarah Jones*, who gave much more than editorial assistance. Acknowledgement of helpful criticism and advice is due also to Drs. *R. Greger, A. Ratner, J. Weiss,* and *S. Wood,* and Prof. *H. Seller*. We are grateful to *Joy Wieser* for her help in checking the proofs. *Wolf-Rüdiger* and *Barbara Gay* are especially recognized, not only for their art work, but for their conceptual contributions as well. The publishers, Georg Thieme Verlag and Deutscher Taschenbuch Verlag, contributed valuable assistance based on extensive experience; an author could wish for no better relationship. Finally, special recognition to Dr. *Walter Kumpmann* for inspiring the project and for his unquestioning confidence in the authors.

Basel and Innsbruck, Summer 1979
Agamemnon Despopoulos
Stefan Silbernagl

From the Preface to the Third Edition

The first German edition of this book was already in press when, on November 2nd, 1979, *Agamennon Despopoulos* and his wife, *Sarah Jones-Despopoulos* put to sea from Bizerta, Tunisia. Their intention was to cross the Atlantic in their sailing boat. This was the last that was ever heard of them and we have had to abandon all hope of seeing them again.

Without the creative enthusiasm of Agamennon Despopoulos, it is doubtful whether this book would have been possible; without his personal support it has not been easy to continue with the project. Whilst keeping in mind our original aims, I have completely revised the book, incorporating the latest advances in the field of physiology as well as the welcome suggestions provided by readers of the earlier edition, to whom I extend my thanks for their active interest.

Würzburg, Fall 1985
Stefan Silbernagl

Dr. Agamemnon Despopoulos

Born 1924 in New York; Professor of Physiology at the University of New Mexico. Albuquerque, USA, until 1971; thereafter scientific adviser to CIBA-GEIGY, Basel.

Table of Contents

13 | **Appendix** | **372**

Further Reading | **391**

Index | **394**

"… If we break up a living organism by isolating its different parts, it is only for the sake of ease in analysis and by no means in order to conceive them separately. Indeed, when we wish to ascribe to a physiological quality its value and true significance, we must always refer it to the whole and draw our final conclusions only in relation to its effects on the whole."

Claude Bernard (1865)

The Body: an Open System with an Internal Environment

The existence of unicellular organisms is the epitome of life in its simplest form. Even simple protists must meet two basic but essentially conflicting demands in order to survive. A unicellular organism must, on the one hand, isolate itself from the seeming disorder of its inanimate surroundings, yet, as an "open system" ($\rightarrow$ p. 40), it is dependent on its environment for the exchange of heat, oxygen, nutrients, waste materials, and information.

"Isolation" is mainly ensured by the cell membrane, the hydrophobic properties of which prevent the potentially fatal mixing of hydrophilic components in watery solutions inside and outside the cell. Protein molecules within the cell membrane ensure the permeability of the membrane barrier. They may exist in the form of *pores* (*channels*) or as more complex transport proteins known as *carriers* ($\rightarrow$ p. 26 ff.). Both types are selective for certain substances, and their activity is usually regulated. The cell membrane is relatively well permeable to hydrophobic molecules such as gases. This is useful for the exchange of O_2 and CO_2 and for the uptake of lipophilic signal substances, yet exposes the cell to poisonous gases such as carbon monoxide (CO) and lipophilic noxae such as organic solvents. The cell membrane also contains other proteins—namely, receptors and enzymes. *Receptors* receive signals from the external environment and convey the information to the interior of the cell (signal transduction), and *enzymes* enable the cell to metabolize extracellular substrates.

Let us imagine the primordial sea as the external environment of the unicellular organism ($\rightarrow$ **A**). This milieu remains more or less constant, although the organism absorbs nutrients from it and excretes waste into it. In spite of its simple structure, the unicellular organism is capable of eliciting motor responses to signals from the environment. This is achieved by moving its pseudopodia or flagella, for example, in response to changes in the food concentration.

The evolution from unicellular organisms to multicellular organisms, the transition from specialized cell groups to organs, the emergence of the two sexes, the coexistence of individuals in social groups, and the transition from water to land have tremendously increased the efficiency, survival, radius of action, and independence of living organisms. This process required the simultaneous development of a complex infrastructure within the organism. Nonetheless, the individual cells of the body still need a milieu like that of the primordial sea for life and survival. Today, the **extracellular fluid** is responsible for providing constant environmental conditions ($\rightarrow$ **B**), but the volume of the fluid is no longer infinite. In fact, it is even smaller than the intracellular volume ($\rightarrow$ p. 168). Because of their metabolic activity, the cells would quickly deplete the oxygen and nutrient stores within the fluids and flood their surroundings with waste products if organs capable of maintaining a **stable internal environment** had not developed. This is achieved through **homeostasis**, a process by which physiologic self-regulatory mechanisms (see below) maintain steady states in the body through coordinated physiological activity. Specialized organs ensure the continuous absorption of nutrients, electrolytes and water and the excretion of waste products via the urine and feces. The *circulating blood* connects the organs to every inch of the body, and the exchange of materials between the blood and the intercellular spaces (*interstices*) creates a stable environment for the cells. Organs such as the digestive tract and liver absorb nutrients and make them available by processing, metabolizing and distributing

$\blacktriangleright$

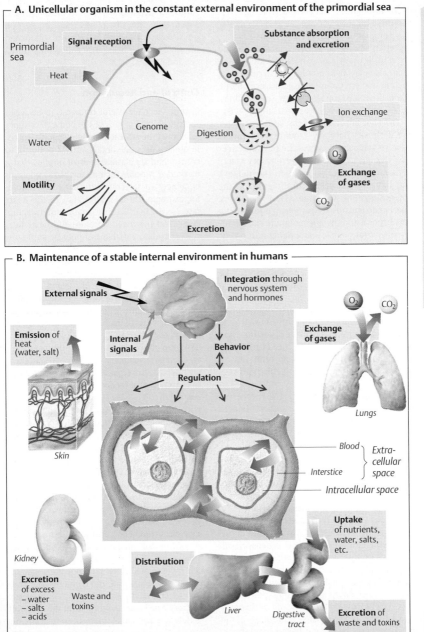

A. Unicellular organism in the constant external environment of the primordial sea

Primordial sea

Signal reception

Heat

Genome

Water

Motility

Substance absorption and excretion

Digestion

Ion exchange

O_2

Exchange of gases

CO_2

Excretion

B. Maintenance of a stable internal environment in humans

External signals

Integration through nervous system and hormones

O_2 CO_2

Exchange of gases

Emission of heat (water, salt)

Internal signals

Behavior

Regulation

Skin

Lungs

Blood
Interstice
Intracellular space

Extra-cellular space

Kidney

Excretion of excess
– water
– salts
– acids

Waste and toxins

Distribution

Uptake of nutrients, water, salts, etc.

Liver

Digestive tract

Excretion of waste and toxins

Plate 1.1 Internal and External Environment

them throughout the body. The lung is responsible for the exchange of gases (O_2 intake, CO_2 elimination), the liver and kidney for the excretion of waste and foreign substances, and the skin for the release of heat. The kidney and lungs also play an important role in regulating the internal environment, e.g., water content, osmolality, ion concentrations, pH (kidney, lungs) and O_2 and CO_2 pressure (lungs) ($\rightarrow$ **B**).

The specialization of cells and organs for specific tasks naturally requires **integration**, which is achieved by convective transport over long distances (circulation, respiratory tract), humoral transfer of information (hormones), and transmission of electrical signals in the nervous system, to name a few examples. These mechanisms are responsible for supply and disposal and thereby maintain a stable internal environment, even under conditions of extremely high demand and stress. Moreover, they control and regulate functions that ensure survival in the sense of **preservation of the species**. Important factors in this process include not only the timely development of reproductive organs and the availability of fertilizable gametes at sexual maturity, but also the control of erection, ejaculation, fertilization, and nidation. Others include the coordination of functions in the mother and fetus during pregnancy and regulation of the birth process and the lactation period.

The **central nervous system** (CNS) processes signals from peripheral sensors (single sensory cells or sensory organs), activates outwardly directed effectors (e.g., **skeletal muscles**), and influences the endocrine **glands**. The CNS is the focus of attention when studying human or animal **behavior**. It helps us to locate food and water and protects us from heat or cold. The central nervous system also plays a role in partner selection, concern for offspring even long after their birth, and integration into social systems. The CNS is also involved in the development, expression, and processing of emotions such as desire, listlessness, curiosity, wishfulness, happiness, anger, wrath, and envy and of traits such as creativeness, inquisitiveness, self-awareness, and responsibility. This goes far beyond the scope of physiology—which in the narrower sense is the study of the functions of the body—and, hence, of this book.

Although behavioral science, sociology, and psychology are disciplines that border on physiology, true bridges between them and physiology have been established only in exceptional cases.

Control and Regulation

In order to have useful cooperation between the specialized organs of the body, their functions must be adjusted to meet specific needs. In other words, the organs must be subject to control and regulation. **Control** implies that a *controlled variable* such as the blood pressure is subject to selective external modification, for example, through alteration of the heart rate ($\rightarrow$ p. 218). Because many other factors also affect the blood pressure and heart rate, the controlled variable can only be kept constant by continuously measuring the current blood pressure, comparing it with the reference signal (*set point*), and continuously correcting any deviations. If the blood pressure drops—due, for example, to rapidly standing up from a recumbent position—the heart rate will increase until the blood pressure has been reasonably adjusted. Once the blood pressure has risen above a certain limit, the heart rate will decrease again and the blood pressure will normalize. This type of *closed-loop control* is called a **negative feedback control system** or a **control circuit** ($\rightarrow$ **C1**). It consists of a *controller* with a programmed *set-point* value (target value) and *control elements* (*effectors*) that can adjust the *controlled variable* to the set point. The system also includes *sensors* that continuously measure the actual value of the controlled variable of interest and report it (feedback) to the controller, which compares the actual value of the controlled variable with the set-point value and makes the necessary adjustments if disturbance-related discrepancies have occurred. The control system operates either from within the organ itself (*autoregulation*) or via a *superordinate organ* such as the central nervous system or hormone glands. Unlike simple control, the elements of a control circuit can work rather imprecisely without causing a deviation from the set point (at least on average). Moreover, control circuits are capable of responding to unexpected dis-

C. Control circuit

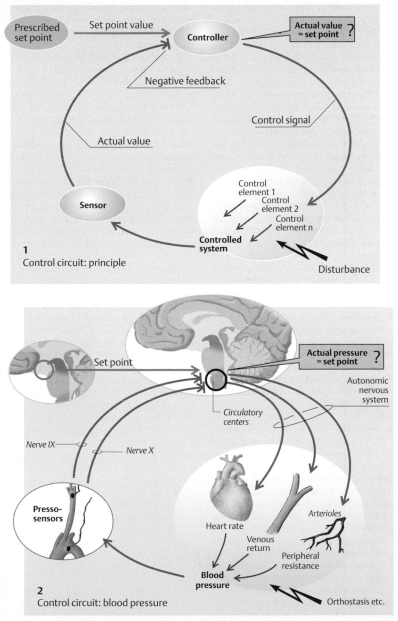

1 Control circuit: principle

- Prescribed set point
- Set point value
- Controller
- Actual value = set point ?
- Negative feedback
- Control signal
- Actual value
- Sensor
- Control element 1
- Control element 2
- Control element n
- Controlled system
- Disturbance

2 Control circuit: blood pressure

- Set point
- Actual pressure = set point ?
- Autonomic nervous system
- Circulatory centers
- Nerve IX
- Nerve X
- Presso-sensors
- Heart rate
- Venous return
- Arterioles
- Peripheral resistance
- Blood pressure
- Orthostasis etc.

Plate 1.2 Control and Regulation I

turbances. In the case of blood pressure regulation (→ **C2**), for example, the system can respond to events such as orthostasis (→ p. 204) or sudden blood loss.

The type of control circuits described above keep the controlled variables constant when **disturbance variables** cause the controlled variable to deviate from the set point (→ **D2**). Within the body, the set point is rarely invariable, but can be "shifted" when requirements of higher priority make such a change necessary. In this case, it is the **variation of the set point** that creates the discrepancy between the nominal and actual values, thus leading to the activation of regulatory elements (→ **D3**). Since the regulatory process is then triggered by variation of the set point (and not by disturbance variables), this is called **servocontrol** or **servomechanism**. Fever (→ p. 224) and the adjustment of muscle length by muscle spindles and γ-motor neurons (→ p. 316) are examples of servocontrol.

In addition to relatively simple variables such as blood pressure, cellular pH, muscle length, body weight and the plasma glucose concentration, the body also regulates complex sequences of events such as fertilization, pregnancy, growth and organ differentiation, as well as sensory stimulus processing and the motor activity of skeletal muscles, e.g., to maintain equilibrium while running. The regulatory process may take parts of a second (e.g., purposeful movement) to several years (e.g., the growth process).

In the control circuits described above, the controlled variables are kept constant on average, with variably large, wave-like deviations. The sudden emergence of a disturbance variable causes larger deviations that quickly normalize in a stable control circuit (→ **E**, test subject no. 1). The **degree of deviation** may be slight in some cases but substantial in others. The latter is true, for example, for the blood glucose concentration, which nearly doubles after meals. This type of regulation obviously functions only to prevent extreme rises and falls (e.g., hyper- or hypoglycemia) or chronic deviation of the controlled variable. More precise maintenance of the controlled variable requires a higher level of regulatory sensitivity (*high amplification factor*). However, this ex-

tends the settling time (→ **E**, subject no. 3) and can lead to regulatory instability, i.e., a situation where the actual value oscillates back and forth between extremes (*unstable oscillation*, → **E**, subject no. 4).

Oscillation of a controlled variable in response to a disturbance variable can be *attenuated* by either of two mechanisms. First, sensors with differential characteristics (*D sensors*) ensure that the intensity of the sensor signal increases in proportion with the **rate of deviation** of the controlled variable from the set point (→ p. 312 ff.). Second, **feedforward control** ensures that information regarding the expected intensity of disturbance is reported to the controller *before* the value of the controlled variable has changed at all. Feedforward control can be explained by example of physiologic thermoregulation, a process in which cold receptors on the skin trigger counterregulation before a change in the controlled value (core temperature of the body) has actually occurred (→ p. 224). The disadvantage of having *only* D sensors in the control circuit can be demonstrated by example of arterial pressosensors (= pressoreceptors) in acute blood pressure regulation. Very slow but steady changes, as observed in the development of arterial hypertension, then escape regulation. In fact, a rapid drop in the blood pressure of a hypertensive patient will even cause a counterregulatory increase in blood pressure. Therefore, other control systems are needed to ensure proper long-term blood pressure regulation.

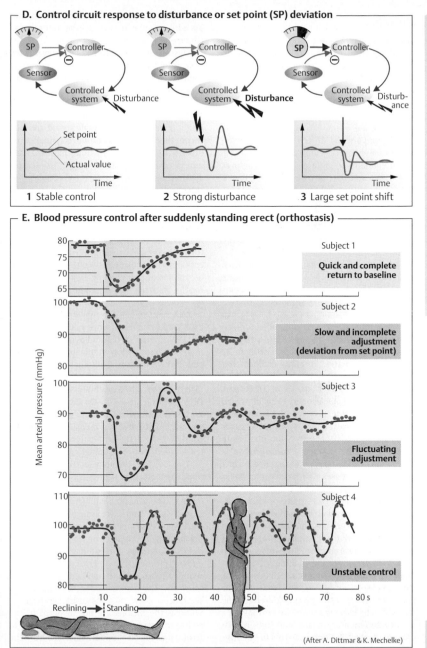

D. Control circuit response to disturbance or set point (SP) deviation

1 Stable control

SP → Controller
Sensor
Controlled system ← Disturbance

Set point
Actual value
Time

2 Strong disturbance

SP → Controller
Sensor
Controlled system ← **Disturbance**

Time

3 Large set point shift

SP → Controller
Sensor
Controlled system ← Disturbance

Time

E. Blood pressure control after suddenly standing erect (orthostasis)

Mean arterial pressure (mmHg)

Subject 1
Quick and complete return to baseline

Subject 2
Slow and incomplete adjustment (deviation from set point)

Subject 3
Fluctuating adjustment

Subject 4
Unstable control

Reclining → Standing

(After A. Dittmar & K. Mechelke)

Plate 1.3 Control and Regulation II

The Cell

The cell is the smallest functional unit of a living organism. In other words, a cell (and no smaller unit) is able to perform essential vital functions such as metabolism, growth, movement, reproduction, and hereditary transmission (W. Roux) ($\rightarrow$ p. 4). Growth, reproduction, and hereditary transmission can be achieved by *cell division*.

Cell components: All cells consist of a cell membrane, cytosol or cytoplasm (ca. 50 vol.%), and membrane-bound subcellular structures known as *organelles* ($\rightarrow$ **A, B**). The organelles of eukaryotic cells are highly specialized. For instance, the genetic material of the cell is concentrated in the cell nucleus, whereas "digestive" enzymes are located in the lysosomes. Oxidative ATP production takes place in the mitochondria.

The **cell nucleus** contains a liquid known as karyolymph, a nucleolus, and chromatin. Chromatin contains deoxyribonucleic acids (**DNA**), the carriers of genetic information. Two strands of DNA forming a *double helix* (up to 7 cm in length) are twisted and folded to form chromosomes 10 μm in length. Humans normally have 46 chromosomes, consisting of 22 autosomal pairs and the chromosomes that determine the sex (XX in females, XY in males). DNA is made up of a strand of three-part molecules called *nucleotides*, each of which consists of a pentose (deoxyribose) molecule, a phosphate group, and a base. Each sugar molecule of the monotonic sugar–phosphate backbone of the strands (…deoxyribose – phosphate–deoxyribose…) is attached to one of four different bases. The sequence of bases represents the **genetic code** for each of the roughly 100 000 different proteins that a cell produces during its lifetime (**gene expression**). In a DNA double helix, each base in one strand of DNA is bonded to its complementary base in the other strand according to the rule: adenine (A) with thymine (T) and guanine (G) with cytosine (C). The base sequence of one strand of the double helix ($\rightarrow$ **E**) is always a "mirror image" of the opposite strand. Therefore, one strand can be used as a template for making a new complementary strand, the information content of which is identical to that of the orig-inal. In cell division, this process is the means by which duplication of genetic information (**replication**) is achieved.

Messenger RNA (**mRNA**) is responsible for code transmission, that is, passage of coding sequences from DNA in the nucleus (base sequence) for protein synthesis in the cytosol (amino acid sequence) ($\rightarrow$ **C1**). mRNA is formed in the nucleus and differs from DNA in that it consists of only a single strand and that it contains ribose instead of deoxyribose, and uracil (U) instead of thymine. In DNA, each amino acid (e.g., glutamate, $\rightarrow$ **E**) needed for synthesis of a given protein is coded by a set of three adjacent bases called a *codon* or *triplet* (C–T–C in the case of glutamate). In order to transcribe the DNA triplet, mRNA must form a complementary codon (e.g., G–A–G for glutamate). The relatively small transfer RNA (**tRNA**) molecule is responsible for reading the codon in the ribosomes ($\rightarrow$ **C2**). tRNA contains a complementary codon called the *anticodon* for this purpose. The anticodon for glutamate is C–U–C ($\rightarrow$ **E**).

RNA synthesis in the nucleus is controlled by *RNA polymerases* (types I–III). Their effect on DNA is normally blocked by a *repressor protein*. Phosphorylation of the polymerase occurs if the repressor is eliminated (de-repression) and the *general transcription factors* attach to the so-called promoter sequence of the DNA molecule (T–A–T–A in the case of polymerase II). Once activated, it separates the two strands of DNA at a particular site so that the code on one of the strands can be read and transcribed to form mRNA (**transcription**, $\rightarrow$ **C1a, D**). The heterogeneous nuclear RNA (*hnRNA*) molecules synthesized by the polymerase have a characteristic "cap" at their 5′ end and a polyadenine "tail" (A–A–A–…) at the 3′ end ($\rightarrow$ **D**). Once synthesized, they are immediately "enveloped" in a protein coat, yielding heterogeneous nuclear ribonucleoprotein (*hnRNP*) particles. The *primary RNA* or *pre-mRNA* of hnRNA contains both coding sequences (*exons*) and non-coding sequences (*introns*). The exons code for amino acid sequences of the proteins to be synthesized, whereas the introns are not involved in the coding process. Introns may contain 100 to 10 000 nucleotides; they are removed from the

▶

A. Cell organelles (epithelial cell)

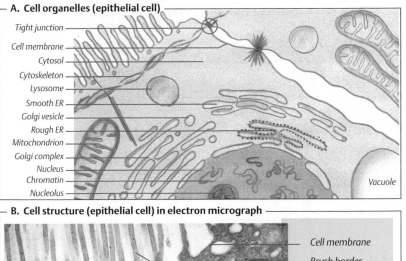

Tight junction
Cell membrane
Cytosol
Cytoskeleton
Lysosome
Smooth ER
Golgi vesicle
Rough ER
Mitochondrion
Golgi complex
Nucleus
Chromatin
Nucleolus

Vacuole

B. Cell structure (epithelial cell) in electron micrograph

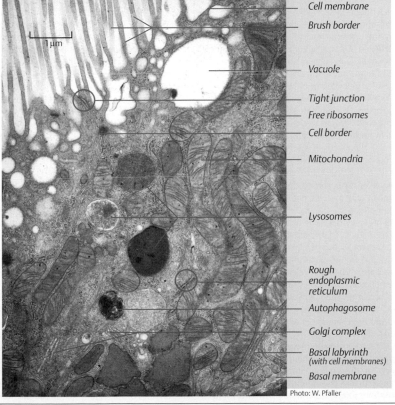

1 µm

Cell membrane
Brush border

Vacuole

Tight junction
Free ribosomes
Cell border

Mitochondria

Lysosomes

Rough endoplasmic reticulum

Autophagosome

Golgi complex

Basal labyrinth (with cell membranes)

Basal membrane

Photo: W. Pfaller

Plate 1.4 The Cell I

primary mRNA strand by **splicing** (→ **C1b, D**) and then degraded. The introns, themselves, contain the information on the exact splicing site. Splicing is ATP-dependent and requires the interaction of a number of proteins within a ribonucleoprotein complex called the *spliceosome*. Introns usually make up the lion's share of pre-mRNA molecules. For example, they make up 95% of the nucleotide chain of coagulation factor VIII, which contains 25 introns. mRNA can also be modified (e.g., through methylation) during the course of **posttranscriptional modification**.

RNA now exits the nucleus through **nuclear pores** (around 4000 per nucleus) and enters the cytosol (→ **C1c**). Nuclear pores are high-molecular-weight protein complexes (125 MDa) located within the nuclear envelope. They allow large molecules such as transcription factors, RNA polymerases or cytoplasmic steroid hormone receptors to pass into the nucleus, nuclear molecules such as mRNA and tRNA to pass out of the nucleus, and other molecules such as ribosomal proteins to travel both ways. The (ATP-dependent) passage of a molecule in either direction cannot occur without the help of a specific signal that guides the molecule into the pore. The abovementioned *5′ cap* is responsible for the exit of mRNA from the nucleus, and one or two specific sequences of a few (mostly cationic) amino acids are required as the signal for the entry of proteins into the nucleus. These sequences form part of the peptide chain of such *nuclear proteins* and probably create a peptide loop on the protein's surface. In the case of the cytoplasmic receptor for glucocorticoids (→ p. 278), the *nuclear localization signal* is masked by a chaperone protein (heat shock protein 90, hsp90) in the absence of the glucocorticoid, and is released only after the hormone binds, thereby freeing hsp90 from the receptor. The "activated" receptor then reaches the cell nucleus, where it binds to specific DNA sequences and controls specific genes.

The **nuclear envelope** consists of *two* membranes (= two phospholipid bilayers) that merge at the nuclear pores. The two membranes consist of different materials. The external membrane is continuous with the membrane of the endoplasmic reticulum (ER), which is described below (→ **F**).

The mRNA exported from the nucleus travels to the **ribosomes** (→ **C1**), which either float freely in the cytosol or are bound to the cytosolic side of the endoplasmic reticulum, as described below. Each ribosome is made up of dozens of proteins associated with a number of structural RNA molecules called *ribosomal RNA* (**rRNA**). The two subunits of the ribosome are first transcribed from numerous rRNA genes in the **nucleolus**, then separately exit the cell nucleus through the nuclear pores. Assembled together to form a ribosome, they now comprise the biochemical "machinery" for **protein synthesis** (**translation**) (→ **C2**). Synthesis of a peptide chain also requires the presence of specific tRNA molecules (at least one for each of the 21 proteinogenous amino acids). In this case, the target amino acid is bound to the C–C–A end of the tRNA molecule (same in all tRNAs), and the corresponding anticodon that recognizes the mRNA codon is located at the other end (→ **E**). Each ribosome has two tRNA binding sites: one for the last incorporated amino acid and another for the one beside it (not shown in **E**). Protein synthesis begins when the *start codon* is read and ends once the *stop codon* has been reached. The ribosome then breaks down into its two subunits and releases the mRNA (→ **C2**). Ribosomes can add approximately 10–20 amino acids per second. However, since an mRNA strand is usually translated simultaneously by many ribosomes (*polyribosomes* or *polysomes*) at different sites, a protein is synthesized much faster than its mRNA. In the bone marrow, for example, a total of around 5×10^{14} hemoglobin copies containing 574 amino acids each are produced per second.

The **endoplasmic reticulum** (**ER**, → **C, F**) plays a central role in the *synthesis of proteins and lipids*; it also serves as an intracellular Ca^{2+} *store* (→ p. 17 A). The ER consists of a net-like system of interconnected branched channels and flat cavities bounded by a membrane. The enclosed spaces (*cisterns*) make up around 10% of the cell volume, and the membrane comprises up to 70% of the membrane mass of a cell. *Ribosomes* can attach to the cytosolic surface of parts of the ER, forming a **rough endo-**

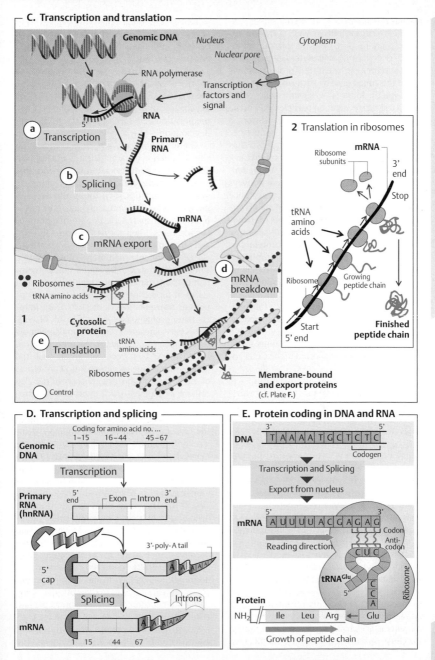

C. Transcription and translation

Genomic DNA — *Nucleus* — *Cytoplasm*

Nuclear pore

RNA polymerase

Transcription factors and signal

RNA

5'

(a) Transcription

Primary RNA

(b) Splicing

mRNA

(c) mRNA export

Ribosomes
tRNA amino acids

(d) mRNA breakdown

1

Cytosolic protein

tRNA amino acids

(e) Translation

Ribosomes

Membrane-bound and export proteins
(cf. Plate **F.**)

(Control)

2 Translation in ribosomes

Ribosome subunits

mRNA

3' end

Stop

tRNA amino acids

Ribosome

Growing peptide chain

Start

5' end

Finished peptide chain

Plate 1.5 **The Cell II**

D. Transcription and splicing

Coding for amino acid no. ...
1–15   16–44   45–67

Genomic DNA

Transcription

Primary RNA (hnRNA)

5' end — Exon — Intron — 3' end

3'-poly-A tail

5' cap   A A A A A

Splicing   Introns

mRNA

1   15   44   67

E. Protein coding in DNA and RNA

DNA
3' | T A A A A T G C T C T C | 5'
Codogen

Transcription and Splicing

Export from nucleus

mRNA
5' | A U U U A C G A G A G | 3'
Codon
Reading direction
Anti-codon
C U C

tRNA^Glu
5'
C C A

Protein
NH₂ — Ile | Leu | Arg ← Glu

Growth of peptide chain

Ribosome

11

▶

plasmic reticulum (RER). These ribosomes synthesize export proteins as well as transmembrane proteins ($\rightarrow$ **G**) for the plasma membrane, endoplasmic reticulum, Golgi apparatus, lysosomes, etc. The start of protein synthesis (at the amino end) by such ribosomes (still unattached) induces a *signal sequence* to which a signal recognition particle (SRP) in the cytosol attaches. As a result, (a) synthesis is temporarily halted and (b) the ribosome (mediated by the SRP and a SRP receptor) attaches to a ribosome receptor on the ER membrane. After that, synthesis continues. In *export protein synthesis*, a translocator protein conveys the peptide chain to the cisternal space once synthesis is completed. *Synthesis of membrane proteins* is interrupted several times (depending on the number of membrane-spanning domains ($\rightarrow$ **G2**) by translocator protein closure, and the corresponding (hydrophobic) peptide sequence is pushed into the phospholipid membrane. The **smooth endoplasmic reticulum** (**SER**) contains no ribosomes and is the *production site of lipids* (e.g., for lipoproteins, $\rightarrow$ p. 254 ff.) and other substances. The ER membrane containing the synthesized membrane proteins or export proteins forms vesicles which are transported to the Golgi apparatus.

The Golgi complex or **Golgi apparatus** ($\rightarrow$ **F**) has sequentially linked functional compartments for further processing of products from the endoplasmic reticulum. It consists of a *cis*-Golgi network (entry side facing the ER), stacked flattened cisternae (Golgi stacks) and a *trans*-Golgi network (sorting and distribution). Functions of the Golgi complex:

◆ polysaccharide synthesis;

◆ protein processing (**posttranslational modification**), e.g., glycosylation of membrane proteins on certain amino acids (in part in the ER) that are later borne as glycocalyces on the external cell surface (see below) and γ-carboxylation of glutamate residues ($\rightarrow$ p. 102);

◆ phosphorylation of sugars of glycoproteins (e.g., to mannose-6-phosphate, as described below);

◆ "packaging" of proteins meant for export into *secretory vesicles* (secretory granules), the contents of which are exocytosed into the extracellular space; see p. 246, for example.

Hence, the Golgi apparatus represents a central **modification, sorting and distribution center** for proteins and lipids received from the endoplasmic reticulum.

Regulation of gene expression takes place on the level of transcription ($\rightarrow$ **C1a**), RNA modification ($\rightarrow$ **C1b**), mRNA export ($\rightarrow$ **C1c**), RNA degradation ($\rightarrow$ **C1d**), translation ($\rightarrow$ **C1e**), modification and sorting ($\rightarrow$ **F,f**), and protein degradation ($\rightarrow$ **F,g**).

The **mitochondria** ($\rightarrow$ **A, B**; p. 17 B) are the site of oxidation of carbohydrates and lipids to CO_2 and H_2O and associated O_2 expenditure. The Krebs cycle (citric acid cycle), respiratory chain and related ATP synthesis also occur in mitochondria. Cells intensely active in metabolic and transport activities are rich in mitochondria—e.g., hepatocytes, intestinal cells, and renal epithelial cells. Mitochondria are enclosed in a double membrane consisting of a smooth outer membrane and an inner membrane. The latter is deeply infolded, forming a series of projections (cristae); it also has important transport functions ($\rightarrow$ p. 17 B). Mitochondria probably evolved as a result of symbiosis between aerobic bacteria and anaerobic cells (*symbiosis hypothesis*). The mitochondrial DNA (mtDNA) of bacterial origin and the double membrane of mitochondria are relicts of their ancient history. Mitochondria also contain ribosomes which synthesize all proteins encoded by mtDNA.

Lysosomes are vesicles ($\rightarrow$ **F**) that arise from the ER (via the Golgi apparatus) and are involved in the intracellular digestion of macromolecules. These are taken up into the cell either by *endocytosis* (e.g., uptake of albumin into the renal tubules; $\rightarrow$ p. 158) or by *phagocytosis* (e.g., uptake of bacteria by macrophages; $\rightarrow$ p. 94 ff.). They may also originate from the degradation of a cell's own organelles (autophagia, e.g., of mitochondria) delivered inside autophagosomes ($\rightarrow$ **B, F**). A portion of the endocytosed membrane material recycles (e.g., receptor recycling in receptor-mediated endocytosis; $\rightarrow$ p. 28). *Early* and *late* **endosomes** are intermediate stages in this *vesicular transport*. Late endosomes and lysosomes contain acidic hydrolases (proteases, nucleases, lipases, glycosidases, phosphatases, etc., that are active only under acidic conditions). The

▶

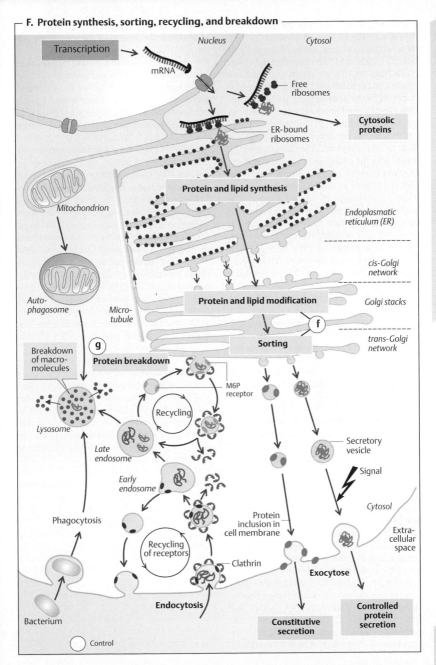

Plate 1.6 The Cell III

13

membrane contains an H^+-ATPase that creates an acidic (pH 5) interior environment within the lysosomes and assorted *transport proteins* that (a) release the products of digestion (e.g., amino acids) into the cytoplasm and (b) ensure charge compensation during H^+ uptake (Cl^- channels). These enzymes and transport proteins are delivered in *primary lysosomes* from the Golgi apparatus. Mannose-6-phosphate (M6P) serves as the "label" for this process; it binds to M6P receptors in the Golgi membrane which, as in the case of receptor-mediated endocytosis ($\rightarrow$ p. 28), cluster in the membrane with the help of a clathrin framework. In the acidic environment of the lysosomes, the enzymes and transport proteins are separated from the receptor, and M6P is dephosphorylated. The M6P receptor returns to the Golgi apparatus (recycling, $\rightarrow$ F). The M6P receptor no longer recognizes the dephosphorylated proteins, which prevents them from returning to the Golgi apparatus.

Peroxisomes are microbodies containing enzymes (imported via a signal sequence) that permit the oxidation of certain organic molecules ($R\text{-}H_2$), such as amino acids and fatty acids: $R\text{-}H_2 + O_2 \rightarrow R + H_2O_2$. The peroxisomes also contain catalase, which transforms $2\,H_2O_2$ into $O_2 + H_2O$ and oxidizes toxins, such as alcohol and other substances.

Whereas the membrane of organelles is responsible for intracellular compartmentalization, the main job of the **cell membrane** ($\rightarrow$ G) is to separate the cell interior from the extracellular space ($\rightarrow$ p. 2). The cell membrane is a phospholipid bilayer ($\rightarrow$ G1) that may be either smooth or deeply infolded, like the brush border or the basal labyrinth ($\rightarrow$ B). Depending on the cell type, the cell membrane contains variable amounts of *phospholipids*, *cholesterol*, and *glycolipids* (e.g., cerebrosides). The phospholipids mainly consist of phosphatidylcholine ($\rightarrow$ G3), phosphatidylserine, phosphatidylethanolamine, and sphingomyelin. The hydrophobic components of the membrane face each other, whereas the hydrophilic components face the watery surroundings, that is, the extracellular fluid or cytosol ($\rightarrow$ G4). The lipid composition of the two layers of the membrane differs greatly. Glycolipids are present only in the external layer, as described below. Cholesterol (present in both layers) reduces both the fluidity of the membrane and its permeability to polar substances. Within the two-dimensionally fluid phospholipid membrane are proteins that make up 25% (myelin membrane) to 75% (inner mitochondrial membrane) of the membrane mass, depending on the membrane type. Many of them span the entire lipid bilayer once ($\rightarrow$ G1) or several times ($\rightarrow$ G2) (*transmembrane proteins*), thereby serving as ion channels, carrier proteins, hormone receptors, etc. The proteins are anchored by their lipophilic amino acid residues, or attached to already anchored proteins. Some proteins can move about freely within the membrane, whereas others, like the anion exchanger of red cells, are anchored to the cytoskeleton. The cell surface is largely covered by the *glycocalyx*, which consists of sugar moieties of glycoproteins and glycolipids in the cell membrane ($\rightarrow$ G1,4) and of the extracellular matrix. The glycocalyx mediates cell–cell interactions (surface recognition, cell docking, etc.). For example, components of the glycocalyx of neutrophils dock onto endothelial membrane proteins, called *selectins* ($\rightarrow$ p. 94).

The **cytoskeleton** allows the cell to maintain and change its shape (during cell division, etc.), make selective movements (migration, cilia), and conduct intracellular transport activities (vesicle, mitosis). It contains *actin filaments* as well as *microtubules* and *intermediate filaments* (e.g., vimentin and desmin filaments, neurofilaments, keratin filaments) that extend from the centrosome.

Plate 1.7 The Cell IV

G. Cell membrane

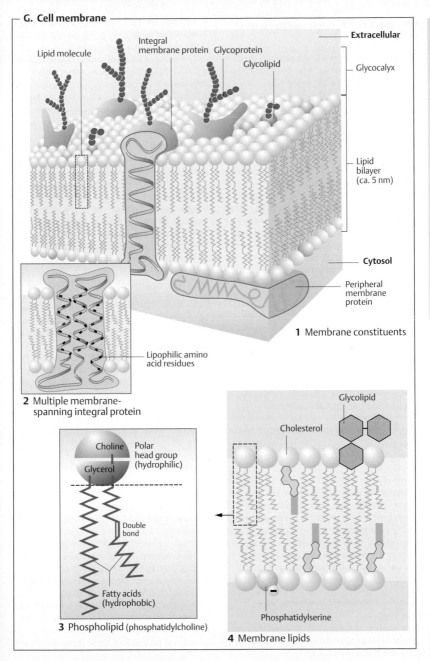

Lipid molecule

Integral membrane protein Glycoprotein

Glycolipid

Extracellular

Glycocalyx

Lipid bilayer (ca. 5 nm)

Cytosol

Peripheral membrane protein

1 Membrane constituents

Lipophilic amino acid residues

2 Multiple membrane-spanning integral protein

Choline Polar head group (hydrophilic)

Glycerol

Double bond

Fatty acids (hydrophobic)

3 Phospholipid (phosphatidylcholine)

Glycolipid

Cholesterol

Phosphatidylserine

4 Membrane lipids

15

Transport In, Through and Between Cells

The lipophilic cell membrane protects the cell interior from the extracellular fluid, which has a completely different composition ($\rightarrow$ p. 2). This is imperative for the creation and maintenance of a *cell's internal environment* by means of metabolic energy expenditure. Channels (pores), carriers, ion pumps ($\rightarrow$ p. 26ff.) and the process of cytosis ($\rightarrow$ p. 28) allow **transmembrane transport** of selected substances. This includes the import and export of metabolic substrates and metabolites and the selective transport of ions used to create or modify the *cell potential* ($\rightarrow$ p. 32), which plays an essential role in excitability of nerve and muscle cells. In addition, the effects of substances that readily penetrate the cell membrane in most cases (e.g., water and CO_2) can be mitigated by selectively transporting certain other substances. This allows the cell to compensate for undesirable changes in the cell volume or pH of the cell interior.

Intracellular Transport

The cell interior is divided into different compartments by the organelle membranes. In some cases, very broad intracellular spaces must be crossed during transport. For this purpose, a variety of specific intracellular transport mechanisms exist, for example:

◆ Nuclear pores in the nuclear envelope provide the channels for RNA export out of the nucleus and protein import into it ($\rightarrow$ p. 11 C);

◆ *Protein transport* from the rough endoplasmic reticulum to the Golgi complex ($\rightarrow$ p. 13 F);

◆ *Axonal transport* in the nerve fibers, in which distances of up to 1 meter can be crossed ($\rightarrow$ p. 42). These transport processes mainly take place along the filaments of the cytoskeleton. Example: while expending ATP, the microtubules set dynein-bound vesicles in motion in the one direction, and kinesin-bound vesicles in the other ($\rightarrow$ p. 13 F).

Intracellular Transmembrane Transport

Main sites:

◆ *Lysosomes*: Uptake of H^+ ions from the cytosol and release of metabolites such as amino acids into the cytosol ($\rightarrow$ p. 12);

◆ *Endoplasmic reticulum (ER)*: In addition to a translocator protein ($\rightarrow$ p. 10), the ER has two other proteins that transport Ca^{2+} ($\rightarrow$ **A**). Ca^{2+} can be pumped from the cytosol into the ER by a Ca^{2+}-ATPase called *SERCA* (sarcoplasmic endoplasmic reticulum Ca^{2+}-transporting ATPase). The resulting Ca^{2+} stores can be released into the cytosol via a *Ca^{2+} channel* (ryanodine receptor, RyR) in response to a triggering signal ($\rightarrow$ p. 36).

◆ *Mitochondria*: The outer membrane contains large pores called *porins* that render it permeable to small molecules ($< 5\,kDa$), and the inner membrane has high concentrations of specific carriers and enzymes ($\rightarrow$ **B**). Enzyme complexes of the *respiratory chain* transfer electrons (e^-) from high to low energy levels, thereby pumping H^+ ions from the matrix space into the intermembrane space ($\rightarrow$ **B1**), resulting in the formation of an *H^+ ion gradient* directed into the matrix. This not only drives ATP synthetase (ATP production; $\rightarrow$ **B2**), but also promotes the inflow of pyruvate$^-$ and anorganic phosphate, P_i^- (symport; $\rightarrow$ **B2b,c** and p. 28). *Ca^{2+} ions* that regulate Ca^{2+}-sensitive mitochondrial enzymes in muscle tissue can be pumped into the matrix space with ATP expenditure ($\rightarrow$ **B2**), thereby allowing the mitochondria to form a sort of Ca^{2+} buffer space for protection against dangerously high concentrations of Ca^{2+} in the cytosol. The inside-negative *membrane potential* (caused by H^+ release) drives the uptake of ADP^{3-} in exchange for ATP^{4-} (potential-driven transport; $\rightarrow$ **B2a** and p. 22).

Transport between Adjacent Cells

In the body, transport between adjacent cells occurs either via diffusion through the extracellular space (e.g., paracrine hormone effects) or through channel-like connecting structures (**connexons**) located within a so-called **gap junction** or **nexus** ($\rightarrow$ **C**). A connexon is a hemichannel formed by six connexin molecules ($\rightarrow$ **C2**). One connexon docks with another connexon on an adjacent cell, thereby forming a common channel through which substances with molecular masses of up to around 1 kDa can pass. Since this applies not only for ions such as Ca^{2+}, but also for a number of organic substances such as ATP, these types of cells are

▶

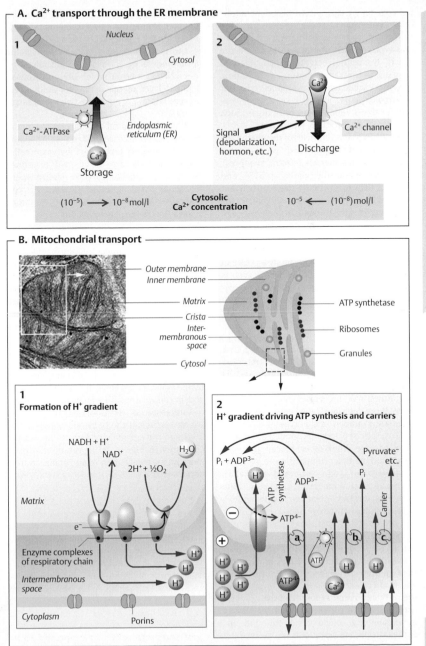

Plate 1.8 Transport In, Through and Between Cells I

A. Ca²⁺ transport through the ER membrane

1

Nucleus

Cytosol

Ca²⁺-ATPase

Endoplasmic reticulum (ER)

Ca²⁺

Storage

2

Ca²⁺

Signal (depolarization, hormon, etc.)

Ca²⁺ channel

Discharge

$(10^{-5}) \longrightarrow 10^{-8}$ mol/l   **Cytosolic Ca²⁺ concentration** $10^{-5} \longleftarrow (10^{-8})$ mol/l

B. Mitochondrial transport

Outer membrane

Inner membrane

Matrix

Crista

Intermembranous space

Cytosol

ATP synthetase

Ribosomes

Granules

1

Formation of H⁺ gradient

$NADH + H^+$

NAD^+

$2H^+ + \frac{1}{2}O_2$

H_2O

Matrix

e^-

Enzyme complexes of respiratory chain

Intermembranous space

H^+

H^+

H^+

Cytoplasm

Porins

2

H⁺ gradient driving ATP synthesis and carriers

$P_i + ADP^{3-}$

Pyruvate⁻ etc.

H^+

ATP synthetase

ADP^{3-}

P_i

$\ominus$

ATP^{4-}

ATP

Carrier

$\oplus$

H^+ H^+ H^+ H^+ H^+

ATP^{4-}

Ca^{2+}

H^+

H^+

a   **b**   **c**

united to form a close electrical and metabolic unit (**syncytium**), as is present in the epithelium, many smooth muscles (single-unit type, → p. 70), the myocardium, and the glia of the central nervous system. Electric coupling permits the transfer of excitation, e.g., from excited *muscle cells* to their adjacent cells, making it possible to trigger a wave of excitation across wide regions of an organ, such as the stomach, intestine, biliary tract, uterus, ureter, atrium, and ventricles of the heart. Certain neurons of the retina and CNS also communicate in this manner (*electric synapses*). Gap junctions in the glia (→ p. 338) and epithelia help to distribute the stresses that occur in the course of transport and barrier activities (see below) throughout the entire cell community. However, the connexons close when the concentration of Ca^{2+} (in an extreme case, due to a hole in cell membrane) or H^+ concentration increases too rapidly (→ **C3**). In other words, the individual (defective) cell is left to deal with its own problems when necessary to preserve the functionality of the cell community.

Transport through Cell Layers

Multicellular organisms have *cell layers* that are responsible for separating the "interior" from the "exterior" of the organism and its larger compartments. The *epithelia* of skin and gastrointestinal, urogenital and respiratory tracts, the *endothelia* of blood vessels, and *neuroglia* are examples of this type of extensive barrier. They separate the immediate extracellular space from other spaces that are greatly different in composition, e.g., those filled with air (skin, bronchial epithelia), gastrointestinal contents, urine or bile (tubules, urinary bladder, gallbladder), aqueous humor of the eye, blood (endothelia) and cerebrospinal fluid (blood–cerebrospinal fluid barrier), and from the extracellular space of the CNS (blood–brain barrier). Nonetheless, certain substances must be able to pass through these cell layers. This requires selective **transcellular transport** with import into the cell followed by export from the cell. Unlike cells with a completely uniform plasma membrane (e.g., blood cells), epi- and endothelial cells are *polar cells,* as defined by

their structure (→ p. 9A and B) and transport function. Hence, the *apical membrane* (facing exterior) of an epithelial cell has a different set of transport proteins from the *basolateral membrane* (facing the blood). Tight junctions (described below) at which the outer phospholipid layer of the membrane folds over, prevent lateral mixing of the two membranes (→ **D2**).

Whereas the apical and basolateral membranes permit transcellular transport, **paracellular transport** takes place between cells. Certain *epithelia* (e.g., in the small intestinal and proximal renal tubules) are relatively permeable to small molecules (leaky), whereas others are less leaky (e.g., distal nephron, colon). The degree of permeability depends on the strength of the *tight junctions* (*zonulae* ■ *occludentes*) holding the cells together (→ **D**). The paracellular pathway and the extent of its permeability (sometimes cation-specific) are essential functional elements of the various epithelia. Macromolecules can cross the barrier formed by the *endothelium* of the vessel wall by transcytosis (→ p. 28), yet paracellular transport also plays an essential role, especially in the fenestrated endothelium. Anionic macromolecules like albumin, which must remain in the bloodstream because of its colloid osmotic action (→ p. 208), are held back by the wall charges at the intercellular spaces and, in some cases, at the fenestra.

Long-distance transport between the various organs of the body and between the body and the outside world is also necessary. *Convection* is the most important transport mechanism involved in long-distance transport (→ p. 24).

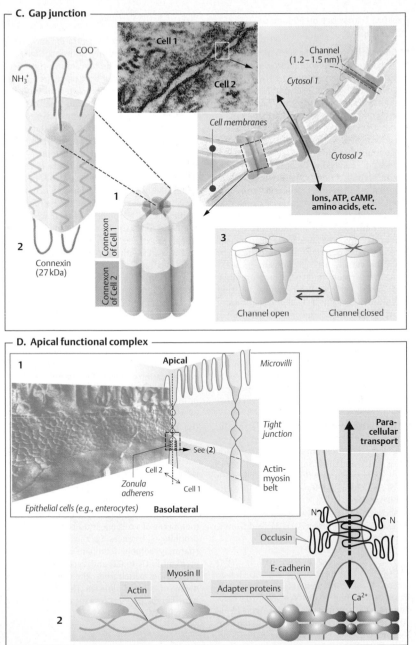

C. Gap junction

NH₃⁺ — COO⁻

Cell 1

Cell 2

Channel (1.2 – 1.5 nm)

Cytosol 1

Cell membranes

Cytosol 2

Ions, ATP, cAMP, amino acids, etc.

2

Connexin (27 kDa)

1

Connexon of Cell 1

Connexon of Cell 2

3

Channel open — Channel closed

D. Apical functional complex

1

Apical

Microvilli

Tight junction

Actin-myosin belt

Zonula adherens

See (2)

Cell 2

Cell 1

Epithelial cells (e.g., enterocytes)

Basolateral

Para-cellular transport

N — N

Occlusin

Ca²⁺

2

Actin

Myosin II

Adapter proteins

E-cadherin

Plate 1.9 Transport In, Through and Between Cells II

Photos: H. Lodish. Reproduced with permission from Scientific American Books, New York, 1995.

Passive Transport by Means of Diffusion

Diffusion is movement of a substance owing to the random thermal motion (brownian movement) of its molecules or ions ($\rightarrow$ **A1**) in all directions throughout a solvent. *Net diffusion* or selective transport can occur only when the solute concentration at the starting point is higher than at the target site. (*Note:* unidirectional fluxes also occur in absence of a concentration gradient—i.e., at equilibrium—but net diffusion is zero because there is equal flux in both directions.) The driving force of diffusion is, therefore, a **concentration gradient**. Hence, diffusion equalizes concentration differences and requires a driving force: **passive transport** (= downhill transport).

Example: When a layer of O_2 gas is placed on water, the O_2 quickly diffuses into the water along the initially high gas pressure gradient ($\rightarrow$ **A2**). As a result, the partial pressure of O_2 (Po_2) rises, and O_2 can diffuse further downward into the next O_2-poor layer of water ($\rightarrow$ **A1**). (*Note:* with gases, partial pressure is used in lieu of concentration.) However, the steepness of the Po_2 profile or gradient (dPo_2/dx) decreases (exponentially) in each subsequent layer situated at distance x from the O_2 source ($\rightarrow$ **A3**). Therefore, diffusion is only feasible for transport across short distances within the body. Diffusion in liquids is slower than in gases.

The diffusion rate, J_{diff} ($mol \cdot s^{-1}$), is the amount of substance that diffuses per unit of time. It is proportional to the area available for diffusion (A) and the absolute temperature (T) and is inversely proportional to the viscosity (η) of the solvent and the radius (r) of the diffused particles.

According to the *Stokes–Einstein equation*, the *coefficient of diffusion* (D) is derived from T, η, and r as

$$D = \frac{R \cdot T}{N_A \cdot 6\pi \cdot r \cdot \eta} \ [m^2 \cdot s^{-1}],$$
[1.1]

where R is the general gas constant ($8.3144\,J \cdot K^{-1} \cdot mol^{-1}$) and N_A Avogadro's constant ($6.022 \cdot 10^{23}\,mol^{-1}$). In Fick's first law of diffusion (Adolf Fick, 1855), the diffusion rate is expressed as

$$J_{diff} = A \cdot D \cdot \left(\frac{dC}{dx}\right) [mol \cdot s^{-1}]$$
[1.2]

where C is the molar concentration and x is the distance traveled during diffusion. Since the *driving "force"*—i.e., the concentration gradient (dC/dx)—decreases with distance, as was explained above, the *time* required for diffusion increases exponentially with the distance traveled ($t \sim x^2$). If, for example, a molecule travels the first μm in 0.5 ms, it will require 5 s to travel $100\,\mu m$ and a whopping 14 h for 1 cm.

Returning to the previous example ($\rightarrow$ **A2**), if the above-water partial pressure of free O_2 diffusion ($\rightarrow$ **A2**) is kept constant, the Po_2 in the water and overlying gas layer will eventually equalize and net diffusion will cease (*diffusion equilibrium*). This process takes place within the body, for example, when O_2 diffuses from the alveoli of the lungs into the bloodstream and when CO_2 diffuses in the opposite direction (→ p. 120).

Let us imagine two spaces, a and b ($\rightarrow$ **B1**) containing different concentrations ($C_a > C_b$) of an uncharged solute. The membrane separating the solutions has pores Δx in length and with total cross-sectional area of A. Since the pores are permeable to the molecules of the dissolved substance, the molecules will diffuse from a to b, with $C^a - C^b = \Delta C$ representing the concentration gradient. If we consider only the spaces a and b (while ignoring the gradients dC/dx in the pore, as shown in B2, for the sake of simplicity), **Fick's first law of diffusion** (Eq. 1.2) can be modified as follows:

$$J_{diff} = A \cdot D \cdot \frac{\Delta C}{\Delta x} [mol \cdot s^{-1}].$$
[1.3]

In other words, the rate of diffusion increases as A, D, and ΔC increase, and decreases as the thickness of the membrane (Δx) decreases.

When diffusion occurs through the **lipid membrane** of a cell, one must consider that hydrophilic substances in the membrane are sparingly soluble (compare intramembrane gradient in **C1** to **C2**) and, accordingly, have a hard time penetrating the membrane by means of *"simple"* diffusion. The *oil-and-water partition coefficient* (k) is a measure of the lipid solubility of a substance ($\rightarrow$ **C**).

Plate 1.10   Passive Transport by Means of Diffusion I

A. Diffusion in homogeneous media

1 Brownian particle movement (~T)

2 Passive transport

P_{O_2} Gas O_2

O_2 O_2 O_2

P_{O_2} ↑

Water

0

X

3 P_{O_2} profile

P_{O_2}

Slope = gradient = dP/dx

P
x

P
x

0 Distance from O_2 source (x)

B. Diffusion through porous membranes

1

Porous membrane

Space a   Space b

C^a C^b

Δx

$C^a - C^b = \Delta C$

2

C^a **Gradient** C^b

Pore

Space a   Membrane   Space b

C. Diffusion through lipid membranes

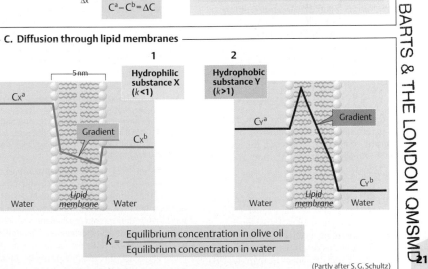

1

5 nm

Hydrophilic substance X ($k < 1$)

Cx^a

Gradient

Cx^b

Water   *Lipid membrane*   Water

2

Hydrophobic substance Y ($k > 1$)

Cy^a Gradient

Cy^b

Water   *Lipid membrane*   Water

$$k = \frac{\text{Equilibrium concentration in olive oil}}{\text{Equilibrium concentration in water}}$$

(Partly after S. G. Schultz)

BARTS & THE LONDON QMSMD

21

The higher the k value, the more quickly the substance will diffuse through a pure phospholipid bilayer membrane. Substitution into Eq. 1.3 gives

$$J_{diff} = k \cdot A \cdot D \cdot \frac{\Delta C}{\Delta x} \, [\text{mol} \cdot s^{-1}]; \qquad [1.4]$$

Whereas the molecular radius r ($\rightarrow$ Eq. 1.1) still largely determines the magnitude of D when k remains constant (cf. diethylmalonamide with ethylurea in **D**), k can vary by many powers of ten when r remains constant (cf. urea with ethanol in **D**) and can therefore have a decisive effect on the permeability of the membrane.

Since the value of the variables k, D, and Δx within the body generally cannot be determined, they are usually summarized as the *permeability coefficient* P, where

$$P = k \cdot \frac{D}{\Delta x} \, [m \cdot s^{-1}]. \qquad [1.5]$$

If the diffusion rate, J_{diff} [mol·s^{-1}], is related to area A, Eq. 1.4 is transformed to yield

$$\frac{J_{diff}}{A} = P \cdot \Delta C \, [\text{mol} \cdot m^{-2} \cdot s^{-1}]. \qquad [1.6]$$

The quantity of substance (net) diffused per unit area and time is therefore proportional to ΔC and P ($\rightarrow$ **E**, blue line with slope P).

When considering the **diffusion of gases**, ΔC in Eq. 1.4 is replaced by $\alpha \cdot \Delta P$ (solubility coefficient times partial pressure difference; $\rightarrow$ p. 126) and J_{diff} [mol · s^{-1}] by $\dot{V}_{diff}$ [m^3· s^{-1}]. $k \cdot \alpha \cdot D$ is then summarized as diffusion conductance, or *Krogh's diffusion coefficient* K [m^2 · s^{-1} · Pa^{-1}]. Substitution into Fick's first diffusion equation yields

$$\frac{\dot{V}_{diff}}{A} = K \cdot \frac{\Delta P}{\Delta x} \, [m \cdot s^{-1}]. \qquad [1.7]$$

Since A and Δx of alveolar gas exchange ($\rightarrow$ p. 120) cannot be determined in living organisms, $K \cdot F/\Delta x$ for O_2 is often expressed as the O_2 *diffusion capacity* of the lung, D_L:

$$\dot{V}_{O_2 diff} = D_L \cdot \Delta P_{O_2} \, [m^3 \cdot s^{-1}]. \qquad [1.8]$$

Nonionic diffusion occurs when the uncharged form of a weak base (e.g., ammonia = NH_3) or acid (e.g., formic acid, HCOOH) passes through a membrane more readily than the charged form ($\rightarrow$ **F**). In this case, the membrane would be more permeable to NH_3 than to NH_4^+

($\rightarrow$ p. 176 ff.). Since the pH of a solution determines whether these substances will be charged or not (pK value; $\rightarrow$ p. 378), the diffusion of weak acids and bases is clearly dependent on the pH.

The previous equations have not made allowances for the diffusion of electrically charged particles (**ions**). In their case, the **electrical potential difference** at cell membranes must also be taken into account. The electrical potential difference can be a driving force of diffusion (*electrodiffusion*). In that case, positively charged ions (cations) will then migrate to the negatively charged side of the membrane, and negatively charged ions (anions) will migrate to the positively charged side. The prerequisite for this type of transport is, of course, that the membrane contain ion channels ($\rightarrow$ p. 32 ff.) that make it permeable to the transported ions. Inversely, every ion diffusing along a concentration gradient carries a charge and thus creates an electric *diffusion potential* ($\rightarrow$ p. 32 ff.).

As a result of the electrical charge of an ion, the permeability coefficient of the ion x (= P_x) can be transformed into the **electrical conductance** of the membrane for this ion, g_x ($\rightarrow$ p. 32):

$$g_x = \cdot P_x \cdot z_x^2 \cdot F^2 \cdot R^{-1} \cdot T^{-1} \cdot \bar{c}_x \, [S \cdot m^{-2}] \qquad [1.9]$$

where R and T have their usual meaning (explained above) and z_x equals the charge of the ion, F equals the *Faraday constant* ($9,65 \cdot 10^4$ A · s · mol^{-1}), and $\bar{c}_x$ equals the mean ionic activity in the membrane. Furthermore,

$$\bar{c} = \frac{c_1 - c_2}{\ln c_1 - \ln c_2}. \qquad [1.10]$$

where index 1 = one side and index 2 = the other side of the membrane. Unlike P, *g is concentration-dependent*. If, for example, the extracellular K$^+$ concentration rises from 4 to 8 mmol/kg H_2O (cytosolic concentration remains constant at 160 mmol/kg H_2O), $\bar{c}$ will rise, and *g* will increase by 20%.

Since most of the biologically important substances are so polar or lipophobic (*small k value*) that simple diffusion of the substances through the membrane would proceed much too slowly, other membrane transport proteins called **carriers** or transporters exist in addition to ion channels. Carriers bind the target molecule (e.g., glucose) on one side of the membrane and detach from it on the other side

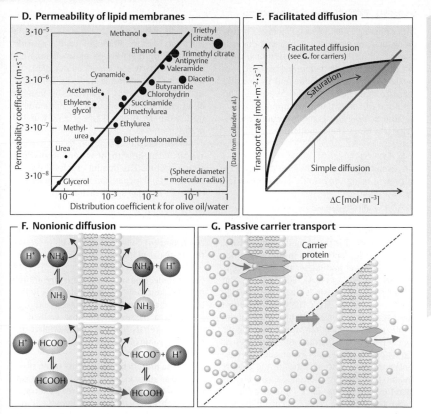

D. Permeability of lipid membranes

Permeability coefficient (m·s⁻¹) vs Distribution coefficient k for olive oil/water

$3 \cdot 10^{-5}$ — Methanol — Triethyl citrate
Ethanol — Trimethyl citrate
Antipyrine
Valeramide
Cyanamide — Diacetin
$3 \cdot 10^{-6}$ — Butyramide
Acetamide — Chlorohydrin
Ethylene glycol — Succinamide
Dimethylurea
Methylurea — Ethylurea
$3 \cdot 10^{-7}$ — Diethylmalonamide
Urea
$3 \cdot 10^{-8}$ — Glycerol

(Sphere diameter = molecular radius)

(Data from Collander et al.)

E. Facilitated diffusion

Transport rate [mol·m⁻²·s⁻¹] vs ΔC [mol·m⁻³]

Facilitated diffusion (see **G.** for carriers)

Saturation

Simple diffusion

F. Nonionic diffusion

$H^+ + NH_4^+$ → $NH_4^+ + H^+$
NH_3 → NH_3

$H^+ + HCOO^-$ → $HCOO^- + H^+$
$HCOOH$ → $HCOOH$

G. Passive carrier transport

Carrier protein

(after a conformational change) ($\rightarrow$ **G**). As in simple diffusion, a concentration gradient is necessary for such carrier-mediated transport (passive transport), e.g., with GLUT uniporters for glucose ($\rightarrow$ p. 158). On the other hand, this type of "facilitated diffusion" is subject to *satu-* *ration* and is *specific* for structurally similar substances that may *competitively inhibit* one another. The carriers in both passive and active transport have the latter features in common ($\rightarrow$ p. 26).

Plate 1.11 Passive Transport by Means of Diffusion II

Osmosis, Filtration and Convection

Water flow or *volume flow* (J_V) across a membrane, in living organisms is achieved through *osmosis* (diffusion of water) or *filtration*. They can occur only if the membrane is water-permeable. This allows osmotic and hydrostatic pressure differences ($\Delta\pi$ and ΔP) across the membrane to drive the fluids through it.

Osmotic flow equals the hydraulic conductivity (K_f) times the osmotic pressure difference ($\Delta\pi$) ($\rightarrow$ **A**):

$$J_V = K_f \cdot \Delta\pi \qquad [1.11]$$

The **osmotic pressure difference** ($\Delta\pi$) can be calculated using *van't Hoff's* law, as modified by *Staverman*:

$$\Delta\pi = \sigma \cdot R \cdot T \cdot \Delta C_{osm}, \qquad [1.12]$$

where σ is the reflection coefficient of the particles (see below), R is the universal gas constant ($\rightarrow$ p. 20), T is the absolute temperature, and ΔC_{osm} [osm $\cdot$ kgH$_2$O^{-1}] is the difference between the lower and higher particle concentrations, $C_{osm}^a - C_{osm}^b$ ($\rightarrow$ **A**). Since ΔC_{osm}, the *driving force for osmosis*, is a negative value, J_V is also negative (Eq. 1.11). The water therefore flows against the concentration gradient of the solute particles. In other words, the higher concentration, C_{osm}^b, *attracts* the water. When the *concentration of water* is considered in osmosis, the H$_2$O concentration in **A,a**, $C_{H_2O}^a$, is greater than that in **A,b**, $C_{H_2O}^b$. $C_{H_2O}^a - C_{H_2O}^b$ is therefore the *driving force for H$_2$O diffusion* ($\rightarrow$ **A**). Osmosis also cannot occur unless the reflection coefficient is greater than zero ($\sigma > 0$), that is, unless the membrane is less permeable to the solutes than to water.

Aquaporins (AQP) are *water channels* that permit the passage of water in many cell membranes. A chief cell in the renal collecting duct contains a total of ca. 107 water channels, comprising AQP2 (regulated) in the luminal membrane, and AQP3 and 4 (permanent?) in the basolateral membrane. The permeability of the epithelium of the renal collecting duct to water ($\rightarrow$ **A**, right panel) is controlled by the *insertion and removal of AQP2*, which is stored in the membrane of intracellular vesicles. In the presence of the antidiuretic hormone ADH (V$_2$ receptors, cAMP; $\rightarrow$ p. 274), water channels are inserted in the luminal membrane within minutes, thereby increasing the water permeability of the membrane to around 1.5×10^{-17} L s^{-1} per channel.

In **filtration** ($\rightarrow$ **B**),

$$J_V = K_f \cdot \Delta P - \Delta\pi \qquad [1.13]$$

Filtration occurs through *capillary walls*, which allow the passage of small ions and molecules ($\sigma = 0$; see below), but not of plasma proteins ($\rightarrow$ **B**, molecule x). Their concentration difference leads to an oncotic pressure difference ($\Delta\pi$) that opposes ΔP. Therefore, filtration can occur only if $\Delta P > \Delta\pi$ ($\rightarrow$ **B**, p. 152, p. 208).

Solvent drag occurs when solute particles are carried along with the water flow of osmosis or filtration. The amount of solvent drag for solute X (J_X) depends mainly on osmotic flow (J_V) and the mean *solute activity* $\bar{a}_x$ ($\rightarrow$ p. 376) at the site of penetration, but also on the degree of particle reflection from the membrane, which is described using the **reflection coefficient (σ)**. Solvent drag for solute X (J_X) is therefore calculated as

$$J_x = J_V (1 - \sigma) \, \bar{a}_x \, [\text{mol} \cdot \text{s}^{-1}] \qquad [1.14]$$

Larger molecules such as proteins are entirely reflected, and $\sigma = 1$ ($\rightarrow$ **B**, molecule X). Reflection of smaller molecules is lower, and $\sigma < 1$. When urea passes through the wall of the proximal renal tubule, for example, $\sigma = 0.68$. The value $(1-\sigma)$ is also called the *sieving coefficient* ($\rightarrow$ p. 154).

Plasma protein binding occurs when small-molecular substances in plasma bind to proteins ($\rightarrow$ **C**). This hinders the free penetration of the substances through the endothelium or the glomerular filter ($\rightarrow$ p. 154 ff.). At a glomerular filtration fraction of 20%, 20% of a freely filterable substance is filtered out. If, however, 9/10 of the substance is bound to plasma proteins, only 2% will be filtered during each renal pass.

Convection functions to transport solutes over *long distances*—e.g., in the circulation or urinary tract. The solute is then carried along like a piece of driftwood. The quantity of solute transported over time (J_{conv}) is the product of volume flow J_V (in m$^3 \cdot$ s^{-1}) and the solute concentration C (mol $\cdot$ m^{-3}):

$$J_{conv} = J_V \cdot C \, [\text{mol} \cdot \text{s}^{-1}]. \qquad [1.15]$$

The flow of gases in the respiratory tract, the transmission of heat in the blood and the release of heat in the form of warmed air occurs through convection ($\rightarrow$ p. 222).

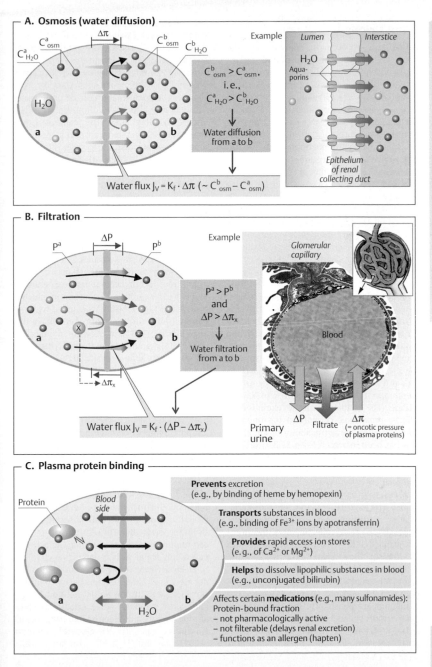

A. Osmosis (water diffusion)

$C^a_{H_2O}$ C^a_{osm} $\Delta\pi$ C^b_{osm} $C^b_{H_2O}$

H_2O

a b

$C^b_{osm} > C^a_{osm}$,
i.e.,
$C^a_{H_2O} > C^b_{H_2O}$

Water diffusion from a to b

Water flux $J_V = K_f \cdot \Delta\pi$ ($\sim C^b_{osm} - C^a_{osm}$)

Example

Lumen *Interstice*

H_2O
Aquaporins

Epithelium of renal collecting duct

B. Filtration

P^a ΔP P^b

a b

X

$\Delta\pi_x$

$P^a > P^b$
and
$\Delta P > \Delta\pi_x$

Water filtration from a to b

Water flux $J_V = K_f \cdot (\Delta P - \Delta\pi_x)$

Example

Glomerular capillary

Blood

ΔP Filtrate $\Delta\pi$
(= oncotic pressure of plasma proteins)

Primary urine

C. Plasma protein binding

Protein

Blood side

a b

H_2O

Prevents excretion
(e.g., by binding of heme by hemopexin)

Transports substances in blood
(e.g., binding of Fe^{3+} ions by apotransferrin)

Provides rapid access ion stores
(e.g., of Ca^{2+} or Mg^{2+})

Helps to dissolve lipophilic substances in blood
(e.g., unconjugated bilirubin)

Affects certain **medications** (e.g., many sulfonamides):
Protein-bound fraction
– not pharmacologically active
– not filterable (delays renal excretion)
– functions as an allergen (hapten)

Plate 1.12 Osmosis, Filtration and Convection

Active Transport

Active transport occurs in many parts of the body when solutes are transported against their concentration gradient (*uphill transport*) and/or, in the case of ions, against an electrical potential (→ p. 22). All in all, active transport occurs against the *electrochemical gradient* or *potential* of the solute. Since passive transport mechanisms represent "downhill" transport (→ p. 20 ff.), they are not appropriate for this task. Active transport requires the **expenditure of energy**. A large portion of chemical energy provided by foodstuffs is utilized for active transport once it has been made readily available in the form of ATP (→ p. 41). The energy created by ATP hydrolysis is used to drive the transmembrane transport of numerous ions, metabolites, and waste products. According to the laws of thermodynamics, the energy expended in these reactions produces *order* in cells and organelles—a prerequisite for survival and normal function of cells and, therefore, for the whole organism (→ p. 38 ff.).

In **primary active transport**, the energy produced by hydrolysis of ATP goes *directly* into ion transport through an ion pump. This type of ion pump is called an **ATPase**. They establish the electrochemical gradients rather slowly, e.g., at a rate of around $1\,\mu mol \cdot s^{-1} \cdot m^{-2}$ of membrane surface area in the case of Na^+-K^+-ATPase. The gradient can be exploited to achieve *rapid ionic currents* in the opposite direction after the permeability of ion channels has been increased (→ p. 32 ff.). Na^+ can, for example, be driven into a nerve cell at a rate of up to $1000\,\mu mol \cdot s^{-1} \cdot m^{-2}$ during an action potential.

ATPases occur ubiquitously in cell membranes (Na^+-K^+-ATPase) and in the endoplasmic reticulum and plasma membrane (Ca^{2+}-ATPase), renal collecting duct and stomach glands (H^+,K^+-ATPase), and in lysosomes (H^+-ATPase). They transport Na^+, K^+, Ca^{2+} and H^+, respectively, by primarily active mechanisms. All except H^+-ATPase consist of 2 α-subunits and 2 β-subunits (P-type ATPases). The α-subunits are phosphorylated and form the ion transport channel (→ **A1**).

Na^+-K^+-ATPase is responsible for maintenance of *intracellular Na^+ and K^+ homeostasis*

and, thus, for maintenance of the *cell membrane potential*. During each transport cycle (→ **A1, A2**), 3 Na^+ and 2 K^+ are "pumped" out of and into the cell, respectively, while 1 ATP molecule is used to phosphorylate the carrier protein (→ **A2b**). Phosphorylation first changes the conformation of the protein and subsequently alters the affinities of the Na^+ and K^+ binding sites. The conformational change is the actual ion transport step since it moves the binding sites to the opposite side of the membrane (→ **A2b–d**). Dephosphorylation restores the pump to its original state (→ **A2e–f**). The Na^+/K^+ pumping rate increases when the cytosolic Na^+ concentration rises—due, for instance, to increased Na^+ influx, or when the extracellular K^+ rises. Therefore, Na^+,K^+-*activatable* ATPase is the full name of the pump. Na-$^+K^+$-ATPase is inhibited by *ouabain* and *cardiac glycosides*.

Secondary active transport occurs when uphill transport of a compound (e.g., glucose) via a carrier protein (e.g., sodium glucose transporter type 2, SGLT2) is coupled with the passive (downhill) transport of an ion (in this example Na^+; → **B1**). In this case, the electrochemical Na^+ gradient into the cell (created by Na^+-K^+-ATPase at another site on the cell membrane; → **A**) provides the driving force needed for secondary active uptake of glucose into the cell. Coupling of the transport of two compounds across a membrane is called **cotransport**, which may be in the form of symport or antiport. **Symport** occurs when the two compounds (i.e., compound and driving ion) are transported across the membrane in the same direction (→ **B1–3**). **Antiport** (countertransport) occurs when they are transported in opposite directions. Antiport occurs, for example, when an electrochemical Na^+ gradient drives H^+ in the opposite direction by secondary active transport (→ **B4**). The resulting H^+ gradient can then be exploited for *tertiary active symport* of molecules such as peptides (→ **B5**).

Electroneutral transport occurs when the net electrical charge remains balanced during transport, e.g., during Na^+/H^+ antiport (→ **B4**) and Na^+-Cl^- symport (→ **B2**). Small charge separation occurs in **electrogenic (rheogenic) transport**, e.g., in Na^+-glucose^0 symport (→ **B1**), Na^+-amino acid^0 symport (→ **B3**),

▶

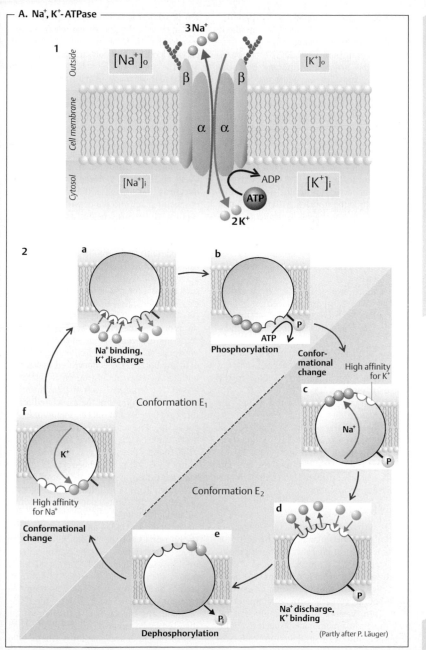

A. Na⁺, K⁺-ATPase

1

Outside — $[Na^+]_o$ — 3 Na⁺ — $[K^+]_o$

β — α — α — β

Cell membrane

Cytosol — $[Na^+]_i$ — ADP — ATP — $[K^+]_i$ — 2 K⁺

2

a Na⁺ binding, K⁺ discharge

b Phosphorylation — ATP — P

Conformational change — High affinity for K⁺

c Na⁺ — P

Conformation E₁

d Na⁺ discharge, K⁺ binding — P

e Dephosphorylation — Pᵢ

Conformation E₂

f Conformational change — K⁺ — High affinity for Na⁺

(Partly after P. Läuger)

Plate 1.13 Active Transport I

2 Na$^+$-amino acid$^-$ symport, or H$^+$-peptide^0 symport ($\rightarrow$ **B5**). The chemical Na$^+$ gradient provides the sole driving force for electroneutral transport (e.g., Na$^+$/H$^+$ antiport), whereas the negative membrane potential ($\rightarrow$ p. 32 ff.) provides an additional driving force for electrogenication–coupled cotransport into the cell. When secondary active transport (e.g., of glucose) is coupled with the influx of not one but two Na$^+$ ions (e.g., SGLT1 symporter), the driving force is doubled. The aid of ATPases is necessary, however, if the required "uphill" concentration ratio is several decimal powers large, e.g., 10^6 in the extreme case of H$^+$ ions across the luminal membrane of parietal cells in the stomach. ATPase-mediated transport can also be electrogenic or electroneutral, e.g., Na$^+$-K$^+$-ATPase (3 Na$^+$/2 K$^+$; cf. p. 46) or H$^+$/K$^+$-ATPase (1 H$^+$/1 K$^+$), respectively.

Characteristics of active transport:

◆ It can be saturated, i.e., it has a limited *maximum capacity* (J$_{max}$).

◆ It is more or less *specific*, i.e., a carrier molecule will transport only certain chemically similar substances which inhibit the transport of each other (*competitive inhibition*).

◆ Variable quantities of the similar substances are transported at a given concentration, i.e., each has a different *affinity* (~1/K$_M$) to the transport system.

◆ Active transport is inhibited when the *energy supply* to the cell is disrupted.

All of these characteristics except the last apply to passive carriers, that is, to uniporter-mediated (facilitated) diffusion ($\rightarrow$ p. 22).

The **transport rate** of saturable transport (J$_{sat}$) is usually calculated according to Michaelis–Menten kinetics:

$$J_{sat} = J_{max} \cdot \frac{C}{K_M + C} \qquad [1.16]$$

where C is the concentration of the substrate in question, J$_{max}$ is its maximum transport rate, and K$_M$ is the substrate concentration that produces one-half J$_{max}$ ($\rightarrow$ p. 383).

Cytosis is a completely different type of active transport involving the formation of membrane-bound vesicles with a diameter of 50–400 nm. **Vesicles** are either pinched off from the plasma membrane (*exocytosis*) or incorporated into it by invagination (*endocytosis*) in conjunction with the expenditure of ATP. In cytosis, the uptake and release of *macromolecules* such as proteins, lipoproteins, polynucleotides, and polysaccharides into and out of a cell occurs by specific mechanisms similar to those involved in intracellular transport ($\rightarrow$ p. 12 ff.).

Endocytosis ($\rightarrow$ p. 13) can be broken down into different types, including pinocytosis, receptor-mediated endocytosis, and phagocytosis. **Pinocytosis** is characterized by the continuous unspecific uptake of extracellular fluid and molecules dissolved in it through relatively small vesicles. **Receptor-mediated endocytosis** ($\rightarrow$ **C**) involves the selective uptake of specific macromolecules with the aid of receptors. This usually begins at small depressions (*pits*) on the plasma membrane surface. Since the insides of the pits are often densely covered with the protein *clathrin*, they are called *clathrin-coated pits*. The **receptors** involved are integral cell membrane proteins such as those for low-density lipoprotein (LPL; e.g., in hepatocytes) or intrinsic factor-bound cobalamin (e.g., in ileal epithelial cells). Thousands of the same receptor type or of different receptors can converge at coated pits ($\rightarrow$ **C**), yielding a tremendous increase in the efficacy of ligand uptake. The endocytosed vesicles are initially coated with clathrin, which is later released. The vesicles then transform into *early endosomes*, and most of the associated receptors circulate back to the cell membrane ($\rightarrow$ **C** and p. 13). The endocytosed ligand is either exocytosed on the opposite side of the cell (*transcytosis*, see below), or is digested by *lysosomes* ($\rightarrow$ **C** and p. 13). **Phagocytosis** involves the endocytosis of particulate matter, such as microorganisms or cell debris, by phagocytes ($\rightarrow$ p. 94 ff.) in conjunction with lysosomes. Small digestion products, such as amino acids, sugars and nucleotides, are transported out of the lysosomes into the cytosol, where they can be used for cellular metabolism or secreted into the extracellular fluid. When certain hormones such as *insulin* ($\rightarrow$ p. 282) bind to receptors on the surface of target cells, hormone-receptor complexes can also enter the coated pits and are endocytosed (*internalized*) and digested by lysosomes. This reduces the density of receptors available for hormone bind-

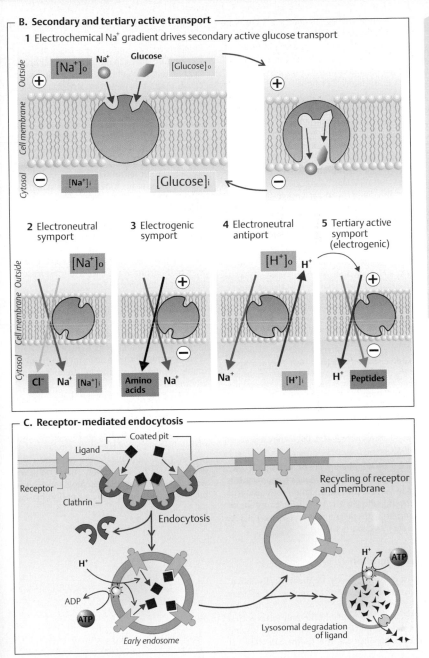

B. Secondary and tertiary active transport

1 Electrochemical Na^+ gradient drives secondary active glucose transport

Outside

$[Na^+]_o$ Na^+ Glucose $[Glucose]_o$

Cell membrane

Cytosol

$[Na^+]_i$ $[Glucose]_i$

2 Electroneutral symport

$[Na^+]_o$

Cl^- Na^+ $[Na^+]_i$

3 Electrogenic symport

Amino acids Na^+

4 Electroneutral antiport

$[H^+]_o$ H^+

Na^+ $[H^+]_i$

5 Tertiary active symport (electrogenic)

H^+ Peptides

Outside

Cell membrane

Cytosol

C. Receptor-mediated endocytosis

Coated pit

Ligand

Receptor

Clathrin

Endocytosis

H^+

ADP

ATP

Early endosome

Recycling of receptor and membrane

H^+

ATP

Lysosomal degradation of ligand

Plate 1.14   Active Transport II

29

ing. In other words, an increased hormone supply *down-regulates* the receptor density.

Exocytosis (→ p. 13) is a method for selective export of macromolecules out of the cell (e.g., pancreatic enzymes; → p. 246 ff.) and for release of many hormones (e.g., posterior pituitary hormone; → p. 280) or neurotransmitters (→ p. 50 ff.). These substances are kept "packed" and readily available in (clathrin-coated) *secretory vesicles*, waiting to be released when a certain signal is received (increase in cytosolic Ca^{2+}). The "packing material" (vesicle membrane) is later re-endocytosed and *recycled*. Exocytotic membrane fusion also helps to insert vesicle-bound proteins into the plasma membrane (→ p. 13).The liquid contents of the vesicle then are automatically emptied in a process called **constitutive exocytosis** (→ **D**).

In constitutive exocytosis, the protein complex *coatomer* (coat assembly protomer) takes on the role of clathrin (see above). Within the Golgi membrane, GNRP (guanine nucleotide-releasing protein) phosphorylates the GDP of the ADP-ribosylation factor (ARF) to GTP → **D1**, resulting in the dispatch of vesicles from the trans-Golgi network. ARF-GTP complexes then anchor on the membrane and bind with coatomer (→ **D2**), thereby producing **coatomer-coated vesicles** (→ **D3**). The membranes of the vesicles contain v-SNAREs (vesicle synaptosome-associated protein receptors), which recognize t-SNAREs (target-SNAREs) in the target membrane (the plasma membrane, in this case). This results in cleavage of ARF-GTP, dissociation of ARF-GDP and coatomer molecules and, ultimately, to membrane fusion and exocytosis (→ **D4, D5**) to the extracellular space (ECS).

Transcytosis is the uptake of macromolecules such as proteins and hormones by endocytosis on one side of the cell, and their release on the opposite side. This is useful for *transcellular transport of the macromolecules* across cell layers such as endothelia.

Cell Migration

Most cells in the body are theoretically able to move from one place to another or *migrate* (→ **E**), but only a few cell species actually do so. The *sperm* are probably the only cells with a special propulsion mechanism. By waving their whip-like tail, the sperm can travel at speeds of up to around 2000 μm/min. Other cells also migrate, but at much slower rates. *Fibroblasts*, for example, move at a rate of around 1.2 μm/min. When an injury occurs, fibroblasts migrate to the wound and aid in the formation of scar tissue. Cell migration also plays a role in *embryonal development*. Chemotactically attracted *neutrophil granulocytes* and *macrophages* can even migrate through vessel walls to attack invading bacteria (→ p. 94ff.). Cells of some tumors can also migrate to various tissues of the body or *metastasize*, thereby spreading their harmful effects.

Cells migrate by "crawling" on a stable surface (→**E1**). The following activities occur during cell migration:

◆ *Back end of the cell*: (a) Depolymerization of actin and tubulin in the cytoskeleton; (b) endocytosis of parts of the cell membrane, which are then propelled forward as endocytotic vesicles to the front of the cell, and (c) release of ions and fluids from the cell.

◆ *Front end of the cell (lamellipodia)*: (a) Polymerization of actin monomers is achieved with the aid of profilin (→**E2**). The monomers are propelled forward with the help of plasma membrane-based myosin I (fueled by ATP); (b) reinsertion of the vesicles in the cell membrane; (c) uptake of ions and fluids from the environment.

Parts of the cell membrane that are not involved in cytosis are conveyed from front to back, as on a track chain. Since the cell membrane is attached to the stable surface (primarily fibronectin of the extracellular matrix in the case of fibroblasts), the cell moves forward relative to the surface. This is achieved with the aid of specific receptors, such as fibronectin receptors in the case of fibroblasts.

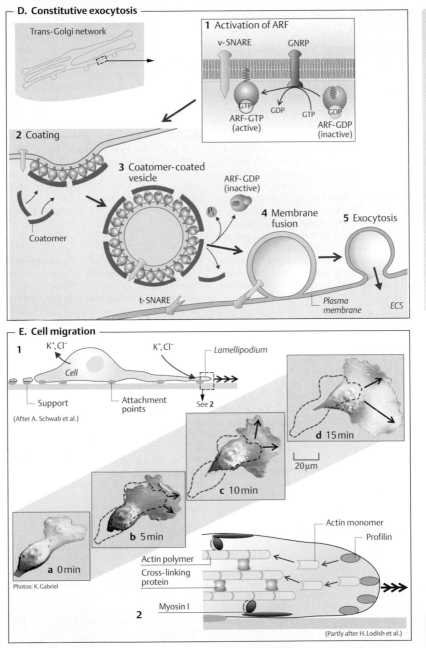

Plate 1.15 Cell Migration

D. Constitutive exocytosis

Trans-Golgi network

1 Activation of ARF

v-SNARE   GNRP

GTP

GDP   GTP

ARF-GTP
(active)

ARF-GDP
(inactive)

GDP

2 Coating

3 Coatomer-coated
vesicle

ARF-GDP
(inactive)

P_i

4 Membrane
fusion

5 Exocytosis

Coatomer

t-SNARE

Plasma
membrane

ECS

E. Cell migration

1

K^+, Cl^- K^+, Cl^-

Lamellipodium

Cell

See 2

Support   Attachment
points

(After A. Schwab et al.)

d 15 min

20 μm

c 10 min

b 5 min

a 0 min

Photos: K. Gabriel

Actin monomer

Profilin

Actin polymer

Cross-linking
protein

Myosin I

2

(Partly after H. Lodish et al.)

Electrical Membrane Potentials and Ion Channels

An **electrical potential difference** occurs due to the *net movement of charge during ion transport*. A **diffusion potential** develops for instance, when ions (e.g., K^+) diffuse (down a chemical gradient; → p. 20ff.) out of a cell, making the cell interior negative relative to the outside. The rising diffusion potential then drives the ions back into the cell (potential-driven ion transport; → p. 22). Outward K^+ diffusion persists until equilibrium is reached. At equilibrium, the two opposing forces become equal and opposite. In other words, the sum of the two or the **electrochemical gradient** (and thereby the electrochemical potential) equals zero, and there is no further net movement of ions (*equilibrium concentration*) at a certain voltage (*equilibrium potential*).

The **equilibrium potential (E_X)** for any species of ion X distributed inside (i) and outside (o) a cell can be calculated using the Nernst equation:

$$E_X = \frac{R \cdot T}{F \cdot z_X} \cdot \ln \frac{[X]_o}{[X]_i} \; [V] \qquad [1.17]$$

where R is the universal gas constant ($= 8.314$ $J \cdot K^{-1} \cdot mol^{-1}$), T is the absolute temperature ($310\,°K$ in the body), F is the Faraday constant or charge per mol ($= 9.65 \times 10^4$ $A \cdot s \cdot mol^{-1}$), z is the valence of the ion in question ($+1$ for K^+, $+2$ for Ca^{2+}, -1 for Cl^-, etc.), ln is the natural logarithm, and [X] is the effective concentration = *activity* of the ion X (→ p. 376). $R \cdot T/F = 0.0267$ V^{-1} at body temperature ($310\,°K$). It is sometimes helpful to convert $\ln([X]_o/[X]_i)$ into $-\ln([X]_i/[X]_o)$, V into mV and ln into log before calculating the equilibrium potential (→ p. 380). After insertion into Eq. 1.17, the **Nernst equation** then becomes

$$E_X = -61 \cdot \frac{1}{z_X} \cdot \log \frac{[X]_i}{[X]_o} \; [mV] \qquad [1.18]$$

If the ion of species X is K^+, and $[K^+]_i = 140$, and $[K^+]_o = 4.5$ mmol/kg H_2O, the equilibrium potential $E_K = -61 \cdot \log 31$ mV or -91 mV. If the cell membrane is permeable only to K^+, the **membrane potential (E_m)** will eventually reach a value of -91 mV, and E_m will equal E_K (→ **A1**).

At equilibrium potential, the chemical gradient will drive just as many ions of species X in the one direction as the electrical potential does in the opposite direction. The **electrochemical potential ($E_m - E_X$)** or so-called electrochemical driving "force", will equal zero, and the sum of ionic inflow and outflow or the **net flux (I_X)** will also equal zero.

Membrane conductance (g_X), a concentration-dependent variable, is generally used to describe the permeability of a cell membrane to a given ion instead of the permeability coefficient P (see Eq. 1.5 on p. 22 for conversion). Since it is relative to membrane surface area, g_X is expressed in siemens (S = $1/\Omega$) per m^2 (→ p. 22, Eq. 1.9). **Ohm's law** defines the net ion current (I_X) per unit of membrane surface area as

$$I_X = g_X \cdot (E_m - E_X) \; [A \cdot m^{-2}] \qquad [1.19]$$

I_X will therefore differ from zero when the prevailing membrane potential, E_m, does not equal the equilibrium potential, E_X. This occurs, for example, after strong transient activation of Na^+-K^+-ATPase (electrogenic; → p. 26): hyperpolarization of the membrane (→ **A2**), or when the cell membrane conducts more than one ion species, e.g., K^+ as well as Cl^- and Na^+: depolarization (→ **A3**). If the membrane is permeable to different ion species, the *total conductance* of the membrane (g_m) equals the sum of all parallel conductances ($g_1 + g_2 + g_3 + ...$). The **fractional conductance** for the ion species X **(f_X)** can be calculated as

$$f_X = g_X/g_m \qquad [1.20]$$

The membrane potential, E_m, can be determined if the fractional conductances and equilibrium potentials of the conducted ions are known (see Eq. 1.18). Assuming K^+, Na^+, and Cl^- are the ions in question,

$$E_m = (E_K \cdot f_K) + (E_{Na} \cdot f_{Na}) + (E_{Cl} \cdot f_{Cl}) \qquad [1.21]$$

Realistic values in *resting nerve cells are*: $f_K = 0.90$, $f_{Na} = 0.03$, $f_{Cl} = 0.07$; $E_K = -90$ mV, $E_{Na} = +70$ mV, $E_{Cl} = -83$ mV. Inserting these values into equation 1.21 results in an E_m of -85 mV. Thus, the driving forces (= electrochemical potentials = $E_m - E_X$), equal $+5$ mV for K^+, -145 mV for Na^+, and -2 mV for Cl^-. The driv-

▶

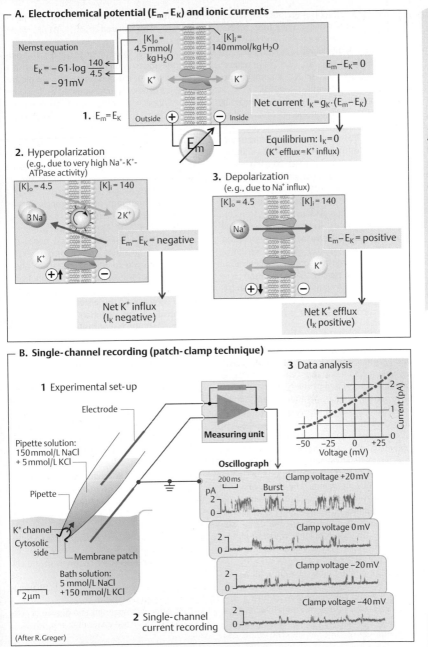

A. Electrochemical potential ($E_m - E_K$) and ionic currents

Nernst equation

$$E_K = -61 \cdot \log \frac{140}{4.5}$$
$$= -91\,mV$$

1. $E_m = E_K$

$[K]_o = 4.5\,mmol/kg\,H_2O$ $[K]_i = 140\,mmol/kg\,H_2O$

$E_m - E_K = 0$

Net current $I_K = g_K \cdot (E_m - E_K)$

Equilibrium: $I_K = 0$
(K^+ efflux = K^+ influx)

Outside ⊕   ⊖ Inside

E_m

2. Hyperpolarization
(e.g., due to very high Na^+-K^+-ATPase activity)

$[K]_o = 4.5$ $[K]_i = 140$

$3\,Na^+$ $2\,K^+$

$E_m - E_K$ = negative

K^+

⊕↑   ⊖

Net K^+ influx
(I_K negative)

3. Depolarization
(e.g., due to Na^+ influx)

$[K]_o = 4.5$ $[K]_i = 140$

Na^+

$E_m - E_K$ = positive

K^+

⊕↓   ⊖

Net K^+ efflux
(I_K positive)

B. Single-channel recording (patch-clamp technique)

1 Experimental set-up

Electrode

Measuring unit

Pipette solution:
150 mmol/L NaCl
+ 5 mmol/L KCl

Pipette

K^+ channel
Cytosolic side

Membrane patch

Bath solution:
5 mmol/L NaCl
+150 mmol/L KCl

2 μm

2 Single-channel current recording

3 Data analysis

Current (pA)

-50 -25 0 $+25$
Voltage (mV)

Oscillograph

200 ms   Burst   Clamp voltage +20 mV
pA
2
0

Clamp voltage 0 mV
2
0

Clamp voltage −20 mV
2
0

Clamp voltage −40 mV
2
0

(After R. Greger)

Plate 1.16   Electrical Potentials I

ing force for K⁺ efflux is therefore low, though g_K is high. Despite a high driving force for Na⁺, Na⁺ influx is low because the g_{Na} and f_{Na} of resting cells are relatively small. Nonetheless, the sodium current, I_{Na}, can rise tremendously when large numbers of Na⁺ channels open during an action potential ($\rightarrow$ p. 46).

Electrodiffusion. The potential produced by the transport of one ion species can also drive other cations or anions through the cell membrane ($\rightarrow$ p. 22), provided it is permeable to them. The K⁺diffusion potential leads to the efflux of Cl⁻, for example, which continues until $E_{Cl} = E_m$. According to Equation 1.18, this means that the cytosolic Cl⁻ concentration is reduced to 1/25 th of the extracellular concentration (*passive distribution* of Cl⁻ between cytosol and extracellular fluid). In the above example, there was a small electrochemical Cl⁻ potential driving Cl⁻ out of the cell ($E_m - E_{Cl} = -2$ mV). This means that the cytosolic Cl⁻ concentration is higher than in passive Cl⁻ distribution ($E_{Cl} = E_m$). Therefore, Cl⁻ ions must have been actively taken up by the cell, e.g., by a Na⁺- Cl⁻ symport carrier ($\rightarrow$ p. 29 B): *active distribution* of Cl⁻.

To achieve ion transport, membranes have a variable number of **channels** (pores) specific for different ion species (Na⁺, Ca²⁺, K⁺, Cl⁻, etc.). The conductance of the cell membrane is therefore determined by the type and number of ion channels that are momentarily open. *Patch–clamp techniques* permit the direct measurement of *ionic currents through single ion channels* ($\rightarrow$ **B**). Patch–clamp studies have shown that the conductance of a membrane does not depend on the change of the pore diameter of its ion channels, but on their average frequency of opening. The ion permeability of a membrane is therefore related to the **open-probability** of the channels in question. Ion channels open in frequent bursts ($\rightarrow$ **B2**). Several ten thousands of ions pass through the channel during each individual burst, which lasts for only a few milliseconds.

During a **patch–clamp recording**, the opening (0.3–3 μm in diameter) of a glass electrode is placed over a cell membrane in such a way that the opening covers only a small part of the membrane (*patch*) containing only one or a small number of ion channels. The whole cell can either be left intact, or a membrane patch can excised for isolated study ($\rightarrow$ **B1**). In *single-channel recording*, the membrane is kept at a preset value (*voltage clamp*). This permits the measurement of ionic current in a single channel. The measurements are plotted ($\rightarrow$ **B3**) as current (I) over voltage (V). The slope of the *I/V curve* corresponds to the conductance of the channel for the respective ion species (see Eq. 1.18). The *zero-current potential* is defined as the voltage at which the I/V curve intercepts the x-axis of the curve (I = 0). The ion species producing current I can be deduced from the zero-current potential. In example **B**, the zero-current potential equals – 90 mV. Under the conditions of this experiment, an electrochemical gradient exists only for Na⁺ and K⁺, but not for Cl⁻ ($\rightarrow$ **B**). At these gradients, $E_K = -90$ mV and $E_{Na} = +90$ mV. As E_K equals the zero-current potential, the channel is exclusively permeable to K⁺ and does not allow other ions like Na⁺ to pass. The channel type can also be determined by adding specific channel blockers to the system.

Control of ion channels ($\rightarrow$ **C**). Channel open-probability is controlled by five main factors:

◆ *Membrane potential*, especially in Na⁺, Ca²⁺ and K⁺ channels in nerve and muscle fibers ($\rightarrow$ **C1**; pp. 46 and 50).

◆ *External ligands* that bind with the channel ($\rightarrow$ **C2**). This includes acetylcholine on the postsynaptic membrane of nicotinic synapses (cation channels), glutamate (cation channels), and glycine or GABA (Cl⁻ channels).

◆ *Intracellular messenger substances* ($\rightarrow$ **C3**) such as:

— cAMP (e.g., in Ca²⁺ channels in myocardial cells and Cl⁻ channels in epithelial cells);

— cGMP (plays a role in muscarinergic effects of acetylcholine and in excitation of the retinal rods);

— IP3 (opening of Ca²⁺ channels of intracellular Ca²⁺ stores);

— Small G-proteins (Ca²⁺ channels of the cell membrane);

— Tyrosine kinases (Cl⁻ and K⁺ channels during apoptosis);

— Ca²⁺ (affects K⁺ channels and degree of activation of rapid Na⁺ channels; $\rightarrow$ p. 46).

Plate 1.17 Electrical Potentials II

C. Control of ion channels

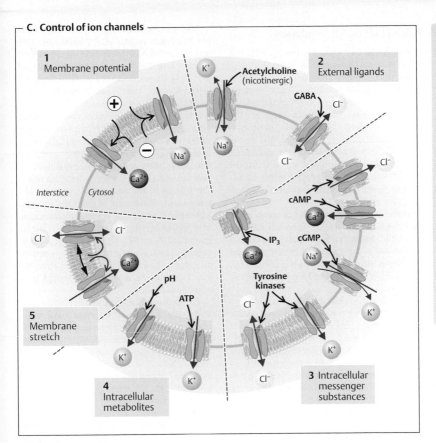

1 Membrane potential

K⁺

Acetylcholine (nicotinergic)

GABA

Cl⁻

2 External ligands

Cl⁻

Cl⁻

Na⁺

Na⁺

cAMP

Cl⁻

Ca²⁺

Interstice Cytosol

Ca²⁺

cGMP

Na⁺

IP₃

Cl⁻

Ca²⁺

Tyrosine kinases

pH

Cl⁻

K⁺

ATP

5 Membrane stretch

Cl⁻

Ca²⁺

K⁺

K⁺

K⁺

K⁺

Cl⁻

4 Intracellular metabolites

3 Intracellular messenger substances

◆ *Intracellular metabolites* (→ **C4**) such as ATP (e.g., in K⁺ channels in the heart and B cells in pancreatic islets) or *H+ ions* (e.g., in K⁺ channels in renal epithelial cells);

◆ *Membrane stretch* (→ **C5**), the direct or indirect (?) effects of which play a role in Ca²⁺ channels of smooth muscle fibers and generally in normal K⁺ and Cl⁻ channels in swelling cells.

Role of Ca²⁺ in Cell Regulation

The **cytosolic Ca²⁺ concentration, [Ca²⁺]ᵢ**, (ca. 0.1 to 0.01 µmol/L) is several decimal powers lower than the extracellular Ca²⁺ concentration **[Ca²⁺]ₒ** (ca. 1.3 mmol/L). This is because Ca²⁺ is continuously pumped from the cytosol into **intracellular Ca²⁺ stores** such as the endoplasmic and sarcoplasmic reticulum (→ p. 17 A), vesicles, mitochondria and nuclei (?) or is *transported out of the cell*. Both processes occur by primary active transport (Ca²⁺-ATPases) and, in the case of efflux, by additional secondary active transport through Ca²⁺/3 Na⁺ antiporters (→ **A1**).

To **increase the cytosolic Ca²⁺ concentration, Ca²⁺ channels** conduct Ca²⁺ from intracellular stores and the extracellular space into the cytosol (→ **A2**). The frequency of Ca²⁺ channel opening in the cell membrane is increased by
- *Depolarization* of the cell membrane (nerve and muscle cells);
- *Ligands* (e.g., via G_o proteins; → p. 274);
- *Intracellular messengers* (e.g., IP₃ and cAMP; → p. 274ff.);
- *Stretching or heating* of the cell membrane.

The Ca²⁺ channels of the endoplasmic and sarcoplasmic reticulum open more frequently in response to signals such as a rise in [Ca²⁺]ᵢ (influx of external Ca²⁺ works as the "spark" or trigger) or inositol *tris*-phosphate (IP₃; → **A2** and p. 276).

A **rise in [Ca²⁺]ᵢ is a signal** for many important cell functions (→ **A**), including myocyte contraction, exocytosis of neurotransmitters in presynaptic nerve endings, endocrine and exocrine hormone secretion, the excitation of certain sensory cells, the closure of gap junctions in various cells (→ p. 19 C), the opening of other types of ion channels, and the migration of leukocytes and tumor cells (→ p. 30) as well as thrombocyte activation and sperm mobilization. Some of these activities are mediated by **calmodulin**. A calmodulin molecule can bind up to 4 Ca²⁺ ions when the [Ca²⁺]ᵢ rises (→ **A2**). The *Ca²⁺-calmodulin complexes* activate a number of different enzymes, including calmodulin-dependent protein kinase II (CaM-kinase II) and myosin light chain kinase (MLCK), which is involved in smooth muscle contraction (→ p. 70).

[Ca²⁺]ᵢ oscillation is characterized by multiple brief and regular [Ca²⁺]ᵢ increases (Ca²⁺ spikes) in response to certain stimuli or hormones (→ **B**). The frequency, not amplitude, of [Ca²⁺]ᵢ oscillation is the quantitative signal for cell response. When low-frequency [Ca²⁺]ᵢ oscillation occurs, CaM-kinase II, for example, is activated and phosphorylates only its target proteins, but is quickly and completely deactivated (→ **B1**, **B3**). High-frequency [Ca²⁺]ᵢ oscillation results in an increasing degree of autophosphorylation and progressively delays the deactivation of the enzyme (→ **B3**). As a result, the activity of the enzyme decays more and more slowly between [Ca²⁺]ᵢ signals, and each additional [Ca²⁺]ᵢ signal leads to a summation of enzyme activity (→ **B2**). As with action potentials (→ p. 46), this frequency-borne, digital *all-or-none* type of signal transmission provides a much clearer message than the [Ca²⁺]ᵢ amplitude, which is influenced by a number of factors.

Ca²⁺ sensors. The extracellular Ca²⁺ concentration [Ca²⁺]ₒ plays an important role in blood coagulation and bone formation as well as in nerve and muscle excitation. [Ca²⁺]ₒ is tightly controlled by hormones such as PTH, calcitriol and calcitonin (→ p. 290), and represents the feedback signal in this control circuit (→ p. 290). The involved Ca²⁺ sensors are membrane proteins that detect **high [Ca²⁺]ₒ levels** on the cell surface and dispatch IP₃ and DAG (diacylglycerol) as intracellular second messengers with the aid of a G_q protein (→ **C1** and p. 274ff.). IP₃ triggers an increase in the [Ca²⁺]ᵢ of parafollicular C cells of the thyroid gland. This induces the exocytosis of calcitonin, a substance that *reduces* [Ca²⁺]ₒ (→ **C2**). In parathyroid cells, on the other hand, a high [Ca²⁺]ₒ reduces the secretion of PTH, a hormone that increases the [Ca²⁺]ₒ. This activity is mediated by DAG and PKC (protein kinase C) and, perhaps, by a (G_i protein-mediated; → p. 274) reduction in the cAMP concentration (→ **C3**). Ca²⁺ sensors are also located on osteoclasts as well as on renal and intestinal epithelial cells.

Plate 1.18 Role of Ca²⁺ in Cell Regulation

A. Role of Ca²⁺ in cell regulation

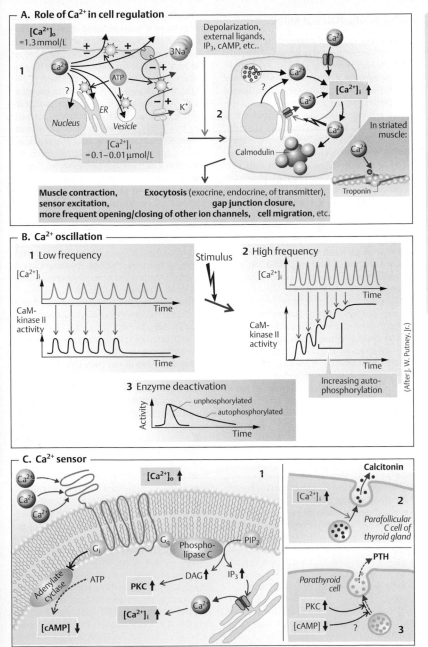

$[Ca^{2+}]_o$ = 1.3 mmol/L

Depolarization, external ligands, IP_3, cAMP, etc..

3Na⁺

ATP

K⁺

1

? Nucleus

ER

Vesicle

$[Ca^{2+}]_i$ = 0.1–0.01 μmol/L

2

Calmodulin

$[Ca^{2+}]_i$ ↑

In striated muscle:

Troponin

Muscle contraction, **Exocytosis** (exocrine, endocrine, of transmitter), **sensor excitation,** **gap junction closure,** **more frequent opening/closing of other ion channels,** **cell migration,** etc.

B. Ca²⁺ oscillation

1 Low frequency

$[Ca^{2+}]_i$

Time

CaM-kinase II activity

Time

Stimulus

2 High frequency

$[Ca^{2+}]_i$

Time

CaM-kinase II activity

Time

Increasing auto-phosphorylation

3 Enzyme deactivation

Activity

unphosphorylated

autophosphorylated

Time

(After J. W. Putney, Jr.)

C. Ca²⁺ sensor

$[Ca^{2+}]_o$ ↑ **1**

G_i

Adenylate Cyclase

ATP

$[cAMP]$ ↓

G_q Phospho-lipase C

PIP_2

DAG ↑

IP_3 ↑

PKC ↑

$[Ca^{2+}]_i$ ↑

Ca²⁺

Calcitonin

$[Ca^{2+}]_i$ ↑

Parafollicular C cell of thyroid gland

2

PTH

Parathyroid cell

PKC ↑

$[cAMP]$ ↓

?

3

Plate 1.18 Role of Ca²⁺ in Cell Regulation

37

Energy Production and Metabolism

Energy is the ability of a system to perform *work*; both are expressed in joules (J). A **potential difference** (potential gradient) is the so-called *driving "force"* that mobilizes the matter involved in the work. Water falling from height X (in meters) onto a power generator, for example, represents the potential gradient in mechanical work. In electrical and chemical work, potential gradients are provided respectively by voltage (V) and a change in free enthalpy ΔG ($J \cdot mol^{-1}$). The amount of work performed can be determined by multiplying the potential difference (*intensity factor*) by the corresponding *capacity factor*. In the case of the water fall, the work equals the height the water falls (m) times the force of the falling water (in N). In the other examples, the amount work performed equals the voltage (V) times the amount of charge (C). Chemical work performed = ΔG times the amount of substance (mol).

Living organisms cannot survive without an adequate supply of energy. Plants utilize solar energy to convert atmospheric CO_2 into oxygen and various organic compounds. These, in turn, are used to fill the energy needs of humans and animals. This illustrates how energy can be converted from one form into another. If we consider such a transformation taking place in a **closed system (**exchange of energy, but not of matter, with the environment), energy can neither appear nor disappear spontaneously. In other words, when energy is converted in a closed system, the total energy content remains constant. This is described in the **first law of thermodynamics**, which states that the change of internal energy (= change of energy content, ΔU) of a system (e.g. of a chemical reaction) equals the sum of the work absorbed (+W) or performed (–W) by a system and the heat lost (–Q) or gained (+Q) by the system. This is described as:

ΔU = heat gained (Q) − work performed
(W) [J] and [1.22]
ΔU = work absorbed (W) − heat lost
(Q) [J]. [1.23]

(By definition, the signs indicate the direction of flow with respect to the system under consideration.)

Heat is transferred in all chemical reactions. The amount of heat produced upon conversion of a given substance into product X is the same, *regardless of the reaction pathway* or whether the system is closed or open, as in a biological system. For caloric values, see p. 228.

Enthalpy change (ΔH) is the heat gained or lost by a system at constant pressure and is related to work, pressure, and volume ($\Delta H = \Delta U + p \cdot \Delta V$). Heat is lost and ΔH is negative in **exothermic** reactions, while heat is gained and ΔH is positive in **endothermic** reactions. The **second law of thermodynamics** states that the total disorder (randomness) or entropy (S) of a closed system increases in any spontaneous process, i.e., **entropy change (ΔS)** > 0. This must be taken into consideration when attempting to determine how much of ΔH is freely available. This free energy or free enthalpy (ΔG) can be used, for example, to drive a chemical reaction. The heat produced in the process is the product of absolute temperature and entropy change ($T \cdot \Delta S$).

Free enthalpy (ΔG) can be calculated using the Gibbs-Helmholtz equation:

$\Delta G = \Delta H - T \cdot \Delta S.$ [1.24]

ΔG and ΔH are approximately equal when ΔS approaches zero. The maximum chemical work of glucose in the body can therefore be determined based on heat transfer, ΔH, measured during the combustion of glucose in a calorimeter (see p. 228 for caloric values). Equation 1.24 also defines the conditions under which chemical reactions can occur. **Exergonic** reactions ($\Delta G < 0$) are characterized by the release of energy and can proceed spontaneously, whereas **endergonic** reactions ($\Delta G > 0$) require the absorption of energy and are not spontaneous. An endothermic reaction ($\Delta H > 0$) can also be exergonic ($\Delta G < 0$) when the entropy change ΔS is so large that $\Delta H - T \cdot \Delta S$ becomes negative. This occurs, for example, in the endothermic dissolution of crystalline NaCl in water.

Free enthalpy, ΔG, is a concentration-dependent variable that can be calculated from the change in **standard free enthalpy (ΔG^0)** and the prevailing concentrations of the substances in question. ΔG^0 is calculated assuming for all reaction partners that concentration = 1 mol/L, pH = 7.0, T = 298 K, and p = 1013 hPa.

▶

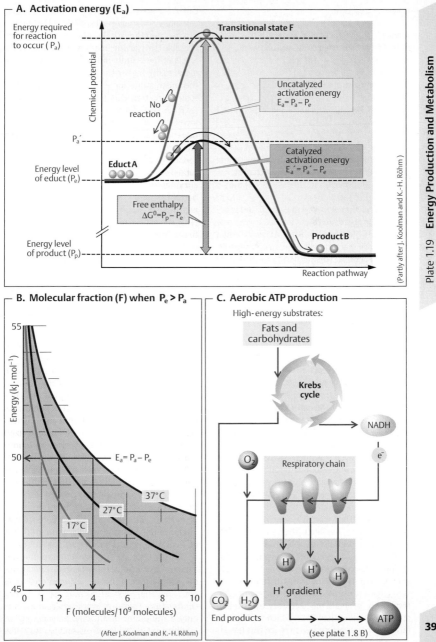

A. Activation energy (E_a)

Energy required for reaction to occur (P_a)

Chemical potential

Transitional state F

Uncatalyzed activation energy
$E_a = P_a - P_e$

No reaction

P_a'

Catalyzed activation energy
$E_a' = P_a' - P_e$

Educt A

Energy level of educt (P_e)

Free enthalpy
$\Delta G^0 = P_p - P_e$

Energy level of product (P_p)

Product B

Reaction pathway

(Partly after J. Koolman and K.-H. Röhm)

B. Molecular fraction (F) when $P_e > P_a$

Energy (kJ · mol^{-1})

$E_a = P_a - P_e$

37°C

27°C

17°C

F (molecules/10^9 molecules)

(After J. Koolman and K.-H. Röhm)

C. Aerobic ATP production

High-energy substrates:

Fats and carbohydrates

Krebs cycle

NADH

e^-

O_2

Respiratory chain

CO_2 H_2O

End products

H$^+$ H$^+$ H$^+$

H$^+$ gradient

ATP

(see plate 1.8 B)

Plate 1.19 **Energy Production and Metabolism**

39

Given the reaction

$$A \rightleftharpoons B + C, \qquad [1.25]$$

where A is the educt and B and C are the products, ΔG^0 is converted to ΔG as follows:

$$\Delta G = \Delta G^0 + R \cdot T \cdot \ln \frac{[B] + [C]}{[A]} \qquad [1.26]$$

or, at a temperature of 37 °C,

$$\Delta G = \Delta G^0 + 8.31 \cdot 310 \cdot 2.3 \cdot \log \frac{[B] + [C]}{[A]} \; [J \cdot mol^{-1}] \qquad [1.27]$$

Assuming the ΔG^0 of a reaction is $+ 20 \; kJ \cdot mol^{-1}$ (endergonic reaction), ΔG will be exergonic (< 0) if [B] · [C] is 10^4 times smaller than A:

$$\Delta G = 20000 + 5925 \cdot \log 10^{-4} = -3.7 \; kJ \cdot mol^{-1}. \qquad [1.28]$$

In this case, A is converted to B and C and reaction 1.25 proceeds to the right.

If $[B] \cdot [C]/[A] = 4.2 \times 10^{-4}$, ΔG will equal zero and the reaction will come to equilibrium (no net reaction). This numeric ratio is called the **equilibrium constant (K_{eq})** of the reaction. K_{eq} can be converted to ΔG^0 and vice versa using Equation 1.26:

$$0 = \Delta G^0 + R \cdot T \cdot \ln K_{eq} \text{ or}$$
$$\Delta G^0 = -R \cdot T \cdot \ln K_{eq} \text{ and} \qquad [1.29]$$

$$K_{eq} = e^{-\Delta G^0/(R \cdot T)}. \qquad [1.30]$$

Conversely, when $[B] \cdot [C]/[A] > 4.2 \times 10^{-4}$, ΔG will be > 0, the net reaction will proceed backwards, and A will arise from B and C.

ΔG is therefore a measure of the direction of a reaction and of its *distance from equilibrium*. Considering the concentration-dependency of ΔG and assuming the reaction took place in an open system (see below) where reaction products are removed continuously, e.g., in subsequent metabolic reactions, it follows that ΔG would be a large negative value, and that the reaction would persist without reaching equilibrium.

The magnitude of ΔG^0, which represents the difference between the energy levels (chemical potentials) of the product P_p and educt P_e ($\rightarrow$ **A**), does not tell us anything about the **rate of the reaction**. A reaction may be very slow, even if $\Delta G^0 < 0$, because the reaction rate also depends on the energy level (P_a) needed *transiently* to create the necessary transitional state. P_a is higher than P_e ($\rightarrow$ **A**). The additional amount of energy required to reach this level is

called the **activation energy (E_a)** : $E_a = P_a - P_e$. It is usually so large ($\approx 50 \; kJ \cdot mol^{-1}$) that only a tiny fraction ($F \approx 10^{-9}$) of the educt molecules are able to provide it ($\rightarrow$ **A, B**). The energy levels of these individual educt molecules are incidentally higher than P_e, which represents the mean value for all educt molecules. The size of **fraction F** is temperature-dependent ($\rightarrow$ **B**). A 10 °C decrease or rise in temperature lowers or raises F (and usually the reaction rate) by a factor of 2 to 4, i.e. the Q_{10} **value** of the reaction is 2 to 4.

Considering the high E_a values of many non-catalyzed reactions, the development of **enzymes** as biological catalysts was a very important step in evolution. Enzymes enormously accelerate reaction rates by lowering the activation energy E_a ($\rightarrow$ **A**). According to the Arrhenius equation, the **rate constant k** (s^{-1}) of a unimolecular reaction is proportional to $e^{-E_a/(R \cdot T)}$. For example, if a given enzyme reduces the E_a of a unimolecular reaction from 126 to 63 $kJ \cdot mol^{-1}$, the rate constant at 310 °K (37 °C) will rise by $e^{-63000/(8.31 \cdot 310)}/e^{-126000/(8.31 \cdot 310)}$, i.e., by a factor of $4 \cdot 10^{10}$. The enzyme would therefore reduce the time required to metabolize 50% of the starting materials ($t^1/_2$) from, say, 10 years to 7 msec! The forward rate of a reaction (mol · $L^{-1} \cdot s^{-1}$) is related to the product of the rate constant (s^{-1}) and the starting substrate concentration (mol · L^{-1}).

The second law of thermodynamics also implies that a continuous loss of free energy occurs as the total disorder or entropy (S) of a closed system increases. A living organism represents an **open system** which, by definition, can absorb energy-rich nutrients and discharge end products of metabolism. While the entropy of a closed system (organism + environment) increases in the process, an open system (organism) can either maintain its entropy level or reduce it using free enthalpy. This occurs, for example, when ion gradients or hydraulic pressure differences are created within the body. A closed system therefore has a maximum entropy, is in a true state of chemical equilibrium, and can perform work only once. An open system such as the body can *continuously* perform work while producing only a minimum of entropy. A true state of equilibrium is achieved in only a very few processes within the body, e.g., in the reaction $CO_2 + H_2O \rightleftharpoons HCO_3^- + H^+$. In most cases (e.g. metabolic pathways, ion gradients), only a *steady state* is reached. Such metabolic pathways are usually *irreversible* due, for example, to excretion of the end products. The thought of reversing the "reaction" germ cell → adult illustrates just how impossible this is.

At **steady state**, the *rate* of a reaction is more important than its equilibrium. The regulation of body functions is achieved by controlling reaction rates. Some reactions are so slow that it is impossible to achieve a sufficient reaction rate with enzymes or by reducing the concentration of the reaction products. These are therefore endergonic reactions that require the input of outside energy. This can involve "activation" of the educt by attachment of a high-energy phosphate group to raise the Pe.

ATP (adenosine triphosphate) is the universal carrier and transformer of free enthalpy within the body. ATP is a nucleotide that derives its chemical energy from energy-rich nutrients (→ **C**). Most ATP is produced by **oxidation** of energy-rich biological molecules such as glucose. In this case, oxidation means the removal of electrons from an electron-rich (reduced) donor which, in this case, is a carbohydrate. CO_2 and H_2O are the end products of the reaction. In the body, oxidation (or electron transfer) occurs in several stages, and a portion of the liberated energy can be simultaneously used for ATP synthesis. This is therefore a **coupled reaction** (→ **C** and p. 17 B). The standard free enthalpy ΔG^0 of **ATP hydrolysis**,

$$ATP \rightleftharpoons ADP + P_i \qquad [1.31]$$

is $-30.5\,kJ \cdot mol^{-1}$. According to Eq. 1.27, the ΔG of reaction 1.31 should increase when the ratio $([ADP] \cdot [P_i])/[ATP]$ falls below the equilibrium constant K_{eq} of ATP hydrolysis. The fact that a high cellular ATP concentration does indeed yield a ΔG of approximately -46 to $-54\,kJ \cdot mol^{-1}$ shows that this also applies in practice.

Some substances have a much higher ΔG^0 of hydrolysis than ATP, e.g., *creatine phosphate* ($-43\,kJ \cdot mol^{-1}$). These compounds react with ADP and P_i to form ATP. On the other hand, the energy of ATP can be used to synthesize other compounds such as UTP, GTP and glucose-6-phosphate. The energy content of these substances is lower than that of ATP, but still relatively high.

The free energy liberated upon hydrolysis of ATP is used to drive hundreds of reactions within the body, including the active transmembrane transport of various substances, protein synthesis, and muscle contraction. According to the laws of thermodynamics, the expenditure of energy in all of these reactions leads to increased order in living cells and, thus, in the organism as a whole. Life is therefore characterized by the continuous reduction of entropy associated with a corresponding increase in entropy in the immediate environment and, ultimately, in the universe.

Neuron Structure and Function

An excitable cell reacts to stimuli by altering its membrane characteristics ($\rightarrow$ p. 32). There are two types of excitable cells: *nerve cells*, which transmit and modify impulses within the nervous system, and *muscle cells*, which contract either in response to nerve stimuli or autonomously ($\rightarrow$ p. 59).

The human nervous system consists of more than 10^{10} nerve cells or neurons. The **neuron** is the *structural and functional unit of the nervous system*. A typical neuron (motor neuron, $\rightarrow$ **A1**) consists of the *soma* or cell body and two types of processes: the *axon* and *dendrites*. Apart from the usual intracellular organelles ($\rightarrow$ p. 8 ff.), such as a nucleus and mitochondria ($\rightarrow$ **A2**), the neuron contains *neurofibrils* and *neurotubules*. The neuron receives *afferent signals* (excitatory and inhibitory) from a few to sometimes several thousands of other neurons via its **dendrites** (usually arborescent) and sums the signals along the cell membrane of the soma (*summation*). The **axon** arises from the axon hillock of the soma and is responsible for the transmission of efferent neural signals to nearby or distant *effectors* (muscle and glandular cells) and adjacent neurons. Axons often have branches (*collaterals*) that further divide and terminate in swellings called *synaptic knobs* or *terminal buttons*. If the summed value of potentials at the axon hillock exceeds a certain threshold, an **action potential** ($\rightarrow$ p. 46) is generated and sent down the axon, where it reaches the next synapse via the **terminal buttons** ($\rightarrow$ **A1,3**) described below.

Vesicles containing materials such as proteins, lipids, sugars, and transmitter substances are conveyed from the Golgi complex of the soma ($\rightarrow$ p. 13 F) to the terminal buttons and the tips of the dendrites by *rapid* **axonal transport** (40 cm/day). This type of *anterograde* transport along the neurotubules is promoted by *kinesin*, a myosin-like protein, and the energy required for it is supplied by ATP ($\rightarrow$ p. 16). Endogenous and exogenous substances such as nerve growth factor (NGF), herpes virus, poliomyelitis virus, and tetanus toxin are conveyed by *retrograde* transport from the peripheral regions to the soma at a rate of ca. 25 cm/day. *Slow* axon transport (ca. 1 mm/day) plays a role in the regeneration of severed neurites.

Along the axon, the plasma membrane of the soma continues as the axolemma ($\rightarrow$ **A1,2**). The axolemma is surrounded by **oligodendrocytes** ($\rightarrow$ p. 338) in the central nervous system (CNS), and by **Schwann cells** in the peripheral nervous system ($\rightarrow$ **A1,2**). A **nerve fiber** consists of an axon plus its sheath. In some neurons, Schwann cells form multiple concentric double phospholipid layers around an axon, comprising the **myelin sheath** ($\rightarrow$ **A1,2**) that insulates the axon from ion currents. The sheath is interrupted every 1.5 mm or so at the **nodes of Ranvier** ($\rightarrow$ **A1**). The *conduction velocity* of myelinated nerve fibers is much higher than that of unmyelinated nerve fibers and increases with the diameter of the nerve fiber ($\rightarrow$ p. 49 C).

A **synapse** ($\rightarrow$ **A3**) is the site where the axon of a neuron communicates with effectors or other neurons (see also p. 50 ff.). With very few exceptions, **synaptic transmissions** in mammals are mediated by chemicals, not by electrical signals. In response to an electrical signal in the axon, vesicles ($\rightarrow$ p. 1.6) on the **presynaptic membrane** release transmitter substances (**neurotransmitters**) by exocytosis ($\rightarrow$ p. 30). The transmitter diffuses across the **synaptic cleft** (10–40 nm) to the postsynaptic membrane, where it binds to receptors effecting new electrical changes ($\rightarrow$ **A3**). Depending on the type of neurotransmitter and postsynaptic receptor involved, the transmitter will either have an excitatory effect (e.g., acetylcholine in skeletal muscle) or inhibitory effect (e.g., glycine in the CNS) on the postsynaptic membrane. Since the postsynaptic membrane normally does not release neurotransmitters (with only few exceptions), nerve impulses can pass the synapse in one direction only. The synapse therefore acts like a *valve* that ensures the orderly transmission of signals. Synapses are also the sites at which neuronal signal transmissions can be *modified* by other (excitatory or inhibitory) neurons.

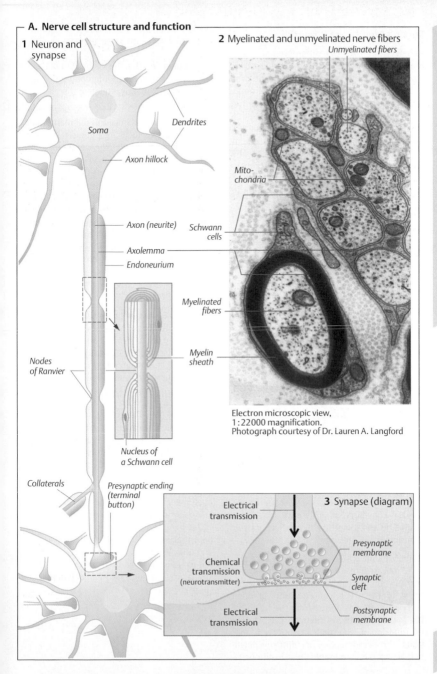

A. Nerve cell structure and function

1 Neuron and synapse

Soma

Dendrites

Axon hillock

Axon (neurite)

Axolemma

Endoneurium

Nodes of Ranvier

Nucleus of a Schwann cell

Collaterals

Presynaptic ending (terminal button)

2 Myelinated and unmyelinated nerve fibers

Unmyelinated fibers

Mito-chondria

Schwann cells

Myelinated fibers

Myelin sheath

Electron microscopic view, 1:22000 magnification.
Photograph courtesy of Dr. Lauren A. Langford

3 Synapse (diagram)

Electrical transmission

Chemical transmission (neurotransmitter)

Presynaptic membrane

Synaptic cleft

Postsynaptic membrane

Electrical transmission

Plate 2.1 Neuron Structure and Function

43

Resting Membrane Potential

An electrical potential difference, or **membrane potential** (E_m), can be recorded across the plasma membrane of living cells. The potential of unstimulated muscle and nerve cells, or **resting potential,** amounts to – 50 to – 100 mV (cell interior is negative). A resting potential is caused by a slightly unbalanced distribution of ions between the intracellular fluid (ICF) and extracellular fluid (ECF) ($\rightarrow$ **B**). The following factors are involved in establishing the membrane potential (see also p. 32 ff.).

◆ **Maintenance of an unequal distribution of ions:** The Na^+-K^+-ATPase ($\rightarrow$ p. 26) continuously "pumps" Na^+ out of the cell and K^+ into it ($\rightarrow$ **A2**). As a result, the intracellular K^+ concentration is around 35 times higher and the intracellular Na^+ concentration is roughly 20 times lower than the extracellular concentration ($\rightarrow$ **B**). As in any active transport, this process requires energy, which is supplied by ATP. Lack of energy or inhibition of the Na^+-K^+-ATPase results in flattening of the ion gradient and breakdown of the membrane potential.

Because anionic proteins and phosphates present in high concentrations in the cytosol are virtually unable to leave the cell, purely passive mechanisms (*Gibbs–Donnan distribution*) could, to a slight extent, contribute to the unequal distribution of diffusable ions ($\rightarrow$ **A1**). For reasons of electroneutrality, $[K^++Na^+]_{ICF} > [K^++Na^+]_{ECF}$ and $[Cl^-]_{ICF} < [Cl^-]_{ECF}$. However, this has practically no effect on the development of resting potentials.

◆ **Low resting Na^+ and Ca^{2+} conductance, g_{Na}, g_{Ca}:** The membrane of a resting cell is only very slightly permeable to Na^+ and Ca^{2+}, and the resting g_{Na} comprises only a small percentage of the total conductance ($\rightarrow$ p. 32 ff.). Hence, the Na^+ concentration difference ($\rightarrow$ **A3–A5**) cannot be eliminated by immediate passive diffusion of Na^+ back into the cell.

◆ **High K^+ conductance, g_K:** It is relatively easy for K^+ ions to diffuse across the cell membrane ($g_K \approx 90\%$ of total conductance; $\rightarrow$ p. 32 ff.). Because of the steep concentration gradient ($\rightarrow$ point 1), K^+ ions diffuse from the ICF to the ECF ($\rightarrow$ **A3**). Because of their positive charge, the diffusion of even small amounts of K^+ ions leads to an electrical potential (*diffusion potential*) across the membrane. This (inside nega-

tive) diffusion potential drives K^+ back into the cell and rises until large enough to almost completely compensate for the K^+ concentration gradient driving the K^+-ions out of the cell ($\rightarrow$ **A4**). As a result, the membrane potential, E_m, is approximately equal to the K^+ equilibrium potential E_K ($\rightarrow$ p. 32).

◆ **Cl^- distribution:** Since the cell membrane is also conductive to Cl^- (g_{Cl} greater in muscle cells than in nerve cells), the membrane potential (electrical driving "force") expels Cl^- ions from the cell ($\rightarrow$ **A4**) until the Cl^- concentration gradient (chemical driving "force") drives them back into the cell at the same rate. The intracellular Cl^- concentration, $[Cl^-]_i$, then continues to rise until the Cl^- equilibrium potential equals E_m ($\rightarrow$ **A5**). $[Cl^-]_i$ can be calculated using the *Nernst equation* ($\rightarrow$ p. 32, Eq. 1.18). Such a "passive" distribution of Cl^- between the intra- and extracellular spaces exists only as long as there is no active Cl^- uptake into the cell ($\rightarrow$ p. 34).

◆ **Why is E_m less negative than E_K?** Although the conductances of Na^+ and Ca^{2+} are very low in resting cells, a few Na^+ and Ca^{2+} ions constantly enter the cell ($\rightarrow$ **A4, 5**). This occurs because the equilibrium potential for both types of ions extends far into the positive range, resulting in a high outside-to-inside electrical and chemical driving "force" for these ions ($\rightarrow$ **B**; $\rightarrow$ p. 32 f.). This cation influx depolarizes the cell, thereby driving K^+ ions out of the cell (1 K^+ for each positive charge that enters). If Na^+-K^+-ATPase did not restore these gradients continuously (Ca^{2+} indirectly via the 3 Na^+/Ca^{2+} exchanger; $\rightarrow$ p. 36), the intracellular Na^+ and Ca^{2+} concentrations would increase continuously, whereas $[K^+]_i$ would decrease, and E_K and E_m would become less negative.

All living cells have a (resting) membrane potential, but only excitable cells such as nerve and muscle cells are able to greatly change the ion conductance of their membrane in response to a stimulus, as in an *action potential* ($\rightarrow$ p. 46).

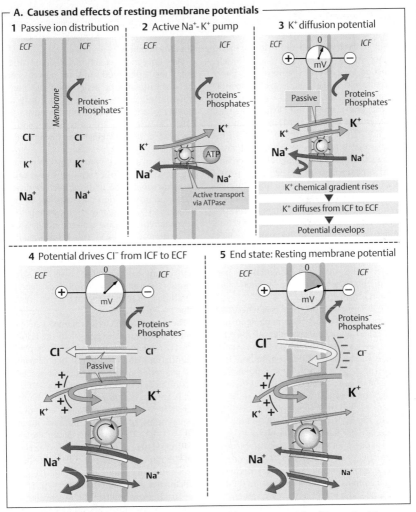

A. Causes and effects of resting membrane potentials

1 Passive ion distribution

ECF   ICF

Proteins⁻
Phosphates⁻

Cl⁻   Cl⁻

K⁺   K⁺

Na⁺   Na⁺

Membrane

2 Active Na⁺-K⁺ pump

ECF   ICF

Proteins⁻
Phosphates⁻

K⁺

K⁺   ATP

Na⁺   Na⁺

Active transport
via ATPase

3 K⁺ diffusion potential

ECF   0 mV   ICF

⊕   ⊖

Proteins⁻
Phosphates⁻

Passive

K⁺   K⁺

Na⁺   Na⁺

K⁺ chemical gradient rises

▼

K⁺ diffuses from ICF to ECF

▼

Potential develops

4 Potential drives Cl⁻ from ICF to ECF

ECF   0 mV   ICF

⊕   ⊖

Proteins⁻
Phosphates⁻

Cl⁻   Cl⁻

Passive

+/ +
+ +

K⁺   K⁺

Na⁺   Na⁺

5 End state: Resting membrane potential

ECF   0 mV   ICF

⊕   ⊖

Proteins⁻
Phosphates⁻

Cl⁻   Cl⁻

+ +
+ +

K⁺   K⁺

Na⁺   Na⁺

B. Typical "effective" concentrations and equilibrium potentials of important ions in skeletal muscle (at 37 °C)

	"Effective" concentration (mmol/kg H₂O)		Equilibrium potential
	Interstice (ECF)	Cell (ICF)	
K⁺	4.5	160	– 95 mV
Na⁺	144	7	+ 80 mV
Ca²⁺	1.3	0.0001 – 0.00001	+125 to +310 mV
H⁺	4·10⁻⁵ (pH 7.4)	10⁻⁴ (pH 7.0)	– 24 mV
Cl⁻	114	7	– 80 mV
HCO₃⁻	28	10	– 27 mV

(After Conway)

Plate 2.2   **Resting Membrane Potential**

45

Action Potential

An action potential is a signal passed on through an axon or along a muscle fiber that influences other neurons or induces muscle contraction. **Excitation** of a neuron occurs if the membrane potential, E_m, on the axon hillock of a motor neuron, for example ($\rightarrow$ p. 42), or on the motor end-plate of a muscle fiber changes from its resting value ($\rightarrow$ p. 44) to a less negative value (slow *depolarization*, $\rightarrow$ **A1**). This depolarization may be caused by neurotransmitter-induced opening of postsynaptic cation channels ($\rightarrow$ p. 50) or by the (electrotonic) transmission of stimuli from the surroundings ($\rightarrow$ p. 48). If the E_m of a stimulated cell comes close to a critical voltage or **threshold potential** ($\rightarrow$ **A1**), "rapid" voltage-gated Na^+ channels are activated ($\rightarrow$ **B4** and **B1** $\Rightarrow$ **B2**). This results in *increased Na^+ conductance*, g_{Na} ($\rightarrow$ p. 32), and the entry of Na^+ into the cell ($\rightarrow$ **A2**). If the threshold potential is not reached, this process remains a *local (subthreshold) response*.

Once the threshold potential is reached, the cell responds with a fast **all-or-none** depolarization called an **action potential, AP** ($\rightarrow$ **A1**). The AP follows a pattern typical of the specific cell type, irregardless of the magnitude of the stimulus that generated it. Large numbers of Na^+ channels are activated, and the influxing Na^+ accelerates depolarization which, in turn, increases g_{Na} and so on (positive feedback). As a result, the E_m rapidly collapses (0.1 ms in nerve cells: fast **depolarization phase** or upsweep) and temporarily reaches positive levels (*overshooting*, + 20 to + 30 mV). The g_{Na} drops before overshooting occurs ($\rightarrow$ **A2**) because the Na^+ channels are *inactivated* within 0.1 ms ($\rightarrow$ **B2** $\Rightarrow$ **B3**). The potential therefore reverses, and restoration of the resting potential, the **repolarization phase** of the action potential, begins. Depolarization increases (relatively slowly) the open-probability of voltage-gated K^+ channels. This increases the potassium conductance, g_K, thereby accelerating repolarization.

In many cases, potassium conductance, g_K is still increased after the original resting potential has been restored ($\rightarrow$ **A2**), and E_m temporarily approaches E_K ($\rightarrow$ pp. 44 and 32 ff.),

resulting in a hyperpolarizing **afterpotential** ($\rightarrow$ **A1**). Increased Na^+-K^+-ATPase pumping rates (electrogenic; $\rightarrow$ p. 28) can contribute to this afterpotential.

Very long trains of action potentials can be generated (up to 1000/s in some nerves) since the quantity of ions penetrating the membrane is very small (only ca. 1/100 000 th the number of intracellular ions). Moreover, the Na^+-K^+-ATPase ($\rightarrow$ p. 26) ensures the continuous restoration of original ion concentrations ($\rightarrow$ p. 46).

During an action potential, the cell remains unresponsive to further stimuli; this is called the **refractory period**. In the *absolute refractory period*, no other action potential can be triggered, even by extremely strong stimuli, since Na^+ channels in depolarized membranes cannot be activated ($\rightarrow$ **B3**). This is followed by a *relative refractory period* during which only action potentials of smaller amplitudes and rates of rise can be generated, even by strong stimuli. The refractory period ends once the membrane potential returns to its resting value ($\rightarrow$ e.g. p. 59 A).

The extent to which Na^+ channels can be activated and, thus, the strength of the Na^+ current, I_{Na}, depends on the pre-excitatory resting potential, not the duration of depolarization. The activation of the Na^+ channels reaches a maximum at resting potentials of ca. – 100 mV and is around 40% lower at – 60 mV. In mammals, Na^+ channels can no longer be activated at potentials of – 50 mV and less negative values ($\rightarrow$ **B3**). This is the reason for the absolute and relative refractory periods (see above) and the non-excitability of cells after the administration of continuously depolarizing substances such as suxamethonium ($\rightarrow$ p. 56). An increased extracellular Ca^{2+} concentration makes it more difficult to stimulate the cell because the threshold potential becomes less negative. On the other hand, excitability increases (lower threshold) in hypocalcemic states, as in muscle spasms in *tetany* ($\rightarrow$ p. 290).

The special features of action potentials in cardiac and smooth muscle fibers are described on pages 192, 70 and 59 A.

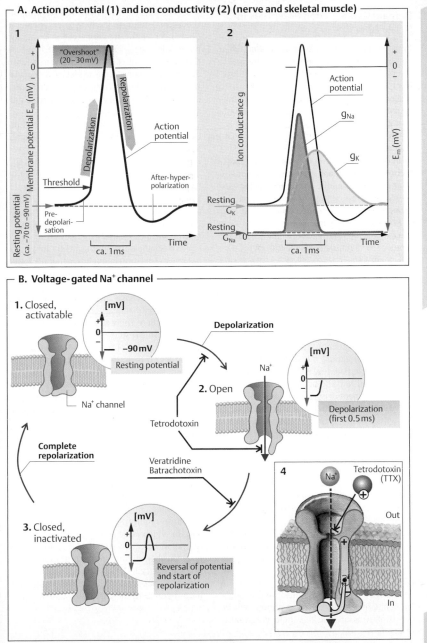

A. Action potential (1) and ion conductivity (2) (nerve and skeletal muscle)

1

"Overshoot" (20–30 mV)

Depolarization

Repolarization

Action potential

Threshold

After-hyperpolarization

Pre-depolarisation

Resting potential (ca. –70 to –90 mV)

Membrane potential E_m (mV)

ca. 1ms

Time

2

Action potential

g_{Na}

g_K

Ion conductance g

Resting G_K

Resting G_{Na}

E_m (mV)

ca. 1ms

Time

B. Voltage-gated Na⁺ channel

1. Closed, activatable

[mV]
–90 mV
Resting potential

Na⁺ channel

Depolarization

2. Open

Na⁺

[mV]
Depolarization (first 0.5ms)

Tetrodotoxin

Complete repolarization

Veratridine
Batrachotoxin

3. Closed, inactivated

[mV]
Reversal of potential and start of repolarization

4

Na⁺

Tetrodotoxin (TTX)

Out

In

Plate 2.3   **Action Potential**

Propagation of Action Potentials in Nerve Fiber

Electrical current flows through a *cable* when voltage is applied to it. The metal wire inside the cable is well insulated and has very low-level resistance, reducing current loss to a minimum. As a result, it can conduct electricity over long distances. *Nerve fibers*, especially unmyelinated ones ($\rightarrow$ p. 42), have a much greater internal longitudinal resistance (R_i) and are not well insulated from their surroundings. Therefore, the cable-like, **electrotonic transmission** of neural impulses dwindles very rapidly, so the conducted impulses must be continuously "refreshed" by generating new **action potentials** ($\rightarrow$ p. 46).

Propagation of action potentials: The start of an action potential is accompanied by a brief influx of Na^+ into the nerve fiber ($\rightarrow$ **A1a**). The cell membrane that previously was inside negative now becomes positive (+20 to +30 mV), thus creating a longitudinal potential difference with respect to the adjacent, still unstimulated nerve segments (internal −70 to −90 mV; $\rightarrow$ p. 44). This is followed by a passive *electrotonic withdrawal of charge* from the adjacent segment of the nerve fiber, causing its *depolarization*. If it exceeds threshold, another action potential is created in the adjacent segment and the action potential in the previous segment dissipates ($\rightarrow$ **A1b**).

Because the membrane acts as a *capacitor*, the withdrawal of charge represents a capacitating (depolarizing) flow of charge that becomes smaller and rises less steeply as the spatial distance increases. Because of the relatively high R_i of nerve fiber, the outward loops of current cross the membrane relatively close to the site of excitation, and the longitudinal current decreases as it proceeds towards the periphery. At the same time, depolarization increases the driving force (= $E_m - E_K$; $\rightarrow$ p. 32) for K^+ outflow. K^+ fluxing out of the cell therefore accelerates repolarization. Hence, distal action potentials are restricted to distances from which the capacitative current suffices to depolarize the membrane quickly and strongly enough. Otherwise, the Na^+ channels will be deactivated before the threshold potential is reached ($\rightarrow$ p. 46).

Action potentials normally run forward (orthodromic) because each segment of nerve fiber becomes refractory when an action potential passes ($\rightarrow$ **A1b** and p. 46). If, however, the impulses are conducted backwards (antidromic) due, for example, to electrical stimulation of nerve fibers from an external source ($\rightarrow$ p. 50), they will terminate at the next synapse (valve-like function, $\rightarrow$ p. 42).

Although the continuous generation of action potentials in the immediately adjacent fiber segment guarantees a refreshed signal, this process is rather time-consuming ($\rightarrow$ **B1**). The **conduction velocity**, θ, in unmyelinated (type C) nerve fibers ($\rightarrow$ **C**) is only around 1 m/s. Myelinated (types A and B) nerve fibers ($\rightarrow$ **C**) conduct much faster (up to 80 m/s = 180 mph in humans). In the *internode regions*, a *myelin sheath* ($\rightarrow$ p. 42) insulates the nerve fibers from the surroundings; thus, longitudinal currents strong enough to generate action potentials can travel further down the axon (ca. 1.5 mm) ($\rightarrow$ **A2**). This results in more rapid conduction because the action potentials are generated only at the unmyelinated *nodes of Ranvier*, where there is a high density of Na^+ channels. This results in rapid, jump-like passage of the action potential from node to node (**saltatory propagation**). The saltatory length is limited since the longitudinal current (1 to 2 nA) grows weaker with increasing distance ($\rightarrow$ **B2**). Before it drops below the threshold level, the signal must therefore be refreshed by a new action potential, with a time loss of 0.1 ms.

Since the internal resistance, R_i, of the nerve fiber limits the spread of depolarization, as described above, the **axon diameter** (2r) also affects the conduction velocity, θ ($\rightarrow$ **C**). R_i is proportional to the cross-sectional area of the nerve fiber (πr^2), i.e., $R_i \sim 1/r^2$. Thick fibers therefore require fewer new APs per unit of length, which is beneficial for θ. Increases in fiber diameter are accompanied by an increase in both fiber circumference ($2\pi r$) and membrane capacity, K ($K \sim r$). Although θ decreases, the beneficial effect of the smaller R_i predominates because of the quadratic relationship.

Plate 2.4 Propagation of Action Potentials in Nerve Fiber

A. Continuous (1a, 1b) and saltatory propagation (2) of action potentials

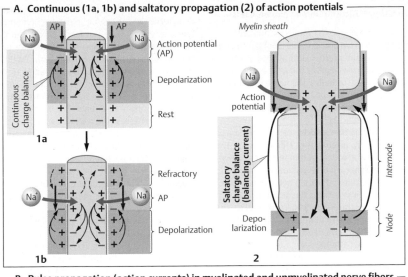

Myelin sheath

Action potential (AP)

Depolarization

Rest

Continuous charge balance

1a

Refractory

AP

Depolarization

1b

Action potential

Saltatory charge balance (balancing current)

Depolarization

Internode

Node

2

B. Pulse propagation (action currents) in myelinated and unmyelinated nerve fibers

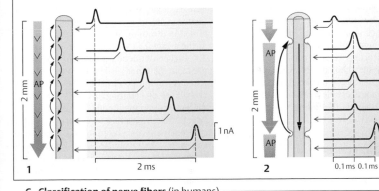

2 mm

AP

[1 nA]

2 ms

1

2 mm

AP

AP

[1 nA]

0.1 ms 0.1 ms

2

C. Classification of nerve fibers (in humans)

Fiber type	Function according to fiber type (Lloyd and Hunt types I–IV)	Diameter (µm)	Conduction velocity (m/s)
Aα	Skeletal muscle efferent, afferents in muscle spindles (Ia) and tendon organs (Ib)	11–16	60–80
Aβ	Mechanoafferents of skin (II)	6–11	30–60
Aγ	Muscle spindle efferents	1–6	2–30
Aδ	Skin afferents (temperature and "fast" pain) (III)		
B	Sympathetic preganglionic; visceral afferents	3	3–15
C	Skin afferents ("slow" pain); sympathetic postganglionic afferents (IV)	0.5–1.5 (unmyelinated)	0.25–1.5

(After Erlanger and Gasser)

Artificial Stimulation of Nerve Cells

When an electrical stimulus is applied to a nerve cell from an external source, current flows from the positive stimulating electrode (*anode*) into the neuron, and exits at the negative electrode (*cathode*). The nerve fiber below the cathode is depolarized and an action potential is generated there if the threshold potential is reached.

The **conduction velocity** of a nerve can be measured by placing two electrodes on the skin along the course of the nerve at a known distance from each other, then stimulating the nerve (containing multiple neurons) and recording the time it takes the summated action potential to travel the known distance. The conduction velocity in humans is normally 40 to 70 m · s^{-1}. Values below 40 m · s^{-1} are considered to be pathological.

Accidental electrification. Exposure of the body to high-voltage electricity, especially low-frequency alternating current (e.g., in an electrical outlet) and low contact resistance (bare feet, bathtub accidents), primarily affects the conduction of impulses in the heart and can cause ventricular fibrillation ($\rightarrow$ p. 200).

Direct current usually acts as a stimulus only when switched on or off: High-frequency alternating current (> 15 kHz), on the other hand, cannot cause depolarization but heats the body tissues. **Diathermy** works on this principle.

Synaptic Transmission

Synapses connect nerve cells to other nerve cells (also applies for certain muscle cells) as well as to sensory and effector cells (muscle and glandular cells).

Electrical synapses are direct, ion-conducting cell–cell junctions through channels (*connexons*) in the region of *gap junctions* ($\rightarrow$ p. 16 f.). They are responsible for the conduction of impulses between neighboring smooth or cardiac muscle fibers (and sometimes between neurons in the retina and in the CNS) and ensure also communication between neighboring epithelial or glial cells.

Chemical synapses utilize **(neuro)transmitters** for the transmission of information and

provide not only simple 1 : 1 connections, but also serve as switching elements for the nervous system. They can facilitate or inhibit the neuronal transmission of information or process them with other neuronal input. At the chemical synapse, the arrival of an action potential (AP) in the axon ($\rightarrow$ **A1,2** and p. 48) triggers the release of the transmitter from the presynaptic axon terminals. The transmitter then diffuses across the narrow synaptic cleft (ca. 30 nm) to bind postsynaptically to **receptors** in the *subsynaptic membrane* of a neuron or of a glandular or muscle cell. Depending on the type of transmitter and receptor involved, the effect on the *postsynaptic membrane* may either be excitatory or inhibitory, as is described below.

Transmitters are **released** by regulated *exocytosis* of so-called *synaptic vesicles* ($\rightarrow$ **A1**). Each vesicle contains a certain *quantum* of neurotransmitters. In the case of the motor end-plate ($\rightarrow$ p. 56), around 7000 molecules of acetylcholine (ACh) are released. Some of the vesicles are already docked on the membrane (*active zone*), ready to exocytose their contents. An incoming action potential functions as the signal for transmitter release ($\rightarrow$ **A1,2**). The higher the action potential *frequency* in the axon the more vesicles release their contents. An action potential increases the open probability of *voltage-gated Ca^{2+} channels* in the presynaptic membrane (sometimes oscillating), thereby leading to an increase in the cytosolic Ca^{2+} concentration, $[Ca^{2+}]_i$ ($\rightarrow$ **A1, 3** and p. 36). Extracellular Mg^{2+} inhibits this process. Ca^{2+} binds to *synaptotagmin* ($\rightarrow$ **A1**), which triggers the interaction of *syntaxin* and *SNAP-25* on the presynaptic membrane with *synaptobrevin* on the vesicle membrane, thereby triggering exocytosis of already docked vesicles (approximately 100 per AP) ($\rightarrow$ **A1, 4**). On the other hand, Ca^{2+} activates calcium-calmodulin-dependent protein kinase-II (*CaM-kinase-II*; $\rightarrow$ **A5**, and p. 36), which activates the enzyme *synapsin* at the presynaptic terminal. As a result, vesicles dock anew on the active zone.

Synaptic facilitation (= **potentiation**). If an action potential should arrive at the presynaptic terminal immediately after another AP (AP frequency > approx. 30 Hz), the cytosolic Ca^{2+}

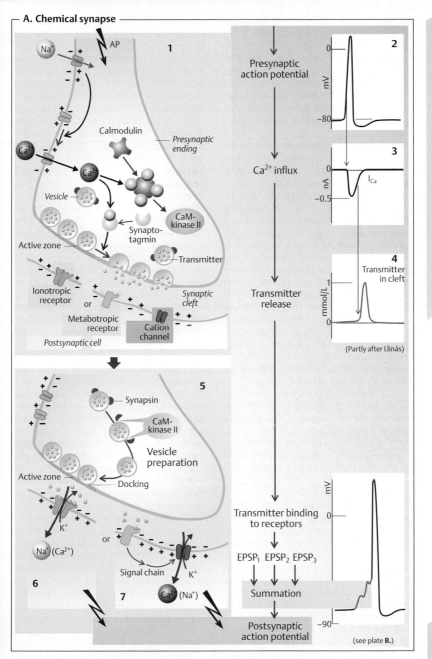

Plate 2.5 Synaptic Transmission I

- A. Chemical synapse

1

Na⁺

AP

Calmodulin

Presynaptic ending

Ca²⁺

Ca²⁺

CaM-kinase II

Vesicle

Synapto-tagmin

Active zone

Transmitter

Ionotropic receptor

or

Metabotropic receptor

Cation channel

Synaptic cleft

Postsynaptic cell

Presynaptic action potential

2

mV

0

−80

Ca²⁺ influx

3

0

nA

−0.5

I_{Ca}

Transmitter release

4

Transmitter in cleft

mmol/L

1

0

(Partly after Llinás)

5

Synapsin

CaM-kinase II

Vesicle preparation

Active zone

Docking

K⁺

Na⁺ (Ca²⁺)

6

or

Signal chain

K⁺

Ca²⁺ (Na⁺)

7

Transmitter binding to receptors

mV

0

EPSP₁ EPSP₂ EPSP₃

Summation

Postsynaptic action potential

−90

(see plate **B.**)

51

concentration will not yet drop to the resting value, and residual Ca^{2+} will accumulate. As a result, the more recent rise in $[Ca^{2+}]_i$ builds on the former one. $[Ca^{2+}]_i$ rises to a higher level after the second stimulus than after the first, and also releases more transmitters. Hence, the first stimulus *facilitates* the response to the second stimulus. Muscle strength increases at high stimulus frequencies for similar reasons ($\rightarrow$ p. 67 A).

Among the many substances that act as **excitatory transmitters** are acetylcholine (ACh) and glutamate (Glu). They are often released together with *co-transmitters* which modulate the transmission of a stimulus (e.g., ACh together with substance P, VIP or galanin; Glu with substance P or enkephalin). If the transmitter's receptor is an ion channel itself (*ionotropic receptor* or *ligand-gated ion channel*; $\rightarrow$ **A6** and **F**), e.g., at the N-cholinergic synapse ($\rightarrow$ p. 82), the channels open more often and allow a larger number of cations to enter (Na^+, sometimes Ca^{2+}) and leave the cell (K^+). Other, so-called *metabotropic receptors* influence the channel via *G proteins* that control channels themselves or by means of "second messengers" ($\rightarrow$ **A7** and **F**). Because of the high electrochemical Na^+ gradient ($\rightarrow$ p. 32), the number of incoming Na^+ ions is much larger than the number of exiting K^+ ions. Ca^{2+} can also enter the cell, e.g., at the glutamate-NMDA receptor ($\rightarrow$ **F**). The net influx of cations leads to depolarization: **excitatory postsynaptic potential (EPSP)** (maximum of ca. 20 mV; $\rightarrow$ **B**). The EPSP begins approx. 0.5 ms after the arrival of an action potential at the presynaptic terminal. This *synaptic delay* (latency) is caused by the relatively slow release and diffusion of the transmitter.

A single EPSP normally is not able to generate a postsynaptic (axonal) action potential (AP_A), but requires the triggering of a large number of local depolarizations in the dendrites. Their depolarizations are transmitted electrotonically across the soma ($\rightarrow$ p. 48) and summed on the axon hillock (**spatial summation**; $\rightarrow$ **B**). Should the individual stimuli arrive at different times (within approx. 50 ms of each other), the prior depolarization will not have dissipated before the next one arrives, and summation will make it easier to reach threshold. This type of **temporal summation** therefore increases the excitability of the postsynaptic neuron ($\rightarrow$ **C**).

Inhibitory transmitters include substances as glycine, GABA (γ-aminobutyric acid), and acetylcholine (at M2 and M3 receptors; $\rightarrow$ p. 82). They increase the conductance, g, of the subsynaptic membrane only to K^+ (e.g., the metabotropic $GABA_B$ receptor.) or Cl^- (e.g., the ionotropic glycine and $GABA_A$ receptors; $\rightarrow$ **F**). The membrane usually becomes hyperpolarized in the process (ca. 4 mV max.). Increases in g_K occur when E_m approaches E_K ($\rightarrow$ p. 44). However, the main effect of this *inhibitory postsynaptic potential* IPSP ($\rightarrow$ **D**) is not hyperpolarization–which works counter to EPSP-related depolarization (the IPSP is sometimes even slightly depolarizing). Instead, the IPSP-related increase in membrane conductance short circuits the electrotonic currents of the EPSP (high g_K or g_{Cl} levels). Since both E_K and E_{Cl} are close to the resting potential ($\rightarrow$ p. 44), stabilization occurs, that is, the EPSP is cancelled out by the high K^+ and Cl^- short-circuit currents. As a result, EPSP-related depolarization is reduced and stimulation of postsynaptic neurons is inhibited ($\rightarrow$ **D**).

Termination of synaptic transmission ($\rightarrow$ **E**) can occur due to *inactivation* of the cation channels due to a conformational change in the channel similar to the one that occurs during an action potential ($\rightarrow$ p. 46). This very rapid process called *desensitization* also functions in the presence of a transmitter. Other terminating pathways include the rapid *enzymatic decay* of the transmitter (e.g., acetylcholine) while still in the synaptic cleft, the *re-uptake* of the transmitter (e.g., noradrenaline) into the presynaptic terminal or *uptake* into extraneuronal cells (e.g., in glial cells of the CNS), endocytotic *internalization* of the receptor ($\rightarrow$ p. 28), and binding of the transmitter to a receptor on the presynaptic membrane (autoceptor). In the latter case, a rise in g_K and a drop in g_{Ca} can occur, thus inhibiting transmitter release, e.g., of GABA via $GABA_B$ receptors or of noradrenaline via α_2-adrenoceptors ($\rightarrow$ **F** and p. 86).

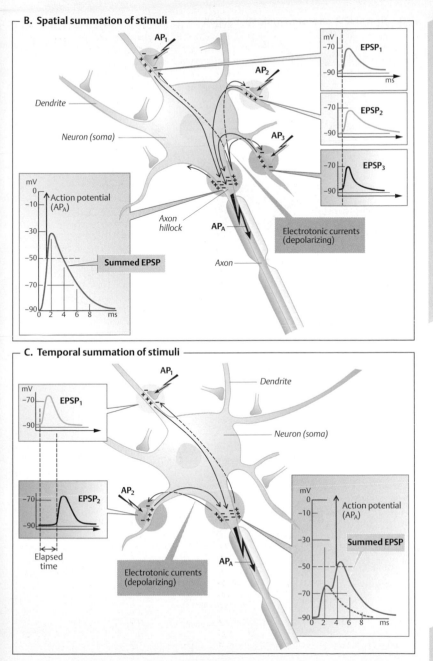

B. Spatial summation of stimuli

AP₁

AP₂

AP₃

mV
−70
−90
EPSP₁
ms

mV
−70
−90
EPSP₂

mV
−70
−90
EPSP₃

Dendrite

Neuron (soma)

mV
0
−10
−30
−50
−70
−90

Action potential (APₐ)

Summed EPSP

0 2 4 6 8 ms

Axon hillock

APₐ

Axon

Electrotonic currents (depolarizing)

C. Temporal summation of stimuli

AP₁

Dendrite

mV
−70
−90
EPSP₁

mV
−70
−90
EPSP₂

Neuron (soma)

AP₂

Elapsed time

Electrotonic currents (depolarizing)

APₐ

mV
0
−10
−30
−50
−70
−90

Action potential (APₐ)

Summed EPSP

0 2 4 6 8 ms

Plate 2.6 Synaptic Transmission II

D. Effect of IPSP on postsynaptic stimulation

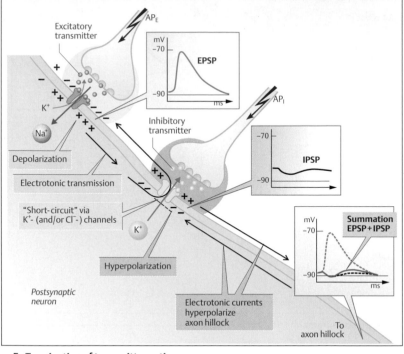

E. Termination of transmitter action

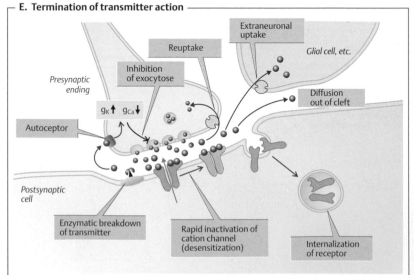

Plate 2.7 u. 2.8   Synaptic Transmission III u. IV

F. Neurotransmitters in the central nervous system

Transmitter	Receptor subtypes	Receptor types	Effect					
			Ion conductance				Second messenger	
			Na⁺	K⁺	Ca²⁺	Cl⁻	cAMP	IP₃/DAG
Acetylcholine	Nicotinic	●	↑	↑	↑			
	Muscarinic: M1,	●						↑
	M2, M3	●		↑			↓	
ADH (= vasopressin)	V1	●						↑
	V2	●					↑	
CCK (= cholecystokinin)	CCK$_{A-B}$	●						↑
Dopamine	D1, D5	●					↑	
	D2	●		↑	↓		↓	
GABA (= γ-aminobutyric acid)	GABA$_A$, GABA$_C$	●				↑		
	GABA$_B$	●		↑	↓		↓	
Glutamate (aspartate)	AMPA	●	↑	↑				
	Kainat	●	↑	↑				
	NMDA	●	↑	↑	↑			
	m-GLU	●					↓	↑
Glycine	–	●				↑		
Histamine	H₁	●						↑
	H₂	●					↑	
Neurotensin	–	●					↓	↑
Norepinephrine, epinephrine	α$_{1(A-D)}$	●		↑				↑
	α$_{2(A-C)}$	●		↑	↓		↓	
	β$_{1-3}$	●					↑	
Neuropeptide Y (NPY)	Y 1–2	●		↑	↓		↓	
Opioid peptides	μ, δ, κ	●		↑	↓		↓	
Oxytocin	–	●						↑
Purines	P₁: A₁	●		↑	↓		↓	
	A$_{2a}$	●					↑	
	P$_{2X}$	●	↑	↑	↑			
	P$_{2Y}$	●						↑
Serotonin (5-hydroxytryptamine)	5-HT₁	●		↓			↓	
	5-HT₂	●						↑
	5-HT₃	●	↑	↑				
	5-HT$_{4-7}$	●					↑	
Somatostatin (= SIH)	SRIF	●		↑	↓		↓	
Tachykinin	NK 1–3	●						↑

Amino acids		
Catecholamines		
Peptides		
Others		

Ionotropic receptor (ligand-gated ion channel)

Metabotropic receptor (G protein-mediated effect)

Inhibits or promotes

cAMP ← ATP

PIP₂ → DAG, IP₃

55

(Modified from F. E. Bloom)

Motor End-plate

The transmission of stimuli from a motor axon to a skeletal muscle fiber occurs at the motor end-plate, MEP ($\rightarrow$ **A**), a type of chemical synapse ($\rightarrow$ p. 50 ff.). The transmitter involved is **acetylcholine** (**ACh**, $\rightarrow$ cf. p. 82), which binds to the N(nicotinergic)-cholinoceptors of the subsynaptic muscle membrane ($\rightarrow$ **A3**). **N-cholinoceptors** are ionotropic, that is, they also function as *ion channels* ($\rightarrow$ **A4**). The N-cholinoceptor of the MEP (type N_M) has 5 subunits (2α, 1β, 1γ, 1δ), each of which contains 4 membrane-spanning α-helices ($\rightarrow$ p. 14).

The channel opens briefly ($\rightarrow$ **B1**) (for approx. 1 ms) when an ACh molecule binds to the two α-subunits of an N-cholinoceptor ($\rightarrow$ **A4**). Unlike voltage-gated Na^+-channels, the open-probability p_o of the N_M-cholinoceptor is not increased by depolarization, but is determined by the *ACh concentration in the synaptic cleft* ($\rightarrow$ p. 50 ff.).

The channel is *specific to cations* such as Na^+, K^+, and Ca^{2+}. Opening of the channel at a resting potential of ca. -90 mV leads mainly to an *influx of Na^+ ions* (and a much lower outflow of K^+; $\rightarrow$ pp. 32 ff. and 44). Depolarization of the subsynaptic membrane therefore occurs: **endplate potential (EPP)**. *Single-channel currents* of 2.7 pA ($\rightarrow$ **B1**) are summated to yield a *miniature end-plate current* of a few nA when spontaneous exocytosis occurs and a vesicle releases a quantum of ACh activating thousands of N_M-cholinoceptors ($\rightarrow$ **B2**). Still, this is not enough for generation of a postsynaptic action potential unless an action potential transmitted by the motor neuron triggers exocytosis of around a hundred vesicles. This opens around 200,000 channels at the same time, yielding a *neurally induced end-plate current (I_{EP})* of ca. 400 nA ($\rightarrow$ **B3**). **End-plate current, I_{EP},** is therefore dependent on:

◆ the number of open channels, which is equal to the total number of channels (n) times the open-probability (p_o), where p_o is determined by the concentration of ACh in the synaptic cleft (up to 1 mmol/L);

◆ the single-channel conductance γ (ca. 30 pS);

◆ and, to a slight extent, the membrane potential, E_m, since the electrical driving "force" ($= E_m - E_{Na,K}$; $\rightarrow$ p. 32 ff.) becomes smaller when E_m is less negative.

$E_{Na,K}$ is the common equilibrium potential for Na^+ and K^+ and amounts to approx. 0 mV. It is also called the **reversal potential** because the direction of I_{EP} ($= I_{Na} + I_K$), which enters the cell when E_m is negative (Na^+ influx $> K^+$ outflow), reverses when E_m is positive (K^+ outflow $> Na^+$ influx). As a result,

$$I_{EP} = n \cdot p_o \cdot \gamma \cdot (E_m - E_{Na,K}) \ [A] \qquad [2.1]$$

Because **neurally induced EPPs** in skeletal muscle are much larger (depolarization by ca. 70 mV) than neuronal EPSPs (only a few mV; $\rightarrow$ p. 50 ff.), single motor axon action potentials are above threshold. The EPP is transmitted electrotonically to the adjacent sarcolemma, where muscle action potentials are generated by means of voltage-gated Na^+ channels, resulting in muscle contraction.

Termination of synaptic transmission in MEPs occurs (1) by rapid degradation of ACh in the synaptic cleft by *acetylcholinesterase* localized at the subsynaptic basal membrane, and (2) by diffusion of ACh out of the synaptic cleft ($\rightarrow$ p. 82).

A motor end-plate can be blocked by certain **poisons** and **drugs**, resulting in muscular weakness and, in some cases, *paralysis*. *Botulinum neurotoxin*, for example, inhibits the discharge of neurotransmitters from the vesicles, and α-*bungarotoxin* in cobra venom blocks the opening of ion channels. Curare-like substances such as (+)-tubocurarine are used as **muscle relaxants** in surgical operations. They displace ACh from its binding site (*competitive inhibition*) but do not have a depolarizing effect of their own. Their inhibitory effect can be reversed by *cholinesterase inhibitors* such as neostigmine (decurarinization). These agents increase the concentration of ACh in the synaptic cleft, thereby displacing curare. Entry of anticholinesterase agents into intact synapses leads to an increase in the ACh concentration and, thus, to *paralysis due to permanent depolarization*. ACh-like substances such as suxamethonium have a similar depolarizing effect, but decay more slowly than ACh. In this case, paralysis occurs because permanent depolarization also permanently inactivates Na^+ channels near the motor end-plate on the sarcolemma ($\rightarrow$ p. 46).

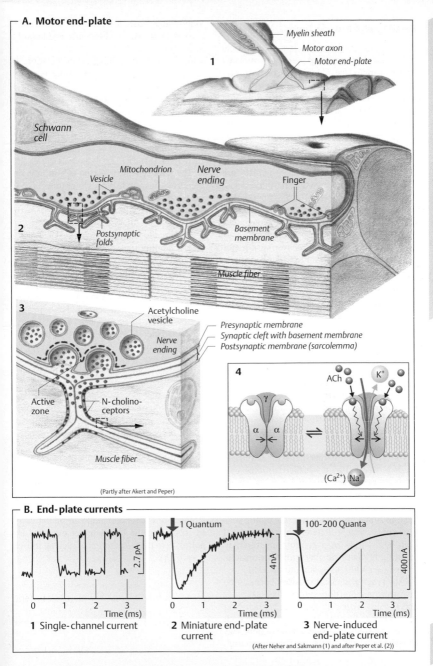

A. Motor end-plate

Myelin sheath
Motor axon
Motor end-plate

1

Schwann cell

Vesicle — *Mitochondrion* — *Nerve ending* — *Finger*

2

Postsynaptic folds — *Basement membrane*

Muscle fiber

3

Acetylcholine vesicle

Nerve ending

Presynaptic membrane
Synaptic cleft with basement membrane
Postsynaptic membrane (sarcolemma)

Active zone — N-cholino-ceptors

Muscle fiber

4 — ACh — K⁺

γ

α α

(Ca²⁺) Na⁺

(Partly after Akert and Peper)

B. End-plate currents

2.7 pA

0   1   2   3
Time (ms)

1 Single-channel current

1 Quantum

4 nA

0   1   2   3
Time (ms)

2 Miniature end-plate current

100–200 Quanta

400 nA

0   1   2   3
Time (ms)

3 Nerve-induced end-plate current

(After Neher and Sakmann (1) and after Peper et al. (2))

Plate 2.9 Motor End-plate

Motility and Muscle Types

Active motility (ability to move) is due to either the **interaction of** energy-consuming **motor proteins** (fueled by ATPase) such as myosin, kinesin and dynein with other proteins such as actin or the **polymerization** and depolymerization of actin and tubulin. Cell division (cytokinesis), cell migration ($\rightarrow$ p. 30), intracellular vesicular transport and cytosis ($\rightarrow$ p. 12f.), sperm motility ($\rightarrow$ p. 306f.), axonal transport ($\rightarrow$ p. 42), electromotility of hair cells ($\rightarrow$ p. 366), and ciliary motility ($\rightarrow$ p. 110) are examples of cell and organelle motility.

The muscles consist of cells (fibers) that contract when stimulated. **Skeletal muscle** is responsible for locomotion, positional change, and the convection of respiratory gases. **Cardiac muscle** ($\rightarrow$ p. 190 ff.) is responsible for pumping the blood, and **smooth muscle** ($\rightarrow$ p. 70) serves as the motor of internal organs and blood vessels. The different muscle types are distinguished by several functional characteristics ($\rightarrow$ **A**).

Motor Unit of Skeletal Muscle

Unlike some types of smooth muscle (single-unit type; $\rightarrow$ p. 70) and cardiac muscle fibers, which pass electric stimuli to each other through gap junctions or nexus ($\rightarrow$ **A**; p. 16f.), **skeletal muscle** fibers are not stimulated by adjacent muscle fibers, but by *motor neurons*. In fact, muscle paralysis occurs if the nerve is severed.

One motor neuron together with all muscle fibers innervated by it is called a **motor unit** (MU). Muscle fibers belonging to a single motor unit can be distributed over large portions (1 cm^2) of the muscle cross-sectional area. To supply its muscle fibers, a motor neuron splits into collaterals with terminal branches ($\rightarrow$ p. 42). A given motor neuron may supply only 25 muscle fibers (mimetic muscle) or well over 1000 (temporal muscle).

Two **types of skeletal muscle fibers** can be distinguished: **S** – slow-twitch fibers (type 1) and **F** – fast-twitch fibers (type 2), including two subtypes, **FR** (2 A) and **FF** (2 B). Since each motor unit contains only one type of fiber, this classification also applies to the motor unit.

Slow-twitch fibers are the least fatigable and are therefore equipped for *sustained performance*. They have high densities of capillaries and mitochondria and high concentrations of fat droplets (high-energy substrate reserves) and the red pigment myoglobin (short-term O_2 storage). They are also rich in oxidative enzymes ($\rightarrow$ p. 72). **Fast-twitch fibers** are mainly responsible for brief and rapid contractions. They are quickly fatigued (FF > FR) and are rich in glycogen (FF > FR) but contain little myoglobin (FF$\ll$FR).

The **fiber type distribution** of a muscle depends on the muscle type. Motor units of the *S type* predominate in "red" muscles such as the soleus muscle, which helps to maintain the body in an upright position, whereas the *F type* predominates in "white" muscles such as the gastrocnemius muscle, which is involved in running activity. Each fiber type can also be *converted* to the other type. If, for example, the prolonged activation of fast-twitch fibers leads to a chronic increase in the cytosolic Ca^{2+} concentration, fast-twitch fibers will be converted to slow-twitch fibers and vice versa.

Graded muscle activity is possible because a variable number of motor units can be recruited as needed. The more motor units a muscle has, the more finely graded its contractions. Contractions are much finer in the external eye muscles, for example, which have around 2000 motor units, than in the lumbrical muscles, which have only around 100 motor units. The larger the **number of motor units recruited**, the stronger the contraction. The number and type of motor units recruited depends on the type of movement involved (fine or coarse movement, intermittent or persistent contraction, reflex activity, voluntary or involuntary movement, etc.). In addition, the strength of each motor unit can be increased by increasing the **frequency of neuronal impulses**, as in the *tetanization* of skeletal muscle ($\rightarrow$ p. 67 A).

A. Structure and function of heart, skeletal and smooth muscle

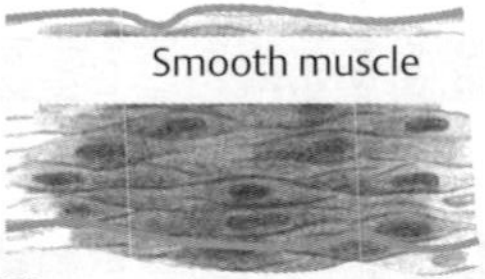

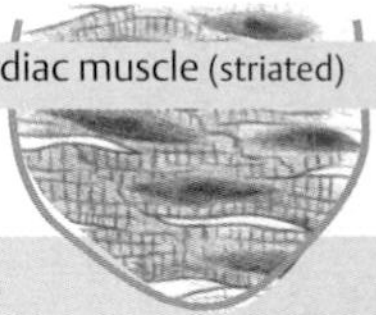

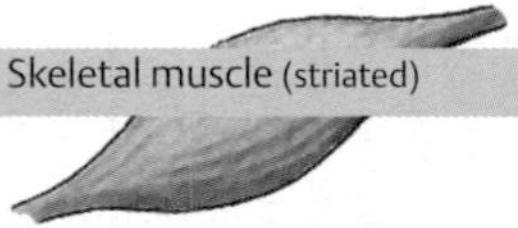

Structure and function	Smooth muscle	Cardiac muscle (striated)	Skeletal muscle (striated)
Motor end-plates	None	None	Yes
Fibers	Fusiform, short (≤ 0.2 mm)	Branched	Cylindrical, long (≤ 15 cm)
Mitochondria	Few	Many	Few (depending on muscle type)
Nuclei per fiber	1	1	Multiple
Sarcomeres	None	Yes, length ≤ 2.6 µm	Yes, length ≤ 3.65 µm
Electr. coupling	Some (single-unit type)	Yes (functional syncytium)	No
Sarcoplasmic reticulum	Little developed	Moderately developed	Highly developed
Ca²⁺ "switch"	Calmodulin/caldesmon	Troponin	Troponin
Pacemaker	Some spontaneous rhythmic activity (1 s⁻¹ – 1 h⁻¹)	Yes (sinus nodes ca. 1 s⁻¹)	No (requires nerve stimulus)
Response to stimulus	Change in tone or rhythm frequency	All or none	Graded
Tetanizable	Yes	No	Yes
Work range	Length-force curve is variable	In rising length-force curve (see 2.15E)	At peak of length-force curve (see 2.15E)

Response to stimulus

Potential ——
Muscle tension ——

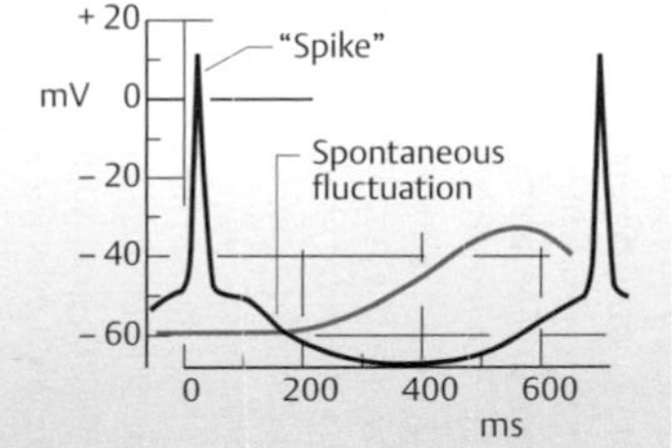

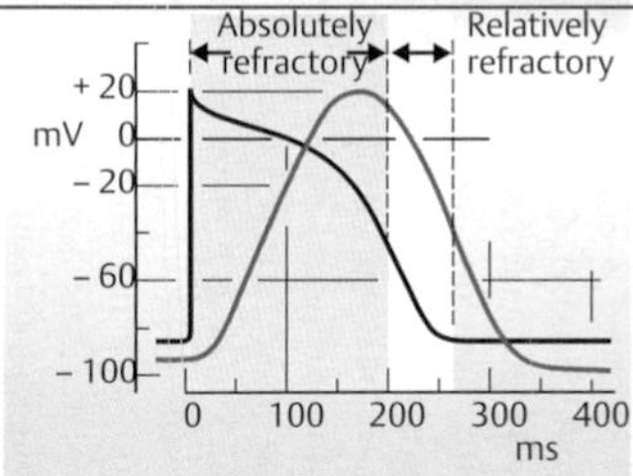

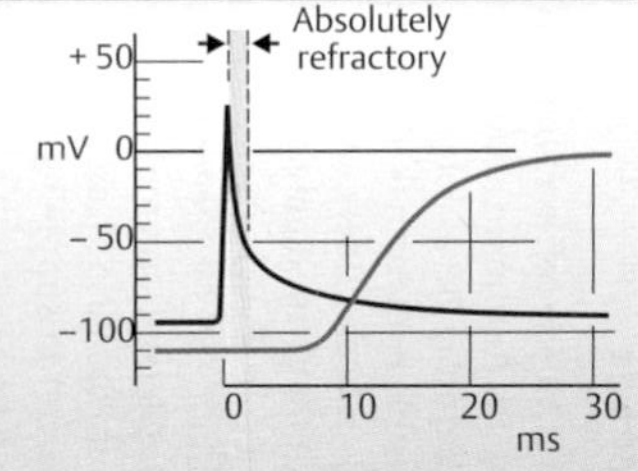

Plate 2.10 **Muscle types, motor unit**

Contractile Apparatus of Striated Muscle

The **muscle cell** is a fiber (→ **A2**) approximately 10 to 100 μm in diameter. Skeletal muscles fibers can be as long as 15 cm. Meat "fibers" visible with the naked eye are actually bundles of muscle fibers that are around 100 to 1000 μm in diameter (→ **A1**). Each striated muscle fiber is invested by a cell membrane called the *sarcolemma*, which surrounds the *sarcoplasm* (cytoplasm), several cell nuclei, mitochondria (*sarcosomes*), substances involved in supplying O_2 and energy (→ p. 72), and several hundreds of *myofibrils*.

So-called Z lines or, from a three-dimensional aspect, *Z plates* (plate-like proteins; → **B**) subdivide each **myofibril** (→ **A3**) into approx. 2 μm long, striated compartments called **sarcomeres** (→ **B**). When observed by (two-dimensional) microscopy, one can identify alternating light and dark bands and lines (hence the name "striated muscle") created by the thick **myosin II filaments** and thin **actin filaments** (→ **B**; for myosin I, see p. 30). Roughly 2000 actin filaments are bound medially to the Z plate. Thus, half of the filament projects into two adjacent sarcomeres (→ **B**). The region of the sarcomere proximal to the Z plate contains only actin filaments, which form a so-called *I band* (→ **B**). The region where the actin and myosin filaments overlap is called the *A band*. The *H zone* solely contains myosin filaments (ca. 1000 per sarcomere), which thicken towards the middle of the sarcomere to form the M line (*M plate*). The (actin) filaments are anchored to the sarcolemma by the protein *dystrophin*.

Each myosin filament consists of a bundle of ca. 300 **myosin-II molecules** (→ **B**). Each molecule has two globular heads connected by flexible necks (head and neck = subfragment S1; formed after proteolysis) to the filamentous tail of the molecule (two intertwined α-helices = subfragment S2) (→ **C**). Each of the heads has a *motor domain* with a *nucleotide binding pocket* (for ATP or ADP + P_i) and an *actin binding site*. Two light protein chains are located on each neck of this heavy molecule (220 kDa): one is regulatory (20 kDa), the other essential (17 kDa). Conformational changes in the head–neck segment allow the myosin head to "tilt" when interacting with actin (*sliding filaments*; → p. 62).

Actin is a globular protein molecule (G-actin). Four hundred such molecules join to form F-actin, a beaded polymer chain. Two of the twisted protein filaments combine to form an **actin filament** (→ **B**), which is positioned by the equally long protein *nebulin*.

Tropomyosin molecules joined end-to-end (40 nm each) lie adjacent to the actin filament, and a **troponin** (TN) molecule is attached every 40 nm or so (→ **B**). Each troponin molecule consists of three subunits: *TN-C*, which has two regulatory bindings sites for Ca^{2+} at the amino end, *TN-I*, which prevents the filaments from sliding when at rest (→ p. 62), and *TN-T*, which interacts with TN-C, TN-I, and actin.

The sarcomere also has another system of filaments (→ **B**) formed by the filamentous protein **titin** (connectin). Titin is more than 1000 nm in length and has some 30 000 amino acids ($M_r > 3000$ kDa). It is the longest known polypeptide chain and comprises 10% of the total muscle mass. Titin is anchored at its carboxyl end to the M plate and, at the amino end, to the Z plate (→ p. 66 for functional description).

The sarcolemma forms a T system with several **transverse tubules** (tube-like invaginations) that run perpendicular to the myofibrils (→ p. 63 A). The endoplasmic reticulum (→ p. 10 ff.) of muscle fibers has a characteristic shape and is called the **sarcoplasmic reticulum** (SR; → p. 63 A). It forms closed chambers without connections between the intra- and extracellular spaces. Most of the chambers run lengthwise to the myofibrils, and are therefore called *longitudinal tubules* (→ p. 63 A). The sarcoplasmic reticulum is more prominently developed in skeletal muscle than in the myocardium and serves as a **Ca^{2+} storage space**. Each T system separates the adjacent longitudinal tubules, forming *triads* (→ p. 63 A, B).

Plate 2.11   Contractile Apparatus of Striated Muscle

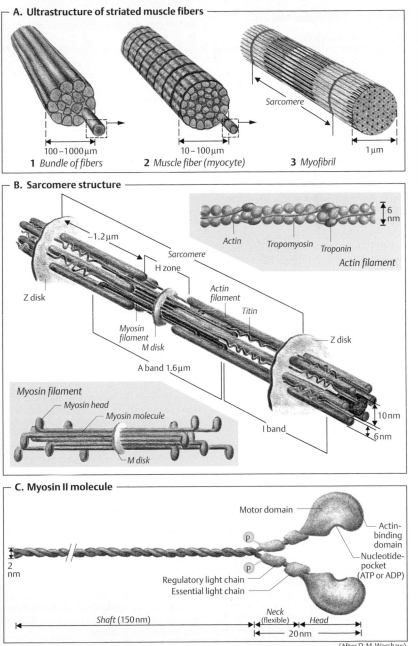

A. Ultrastructure of striated muscle fibers

1 Bundle of fibers
100–1000 µm

2 Muscle fiber (myocyte)
10–100 µm

3 Myofibril
Sarcomere
1 µm

B. Sarcomere structure

Actin filament
Actin   Tropomyosin   Troponin
6 nm

~1.2 µm
Sarcomere
H zone
Actin filament
Titin
Z disk
Myosin filament
M disk
A band 1.6 µm
Z disk
I band
10 nm
6 nm

Myosin filament
Myosin head
Myosin molecule
M disk

C. Myosin II molecule

Motor domain
Actin-binding domain
Nucleotide-pocket (ATP or ADP)
P
P
Regulatory light chain
Essential light chain
2 nm
Shaft (150 nm)
Neck (flexible)
Head
20 nm

(After D. M. Warshaw)

61

Contraction of Striated Muscle

Stimulation of muscle fibers. The release of *acetylcholine* at the motor end-plate of skeletal muscle leads to an *end-plate current* that spreads electrotonically and activates voltage-gated Na^+ channels in the sarcolemma ($\rightarrow$ p. 56). This leads to the firing of **action potentials** (**AP**) that travel at a rate of 2 m/s along the sarcolemma of the entire muscle fiber, and penetrate rapidly into the depths of the fiber along the T system ($\rightarrow$ **A**).

The conversion of this excitation into a contraction is called **electromechanical coupling** ($\rightarrow$ **B**). In the *skeletal muscle*, this process begins with the action potential exciting voltage-sensitive *dihydropyridine receptors* (DHPR) of the sarcolemma in the region of the triads. The DHPR are arranged in rows, and directly opposite them in the adjacent membrane of the sarcoplasmic reticulum (SR) are rows of Ca^{2+} channels called *ryanodine receptors* (type 1 in skeletal muscle: RYR1). Every other RYR1 is associated with a DHPR ($\rightarrow$ **B2**). RYR1 open when they directly "sense" by mechanical means an AP-related conformational change in the DHPR. In the *myocardium*, on the other hand, each DHPR is part of a *voltage-gated Ca^{2+} channel* of the sarcolemma that opens in response to an action potential. Small quantities of extracellular Ca^{2+} enter the cell through this channel, leading to the opening of myocardial RYR2 (so-called *trigger effect* of Ca^{2+} or Ca^{2+} *spark*; $\rightarrow$ **B3**). Ca^{2+} ions stored in the SR now flow through the opened RYR1 or RYR2 into the cytosol, increasing the cytosolic Ca^{2+} concentration $[Ca^{2+}]_i$ from a resting value of ca. 0.01 μmol/L to over 1 μmol/L ($\rightarrow$ **B1**). In skeletal muscle, DHPR stimulation at a single site is enough to trigger the coordinated opening of an entire group of RYR1, thereby increasing the reliability of impulse transmission. The increased cytosolic Ca^{2+} concentration saturates the Ca^{2+} binding sites on **troponin-C**, thereby canceling the troponin-mediated inhibitory effect of tropomyosin on filament sliding ($\rightarrow$ **D**). It is still unclear whether this type of disinhibition involves actin–myosin binding or the detachment of ADP and P_i, as described below.

ATP ($\rightarrow$ p. 72) is essential for **filament sliding** and, hence, for muscle contraction. Due to their ATPase activity, the myosin heads ($\rightarrow$ p. 60) act as the *motors* (motor proteins) of this process. The myosin-II and actin filaments of a sarcomere ($\rightarrow$ p. 60) are arranged in such a way that they can slide past each other. The myosin heads connect with the actin filaments at a particular angle, forming so-called *cross-bridges* ($\rightarrow$ **C1**). Due to a *conformational change* in the region of the nucleotide binding site of myosin-II ($\rightarrow$ p. 61 C), the spatial extent of which is increased by concerted movement of the neck region, the myosin head tilts down, drawing the thin filament a length of roughly 4 nm ($\rightarrow$ **C2**). The second myosin head may also move an adjacent actin filament. The head then detaches and "tenses" in preparation for the next "oarstroke" when it binds to actin anew ($\rightarrow$ **C3**).

Kinesin, another motor protein ($\rightarrow$ pp. 42 u. 58), independently advances on the microtubule by incremental movement of its two heads (8 nm increments), as in tug-of-war. In this case, fifty percent of the cycle time is "work time" (**duty ratio** = 0.5). Between two consecutive interactions with actin in *skeletal muscle*, on the other hand, myosin-II "jumps" 36 nm (or multiples of 36, e.g. 396 nm or more in rapid contractions) to reach the next (or the 11th) suitably located actin binding site ($\rightarrow$ **C3**, jump from **a** to **b**). Meanwhile, the other myosin heads working on this particular actin filament must make at least another 10 to 100 oarstrokes of around 4 nm each. The duty ratio of a myosin-II head is therefore 0.1 to 0.01. This division of labor by the myosin heads ensures that a certain percentage of the heads will always be ready to generate rapid contractions.

When filament sliding occurs, the Z plates approach each other and the overlap region of thick and thin filaments becomes larger, but the length of the filaments remains unchanged. This results in shortening of the I band and H zone ($\rightarrow$ p. 60). When the ends of the thick filaments ultimately bump against the Z plate, maximum muscle shortening occurs, and the ends of the thin filaments overlap ($\rightarrow$ p. 67 C). Shortening of the sarcomere therefore occurs at both ends of the myosin bundle, but in opposite directions.

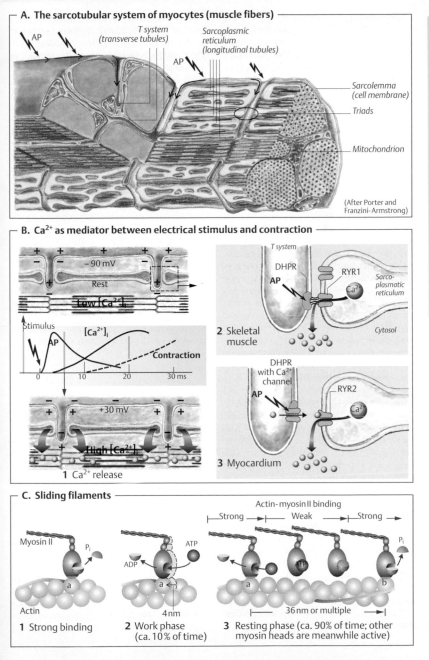

Plate 2.12 Contraction of Striated Muscle I

A. The sarcotubular system of myocytes (muscle fibers)

AP

T system (transverse tubules)

Sarcoplasmic reticulum (longitudinal tubules)

AP

AP

Sarcolemma (cell membrane)

Triads

Mitochondrion

(After Porter and Franzini-Armstrong)

B. Ca²⁺ as mediator between electrical stimulus and contraction

-90 mV

Rest

Low $[Ca^{2+}]_i$

Stimulus

AP

$[Ca^{2+}]_i$

Contraction

0   10   20   30 ms

$+30$ mV

High $[Ca^{2+}]_i$

1 Ca²⁺ release

T system

DHPR

AP

RYR1

Ca²⁺

Sarcoplasmatic reticulum

Cytosol

2 Skeletal muscle

DHPR with Ca²⁺ channel

AP

RYR2

Ca²⁺

3 Myocardium

C. Sliding filaments

Actin-myosin II binding

Strong → ← Weak → ← Strong →

Myosin II

P_i

ATP

ADP

a

a

4 nm

a

ATP

b

P_i

Actin

36 nm or multiple

1 Strong binding

2 Work phase (ca. 10% of time)

3 Resting phase (ca. 90% of time; other myosin heads are meanwhile active)

63

▶ **Contraction cycle** (→ **C** and **D**). Each of the two myosin heads (M) of a myosin-II molecule bind one ATP molecule in their nucleotide binding pocket. The resulting M-ATP complex lies at an approx. 90° angle to the rest of the myosin filament (→ **D4**). In this state, myosin has only a weak affinity for actin binding. Due to the influence of the *increased cytosolic Ca^{2+} concentration* on the troponin – tropomyosin complex, *actin (A) activates myosin's ATPase*, resulting in hydrolysis of ATP (ATP → ADP + P_i) and the formation of an A-M-ADP-P_i complex (→ **D1**). Detachment of P_i (inorganic phosphate) from the complex results in a conformational change of myosin that increases the actin–myosin association constant by four powers of ten (binding affinity now strong). The myosin heads consequently tilt to a 40° angle (→ **D2a**), causing the actin and myosin filaments to slide past each other. The release of ADP from myosin ultimately brings the myosin heads to their final position, a 45° angle (→ **D2b**). The remaining A-M complex (*rigor complex*) is stable and can again be transformed into a weak bond when the myosin heads bind ATP anew (*"softening effect" of ATP*). The high flexibility of the muscle at rest is important for processes such as cardiac filling or the relaxing of the extensor muscles during rapid bending movement. If a new ATP is bound to myosin, the subsequent weakening of the actin–myosin bond allows the realignment of the myosin head from 45° to 90° (→ **D3, 4**), the position preferred by the M-ATP complex. If the cytosolic Ca^{2+} concentration remains > 10^{-6} mol/L, the **D1** to **D4** cycle will begin anew. This depends mainly on whether subsequent action potentials arrive. Only a portion of the myosin heads that pull actin filaments are "on duty" (low duty ratio; see p. 62) to ensure the smoothness of contractions.

The **Ca^{2+} ions** released from the sarcoplasmic reticulum (SR) are continuously pumped back to the SR due to active transport by **Ca^{2+}-ATPase** (→ pp. 17 A and 26), also called SERCA (→ p. 16). Thus, if the RYR-mediated release of Ca^{2+} from the SR is interrupted, the cytosolic Ca^{2+} concentration rapidly drops below 10^{-6} mol/L and filament sliding ceases (*resting position;* → **D**, upper left corner).

Parvalbumin, a protein that occurs in the cytosol of fast-twitch muscle fibers (→ type F; p. 58), accelerates muscle relaxation after short contractions by binding cytosolic Ca^{2+} in exchange for Mg^{2+}. Parvalbumin's binding affinity for Ca^{2+} is higher than that of troponin, but lower than that of SR's Ca^{2+}-ATPase. It therefore functions as a "slow" Ca^{2+} buffer.

The course of the filament sliding cycle as described above mainly applies to **isotonic contractions**, that is, to contractions where muscle shortening occurs. During strictly **isometric contractions** where muscular tension increases but the muscle length remains unchanged, the sliding process tenses elastic components of a muscle, e.g. titin (→ p. 66), and then soon comes to a halt. Afterwards, the A-M-ATP complex (→ **D3**) probably transforms directly into A-M-ADP-P_i (→ **D1**).

The muscle fibers of a dead body do not produce any ATP. This means that, after death, Ca^{2+} is no longer pumped back into the SR, and the ATP reserves needed to break down stable A-M complexes are soon depleted. This results in stiffening of the dead body or **rigor mortis**, which passes only after the actin and myosin molecules in the muscle fibers decompose.

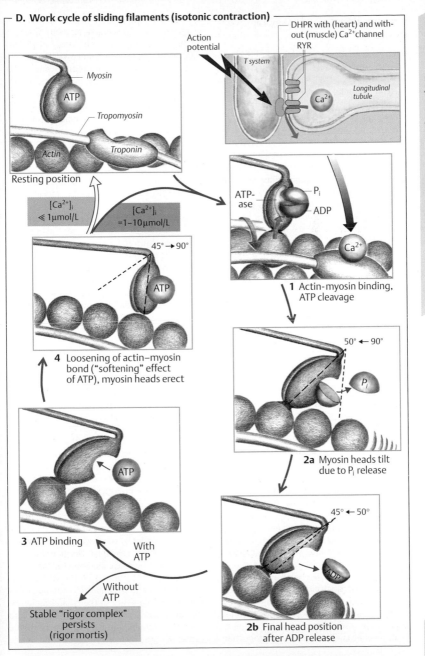

D. Work cycle of sliding filaments (isotonic contraction)

Action potential

DHPR with (heart) and without (muscle) Ca^{2+} channel
RYR

T system

Longitudinal tubule

Ca^{2+}

Myosin

ATP

Tropomyosin

Actin

Troponin

Resting position

$[Ca^{2+}]_i \ll 1\,\mu mol/L$

$[Ca^{2+}]_i = 1{-}10\,\mu mol/L$

ATP-ase

P_i

ADP

Ca^{2+}

1 Actin-myosin binding, ATP cleavage

$45° \rightarrow 90°$

ATP

4 Loosening of actin–myosin bond ("softening" effect of ATP), myosin heads erect

$50° \leftarrow 90°$

P_i

2a Myosin heads tilt due to P_i release

ATP

3 ATP binding

With ATP

Without ATP

$45° \leftarrow 50°$

ADP

2b Final head position after ADP release

Stable "rigor complex" persists (rigor mortis)

Plate 2.13 Contraction of Striated Muscle II

65

Mechanical Features of Skeletal Muscle

Action potentials generated in muscle fibers increase the cytosolic Ca^{2+} concentration $[Ca^{2+}]_i$, thereby triggering a contraction (skeletal muscle; $\rightarrow$ p. 36 B; myocardium; $\rightarrow$ p. 194). In skeletal muscles, **gradation of contraction force** is achieved by variable *recruitment* of motor units ($\rightarrow$ p. 58) and by changing the *action potential frequency*. A single stimulus always leads to maximum Ca^{2+} release and, thus, to a maximum single twitch of skeletal muscle fiber if above threshold (**all-or-none response**). Nonetheless, a single stimulus does not induce maximum shortening of muscle fiber because it is too brief to keep the sliding filaments in motion long enough for the end position to be reached. Muscle shortening continues only if a second stimulus arrives before the muscle has completely relaxed after the first stimulus. This type of stimulus repetition leads to incremental **mechanical summation** or **superposition** of the individual contractions ($\rightarrow$ A). Should the frequency of stimulation become so high that the muscle can no longer relax at all between stimuli, sustained maximum contraction of the motor units or **tetanus** will occur ($\rightarrow$ A). This occurs, for example, at 20 Hz in slow-twitch muscles and at 60–100 Hz in fast-twitch muscles ($\rightarrow$ p. 58). The muscle force during tetanus can be as much as four times larger than that of single twitches. The Ca^{2+} concentration, which decreases to some extent between superpositioned stimuli, remains high in tetanus.

Rigor ($\rightarrow$ p. 2.13) as well as **contracture**, another state characterized by persistent muscle shortening, must be distinguished from tetanus. Contracture is not caused by action potentials, but by *persistent local depolarization* due, for example, to increased extracellular K^+ concentrations (K^+ contracture) or drug-induced intracellular Ca^{2+} release, e.g., in response to caffeine. The contraction of so-called **tonus fibers** (specific fibers in the external eye muscles and in muscle spindles; $\rightarrow$ p. 318) is also a form of contracture. Tonus fibers do not respond to stimuli according to the all-or-none law, but contract in proportion with the magnitude of depolarization. The magnitude of contraction of tonus fibers is regulated by *variation of the cytosolic Ca^{2+} concentration* (not by action potentials!)

In contrast, the general **muscle tone** (*reflex tone*), or the tension of skeletal muscle at rest, is attributable to the arrival of normal action potentials at the individual motor units. The individual contractions cannot be detected because the motor units are alternately (asynchronously) stimulated. When apparently at rest, muscles such as the postural muscles are in this involuntary state of tension. Resting muscle tone is regulated by reflexes ($\rightarrow$ p. 318 ff.) and increases as the state of attentiveness increases.

Types of contractions ($\rightarrow$ B). There are different types of muscle contractions. In **isometric contractions**, muscle force ("tension") varies while the length of the muscle remains constant. (In cardiac muscle, this also represents *isovolumetric contraction*, because the muscle length determines the atrial or ventricular volume.) In **isotonic contractions**, the length of the muscle changes while muscle force remains constant. (In cardiac muscle, this also represents *isobaric contraction*, because the muscle force determines the atrial or ventricular pressure.) In **auxotonic contractions**, muscle length and force both vary simultaneously. An isotonic or auxotonic contraction that builds on an isometric one is called an **afterloaded contraction**.

Muscle extensibility. A resting muscle containing ATP can be stretched like a rubber band. The force required to start the stretching action ($\rightarrow$ D, E; extension force at rest) is very small, but increases exponentially when the muscle is under high elastic strain (see *resting tension curve*, $\rightarrow$ D). A muscle's *resistance to stretch*, which keeps the sliding filaments in the sarcomeres from separating, is influenced to a small extent by the fascia (fibrous tissue). The main factor, however, is the giant filamentous elastic molecule called **titin** (or *connectin*; 1000 nm long, M_r = 3 to 3.7 MDa) which is incorporated in the sarcomere (6 titin molecules per myosin filament). In the A band region of the sarcomere ($\rightarrow$ p. 61 B), titin lies adjacent to a myosin filament and helps to keep it in the center of the sarcomere. Titin molecules in the

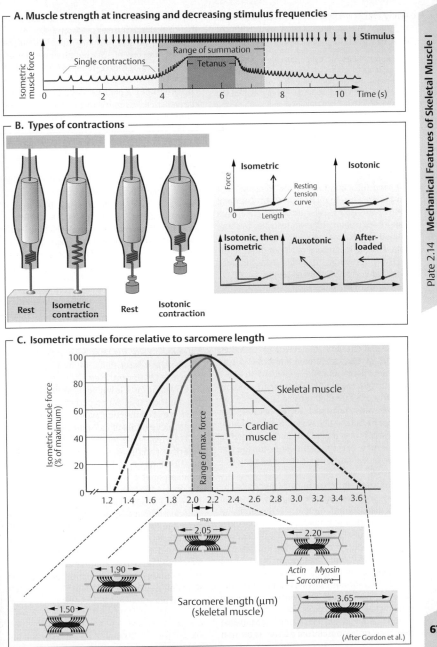

Plate 2.14

A. Muscle strength at increasing and decreasing stimulus frequencies

Stimulus

Isometric muscle force

Single contractions

Range of summation

Tetanus

0   2   4   6   8   10   Time (s)

B. Types of contractions

Rest   Isometric contraction

Rest   Isotonic contraction

Isometric

Force

Resting tension curve

0   Length

Isotonic

Isotonic, then isometric

Auxotonic

After-loaded

C. Isometric muscle force relative to sarcomere length

Isometric muscle force (% of maximum)

100
80
60
40
20
0

1.2 1.4 1.6 1.8 2.0 2.2 2.4 2.6 2.8 3.0 3.2 3.4 3.6

L_{max}

Range of max. force

Skeletal muscle

Cardiac muscle

2.05

2.20

1.90

Actin Myosin
Sarcomere

1.50

3.65

Sarcomere length (µm)
(skeletal muscle)

(After Gordon et al.)

I band region are flexible and function as "elastic bands" that counteract passive stretching of a muscle and influence its shortening velocity.

The **extensibility of titin molecules**, which can stretch to up to around ten times their normal length in skeletal muscle and somewhat less in cardiac muscle, is mainly due to frequent repetition of the *PEVK motif* (proline-glutamate-valine-lysine). In very strong muscle extension, which represents the steepest part of the resting extensibility curve ($\rightarrow$ **D**), globular chain elements called immunoglobulin C2 domains also unfold. The quicker the muscle stretches, the more sudden and crude this type of "shock absorber" action will be.

The **length** (L) and **force** (F) or "tension" of a muscle are closely related ($\rightarrow$ **C, E**). The total force of a muscle is the sum of its active force and its extension force at rest, as was explained above. Since the active force was determined by the magnitude of all potential actin-myosin interactions, it varies in accordance with the initial sarcomere length ($\rightarrow$ **C, D**). Skeletal muscle can develop maximum active (isometric) force (F_0) from its resting length (L_{max}; sarcomere length ca. 2 to 2.2 µm; $\rightarrow$ **C**). When the sarcomeres shorten ($L < L_{max}$), part of the thin filaments overlap, allowing only forces smaller than F_0 to develop ($\rightarrow$ **C**). When L is 70% of L_{max} (sarcomere length: 1.65 µm), the thick filaments make contact with the Z disks, and F becomes even smaller. In addition, a greatly pre-extended muscle ($L > L_{max}$) can develop only restricted force, because the number of potentially available actin–myosin bridges is reduced ($\rightarrow$ **C**). When extended to 130% or more of the L_{max}, the extension force at rest becomes a major part of the total muscle force ($\rightarrow$ **E**).

The length–force curve corresponds to the **cardiac pressure–volume diagram** in which ventricular filling volume corresponds to muscle length, and ventricular pressure corresponds to muscle force; $\rightarrow$ p. 202. Changes in the cytosolic Ca^{2+} concentration can modify the pressure–volume relationship by causing a change in *contractility* ($\rightarrow$ p. 203 B2).

Other important functional **differences between cardiac muscle and skeletal muscle** are listed below (see also p. 59 A):

Since skeletal muscle is more extensible than the cardiac muscle, the passive extension force of cardiac muscle at rest is greater than that of skeletal muscle ($\rightarrow$ **E1, 2**).

Skeletal muscle normally functions in the plateau region of its length–force curve, whereas cardiac muscle tends to operate in the ascending limb (below L_{max}) of its length–force curve without a plateau ($\rightarrow$ **C, E1, 2**). Hence, the ventricle responds to increased diastolic filling loads by increasing its force development (**Frank–Starling mechanism**; $\rightarrow$ p. 204). In cardiac muscle, extension also affects troponin's sensitivity to Ca^{2+}, resulting in a steeper curve ($\rightarrow$ **E2**).

Action potentials in cardiac muscle are of much longer duration than those in skeletal muscle ($\rightarrow$ p. 59 A) because g_K temporarily decreases and g_{Ca} increases for 200 to 500 ms after rapid inactivation of Na^+ channels. This allows the slow influx of Ca^{2+}, causing the action potential to reach a *plateau*. As a result, the refractory period does not end until a contraction has almost subsided ($\rightarrow$ p. 59 A). Therefore, *tetanus cannot be evoked in cardiac muscle*.

Unlike skeletal muscle, *cardiac muscle has no motor units*. Instead, the stimulus spreads across all myocardial fibers of the atria and subsequently of the ventricles generating an *all-or-none contraction* of both atria and, thereafter, both ventricles.

In cardiac muscle but not in skeletal muscle, the duration of an action potential can change the force of contraction, which is controlled by the variable influx of Ca^{2+} into the cell.

The greater the force (load), the lower the **velocity** of an (isotonic) contraction (see velocity–force diagram, **F1**). Maximal force and a small amount of heat will develop if shortening does not occur. The maximal velocity (biceps: ca. 7 m/s) and a lot of heat will develop in muscle without a stress load. Light loads can therefore be picked up more quickly than heavy loads ($\rightarrow$ **F2**). The total amount of energy consumed for work and heat is greater in isotonic contractions than in isometric ones. **Muscle power** is the product of force and the shortening velocity: $N \cdot m \cdot s^{-1} = W$ ($\rightarrow$ **F1**, colored areas).

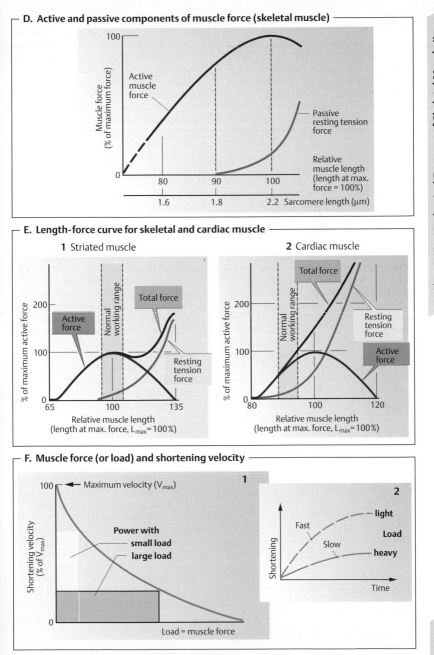

Plate 2.15 Mechanical Features of Skeletal Muscle II

D. Active and passive components of muscle force (skeletal muscle)

- Muscle force (% of maximum force)
- Active muscle force
- Passive resting tension force
- Relative muscle length (length at max. force = 100%)
- 80 — 90 — 100
- 1.6 — 1.8 — 2.2   Sarcomere length (μm)

E. Length-force curve for skeletal and cardiac muscle

1 Striated muscle

- % of maximum active force
- Active force
- Normal working range
- Total force
- Resting tension force
- 200 — 100 — 0
- 65 — 100 — 135
- Relative muscle length (length at max. force, L_{max} = 100%)

2 Cardiac muscle

- % of maximum active force
- Normal working range
- Total force
- Resting tension force
- Active force
- 200 — 100 — 0
- 80 — 100 — 120
- Relative muscle length (length at max. force, L_{max} = 100%)

F. Muscle force (or load) and shortening velocity

1

- Shortening velocity (% of V_{max})
- Maximum velocity (V_{max})
- 100 — 0
- Power with small load
- large load
- Load = muscle force

2

- Shortening
- Fast
- Slow
- light
- Load
- heavy
- Time

Smooth Muscle

Smooth muscle (SmM) consists of multiple layers of spindle-shaped cells. It is involved in the function of many organs (stomach, intestine, gall bladder, urinary bladder, uterus, bronchi, eyes, etc.) and the blood vessels, where it plays an important role in circulatory control. SmM contains a special type of F-actin-tropomyosin and myosin II filaments ($\rightarrow$ p. 60), but lacks troponin and myofibrils. Furthermore, it has no distinct tubular system and no sarcomeres (nonstriated). It is therefore called smooth muscle because of this lack of striation (see p. 59 A for further differences in the muscle types). SmM filaments form a loose contractile apparatus arranged approximately longitudinally within the cell and attached to discoid plaques (see **B** for model), which also provide a mechanical means for cell–cell binding of SmM. Smooth muscle can shorten much more than striated muscle.

The **membrane potential** of the SmM cells of many organs (e.g., the intestine) is not constant, but fluctuates rhythmically at a low frequency (3 to 15 min^{-1}) and amplitude (10 to 20 mV), producing *slow waves*. These waves trigger a burst of action potentials (spikes) when they exceed a certain threshold potential. The longer the slow wave remains above the threshold potential, the greater the number and frequency of the action potentials it produces. A relatively sluggish contraction occurs around 150 ms after a spike ($\rightarrow$ p. 59 A, left panel). *Tetanus* occurs at relatively low spike frequencies ($\rightarrow$ p. 66). Hence, SmM is constantly in a state of a more or less strong contraction (**tonus** or tone). The action potential of SmM cells of some organs has a plateau similar to that of the cardiac action potential ($\rightarrow$ p. 59 A, middle panel).

There are two types of smooth muscles ($\rightarrow$ **A**). The cells of **single-unit SmM** are electrically coupled with each other by *gap junctions* ($\rightarrow$ pp. 18 and 50). Stimuli are passed along from cell to cell in organs such as the stomach, intestine, gallbladder, urinary bladder, ureter, uterus, and some types of blood vessels. Stimuli are generated autonomously from within the SmM, partly by pacemaker cells). In other words, the stimulus is innervation-inde-

pendent and, in many cases, spontaneous (*myogenic tonus*). The second type, **multi-unit SmM**, contracts primarily due to stimuli from the autonomic nervous system (*neurogenic tonus*). This occurs in structures such as the arterioles, spermatic ducts, iris, ciliary body, and the muscles at the roots of the hair. Since these SmM cells generally are not connected by gap junctions, stimulation remains localized, as in the motor units of the skeletal muscle.

Smooth muscle tonus is regulated by the degree of depolarization (e.g., through stretch or pacemaker cells) as well as by transmitter substances (e.g., acetylcholine or noradrenaline) and numerous hormones (e.g., estrogens, progesterone and oxytocin in the uterus and histamine, angiotensin II, adiuretin, serotonin and bradykinin in vascular muscle). An increase in tonus will occur if any of these factors directly or indirectly increases the cytosolic Ca^{2+} concentration to more than 10^{-6} mol/L. The Ca^{2+} influx comes mainly from extracellular sources, but a small portion comes from intracellular stores ($\rightarrow$ **B1**). Ca^{2+} ions bind to calmodulin (CM) ($\rightarrow$ **B2**), and Ca^{2+}-CM promotes contraction in the following manner.

Regulation at **myosin II** ($\rightarrow$ **B3**): The Ca^{2+}-CM complex activates myosin light chain kinase (MLCK), which phosphorylates myosin's regulatory light chain (RLC) in a certain position, thereby enabling the myosin head to interact with actin ($\rightarrow$ **B6**).

Regulation at the **actin** *level* ($\rightarrow$ **B4**). The Ca^{2+}-CM complex also binds with caldesmon (CDM), which then detaches from the actin–tropomyosin complex, thus making it available for filament sliding ($\rightarrow$ **B6**). Phosphorylation of CDM by protein kinase C (PK-C) also seems to be able to induce filament sliding ($\rightarrow$ **B5**).

Factors that lead to a **reduction of tonus** are: reduction of the cytosolic Ca^{2+} concentration to less than 10^{-6} mol/L ($\rightarrow$ **B7**), phosphatase activity ($\rightarrow$ **B8**), and PK-C if it phosphorylates another position on the RLC ($\rightarrow$ **B9**).

When length–force curves are recorded for smooth muscle, the curve shows that muscle force decreases continuously while muscle length remains constant. This property of a muscle is called **plasticity**.

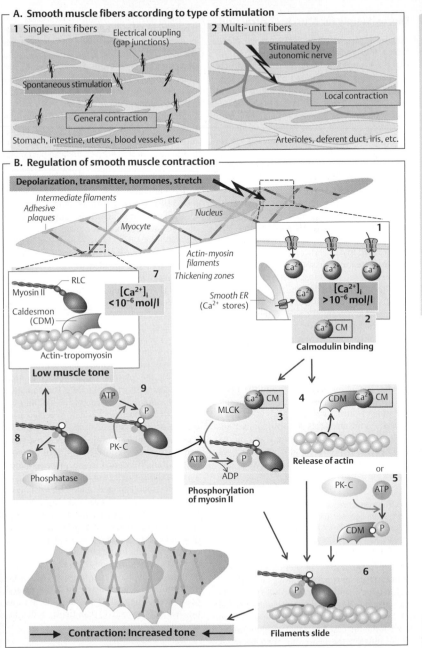

A. Smooth muscle fibers according to type of stimulation

1 Single-unit fibers

Electrical coupling (gap junctions)

Spontaneous stimulation

General contraction

Stomach, intestine, uterus, blood vessels, etc.

2 Multi-unit fibers

Stimulated by autonomic nerve

Local contraction

Arterioles, deferent duct, iris, etc.

B. Regulation of smooth muscle contraction

Depolarization, transmitter, hormones, stretch

Intermediate filaments
Adhesive plaques
Myocyte
Nucleus
Actin-myosin filaments
Thickening zones

1
$[Ca^{2+}]_i$ > 10^{-6} mol/l

Smooth ER (Ca^{2+} stores)

2 Ca^{2+} CM
Calmodulin binding

7
Myosin II — RLC
Caldesmon (CDM)
$[Ca^{2+}]_i$ < 10^{-6} mol/l
Actin-tropomyosin

Low muscle tone

8
P
Phosphatase

9
ATP
P
PK-C

3
MLCK
Ca^{2+} CM
ATP → P
ADP
Phosphorylation of myosin II

4
CDM Ca^{2+} CM
Release of actin

or
5
PK-C ATP
CDM P

6
P
Filaments slide

➤ **Contraction: Increased tone** ◄

Plate 2.16 Smooth Muscle

Energy Supply for Muscle Contraction

Adenosine triphosphate (ATP) is a direct source of chemical energy for muscle contraction (→ A, pp. 40 and 64). However, a muscle cell contains only a limited amount of ATP–only enough to take a sprinter some 10 to 20 m or so. Hence, spent ATP is continuously regenerated to keep the intracellular ATP concentration constant, even when large quantities of it are needed. The three routes of ATP regeneration are (→ B):

1. Dephosphorylation of creatine phosphate
2. Anaerobic glycolysis
3. Aerobic oxidation of glucose and fatty acids.

Routes 2 and 3 are relatively slow, so creatine phosphate (CrP) must provide the chemical energy needed for rapid ATP regeneration. ADP derived from metabolized ATP is immediately transformed to ATP and creatine (Cr) by mitochondrial creatine kinase (→ B1 and p. 40). The CrP reserve of the muscle is sufficient for short-term high-performance bursts of 10 – 20 s (e.g., for a 100-m sprint).

Anaerobic glycolysis occurs later than CrP dephosphorylation (after a maximum of ca. 30 s). In anaerobic glycolysis, muscle glycogen is converted via glucose-6-phosphate to lactic acid (→ lactate + H^+), yielding 3 ATP molecules for each glucose residue (→ B2). During light exercise, lactate$^-$ is broken down in the heart and liver whereby H^+ ions are used up. Aerobic oxidation of glucose and fatty acids takes place approx. 1 min after this less productive anaerobic form of ATP regeneration. If aerobic oxidation does not produce a sufficient supply of ATP during strenuous exercise, anaerobic glycolysis must also be continued.

In this case, however, glucose must be imported from the liver where it is formed by glycogenolysis and gluconeogenesis (see also p. 282f.). Imported glucose yields only two ATP for each molecule of glucose, because one ATP is required for 6-phosphorylation of glucose.

Aerobic regeneration of ATP from glucose (2 + 34 ATP per glucose residue) or fatty acids is required for sustained exercise (→ B3). The cardiac output (= heart rate × stroke volume) and total ventilation must therefore be increased to meet the increased metabolic requirements of the muscle; the heart rate then becomes constant (→ p. 75 B). The several minutes that pass before this steady state is achieved are bridged by anaerobic energy production, increased O_2 extraction from the blood and depletion of short-term O_2 reserves in the muscle (myoglobin). The interim between the two phases is often perceived as the "low point" of physical performance.

The O_2 affinity of myoglobin is higher than that of hemoglobin, but lower than that of respiratory chain enzymes. Thus, myoglobin is normally saturated with O_2 and can pass on its oxygen to the mitochondria during brief arterial oxygen supply deficits.

The endurance limit, which is some 370 W ($\approx$ 0.5 HP) in top athletes, is mainly dependent on the speed at which O_2 is supplied and on how fast aerobic oxidation takes place. When the endurance limit is exceeded, steady state cannot occur, the heart rate then rises continuously (→ p. 75 B). The muscles can temporarily compensate for the energy deficit (see above), but the H^+-consuming lactate metabolism cannot keep pace with the persistently high level of anaerobic ATP regeneration. An excess of lactate and H^+ ions, i.e. lactacidosis, therefore develops. If an individual exceeds his or her endurance limit by around 60%, which is about equivalent to maximum O_2 consumption (→ p. 74), the plasma lactate concentration will increase sharply, reaching the so-called anaerobic threshold at 4 mmol/L. No significant increase in performance can be expected after that point. The systemic drop in pH results in increasing inhibition of the chemical reactions needed for muscle contraction. This ultimately leads to an ATP deficit, rapid muscle fatigue and, finally, a stoppage of muscle work.

CrP metabolism and anaerobic glycolysis enable the body to achieve three times the performance possible with aerobic ATP regeneration, albeit for only about 40 s. However, these processes result in an O_2 deficit that must be compensated for in the post-exercise recovery phase (O_2 debt). The body "pays off" this debt by regenerating its energy reserves and breaking down the excess lactate in the liver and heart. The O_2 debt after strenuous exercise is much larger (up to 20 L) than the O_2 deficit for several reasons.

Plate 2.17 Energy Supply for Muscle Contraction

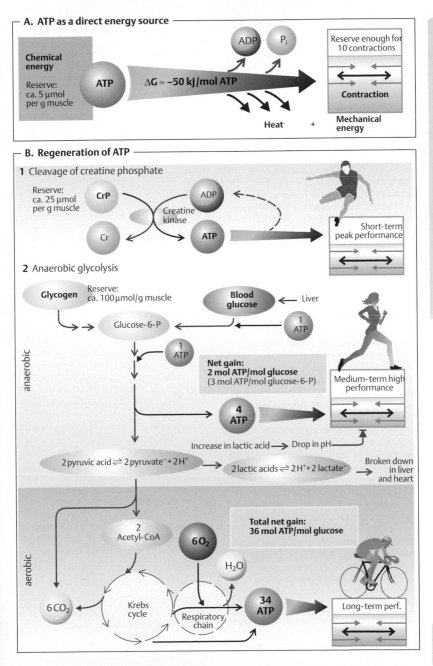

A. ATP as a direct energy source

Chemical energy

Reserve: ca. 5 μmol per g muscle

ATP → ADP + P_i

$\Delta G \approx -50$ kJ/mol ATP

Heat + Mechanical energy

Reserve enough for 10 contractions

Contraction

B. Regeneration of ATP

1 Cleavage of creatine phosphate

Reserve: ca. 25 μmol per g muscle

CrP → Cr

Creatine kinase

ADP → **ATP**

Short-term peak performance

2 Anaerobic glycolysis

anaerobic

Glycogen Reserve: ca. 100 μmol/g muscle

Blood glucose ← Liver

Glucose-6-P ← 1 ATP

1 ATP

Net gain: 2 mol ATP/mol glucose (3 mol ATP/mol glucose-6-P)

4 ATP

Medium-term high performance

Increase in lactic acid → Drop in pH

2 pyruvic acid ⇌ 2 pyruvate⁻ + 2H⁺ → 2 lactic acids ⇌ 2H⁺ + 2 lactate⁻ → Broken down in liver and heart

aerobic

2 Acetyl-CoA

$6O_2$

H_2O

Total net gain: 36 mol ATP/mol glucose

$6 CO_2$ ← Krebs cycle ⇌ Respiratory chain → **34 ATP** → Long-term perf.

Plate 2.17 Energy Supply for Muscle Contraction

73

Physical Work

There are three types of muscle work:

◆ *Positive dynamic work*, which requires to muscles involved to alternately contract and relax (e.g., going uphill).

◆ *Negative dynamic work*, which requires the muscles involved to alternately extend while braking (*braking work*) and contract without a load (e.g., going downhill).

◆ *Static postural work*, which requires continuous contraction (e.g., standing upright).

Many activities involve a combination of two or three types of muscle work. Outwardly directed mechanical work is produced in dynamic muscle activity, but not in purely postural work. In the latter case, force × distance = 0. However, chemical energy is still consumed and completely transformed into a form of heat called *maintenance heat* (= muscle force times the duration of postural work).

In *strenuous exercise*, the muscles require up to 500 times more O_2 than when at rest. At the same time, the muscle must rid itself of metabolic products such as H^+, CO_2, and **lactate** ($\rightarrow$ p. 72). Muscle work therefore requires drastic cardiovascular and respiratory changes.

In untrained subjects (UT), the cardiac output (CO; $\rightarrow$ p. 186) rises from 5–6 L/min at rest to a maximum of 15–20 L/min during exercise ($\rightarrow$ p. 77 C). Work-related **activation of the sympathetic nervous system** increases the heart rate up to ca. 2.5 fold and the stroke volume up to ca. 1.2 fold (UT). In *light to moderate exercise*, the heart rate soon levels out at a new constant level, and no fatigue occurs. *Very strenuous exercise*, on the other hand, must soon be interrupted because the heart cannot achieve the required long-term performance ($\rightarrow$ **B**). The increased CO provides more blood for the muscles ($\rightarrow$ **A**) and the skin (*heat loss;* $\rightarrow$ p. 222.). The blood flow in the kidney and intestine, on the other hand, is reduced by the sympathetic tone below the resting value ($\rightarrow$ **A**). The systolic **blood pressure** ($\rightarrow$ p. 206) rises while the diastolic pressure remains constant, yielding only a moderate increase in the mean pressure.

The smaller the muscle mass involved in the work, the higher the increase in blood pressure. Hence, the blood pressure increase in arm activity (cutting hedges) is higher than that in leg activity (cycling). In patients with coronary artery disease or cerebrovascular sclerosis, arm activity is therefore more dangerous than leg activity due to the risk of myocardial infarction or brain hemorrhage.

Muscular blood flow. At the maximum work level, the blood flow in 1 kg of active muscle rises to as much as 2.5 L/min ($\rightarrow$ p. 213 A), equivalent to 10% of the maximum cardiac output. Hence, no more than 10 kg of muscle ($< 1/3$ the total muscle mass) can be fully active at any one time. Vasodilatation, which is required for the higher blood flow, is mainly achieved through *local chemical influences* ($\rightarrow$ p. 212). In purely postural work, the increase in blood flow is prevented in part by the fact that the continuously contracted muscle squeezes its own vessels. The muscle then *fatigues* faster than in rhythmic dynamic work.

During physical exercise ($\rightarrow$ **C1**), the **ventilation** ($\dot{V}_E$) increases from a resting value of ca. 7.5 L/min to a maximum of 90 to 120 L/min ($\rightarrow$ **C3**). Both the respiratory rate (40–60 min^{-1} max; $\rightarrow$ **C2**) and the tidal volume (ca. 2 L max.) contribute to this increase. Because of the high $\dot{V}_E$ and increased CO, oxygen consumption ($\dot{V}_{O_2}$) can increase from ca. 0.3 L/min at rest to a maximum ($\dot{V}_{O_2}$ max) of ca. 3 L/min in UT ($\rightarrow$ **C4** and p. 76). Around 25 L of air has to be ventilated to take up 1 L of O_2 at rest, corresponding to a *respiratory equivalent* ($\dot{V}_E/\dot{V}_{O_2}$) of 25. During physical exercise, $\dot{V}_E/\dot{V}_{O_2}$ rises beyond the endurance limit to a value of 40–50.

Increased **O_2 extraction** in the tissues also contributes to the large increase in $\dot{V}_{O_2}$ during exercise. The decreasing pH and increasing temperature shift the O_2 binding curve towards the right ($\rightarrow$ p. 129 B). O_2 extraction is calculated as the arteriovenous difference in O_2 concentration (avDo$_2$ in L/L blood) times the blood flow in L/min. The **maximum O_2 consumption ($\dot{V}_{O_2}$ max)** is therefore defined as:

$$\dot{V}_{O_2} \max = HR\max \cdot SV\max \cdot avD_{O_2}\max$$

where HR is the heart rate and SV is the stroke volume. $\dot{V}_{O_2}$ max per body weight is an ideal measure of *physical exercise capacity* ($\rightarrow$ p. 76).

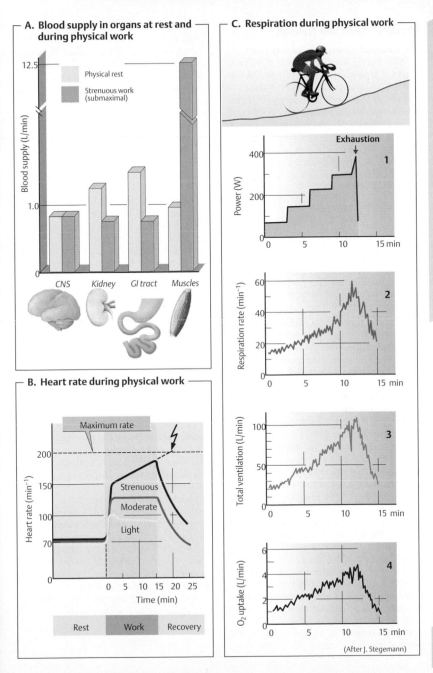

A. Blood supply in organs at rest and during physical work

Physical rest

Strenuous work (submaximal)

Blood supply (L/min)

12.5

1.0

0

CNS   Kidney   GI tract   Muscles

B. Heart rate during physical work

Maximum rate

Heart rate (min⁻¹)

200

150

100

70

0

Strenuous

Moderate

Light

0   5   10   15   20   25

Time (min)

Rest   Work   Recovery

C. Respiration during physical work

Exhaustion

Power (W)

400

200

0

0   5   10   15 min

1

Respiration rate (min⁻¹)

60

40

20

0

0   5   10   15 min

2

Total ventilation (L/min)

100

50

0

0   5   10   15 min

3

O₂ uptake (L/min)

6

4

2

0

0   5   10   15 min

4

(After J. Stegemann)

Plate 2.18   Physical Work

Physical Fitness and Training

The **physical exercise capacity** can be measured using simple yet standardized techniques of **ergometry**. This may be desirable in athletes, for example, to assess the results of training, or in patients undergoing rehabilitation therapy. Ergometry assesses the effects of exercise on physiological parameters such as O_2 consumption ($\dot{V}_{O_2}$), respiration rate, heart rate ($\rightarrow$ p. 74), and the plasma lactate concentration ($\rightarrow$ **A**). The measured physical power (performance) is expressed in watts (W) or W/kg body weight (BW).

In *bicycle ergometry*, a brake is used to adjust the watt level. In "uphill" ergometry on a *treadmill* set at an angle α, exercise performance in watts is calculated as a factor of body mass (kg) $\times$ gravitational acceleration g (m $\cdot$ s^{-2}) $\times$ distance traveled (m) $\times$ sin $\alpha \times 1/$ time required (s^{-1}). In the *Margaria step test*, the test subject is required to run up a staircase as fast as possible after a certain starting distance. Performance is then measured as body mass (kg) $\times$ g (m $\cdot$ s^{-2}) $\times$ height/time (m $\cdot$ s^{-1}).

Short-term performance tests (10–30 s) measure performance achieved through the rapidly available energy reserves (creatine phosphate, glycogen). *Medium-term performance tests* measure performance fueled by anaerobic glycolysis ($\rightarrow$ p. 72). The **maximum O_2 consumption** ($\dot{V}_{O_2}$ **max**) is used to measure longer term aerobic exercise performance achieved through oxidation of glucose and free fatty acids ($\rightarrow$ p. 74).

In strenuous exercise (roughly 2/3 the maximum physical capacity or more), the aerobic mechanisms do not produce enough energy, so anaerobic metabolism must continue as a parallel energy source. This results in lactacidosis and a sharp increase in the plasma lactate concentration ($\rightarrow$ **A**). **Lactate concentrations** of up to 2 mmol/L (*aerobic threshold*) can be tolerated for prolonged periods of exercise. Lactate concentrations above 4 mmol/L (*anaerobic threshold*) indicate that the performance limit will soon be reached. Exercise must eventually be interrupted, not because of the increasing lactate concentration, but because of the increasing level of acidosis ($\rightarrow$ p. 74).

Physical training raises and maintains the physical exercise capacity. There are three types of physical training strategies, and most training programs use a combination of them.

Motor learning, which increases the rate and accuracy of motor skills (e.g., typewriting). These activities primarily involve the CNS.

Endurance training, which improves submaximal long-term performance (e.g., running a marathon). The main objectives of endurance training are to increase the oxidative capacity of slow-twitch motor units ($\rightarrow$ p. 58), e.g., by increasing the mitochondrial density, increase the cardiac output and, consequently, to increase $\dot{V}_{O_2}$ max ($\rightarrow$ **B, C**). The resulting increase in heart weight allows higher stroke volumes ($\rightarrow$ **C**) as well as higher tidal volumes, resulting in very low resting heart rates and respiratory rates. Trained athletes can therefore achieve larger increases in cardiac output and ventilation than untrained subjects ($\rightarrow$ **C**). The $\dot{V}_{O_2}$ max of a healthy individual is limited by the cardiovascular capacity, not the respiratory capacity. In individuals who practice endurance training, the exercise-related rise in the lactate concentraton is also lower and occurs later than in untrained subjects ($\rightarrow$ **A**).

Strength training improves the maximum short-term performance level (e.g., in weight lifting). The main objectives are to increase the muscle mass by increasing the size of the muscle fibers (hypertrophy) and to increase the glycolytic capacity of type motor units ($\rightarrow$ p. 58).

Excessive physical exercise causes **muscle soreness and stiffness.** The underlying cause is not lactic acid accumulation, but sarcomere microtrauma, which leads to muscle swelling and pain. The muscle ache, is a sign of micro-inflammation ($\rightarrow$ **D**).

Muscle fatigue may be peripheral or central. *Peripheral fatigue* ist caused by the exhaustion of energy reserves and the accumulation of metabolic products in the active muscle. This is particularly quick to occur during postural work ($\rightarrow$ p. 66). *Central fatigue* is characterized by work-related pain in the involved muscles and joints that prevents the continuation of physical exercise or decreased the individual's motivation to continue the exercise.

Plate 2.19 Physical Fitness and Training

A. Work load and plasma lactate

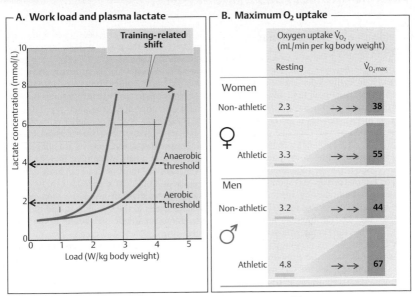

Lactate concentration (mmol/L)

Training-related shift

Anaerobic threshold

Aerobic threshold

Load (W/kg body weight)

B. Maximum O₂ uptake

Oxygen uptake $\dot{V}_{O_2}$ (mL/min per kg body weight)

	Resting		$\dot{V}_{O_2 max}$
Women			
Non-athletic	2.3	→ →	38
♀ Athletic	3.3	→ →	55
Men			
Non-athletic	3.2	→ →	44
♂ Athletic	4.8	→ →	67

C. Comparison of non-athletic individuals and endurance athletes

Physiological parameters (2 men, age 25, 70 kg)

	Non-athletes		Endurance athletes	
	Resting	Maximum	Resting	Maximum
Heart weight (g)	300		500	
Blood volume (L)	5.6		5.9	
Heart rate (min⁻¹)	80 → →	180	40 → →	180
Stroke volume (mL)	70 → →	100	140 → →	190
Cardiac output (L/min)	5.6 → →	18	5.6 → →	35
Total ventilation (L/min)	8.0 → →	100	8.0 → →	200
O₂ uptake (L/min)	0.3 → →	2.8	0.3 → →	5.2

(Data partly from H.-J. Ulmer)

D. Post-exercise muscle ache

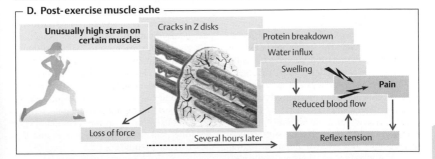

Unusually high strain on certain muscles

Cracks in Z disks

Protein breakdown

Water influx

Swelling

Pain

Reduced blood flow

Loss of force

Reflex tension

Several hours later

77

Organization of the Autonomic Nervous System

In the *somatic nervous system*, nerve fibers extend to and from the skeletal muscles, skin and sense organs. They usually emit impulses in response to stimuli from the outside environment, as in the *withdrawal reflex* (→ p. 320). Much somatic nervous activity occurs *consciously* and under *voluntary control*. In contrast, the **autonomic nervous system** (**ANS**) is mainly concerned with **regulation of circulation and internal organs**. It responds to changing outside conditions by triggering orthostatic responses, work start reactions, etc. to regulate the body's **internal environment** (→ p. 2). As the name implies, most activities of the ANS are not subject to voluntary control.

For the most part, the autonomic and somatic nervous systems are anatomically and functionally separate in the periphery (→ **A**), but closely connected in the central nervous system, CNS (→ p. 266). The *peripheral ANS is efferent*, but most of the nerves containing ANS fibers hold also afferent neurons. These are called *visceral afferents* because their signals originate from visceral organs, such as the esophagus, gastrointestinal (GI) tract, liver, lungs, heart, arteries, and urinary bladder. Some are also named after the nerve they accompany (e.g., vagal afferents).

Autonomic nervous activity is usually regulated by the **reflex arc**, which has an afferent limb (visceral and/or somatic afferents) and an efferent limb (autonomic and/or somatic efferents). The *afferent fibers* convey stimuli from the skin (e.g. nociceptive stimuli; → p. 316) and nocisensors, mechanosensors and chemosensors in organs such as the lungs, gastrointestinal tract, bladder, vascular system and genitals. The ANS provides the autonomic *efferent fibers* that convey the reflex response to such afferent information, thereby inducing smooth muscle contraction (→ p. 70) in organs such as the eye, lung, digestive tract and bladder, and influencing the function of the heart (→ p. 194) and glands. Examples of *somatic nervous system* involvement are afferent stimuli from the skin and sense organs (e.g., light stimuli) and efferent impulses to the skeletal muscles (e.g., coughing and vomiting).

Simple reflexes can take place *within an organ* (e.g., in the gut, → p. 244), but complex reflexes are controlled by **superordinate autonomic centers** in the CNS, primarily in the *spinal cord* (→ **A**). These centers are controlled by the *hypothalamus*, which incorporates the ANS in the execution of its programs (→ p. 330). The *cerebral cortex* is an even higher-ranking center that integrates the ANS with other systems.

The **peripheral ANS** consists of a *sympathetic division* and a *parasympathetic division* (→ **A**) which, for the most part, are separate entities (→ also p. 80ff.). The autonomic centers of the sympathetic division lie in the *thoracic and lumbar* levels of the spinal cord, and those of the parasympathetic division lie in the brain stem (eyes, glands, and organs innervated by the vagus nerve) and sacral part of the spinal cord (bladder, lower parts of the large intestine, and genital organs). (→ **A**). *Preganglionic fibers* of both divisions of the ANS extend from their centers to the **ganglia**, where they terminate at the *postganglionic neurons*.

Preganglionic **sympathetic neurons** arising from the spinal cord terminate either in the *paravertebral ganglionic chain*, in the *cervical or abdominal ganglia* or in so-called terminal ganglia. Transmission of stimuli from preganglionic to postganglionic neurons is *cholinergic*, that is, mediated by release of the neurotransmitter *acetylcholine* (→ p. 82). Stimulation of all effector organs except sweat glands by the postganglionic sympathetic fibers is *adrenergic*, i.e., mediated by the release of *norepinephrine* (→ **A** and p. 84ff.).

Parasympathetic ganglia are situated near or within the effector organ. Synaptic transmissions in the parasympathetic ganglia and at the effector organ are *cholinergic* (→ **A**).

Most organs are innervated by sympathetic and parasympathetic nerve fibers. Nonetheless, the organ's response to the two systems can be either antagonistic (e.g., in the heart) or complementary (e.g., in the sex organs).

The **adrenal medulla** is a ganglion and hormone gland combined. Preganglionic sympathetic fibers in the adrenal medulla release acetylcholine, leading to the secretion of *epinephrine* (and some norepinephrine) *into the bloodstream* (→ p. 86).

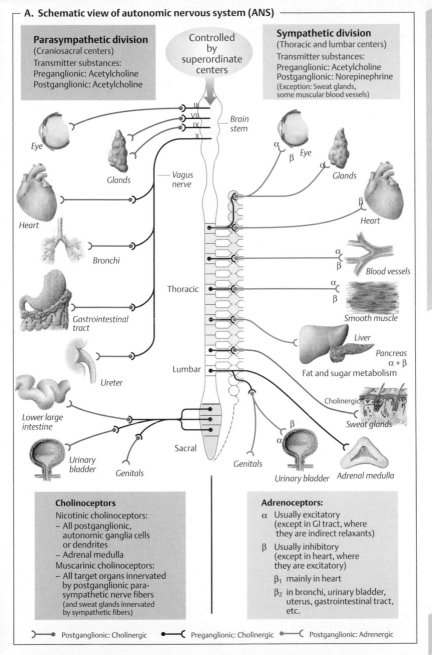

A. Schematic view of autonomic nervous system (ANS)

Parasympathetic division
(Craniosacral centers)
Transmitter substances:
Preganglionic: Acetylcholine
Postganglionic: Acetylcholine

Controlled
by
superordinate
centers

Sympathetic division
(Thoracic and lumbar centers)
Transmitter substances:
Preganglionic: Acetylcholine
Postganglionic: Norepinephrine
(Exception: Sweat glands,
some muscular blood vessels)

III
VII
IX
X

*Brain
stem*

Eye

Glands

*Vagus
nerve*

Heart

Bronchi

Thoracic

Gastrointestinal
tract

Lumbar

Ureter

Lower large
intestine

Sacral

Urinary
bladder

Genitals

Eye
α
β

Glands
α

Heart
β

Blood vessels
α
β

Smooth muscle
α
β

Liver

Pancreas
α + β
Fat and sugar metabolism

Cholinergic

Sweat glands

Genitals
β
α

Urinary bladder

Adrenal medulla

Cholinoceptors
Nicotinic cholinoceptors:
– All postganglionic,
  autonomic ganglia cells
  or dendrites
– Adrenal medulla
Muscarinic cholinoceptors:
– All target organs innervated
  by postganglionic para-
  sympathetic nerve fibers
  (and sweat glands innervated
  by sympathetic fibers)

Adrenoceptors:
α Usually excitatory
   (except in GI tract, where
   they are indirect relaxants)

β Usually inhibitory
   (except in heart, where
   they are excitatory)

β_1 mainly in heart

β_2 in bronchi, urinary bladder,
    uterus, gastrointestinal tract,
    etc.

Postganglionic: Cholinergic   Preganglionic: Cholinergic   Postganglionic: Adrenergic

Plate 3.1 Organization of ANS

79

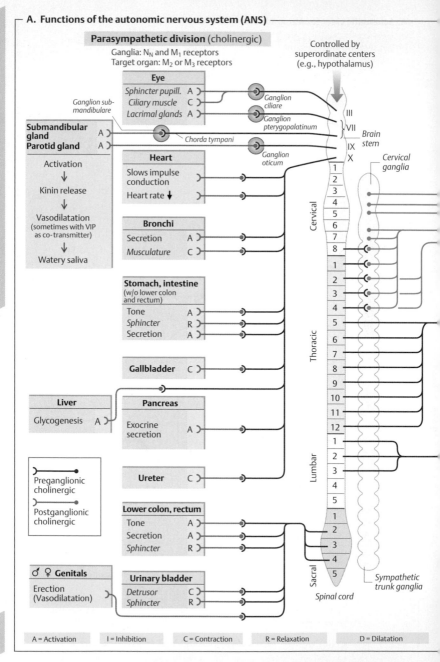

A. Functions of the autonomic nervous system (ANS)

Parasympathetic division (cholinergic)

Ganglia: N_N and M_1 receptors
Target organ: M_2 or M_3 receptors

Controlled by superordinate centers (e.g., hypothalamus)

Ganglion submandibulare

Eye
Sphincter pupill. A
Ciliary muscle C
Lacrimal glands A

Ganglion ciliare
Ganglion pterygopalatinum

Submandibular gland A
Parotid gland A

Chorda tympani
Ganglion oticum

III
VII *Brain stem*
IX
X

Activation
↓
Kinin release
↓
Vasodilatation
(sometimes with VIP as co-transmitter)
↓
Watery saliva

Heart
Slows impulse conduction
Heart rate ↓

Bronchi
Secretion A
Musculature C

Stomach, intestine
(w/o lower colon and rectum)
Tone A
Sphincter R
Secretion A

Gallbladder C

Liver
Glycogenesis A

Pancreas
Exocrine secretion A

Ureter C

Preganglionic cholinergic
Postganglionic cholinergic

Lower colon, rectum
Tone A
Secretion A
Sphincter R

♂ ♀ Genitals
Erection (Vasodilatation)

Urinary bladder
Detrusor C
Sphincter R

Cervical 1–8
Thoracic 1–12
Lumbar 1–5
Sacral 1–5

Cervical ganglia

Sympathetic trunk ganglia

Spinal cord

A = Activation   I = Inhibition   C = Contraction   R = Relaxation   D = Dilatation

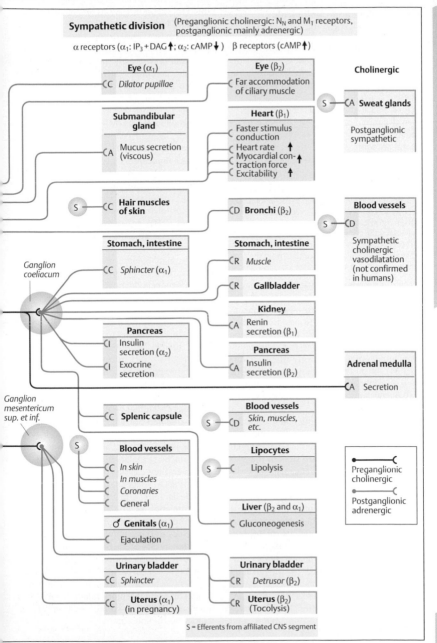

Sympathetic division (Preganglionic cholinergic: N_N and M_1 receptors, postganglionic mainly adrenergic)

α receptors (α_1: IP_3 + DAG ↑; α_2: cAMP ↓) β receptors (cAMP ↑)

Eye (α_1)
C C *Dilator pupillae*

Eye (β_2)
C Far accommodation of ciliary muscle

Cholinergic

S — C A **Sweat glands**

Postganglionic sympathetic

Submandibular gland
C A Mucus secretion (viscous)

Heart (β_1)
C Faster stimulus conduction
C Heart rate ↑
C Myocardial con-traction force ↑
C Excitability ↑

S — C C **Hair muscles of skin**

C D **Bronchi (β_2)**

Blood vessels
S — C D

Sympathetic cholinergic vasodilatation (not confirmed in humans)

Ganglion coeliacum

Stomach, intestine
C C *Sphincter (α_1)*

Stomach, intestine
C R *Muscle*

C R **Gallbladder**

Kidney
C A Renin secretion (β_1)

Pancreas
C I Insulin secretion (α_2)
C I Exocrine secretion

Pancreas
C A Insulin secretion (β_2)

Adrenal medulla
C A Secretion

Ganglion mesentericum sup. et inf.

C C **Splenic capsule**

Blood vessels
S — C D *Skin, muscles, etc.*

S **Blood vessels**
C C *In skin*
C *In muscles*
C *Coronaries*
C General

Lipocytes
S — C Lipolysis

♂ **Genitals (α_1)**
C Ejaculation

Liver (β_2 and α_1)
C Gluconeogenesis

Urinary bladder
C C *Sphincter*

Urinary bladder
C R *Detrusor (β_2)*

Uterus (α_1)
C C (in pregnancy)

Uterus (β_2)
C R (Tocolysis)

●———C Preganglionic cholinergic
●———C Postganglionic adrenergic

S = Efferents from affiliated CNS segment

Plate 3.2 u. 3.3 **Functions of ANS**

81

Acetylcholine and Cholinergic Transmission

Acetylcholine (**ACh**) serves as a neurotransmitter not only at motor end plates ($\to$ p. 56) and in the central nervous system, but also in the **autonomic nervous system, ANS** ($\to$ p. 78ff.), where it is active

◆ in all preganglionic fibers of the ANS;

◆ in all parasympathetic postganglionic nerve endings;

◆ and in some sympathetic postganglionic nerve endings (sweat glands).

Acetylcholine synthesis. ACh is synthesized in the cytoplasm of nerve terminals, and acetyl coenzyme A (acetyl-CoA) is synthesized in mitochondria. The reaction acetyl-CoA + choline is catalyzed by *choline acetyltransferase*, which is synthesized in the soma and reaches the nerve terminals by axoplasmic transport ($\to$ p. 42). Since choline must be taken up from extracellular fluid by way of a carrier, this is the rate-limiting step of ACh synthesis.

Acetylcholine release. Vesicles on presynaptic nerve terminals empty their contents into the synaptic cleft when the cytosolic Ca^{2+} concentration rises in response to incoming action potentials (AP) ($\to$ **A**, p. 50ff.). Epinephrine and norepinephrine can *inhibit ACh release* by stimulating presynaptic α_2-adrenoceptors ($\to$ p. 84). In postganglionic parasympathetic fibers, ACh blocks its own release by binding to presynaptic autoreceptors (M-receptors; see below), as shown in **B**.

ACh binds to postsynaptic **cholinergic receptors** or **cholinoceptors** in autonomic ganglia and organs innervated by parasympathetic fibers, as in the heart, smooth muscles (e.g., of the eye, bronchi, ureter, bladder, genitals, blood vessels, esophagus, and gastrointestinal tract), salivary glands, lacrimal glands, and (sympathetically innervated) sweat glands ($\to$ p. 80ff.). Cholinoceptors are nicotinic (N) or muscarinic (M). **N-cholinoceptors** *(nicotinic)* can be stimulated by the alkaloid nicotine, whereas *M-cholinoceptors (muscarinic)* can be stimulated by the alkaloid mushroom poison muscarine.

Nerve-specific N_N-cholinoceptors on autonomic ganglia ($\to$ **A**) differ from muscle-specific N_M-cholinoceptors on motor end plates ($\to$ p. 56) in that they are formed by different subunits. They are similar in that they are both *ionotropic receptors*, i.e., they act as cholinoceptors and cation channels at the same time. ACh binding leads to rapid Na^+ and Ca^{2+} influx and in early (rapid) excitatory postsynaptic potentials (EPSP; $\to$ p. 50ff.), which trigger postsynaptic action potentials (AP) once they rise above threshold ($\to$ **A, left panel**).

M-cholinoceptors (M_1–M_5) indirectly affect synaptic transmission through G-proteins (*metabotropic receptors*).

M_1-cholinoceptors occur mainly on *autonomic ganglia* ($\to$ **A**), *CNS*, and *exocrine gland cells*. They activate phospholipase Cβ (PLCβ) via G_q protein in the postganglionic neuron. and inositol *tris*-phosphate (IP_3) and diacylglycerol (DAG) are released as second messengers ($\to$ p. 276) that stimulate Ca^{2+} influx and a *late EPSP* ($\to$ **A, middle panel**). Synaptic signal transmission is modulated by the late EPSP as well as by co-transmitting peptides that trigger *peptidergic EPSP or IPSP* ($\to$ **A, right panel**).

M_2-cholinoceptors occur in the *heart* and function mainly via a G_i protein ($\to$ p. 274 ff.). The G_i protein *opens specific K^+ channels* located mainly in the sinoatrial node, atrioventricular (AV) node, and atrial cells, thereby exerting negative chronotropic and dromotropic effects on the heart ($\to$ **B**). The G_i protein also *inhibits adenylate cyclase*, thereby reducing Ca^{2+} influx ($\to$ **B**).

M_3-cholinoceptors occur mainly in *smooth muscles*. Similar to M_1-cholinoceptors ($\to$ **A**, middle panel), M_3-cholinoceptors trigger contractions by stimulating Ca^{2+} influx ($\to$ p. 70). However, they can also induce relaxation by activating Ca^{2+}-dependent NO synthase, e.g., in *endothelial cells* ($\to$ p. 278).

Termination of ACh action is achieved by *acetylcholinesterase*-mediated cleavage of ACh molecules in the synaptic cleft ($\to$ p. 56). Approximately 50% of the liberated choline is reabsorbed by presynaptic nerve endings ($\to$ **B**).

Antagonists. *Atropine* blocks all M-cholinoceptors, whereas *pirenzepine* selectively blocks M_1-cholinoceptors, *tubocurarine* blocks N_M-cholinoceptors ($\to$ p. 56), and *trimetaphan* blocks N_N-cholinoceptors.

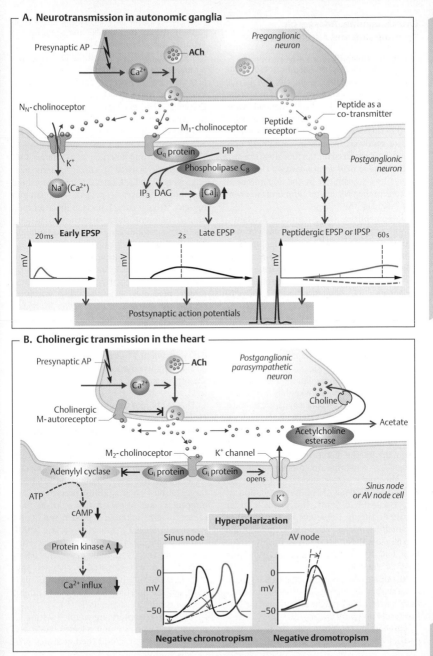

A. Neurotransmission in autonomic ganglia

Presynaptic AP

Ca²⁺ → ACh

Preganglionic neuron

N_N-cholinoceptor

M₁-cholinoceptor

Peptide as a co-transmitter

Peptide receptor

K⁺

G_q protein — PIP

Phospholipase C_β

Postganglionic neuron

Na⁺ (Ca²⁺)

IP₃ DAG → [Ca]_i ↑

20 ms **Early EPSP**

2 s Late EPSP

Peptidergic EPSP or IPSP 60 s

mV

Postsynaptic action potentials

B. Cholinergic transmission in the heart

Presynaptic AP

Ca²⁺ → ACh

Postganglionic parasympathetic neuron

Choline

Cholinergic M-autoreceptor

Acetate

Acetylcholine esterase

M₂-cholinoceptor K⁺ channel

Adenylyl cyclase G_i protein G_i protein opens

Sinus node or AV node cell

ATP

cAMP ↓

K⁺

Hyperpolarization

Protein kinase A ↓

Ca²⁺ influx ↓

Sinus node

AV node

0
mV
−50

0
mV
−50

Negative chronotropism

Negative dromotropism

Plate 3.4 Acetylcholine and Cholinergic Transmission

83

Catecholamine, Adrenergic Transmission and Adrenoceptors

Certain neurons can enzymatically produce *L-dopa* (L-**d**ihydr**o**xy**p**henyl**a**lanine) from the amino acid L-tyrosine. L-dopa is the parent substance of dopamine, norepinephrine, and epinephrine—the three natural **catecholamines**, which are enzymatically synthesized in this order. Dopamine (**DA**) is the final step of synthesis in neurons containing only the enzyme required for the first step (the *aromatic L-amino acid decarboxylase*). Dopamine is used as a transmitter by the dopaminergic neurons in the *CNS* and by autonomic neurons that innervate the *kidney*.

Norepinephrine (NE) is produced when a second enzyme (*dopamine-β-hydroxylase*) is also present. In most *sympathetic postganglionic nerve endings* and noradrenergic central neurons, NE serves as the neurotransmitter along with the *co-transmitters* adenosine triphosphate (ATP), somatostatin (SIH), or neuropeptide Y (NPY).

Within the adrenal medulla (see below) and adrenergic neurons of the medulla oblongata, *phenylethanolamine N-methyltransferase* transforms norepinephrine (NE) into **epinephrine (E)**.

The endings of unmyelinated sympathetic postganglionic neurons are knobby or *varicose* (→ **A**). These knobs establish synaptic contact, albeit not always very close, with the effector organ. They also serve as sites of **NE synthesis** and **storage**. L-tyrosine (→ **A1**) is actively taken up by the nerve endings and transformed into dopamine. In adrenergic stimulation, this step is accelerated by protein kinase A-mediated (PKA; → **A2**) phosphorylation of the responsible enzyme. This yields a larger dopamine supply. Dopamine is transferred to chromaffin vesicles, where it is transformed into NE (→ **A3**). Norepinephrine, the end product, inhibits further dopamine synthesis (negative feedback).

NE release. NE is exocytosed into the synaptic cleft after the arrival of action potentials at the nerve terminal and the initiation of Ca^{2+} influx (→ **A4** and p. 50).

Adrenergic receptors or **adrenoceptors** (→ **B**). Four main types of adrenoceptors (α_1, α_2, β_1 and β_2) can be distinguished according to their affinity to E and NE and to numerous agonists and antagonists. All adrenoceptors respond to E, but NE has little effect on β_2-adrenoceptors. Isoproterenol (isoprenaline) activates only β-adrenoceptors, and phentolamine only blocks α-adrenoceptors. The activities of all adrenoceptors are mediated by G proteins (→ p. 55).

Different subtypes (α_{1A}, α_{1B}, α_{1D}) of **α_1-adrenoceptors** can be distinguished (→ **B1**). Their location and function are as follows: CNS (sympathetic activity ↑), salivary glands, liver (glycogenolysis ↑), kidneys (alters threshold for renin release; → p. 184), and smooth muscles (trigger contractions in the arterioles, uterus, deferent duct, bronchioles, urinary bladder, gastrointestinal sphincters, and dilator pupillae).

Activation of α_1-adrenoceptors (→ **B1**), mediated by G_q *proteins* and *phospholipase C*β (PLCβ), leads to formation of the second messengers *inositol tris-phosphate* (IP_3), which increases the cytosolic Ca^{2+} concentration, and *diacylglycerol* (DAG), which activates protein kinase C (PKC; see also p. 276). G_q protein-mediated α_1-adrenoceptor activity also activates Ca^{2+}-dependent K^+ channels. The resulting K^+ outflow hyperpolarizes and relaxes target smooth muscles, e.g., in the gastrointestinal tract.

Three subtypes (α_{2A}, α_{2B}, α_{2C}) of **α_2-adrenoceptors** (→ **B2**) can be distinguished. Their location and action are as follows: CNS (sympathetic activity ↓, e.g., use of the α_2 agonist clonidine to lower blood pressure), salivary glands (salivation ↓), pancreatic islets (insulin secretion ↓), lipocytes (lipolysis ↓), platelets (aggregation ↑), and neurons (presynaptic autoreceptors, see below). Activated α_2-adrenoceptors (→ **B2**) link with G_i *protein* and inhibit (via α_i subunit of G_i) adenylate cyclase (*cAMP synthesis* ↓, → p. 274) and, at the same time, increase (via the βγ subunit of G_i) the open-probability of voltage-gated K^+ channels (*hyperpolarization*). When coupled with G_0 proteins, activated α_2-adrenoceptors also inhibit voltage-gated Ca^{2+} channels ($[Ca^{2+}]_i$ ↓).

All **β-adrenoceptors** are coupled with a G_S protein, and its α_S subunit releases cAMP as a second messenger. cAMP then activates pro-

▶

Plate 3.5   Adrenergic Transmission I

A. Adrenergic transmission

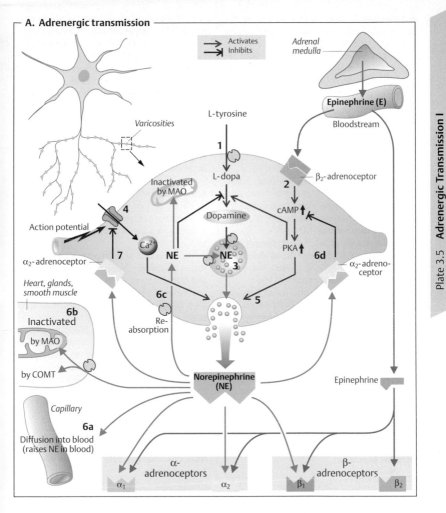

tein kinase A (PKA), which phosphorylates different proteins, depending on the target cell type (→ p. 274).

NE and E work via **β₁-adrenoceptors** (→ **B3**) to open L-type Ca^{2+} channels in *cardiac cell* membranes. This increases the $[Ca^{2+}]_i$ and therefore produces *positive chronotropic, dromotropic, and inotropic effects*. Activated G_s protein can also directly increase the open-

probability of voltage-gated Ca^{2+} channels in the heart. In the *kidney*, the basal renin secretion is increased via β₁-adrenoceptors.

Activation of **β₂-adrenoceptors** by epinephrine (→ **B4**) increases cAMP levels, thereby lowering the $[Ca^{2+}]_i$ (by a still unclear mechanism). This *dilates* the bronchioles and blood vessels of skeletal muscles and *relaxes* the muscles of the uterus, deferent duct, and gastrointestinal tract. Further effects of β₂-adrenoceptor activation are *increased insulin secretion* and *glycogenolysis* in liver and muscle and *decreased platelet aggregation*. Epinephrine also enhances NE release in noradrenergic fibers by way of presynaptic β₂-adrenoceptors (→ **A2, A5**).

Heat production is increased via **β₃-adrenoceptors** on brown lipocytes (→ p. 222).

NE in the synaptic cleft **is deactivated by** (→ **A6 a – d**):

♦ *diffusion* of NE from the synaptic cleft *into the blood*;

♦ *extraneuronal NE uptake* (in the heart, glands, smooth muscles, glia, and liver), and subsequent intracellular degradation of NE by catecholamine-O-methyltransferase (COMT) and monoamine oxidase (MAO);

♦ *active re-uptake of NE* (70%) by the presynaptic nerve terminal. Some of the absorbed NE enters intracellular vesicles (→ **A3**) and is re-used, and some is inactivated by MAO;

♦ stimulation of presynaptic α₂-adrenoceptors (*autoreceptors*; → **A 6d, 7**) by NE in the synaptic cleft, which inhibits the further release of NE.

Presynaptic α₂-adrenoceptors can also be found on cholinergic nerve endings, e.g., in the gastrointestinal tract (motility ↓) and cardiac atrium (negative dromotropic effect), whereas presynaptic M-cholinoceptors are present on noradrenergic nerve terminals. Their mutual interaction permits a certain degree of peripheral ANS regulation.

Adrenal Medulla

After stimulation of preganglionic sympathetic nerve fibers (cholinergic transmission; → p. 81), 95% of all cells in the adrenal medulla secrete the endocrine hormone **epinephrine (E)** into the blood by exocytosis, and another 5% release norepinephrine (NE). Compared to noradrenergic neurons (see above), *NE synthesis* in the adrenal medulla is similar, but most of the NE leaves the vesicle and is enzymatically metabolized into E in the cytoplasm. Special vesicles called *chromaffin bodies* then actively store E and get ready to release it and co-transmitters (enkephalin, neuropeptide Y) by exocytosis.

In **alarm reactions**, secretion of E (and some NE) from the adrenal medulla increases substantially in response to physical and mental or emotional stress. Therefore, cells not sympathetically innervated are also activated in such stress reactions. E also increases neuronal NE release via presynaptic β₂-adrenoceptors (→ **A2**). Epinephrine secretion from the adrenal medulla (mediated by increased sympathetic activity) is stimulated by certain **triggers**, e.g., *physical work*, cold, heat, anxiety, anger (stress), *pain, oxygen deficiency*, or a *drop in blood pressure*. In severe hypoglycemia (< 30 mg/dL), for example, the plasma epinephrine concentration can increase by as much as 20-fold, while the norepinephrine concentration increases by a factor of only 2.5, resulting in a corresponding rise in the E/NE ratio.

The **main task of epinephrine** is to mobilize stored chemical energy, e.g., through *lipolysis* and *glycogenolysis*. Epinephrine enhances the uptake of glucose into skeletal muscle (→ p. 282) and activates enzymes that accelerate glycolysis and lactate formation (→ p. 72ff.). To enhance the blood flow in the muscles involved, the body increases the cardiac output while curbing gastrointestinal blood flow and activity (→ p. 75 A). Adrenal epinephrine and neuronal NE begin to stimulate the secretion of hormones responsible for replenishing the depleted energy reserves (e.g., ACTH; → p. 297 A) while the alarm reaction is still in process.

Non-cholinergic, Non-adrenergic Transmitters

In humans, gastrin-releasing peptide (GRP) and vasoactive intestinal peptide (VIP) serve as co-transmitters in *preganglionic* **sympathetic fibers**; neuropeptide Y (NPY) and somatostatin (SIH) are the ones involved in *postganglionic* fibers. Postganglionic **parasympathetic fibers** utilize the peptides enkephalin, substance P (SP) and/or NPY as co-transmitters.

Modulation of postsynaptic neurons seems to be the primary goal of preganglionic peptide secretion. There is substantial evidence demonstrating that **ATP** (adenosine triphosphate), **NPY** and **VIP** also function as independent neu-

Plate 3.6 Adrenergic Transmission II

B. Adrenoceptors

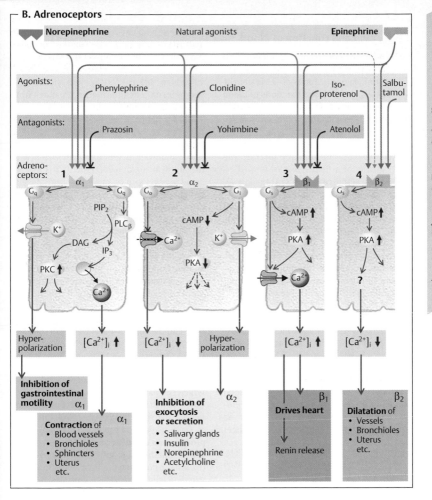

Norepinephrine   Natural agonists   Epinephrine

Agonists:
Phenylephrine   Clonidine   Iso-proterenol   Salbu-tamol

Antagonists:
Prazosin   Yohimbine   Atenolol

Adreno-ceptors: 1 α_1 2 α_2 3 β_1 4 β_2

G_q G_q G_o G_i G_s G_s

PIP$_2$
PLC$_\beta$
cAMP ↓   cAMP ↑   cAMP ↑
K$^+$
DAG
IP$_3$ Ca^{2+} K$^+$ PKA ↑   PKA ↑
PKC ↑   PKA ↓
Ca^{2+} Ca^{2+} ?

Hyper-polarization $[Ca^{2+}]_i$ ↑ $[Ca^{2+}]_i$ ↓   Hyper-polarization $[Ca^{2+}]_i$ ↑ $[Ca^{2+}]_i$ ↓

Inhibition of gastrointestinal motility α_1

Contraction of α_1
• Blood vessels
• Bronchioles
• Sphincters
• Uterus
  etc.

Inhibition of exocytosis or secretion α_2
• Salivary glands
• Insulin
• Norepinephrine
• Acetylcholine
  etc.

Drives heart β_1

Renin release

Dilatation of β_2
• Vessels
• Bronchioles
• Uterus
  etc.

rotransmitters in the autonomic nervous system. VIP and acetylcholine often occur jointly (but in separate vesicles) in the parasympathetic fibers of blood vessels, exocrine glands, and sweat glands. Within the gastrointestinal tract, VIP (along with nitric oxide) induces the slackening of the circular muscle layer and sphincter muscles and (with the co-transmitters dynorphin and galanin) enhances intestinal secretion. **Nitric oxide (NO)** is liberated from nitrergic neurons (→ p. 278)

4 | Blood

Composition and Function of Blood

The **blood volume** of an adult correlates with his or her (fat-free) body mass and amounts to ca. 4–4.5 L in women (♀) and 4.5–5 L in men of 70 kg BW (♂; → table). The **functions of blood** include the *transport* of various molecules (O_2, CO_2, nutrients, metabolites, vitamins, electrolytes, etc.), heat (regulation of body temperature) and *transmission of signals* (hormones) as well as *buffering* and immune *defense*. The blood consists of a fluid (*plasma*) *formed elements*: **Red blood cells** (RBCs) transport O_2 and play an important role in pH regulation. **White blood cells** (WBCs) can be divided into *neutrophilic, eosinophilic* and *basophilic granulocytes, monocytes*, and *lymphocytes*. Neutrophils play a role in nonspecific immune defense, whereas monocytes and lymphocytes participate in specific immune responses. **Platelets (thrombocytes)** are needed for hemostasis. *Hematocrit* (Hct) is the volume ratio of red cells to whole blood (→ **C** and Table). **Plasma** is the fluid portion of the blood in which electrolytes, nutrients, metabolites, vitamins, hormones, gases, and proteins are dissolved.

Plasma proteins (→ Table) are involved in humoral immune defense and maintain oncotic pressure, which helps to keep the blood volume constant. By *binding to plasma proteins*, compounds insoluble in water can be transported in blood, and many substances (e.g., heme) can be protected from breakdown and renal excretion. The binding of small molecules to plasma proteins reduces their osmotic efficacy. Many plasma proteins are involved in blood clotting and fibrinolysis. **Serum** forms when fibrinogen separates from plasma in the process of blood clotting.

The **formation of blood cells** occurs in the red bone marrow of flat bone in adults and in the spleen and liver of the fetus. Hematopoietic tissues contain pluripotent **stem cells** which, with the aid of hematopoietic growth factors (see below), develop into myeloid, erythroid and lymphoid precursor cells. Since pluripotent stem cells are autoreproductive, their existence is ensured throughout life. In *lymphocyte development,* lymphocytes arising from lymphoid precursor cells first undergo special differentiation (in the thymus or bone marrow) and are later formed in the spleen and lymph nodes as well as in the bone marrow. All other precursor cells are produced by *myelocytopoiesis*, that is, the entire process of proliferation, maturation, and release into the bloodstream occurs in the bone marrow. Two hormones, erythropoietin and thrombopoietin, are involved in myelopoiesis. *Thrombopoietin* (formed mainly in the liver) promotes the maturation and development of megakaryocytes from which the platelets are split off. A number of other growth factors affect blood cell formation in bone marrow via paracrine mechanisms.

Erythropoietin promotes the **maturation and proliferation of red blood cells**. It is secreted by the liver in the fetus, and chiefly by the kidney (ca. 90%) in postnatal life. In response to an oxygen deficiency (due to high altitudes, hemolysis, etc.; → **A**), erythropoietin secretion increases, larger numbers of red blood cells are produced, and the fraction of reticulocytes (young erythrocytes) in the blood rises. The **life span** of a red blood cell is around 120 days. Red blood cells regularly exit from arterioles in the splenic pulp and travel through small pores to enter the splenic sinus (→ **B**), where old red blood cells are sorted out and destroyed (*hemolysis*). Macrophages in the spleen, liver, bone marrow, etc. engulf and break down the cell fragments. *Heme*, the iron-containing group of hemoglobin (Hb) released during hemolysis, is broken down into bilirubin (→ p. 250), and the iron is recycled (→ p. 90).

Blood volume in liters relative to body weight (BW)
♂ 0.041 × BW (kg) + 1.53, ♀ 0.047 × BW (kg) + 0.86

Hematocrit (cell volume/ blood volume):
♂ 0.40–0.54   Females: 0.37–0.47

Erythrocytes (10^{12}/L of blood = 10^6/ µL of blood):
♂ 4.6–5.9   ♀ 4.2–5.4

Hemoglobin (g/L of blood):
♂140–180   ♀ 120–160

MCH, MCV, MCHC—mean corpuscular (MC), hemoglobin (Hb), MC volume, MC Hb concentration → **C**

Leukocytes (10^9/L of blood = 10^3/ µL of blood):
3–11 (64% granulocytes, 31% lymphocytes, 6% monocytes)

Platelets (10^9/L of blood = 10^3/ µL of blood):
♂ 170–360   ♀180–400

Plasma proteins (g/L of serum):
66–85 (including 55–64% albumin)

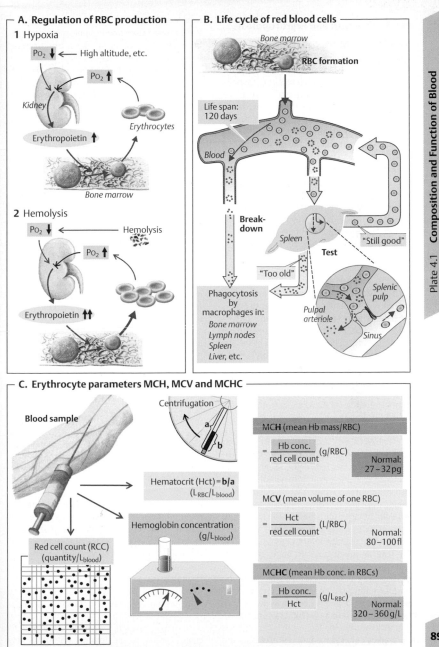

A. Regulation of RBC production

1 Hypoxia

P_{O_2} ↓ ← High altitude, etc.

Kidney

P_{O_2} ↑

Erythrocytes

Erythropoietin ↑

Bone marrow

2 Hemolysis

P_{O_2} ↓ ← Hemolysis

P_{O_2} ↑

Erythropoietin ↑↑

B. Life cycle of red blood cells

Bone marrow

RBC formation

Life span: 120 days

Blood

Break-down

Spleen

"Still good"

Test

"Too old"

Splenic pulp

Pulpal arteriole

Sinus

Phagocytosis by macrophages in:
Bone marrow
Lymph nodes
Spleen
Liver, etc.

C. Erythrocyte parameters MCH, MCV and MCHC

Blood sample

Centrifugation

a.
b.

Hematocrit (Hct) = **b/a**
(L_{RBC}/L_{blood})

Hemoglobin concentration
(g/L_{blood})

Red cell count (RCC)
(quantity/L_{blood})

MC**H** (mean Hb mass/RBC)

$$= \frac{\text{Hb conc.}}{\text{red cell count}} \text{ (g/RBC)}$$

Normal: 27–32 pg

MC**V** (mean volume of one RBC)

$$= \frac{\text{Hct}}{\text{red cell count}} \text{ (L/RBC)}$$

Normal: 80–100 fl

MC**HC** (mean Hb conc. in RBCs)

$$= \frac{\text{Hb conc.}}{\text{Hct}} \text{ (g/L_{RBC})}$$

Normal: 320–360 g/L

Plate 4.1 **Composition and Function of Blood**

89

Iron Metabolism and Erythropoiesis

Roughly $^2/_3$ of the body's **iron pool** (ca. 2 g in women and 5 g in men) is bound to *hemoglobin* (Hb). About $^1/_4$ exists as *stored iron* (ferritin, hemosiderin), the rest as *functional iron* (myoglobin, iron-containing enzymes). **Iron losses** from the body amount to about 1 mg/day in men and up to 2 mg/day in women due to menstruation, birth, and pregnancy. **Iron absorption** occurs mainly in the duodenum and *varies according to need*. The absorption of iron supplied by the diet usually amounts to about 3 to 15% in healthy individuals, but can increase to over 25% in individuals with iron deficiency ($\rightarrow$ **A1**). A minimum daily **iron intake** of at least 10–20 mg/day is therefore recommended (women > children > men).

Iron absorption ($\rightarrow$ **A2**). Fe(II) supplied by the diet (hemoglobin, myoglobin found chiefly in meat and fish) is absorbed relatively efficiently as a **heme-Fe(II)** upon protein cleavage. With the aid of *heme oxygenase*, Fe in mucosal cells cleaves from heme and oxidizes to Fe(III). The triferric form either remains in the mucosa as a ferritin-Fe(III) complex and returns to the lumen during cell turnover or enters the bloodstream. **Non-heme-Fe** can only be absorbed as Fe^{2+}. Therefore, non-heme Fe(III) must first be reduced to Fe^{2+} by *ferrireductase* (FR; $\rightarrow$ **A2**) and ascorbate on the surface of the luminal mucosa ($\rightarrow$ **A2**). Fe^{2+} is probably absorbed through secondary active transport via an Fe^{2+}-H^+ symport carrier (DCT1) (competition with Mn^{2+}, Co^{2+}, Cd^{2+}, etc.). A *low chymous pH* is important since it (a) increases the H^+ gradient that drives Fe^{2+} via DCT1 into the cell and (b) frees dietary iron from complexes. The absorption of iron into the bloodstream is *regulated by the intestinal mucosa*. When an iron deficiency exists, *aconitase* (an iron-regulating protein) in the cytosol binds with ferritin-mRNA, thereby inhibiting mucosal ferritin translation. As a result, larger quantities of absorbed Fe(II) can enter the bloodstream. Fe(II) in the blood is oxidized to Fe(III) by ceruloplasmin (and copper). It then binds to *apotransferrin*, a protein responsible for **iron transport in plasma** ($\rightarrow$ **A2, 3**). *Transferrin* (= apotransferrin loaded with 2 Fe(III)), is taken up by endocytosis into erythroblasts and cells of the liver, placenta, etc. with the aid of *transferrin receptors*. Once iron has been released to the target cells, apotransferrin again becomes available for uptake of iron from the intestine and macrophages (see below).

Iron storage and recycling ($\rightarrow$ **A3**). *Ferritin*, one of the chief forms in which iron is stored in the body, occurs mainly in the intestinal mucosa, liver, bone marrow, red blood cells, and plasma. It contains binding pockets for up to 4500 Fe^{3+} ions and provides rapidly available stores of iron (ca. 600 mg), whereas iron mobilization from *hemosiderin* is much slower (250 mg Fe in macrophages of the liver and bone marrow). Hb-Fe and heme-Fe released from malformed erythroblasts (so-called inefficient erythropoiesis) and hemolyzed red blood cells bind to haptoglobin and hemopexin, respectively. They are then engulfed by macrophages in the bone marrow or in the liver and spleen, respectively, resulting in 97% iron recycling ($\rightarrow$ **A3**).

An **iron deficiency** inhibits Hb synthesis, leading to hypochromic microcytic anemia: MCH < 26 pg, MCV < 70 fL, Hb < 110 g/L. The primary causes are:

◆ blood loss (most common cause); 0.5 mg Fe are lost with each mL of blood;

◆ insufficient iron intake or absorption;

◆ increased iron requirement due to growth, pregnancy, breast-feeding, etc.;

◆ decreased iron recycling (due to chronic infection);

◆ apotransferrin defect (rare cause).

Iron overload most commonly damages the liver, pancreas and myocardium (hemochromatosis). If the iron supply bypasses the intestinal tract (iron injection), the transferrin capacity can be exceeded and the resulting quantities of free iron can induce iron poisoning.

B_{12} **vitamin (cobalamins)** and **folic acid** are also required for erythropoiesis ($\rightarrow$ **B**). Deficiencies lead to *hyperchromic anemia* (decreased RCC, increased MCH). The main causes are lack of intrinsic factor (required for cobalamin resorption) and decreased folic acid absorption due to malabsorption (see also p. 260) or an extremely unbalanced diet. Because of the large stores available, decreased cobalamin absorption does not lead to symptoms of deficiency until many years later, whereas folic acid deficiency leads to symptoms within a few months.

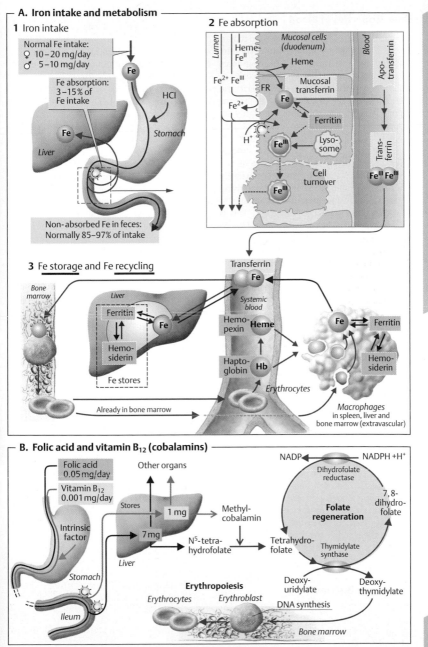

A. Iron intake and metabolism

1 Iron intake

Normal Fe intake:
♀ 10 – 20 mg/day
♂ 5 – 10 mg/day

Fe absorption:
3 – 15 % of
Fe intake

HCl

Stomach

Liver

Fe

Fe

Non-absorbed Fe in feces:
Normally 85 – 97% of intake

2 Fe absorption

Lumen — Mucosal cells (duodenum) — Blood

Heme-Fe^{II}

Heme

Fe^{2+} Fe^{III}

FR

Mucosal transferrin

Apo-transferrin

Fe^{2+}

Fe

Ferritin

H^+

Fe^{III}

Lyso-some

Transferrin

Cell turnover

Fe^{III}

Fe^{III}

Fe^{III} Fe^{III}

3 Fe storage and Fe recycling

Transferrin

Fe

Bone marrow

Liver

Ferritin

Fe

Hemo-siderin

Fe stores

Systemic blood

Hemo-pexin

Heme

Hapto-globin

Hb

Erythrocytes

Fe

Ferritin

Hemo-siderin

Already in bone marrow

Macrophages
in spleen, liver and
bone marrow (extravascular)

B. Folic acid and vitamin B₁₂ (cobalamins)

Folic acid
0.05 mg/day

Vitamin B₁₂
0.001 mg/day

Intrinsic factor

Stomach

Ileum

Liver

Stores

1 mg

7 mg

Other organs

NADP

NADPH +H⁺

Dihydrofolate reductase

Methyl-cobalamin

N^5-tetra-hydrofolate

Tetrahydro-folate

Folate regeneration

7, 8-dihydro-folate

Thymidylate synthase

Deoxy-uridylate

Deoxy-thymidylate

Erythropoiesis

Erythrocytes

Erythroblast

DNA synthesis

Bone marrow

Plate 4.2 Iron Metabolism and Erythropoiesis

Flow Properties of Blood

The **viscosity** (η) of blood is higher than that of plasma due to its erythrocyte (RBC) content. Viscosity (η) = 1/fluidity = shearing force (τ)/shearing action (γ) [Pa·s]. The viscosity of blood rises with increasing *hematocrit* and decreasing *flow velocity*. Erythrocytes lack the major organelles and, therefore, are highly deformable. Because of the low viscosity of their contents, the liquid film-like characteristics of their membrane, and their high surface/volume ratio, the blood behaves more like an emulsion than a cell suspension, especially when it flows rapidly. The viscosity of *flowing* blood (η_{blood}) passing through small arteries ($\varnothing$ 20 μm) is about 4 relative units (RU). This is twice as high as the viscosity of plasma (η_{plasma} = 2 RU; water: 1 RU = 0.7 mPa·s at 37 °C).

Because they are *highly deformable*, normal RBCs normally have no problem passing through capillaries or pores in the splenic vessels (see p. 89 B), although their diameter ($\varnothing$ < 5 μm) is smaller than that of freely mobile RBCs (7 μm). Although the slowness of flow in small vessels causes the blood viscosity to increase, this is partially compensated for ($\eta_{blood} \downarrow$) by the passage of red cells in single file through the center of small vessels (diameter < 300 μm) due to the **Fåhraeus–Lindqvist effect** ($\rightarrow$ **A**). Blood viscosity is only slightly higher than plasma viscosity in arterioles ($\varnothing \approx 7$ μm), but rises again in capillaries ($\varnothing \approx 4$ μm). A *critical increase* in blood viscosity can occur a) if blood flow becomes too sluggish and/or b) if the fluidity of red cells decreases due to hyperosmolality (resulting in crenation), cell inclusion, hemoglobin malformation (e.g., sickle-cell anemia), changes in the cell membrane (e.g., in old red cells), and so forth. Under such circumstances, the RBCs undergo aggregation (*rouleaux formation*), increasing the blood viscosity tremendously (up to 1000 RU). This can quickly lead to the cessation of blood flow in small vessels ($\rightarrow$ p. 218).

Plasma, Ion Distribution

Plasma is obtained by preventing the blood from clotting and extracting the formed elements by centrifugation ($\rightarrow$ p. 89 C). High molecular weight proteins ($\rightarrow$ **B**) as well as ions and non-charged substances with low molecular weights are dissolved in plasma. The sum of the concentrations of these particles yields a *plasma osmolality* of 290 mOsm/kgH$_2$O ($\rightarrow$ pp. 164, 377). The most abundant cation in plasma is Na$^+$, and the most abundant anions are Cl$^-$ and HCO$_3^-$. Although plasma proteins carry a number of anionic net charges ($\rightarrow$ **C**), their osmotic efficacy is smaller because the number of particles, not the ionic valency, is the determining factor.

The fraction of proteins able to leave the blood vessels is small and varies from one organ to another. Capillaries in the liver, for example, are much more permeable to proteins than those in the brain. The composition of *interstitial fluid* therefore differs significantly from that of plasma, especially with respect to protein content ($\rightarrow$ **C**). A completely different composition is found in the *cytosol*, where K$^+$ is the prevailing cation, and where phosphates, proteins and other organic anions comprise the major fraction of anions ($\rightarrow$ **C**). These fractions vary depending on cell type.

Sixty percent of all **plasma protein** ($\rightarrow$ **B**) is albumin (35–46 g/L). Albumin serves as a vehicle for a number of substances in the blood. They are the main cause of colloidal osmotic pressure or, rather, *oncotic pressure* ($\rightarrow$ pp. 208, 378), and they provide a protein reserve in times of protein deficiency. The α_1, α_2 and β globulins mainly serve to transport lipids (apolipoproteins), hemoglobin (haptoglobin), iron (apotransferrin), cortisol (transcortin), and cobalamins (transcobalamin). Most plasma factors for coagulation and fibrinolysis are also proteins. Most plasma *immunoglobulins* (*Ig*, $\rightarrow$ **D**) belong to the group of γ globulins and serve as defense proteins (antibodies). IgG, the most abundant immunoglobulin (7–15 g/L), can cross the placental barrier (maternofetal transmission; $\rightarrow$ **D**). Each Ig consists of two group-specific, heavy protein chains (IgG: γ chain, IgA: α chain, IgM: μ chain, IgD: δ chain, IgE: ε chain) and two light protein chains (λ or $\varkappa$ chain) linked by disulfide bonds to yield a characteristic Y-shaped configuration (see p. 95 A).

A. Fåhraeus-Lindqvist effect

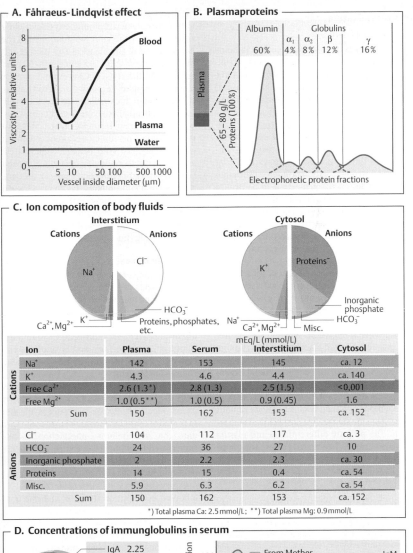

Viscosity in relative units

Blood
Plasma
Water

Vessel inside diameter (μm)

B. Plasmaproteins

| Albumin | | | Globulins | |
| 60% | α₁ 4% | α₂ 8% | β 12% | γ 16% |

Plasma

65–80 g/L Proteins (100%)

Electrophoretic protein fractions

C. Ion composition of body fluids

Interstitium

Cations   Anions

Na⁺   Cl⁻

Ca²⁺,Mg²⁺ K⁺   HCO₃⁻
Proteins, phosphates, etc.

Cytosol

Cations   Anions

K⁺   Proteins⁻

Inorganic phosphate

Na⁺ Ca²⁺,Mg²⁺ Misc.   HCO₃⁻

	Ion	Plasma	Serum	Interstitium	Cytosol
				mEq/L (mmol/L)	
Cations	Na⁺	142	153	145	ca. 12
	K⁺	4.3	4.6	4.4	ca. 140
	Free Ca²⁺	2.6 (1.3*)	2.8 (1.3)	2.5 (1.5)	<0.001
	Free Mg²⁺	1.0 (0.5**)	1.0 (0.5)	0.9 (0.45)	1.6
	Sum	150	162	153	ca. 152
Anions	Cl⁻	104	112	117	ca. 3
	HCO₃⁻	24	36	27	10
	Inorganic phosphate	2	2.2	2.3	ca. 30
	Proteins	14	15	0.4	ca. 54
	Misc.	5.9	6.3	6.2	ca. 54
	Sum	150	162	153	ca. 152

*) Total plasma Ca: 2.5 mmol/L; **) Total plasma Mg: 0.9 mmol/L

D. Concentrations of immunglobulins in serum

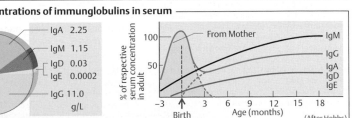

IgA 2.25
IgM 1.15
IgD 0.03
IgE 0.0002
IgG 11.0
g/L

% of respective serum concentration in adult

From Mother

IgM
IgG
IgA
IgD
IgE

Birth

Age (months)

(After Hobbs)

Immune System

Fundamental Principles

The body has *nonspecific (innate) immune defenses* linked with *specific (acquired) immune defenses* that counteract bacteria, viruses, fungi, parasites and foreign (non-self) macromolecules. They all function as **antigens**, i.e., substances that stimulate the specific immune system, resulting in the activation of antigen-specific **T lymphocytes** (**T cells**) and **B lymphocytes** (**B cells**). In the process, B lymphocytes differentiate into *plasma cells* that secrete antigen-specific **antibodies** (**immunoglobulins, Ig**) (→ **C**). Ig function to neutralize and opsonize antigens and to activate the complement system (→ p. 96). These mechanisms ensure that the respective antigen is *specifically recognized*, then eliminated by relatively nonspecific means. Some of the T and B cells have an *immunologic memory*.

Precursor lymphocytes without an antigen-binding receptor are *preprocessed* within the *thymus (T)* or *bone marrow (B)*. These organs produce up to 10^8 monospecific T or B cells, each of which is directed against a specific antigen. *Naive* T and B cells which have not previously encountered antigen circulate through the body (blood → peripheral lymphatic tissue → lymph → blood) and undergo **clonal expansion and selection** after contact with its specific antigen (usually in lymphatic tissue). The lymphocyte then begins to divide rapidly, producing numerous monospecific daughter cells. The progeny differentiates into plasma cells or "armed" T cells that initiate the elimination of the antigen.

Clonal deletion is a mechanism for eliminating lymphocytes with receptors directed against autologous (self) tissue. After first contact with their specific self-antigen, these lymphocytes are eliminated during the early stages of development in the thymus or bone marrow. Clonal deletion results in **central immunologic tolerance**. The ability of the immune system to distinguish between endogenous and foreign antigens is called **self/nonself recognition**. This occurs around the time of birth. All substances encountered by that time are recognized as endogenous (self); others are identified as foreign (nonself). The inability to distinguish self from nonself results in *autoimmune disease*.

At first contact with a virus (e.g., measles virus), the nonspecific immune system usually cannot prevent the viral proliferation and the development of the measles. The specific immune system, with its T-killer cells (→ **B2**) and Ig (first IgM, then IgG; → **C3**), responds slowly: **primary immune response** or **sensitization**. Once activated, it can eliminate the pathogen, i.e., the individual recovers from the measles. **Secondary immune response**: When infected a second time, specific IgG is produced much more rapidly. The virus is quickly eliminated, and the disease does not develop a second time. This type of protection against infectious disease is called **immunity**. It can be achieved by vaccinating the individual with a specific antigen (*active immunization*). *Passive immunization* can be achieved by administering ready-made Ig (immune serum).

Nonspecific Immunity

Lysozyme and *complement factors* dissolved in plasma (→ **A1**) as well as *natural killer cells* (*NK cells*) and phagocytes, especially *neutrophils* and *macrophages* that arise from *monocytes* that migrate into the tissues (→ **A2**) play an important role in nonspecific immunity. Neutrophils, monocytes, and eosinophils circulate throughout the body. They have chemokine receptors (e.g., CXCR1 and 2 for IL-8) and are attracted by various *chemokines* (e.g., IL-8) to the sites where microorganisms have invaded (*chemotaxis*). These cells are able to migrate. With the aid of selectins, they dock onto the endothelium (*margination*), penetrate the endothelium (*diapedesis*), and engulf and damage the microorganism with the aid of *lysozyme*, *oxidants* such as H_2O_2, oxygen radicals (O_2^-, $OH\cdot$, $^1O_2$), and *nitric oxide* (NO). This is followed by digestion (lysis) of the microorganism with the aid of lysosomal enzymes. If the antigen (parasitic worm, etc.) is too large for digestion, other substances involved in nonspecific immunity (e.g., proteases and cytotoxic proteins) are also exocytosed by these cells.

Reducing enzymes such as catalase and superoxide dismutase usually keep the oxidant concentration low. This is often discontinued, especially when macrophages are activated (→ below and **B3**), to fully exploit the bactericidal effect of the oxidants. However,

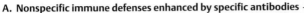

A. Nonspecific immune defenses enhanced by specific antibodies

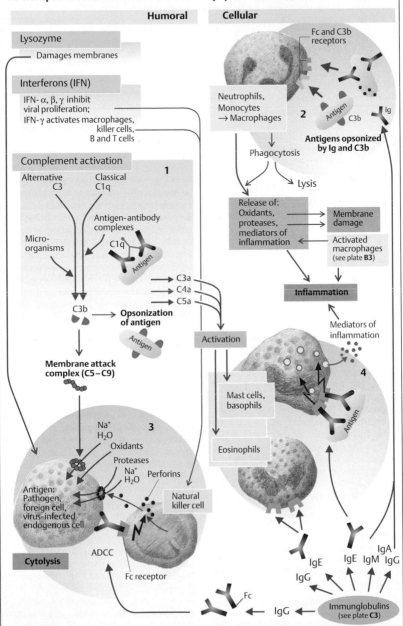

Humoral   **Cellular**

Fc and C3b receptors

Lysozyme
Damages membranes

Interferons (IFN)
IFN- α, β, γ inhibit
viral proliferation;
IFN-γ activates macrophages,
killer cells,
B and T cells

Neutrophils,
Monocytes
→ Macrophages

Antigen   C3b   Ig

2

**Antigens opsonized
by Ig and C3b**

Phagocytosis

Complement activation   1

Alternative   Classical
C3   C1q

Antigen-antibody
complexes

Micro-
organisms

C1q

Antigen

Lysis

Release of:
Oxidants,
proteases,
mediators
of
inflammation

→ Membrane
damage

Activated
macrophages
(see plate **B3**)

→ C3a
→ C4a
→ C5a

Inflammation

C3b → **Opsonization
of antigen**

Antigen

Activation

Mediators
of
inflammation

4

**Membrane attack
complex (C5 – C9)**

Na$^+$
H$_2$O

3

Antigen

Oxidants

Mast cells,
basophils

Proteases
Na$^+$
H$_2$O

Perforins

Eosinophils

Antigen:
Pathogen,
foreign cell,
virus-infected
endogenous cell

Natural
killer cell

Cytolysis

ADCC

Fc receptor

IgE   IgE IgM IgA
             IgG

IgG

Immunglobulins
(see plate **C3**)

Fc

IgG ← IgG

Plate 4.4 **Immune System I**

95

▷

the resulting *inflammation* (→ **A2, 4**) also damages cells involved in nonspecific defense and, in some cases, even other endogenous cells.

Opsonization (→ **A1, 2**) involves the binding of *opsonins*, e.g., IgG or complement factor C3b, to specific domains of an antigen, thereby enhancing phagocytosis. It is the only way to make bacteria with a polysaccharide capsule phagocytable. The phagocytes have **receptors** on their surface for the (antigen-independent) Fc segment of IgG as well as for C3b. Thus, the antigen-bound IgG and C3b bind to their respective receptors, thereby linking the rather unspecific process of phagocytosis with the specific immune defense system. Carbohydrate-binding proteins (lectins) of plasma, called *collectins* (e.g. mannose-*binding protein*), which dock onto microbial cell walls, also acts as unspecific opsonins.

The **complement cascade** is activated by antigens opsonized by Ig (classical pathway) as well as by non-opsonophilic antigens (alternative pathway) (→ **A1**). Complement components C3a, C4a and C5a activate basophils and eosinophils (→ **A4**). Complement components C5 –C9 generate the *membrane-attack complex* (MAC), which perforates and kills (Gram-negative) bacteria by **cytolysis** (→ **A3**). This form of defense is assisted by **lysozyme** (= muramidase), an enzyme that breaks down murein-containing bacterial cell walls. It occurs in granulocytes, plasma, lymph, and secretions.

Natural killer (NK) cells are large, granular lymphocytes specialized in nonspecific defense against viruses, mycobacteria, tumor cells etc. They recognize infected cells and tumor cells on "foreign surfaces" and dock via their Fc receptors on IgG-opsonized surface antigens (*antibody-dependent cell-mediated cytotoxicity,* **ADCC**; → **A3**). *Perforins* exocytosed by NK cells form pores in target cell walls, thereby allowing their subsequent lysis (*cytolysis*). This not only makes the virus unable to proliferate (enzyme apparatus of the cell), but also makes it (and other intracellular pathogens) subject to attack from other defense mechanisms.

Various **interferons (IFNs)** stimulate NK cell activity: IFN-α, IFN-β and, to a lesser degree, IFN-γ. IFN-α and IFN-β are released mainly from leukocytes and fibro-

blasts, while IFN-γ is liberated from activated T cells and NK cells. Virus-infected cells release large quantities of IFNs, resulting in heightened viral resistance in non-virus-infected cells. **Defensins** are cytotoxic peptides released by phagocytes. They can exert unspecific cytotoxic effects on pathogens resistant to NK cells (e.g., by forming ion channels in the target cell membrane).

Macrophages arise from monocytes that migrate into the tissues. Some macrophages are freely mobile (*free macrophages*), whereas others (*fixed macrophages*) remain restricted to a certain area, such as the hepatic sinus (Kupffer cells), the pulmonary alveoli, the intestinal serosa, the splenic sinus, the lymph nodes, the skin (Langerhans cells), the synovia (synovial A cells), the brain (microglia), or the endothelium (e.g., in the renal glomeruli). The **mononuclear phagocytic system (MPS)** is the collective term for the circulating monocytes in the blood and macrophages in the tissues. Macrophages recognize relatively unspecific carbohydrate components on the surface of bacteria and ingest them by phagocytosis. The macrophages have to be activated if the pathogens survive within the phagosomes (→ below and **B3**).

Specific Immunity: Cell-Mediated Immune Responses

Since specific cell-mediated immune responses through "armed" T effector cells need a few days to become effective, this is called *delayed-type immune* response. It requires the participation of **professional antigen-presenting cells (APCs)**: dendritic cells, macrophages and B cells. APCs process and present antigenic peptides to the T cells in association with *MHC-I or MHC-II proteins*, thereby delivering the co-stimulatory signal required for activation of naive T cells. (The gene loci for these proteins are the class I (MHC-I) and class II (MHC-II) major histocompatibility complexes (MHC)), **HLA** *(human leukocyte antigen)* is the term for MHC proteins in humans. Virus-infected dendritic cells, which are mainly located in lymphatic tissue, most commonly serve as APCs. Such HLA-restricted antigen presentation (→ **B1**) involves the insertion of an antigen in the binding pocket of an HLA protein. An *ICAM* (intercellular adhesion molecule) on the surface of the APC then binds to

▶

Plate 4.5 Immune System II

B. Specific immunity: T-cell activation

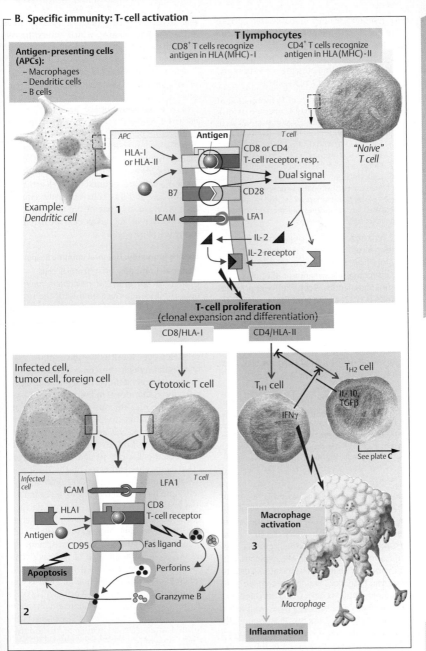

T lymphocytes

CD8$^+$ T cells recognize antigen in HLA (MHC)-I

CD4$^+$ T cells recognize antigen in HLA (MHC)-II

Antigen-presenting cells (APCs):
- Macrophages
- Dendritic cells
- B cells

Example: *Dendritic cell*

"Naive" T cell

APC — **Antigen** — *T cell*

HLA-I or HLA-II

CD8 or CD4 T-cell receptor, resp.

Dual signal

B7 — CD28

ICAM — LFA1

IL-2

IL-2 receptor

1

T-cell proliferation
(clonal expansion and differentiation)

CD8/HLA-I   CD4/HLA-II

Infected cell, tumor cell, foreign cell   Cytotoxic T cell

T_{H1} cell T_{H2} cell

IL-10, TGFβ

IFNγ

See plate **C**

Infected cell   *T cell*

ICAM — LFA1

HLA1

Antigen

CD8 T-cell receptor

CD95 — Fas ligand

Apoptosis

Perforins

Granzyme B

2

Macrophage activation

3

Macrophage

Inflammation

LFA1 (lymphocyte function-associated antigen 1) on the T cell membrane. When a T cell specific for the antigen in question docks onto the complex, the bond is strengthened and the *APC dual signal* stimulates the activation and clonal selection of the T cell (→ **B1**).

The **APC dual signal** consists of **1) recognition of the antigen** (class I or class II HLA-restricted antigen) by the *T cell receptor* and its *co-receptor* and **2) a co-stimulatory signal**, that is, the binding of the B7 protein (on the APC) with the CD28 protein on the T cell (→ **B1**). CD8 molecules on T cytotoxic cells (T$_C$ cells = T-killer cells) and CD4 molecules on T helper cells (T$_H$ cells) function as the co-receptors. When antigen binding occurs without co-stimulation (e.g., in the liver, where there are no APCs), the lymphocyte is inactivated, i.e., it becomes **anergic**, and *peripheral immunologic tolerance* develops.

The T cell can receive APC dual signals from infected macrophages or B cells, provided their receptors have bound the antigen in question (e.g., insect or snake venom or allergens). The APC dual signal induces the T cell to express *interleukin-2* (**IL-2**) and to bind the respective *IL-2 receptor* (→ **B1**). IL-2 is the actual *signal for clonal expansion* of these monospecific T cells. It functions through autocrine and paracrine mechanisms. Potent *immunosuppression,* e.g., for organ transplantation, can be achieved with IL-2 inhibitors like *cyclosporin A.*

During clonal expansion, the T cells differentiate into three "armed" subtypes, i.e., *T cytotoxic cells* (**T$_C$ cells** or **T killer cells**) and *T helper cells type 1* (**T$_{H1}$ cells**) and type 2 (**T$_{H2}$ cells**). These cells no longer require costimulation and express a different type of adhesion molecule (VLA-4 instead of L-selectins) by which they now dock onto the endothelium of inflamed tissues (rather than to lymphatic tissue like their naïve precursors).

T killer cells develop from naïve CD8-containing (CD8$^+$) T cells after HLA-I-restricted antigen presentation (→ **B2**). **Endogenous antigen presentation** occurs when the HLA-I protein takes up the antigen (virus, cytosolic protein) from the cytosol of the APC, which is usually the case. With its CD8-associated T-cell receptor, the T killer cell is able to recognize HLA-I-restricted antigens on (virus) infected endogenous cells and tumor cells as well as on cells of transplanted organs. It subsequently drives the cells into **apoptosis** (programmed cell death) or **necrosis**. Binding of the *Fas lig-*

and to CD95 (= Fas) plays a role, as does granzyme B (protease), which enters the cell through pores created by exocytosed *perforins* (→ **B2**).

Once **HLA-II-restricted presentation** (→ **B1**) of antigens from intracellular vesicles (e.g., phagocytosed bacteria or viral envelope proteins = **exogenous antigen presentation**) has occurred, naive CD4$^+$ T cells transform into immature T helper cells (T$_{H0}$), which differentiate into T$_{H1}$ or T$_{H2}$ cells. **T$_{H1}$ cells** induce *inflammatory responses* and promote the activation of macrophages with the aid of IFN-γ (→ **B3**), while **T$_{H2}$ cells** are required for *B-cell activation* (→ **C2**). T$_{H1}$ and T$_{H2}$ cells mutually *suppress* each other, so only one of the two types will predominate in any given cell-mediated immune response (→ **B3**).

Specific Immunity: Humoral Immune Responses

Humoral immunity arise from **B cells** (→ **C1**). Numerous *IgD* and *IgM monomers* anchored onto the B-cell surface bind with the respective antigen (dissolved IgM occurs in pentameric form). A resulting network of antigen-bound Ig leads to *internalization* and *processing of the antigen-antibody complex* in B cells. However, **B-cell activation** requires a second signal, which can come directly from a thymus-independent (TI) antigen (e.g., bacterial polysaccharide) or indirectly from a **T$_{H2}$ cell** in the case of a thymus-dependent (TD) antigen. In the latter case, the B cell presents the HLA-II-restricted TD antigen to the T$_{H2}$ cell (→ **C2**). If the *CD4-associated T-cell receptor* (TCR) of the T$_{H2}$ cell recognizes the antigen, *CD40 ligands* are expressed on the T$_{H2}$ surface (CD40 ligands bind with CD40 proteins on B cells) and **IL-4** is secreted. The CD40 ligand and IL-4 (later also IL-5 and IL-6) stimulate the B cell to undergo *clonal selection,* IgM secretion, and differentiation into **plasma cells** (→ **C3**). Before differentiation, **class switching** can occur, i.e., a different type of Ig heavy chain (→ p. 92) can be expressed by altered DNA splicing (→ p. 8f.). In this manner, IgM is converted into IgA, IgG or IgE (→ p. 92). All Ig types arising from a given B-cell clone remain monospecific for the same antigen. The plasma cells formed after class switching produce only a single type of Ig.

4 Blood

Plate 4.6   Immune System III

C. Specific immunity: B-cell activation

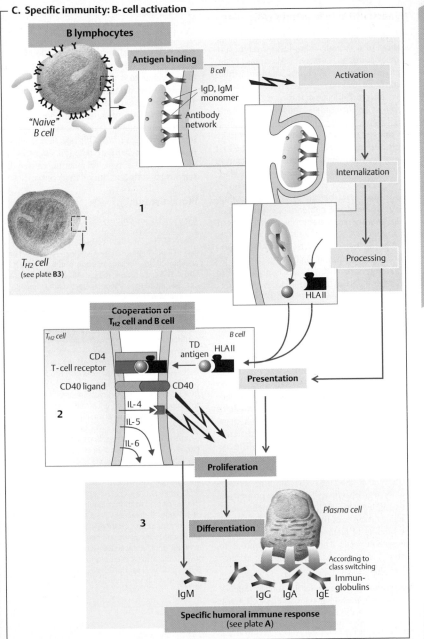

B lymphocytes

"Naive" B cell

Antigen binding

B cell

IgD, IgM monomer

Antibody network

Activation

Internalization

Processing

T_{H2} cell
(see plate **B3**)

1

HLA II

Cooperation of T_{H2} cell and B cell

T_{H2} cell

B cell

CD4 T-cell receptor

CD40 ligand

TD antigen

HLA II

CD40

IL-4

IL-5

IL-6

Presentation

2

Proliferation

3

Differentiation

Plasma cell

According to class switching
Immun-globulins

IgM

IgG   IgA   IgE

Specific humoral immune response
(see plate **A**)

Hypersensitivity Reactions (Allergies)

Allergy is a specific, exaggerated immune response to a (usually harmless) foreign substance or antigen (→ p. 94ff.). **Allergens** are antigens that induce allergies. Small molecules conjugated to endogenous proteins can also have antigenic effects. In this case, they are referred to as incomplete antigens or **haptens**. The heightened immune response to secondary antigen contact (→ p. 94ff.) normally has a protective effect. In allergies, however, the first contact with an antigen induces sensitization (allergization), and subsequent exposure leads to the destruction of healthy cells and intact tissue. This can also result in damage to endogenous proteins and *autoantibody* production. *Inflammatory reactions* are the main causes of damage.

Types of hypersensitivity reactions: **Type I** reactions are common. On first contact, the allergen internalized by B cells is presented to T_{H2} cells. The B cell then proliferates and differentiates into plasma cells (see p. 98), which release *immunoglobulin E* (**IgE**). The Fc fragment of IgE binds to *mast cells* and *basophils*. On subsequent contact, the antigens bind to the already available IgE-linked mast cells (→ **A**). Due to the rapid release of mostly vasoactive mediators of inflammation such as histamine, leukotrienes and platelet-activating factor (PAF), an immediate reaction (*anaphylaxis*) occurs within seconds or minutes: **immediate type hypersensitivity**. This is the mechanism by which allergens breathed into the lungs trigger hay fever and asthma attacks. The vasodilatory effect of a generalized type I reaction can lead to *anaphylactic shock* (see p. 218).

In **type II** reactions, the immune system mainly attacks *cells* with antigenic properties. This can be attributable to the transfusion of the erythrocytes of the wrong blood group or the binding of *haptens* (e.g., medications) to endogenous cells. The binding of haptens to platelets can, for example, result in thrombocytopenia.

Type III reactions are caused by antigen-antibody complexes. If more antigen than antibody is available, *soluble antigen-antibody complexes* circulate in blood for a long time (→ **B**) and settle mainly in the capillaries, making the capillary wall subject to attack by the complement system. This leads to the development of **serum sickness** (→ **B**), the main symptoms of which are joint pain and fever.

Type IV reactions are mainly mediated by T_{H1} cells, T_C cells, and macrophages. Since symptoms peak 2 to 4 days after antigen contact, this is called **delayed type hypersensitivity**. The main triggers are mycobacteria (e.g. Tbc), other foreign proteins, and haptens, such as medications and plant substances, such as poison ivy. Primary *transplant rejection* is also a type IV hypersensitivity reaction. *Contact dermatitis* is also a type IV reaction caused by various haptens (e.g., nickel in jewelry).

Blood Groups

A person's blood group is determined by the type of antigen (certain glycolipids) present on the red blood cells (RBCs). In the **ABO system**, the antigens are A and B (→ **C**). In blood type A, antigen A (on RBC) and anti-B antibody (in serum) are present; in type B, B and anti-A are present; in type AB, A and B are present, no antibody; in type 0 (zero), no antigen but anti-A and anti-B are present.

When giving a **blood transfusion**, it is important that the blood groups of donor and recipient match, i.e. that the RBCs of the donor (e.g. A) do not come in contact with the respective antibodies (e.g. anti-A) in the recipient. If the donor's blood is the wrong type, *agglutination* (cross-linking by IgM) and *hemolysis* (bursting) of the donor's RBCs will occur (→ **C1**). Donor and recipient blood types must therefore be determined and *cross-matched* (→ **C2**) prior to a blood transfusion. Since ABO antibodies belong to the IgM class, they usually do not cross the placenta.

In the **Rh system**, antibodies against rhesus antigens (C, D, E) on RBCs do not develop unless *prior sensitization* has occurred. D is by far the most antigenic. A person is Rh-positive (Rh+) when D is present on their RBCs (most people), and Rh-negative (Rh−) when D is absent. Anti-D antibodies belong to the IgG class of immunoglobulins, which are capable of crossing the placenta (→ p. 93 D). Rh− individuals can form anti-Rh+ (= anti-D) antibodies, e.g., after sensitization by a mismatched blood transfusion or of an Rh− mother by an Rh+ fetus. Subsequent exposure to the mismatched blood leads to a severe antigen-antibody reaction characterized by intravascular agglutination and hemolysis (→ **D**).

Plate 4.7 **Blood Groups**

A. Anaphylaxis

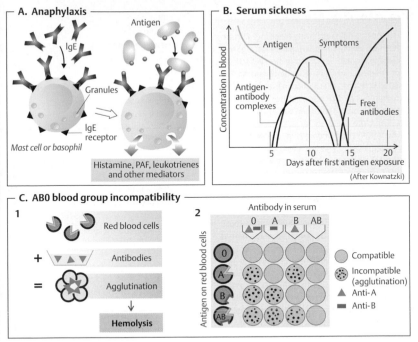

Antigen

IgE

Granules

IgE receptor

Mast cell or basophil

Histamine, PAF, leukotrienes and other mediators

B. Serum sickness

Concentration in blood

Antigen

Symptoms

Antigen-antibody complexes

Free antibodies

5   10   15   20
Days after first antigen exposure

(After Kownatzki)

C. AB0 blood group incompatibility

1

Red blood cells

+ Antibodies

= Agglutination

↓

Hemolysis

2

Antibody in serum

Antigen on red blood cells

0   A   B   AB

0
A
B
AB

○ Compatible

◉ Incompatible (agglutination)

▲ Anti-A

▬ Anti-B

D. Rh sensitization of mother by child or by Rh-mismatched transfusion

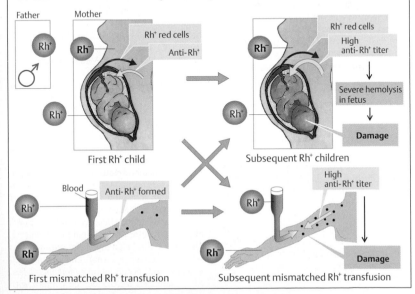

Father

Mother

Rh⁺

Rh⁻

Rh⁺ red cells

Anti-Rh⁺

Rh⁺

First Rh⁺ child

Rh⁻

Rh⁺ red cells

High anti-Rh⁺ titer

Severe hemolysis in fetus

↓

Damage

Rh⁺

Subsequent Rh⁺ children

Blood

Rh⁺

Rh⁻

Anti-Rh⁺ formed

First mismatched Rh⁺ transfusion

Rh⁺

Rh⁻

High anti-Rh⁺ titer

↓

Damage

Subsequent mismatched Rh⁺ transfusion

Hemostasis

The hemostatic system stops bleeding. *Thrombocytes (platelets), coagulation* (or *clotting*) *factors in plasma* and *vessel walls* interact to seal leaks in blood vessels. The damaged vessel constricts (release of endothelin), and platelets aggregate at the site of puncture (and attract more platelets) to seal the leak by a *platelet plug*. The time required for sealing (ca. 2 to 4 min) is called the *bleeding time*. Subsequently, the coagulation system produces a fibrin meshwork. Due to covalent cross-linking of fibrin, it turns to a *fibrin clot* or *thrombus* that retracts afterwards, thus reinforcing the seal. Later recanalization of the vessel can be achieved by fibrinolysis.

Platelets ($170-400 \cdot 10^3$ per µL of blood; half-life ≈ 10 days) are small non-nucleated bodies that are pinched off from megakaryocytes in the bone marrow. When an endothelial injury occurs, platelets adhere to subendothelial collagen fibers ($\rightarrow$ **A1**) bridged by **von Willebrand's factor** (vWF), which is formed by endothelial cells and circulates in the plasma complexed with factor VIII. Glycoprotein complex GP Ib/IX on the platelets are vWF receptors. This *adhesion* activates **platelets** ($\rightarrow$ **A2**). They begin to release substances ($\rightarrow$ **A3**), some of which promote platelet adhesiveness (vWF). Others like serotonin, platelet-derived growth factor (PDGF) and thromboxane A$_2$ (TXA$_2$) promote vasoconstriction. Vasoconstriction and platelet *contraction* slow the blood flow. Mediators released by platelets enhance platelet activation and attract and activate more platelets: ADP, TXA$_2$, platelet-activating factor (PAF). The shape of activated platelets change drastically ($\rightarrow$ **A4**). Discoid platelets become spherical and exhibit pseudopodia that intertwine with those of other platelets. This *platelet aggregation* ($\rightarrow$ **A5**) is further enhanced by thrombin and stabilized by GP IIb/IIIa. Once a platelet changes its shape, GP IIb/IIIa is expressed on the platelet surface, leading to fibrinogen binding and platelet aggregation. GP IIb/IIIa also increases the adhesiveness of platelets, which makes it easier for them to stick to subendothelial fibronectin.

			Half-life (h):	
I	Fibrinogen			96
II ^{K}	Prothrombin			72
III	Tissue thromboplastin			
IV	Ionized calcium (Ca^{2+})			
V	Proaccelerin			20
VII ^{K}	Proconvertin			5
VIII	Antihemophilic factor A			12
IX ^{K}	Antihemophilic factor B; plasma thromboplastin component (PTC); Christmas factor			24
X ^{K}	Stuart–Prower factor			30
XI	Plasma thromboplastin antecedent (PTA)			48
XII	Hageman factor			50
XIII	Fibrin-stabilizing factor (FSF)			250
–	Prekallikrein (PKK); Fletcher factor			
–	High-molecular-weight kininogen (HMK); Fitzgerald factor			

Several **coagulation factors** are involved in the clotting process. Except for Ca^{2+}, they are proteins formed in the *liver* ($\rightarrow$ **B** and Table). Factors labeled with a "K" in the table (as well as protein C and protein S, see below) are produced with **vitamin K**, an essential cofactor in posttranslational γ-carboxylation of glutamate residues of the factors. These γ-carboxyglutamyl groups are chelators of Ca^{2+}. They are required for Ca^{2+}-mediated complex formation of factors on the surface of phospholipid layers (PL), particularly on the platelet membrane (platelet factor 3). Vitamin K is oxidized in the reaction and has to be re-reduced by liver epoxide reductase (*vitamin K recycling*). **Ca^{2+} ions** are required for several steps in the clotting process ($\rightarrow$ **B**). When added to blood samples in vitro, citrate, oxalate, and EDTA bind with Ca^{2+} ions, thereby preventing the blood from clotting. This effect is desirable when performing various blood tests.

Activation of blood clotting ($\rightarrow$ **B**). Most coagulation factors normally are not active, or *zymogenic*. Their activation requires a cascade of events (An "a" added to the factor number means "activated"). Thus, even small amounts of a trigger factor lead to rapid blood clotting. The trigger can be endogenous (within a vessel) or exogenous (external). **Endogenous activation** ($\rightarrow$ **B2**) occurs at an *endothelial defect*. XII is activated to XIIa by the contact with

▶

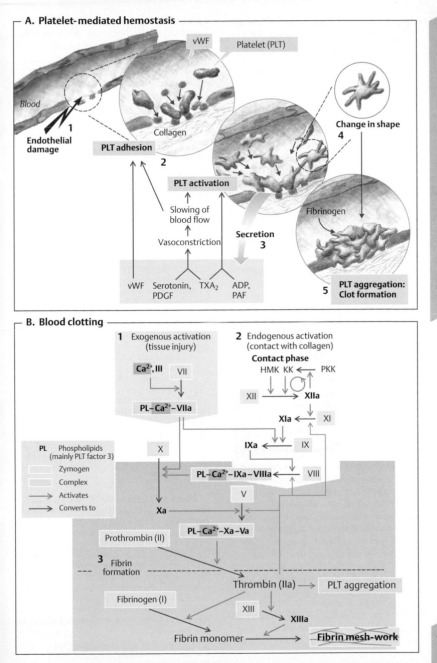

A. Platelet-mediated hemostasis

vWF

Platelet (PLT)

Blood

1 **Endothelial damage**

Collagen

2 **PLT adhesion**

PLT activation

Slowing of blood flow

Vasoconstriction

3 **Secretion**

vWF | Serotonin, PDGF | TXA₂ | ADP, PAF

Change in shape 4

Fibrinogen

5 **PLT aggregation: Clot formation**

B. Blood clotting

1 Exogenous activation (tissue injury)

Ca^{2+}, III | VII

$PL - Ca^{2+} - VIIa$

2 Endogenous activation (contact with collagen)

Contact phase

HMK KK ← PKK

XII → XIIa

XIa ← XI

IXa ← IX

PL Phospholipids (mainly PLT factor 3)

  Zymogen
  Complex
  → Activates
  → Converts to

X

$PL - Ca^{2+} - IXa - VIIIa$ ← VIII

Xa

V

$PL - Ca^{2+} - Xa - Va$

Prothrombin (II)

3 Fibrin formation

Thrombin (IIa) → PLT aggregation

Fibrinogen (I)

XIII

XIIIa

Fibrin monomer → **Fibrin mesh-work**

Plate 4.8 Hemostasis

103

negative charges of subendothelial collagen and sulfatide groups. This stimulates the conversion of prekallikrein (PKK) to kallikrein (KK), which enhances XII activation (positive feedback). Next, XIIa activates XI to XIa, which converts IX to IXa and, subsequently, VIII to VIIIa. Complexes formed by conjugation of IXa and VIIIa with Ca^{2+} on phospholipid (PL) layers activate X. **Exogenous activation** now merges with endogenous activation ($\rightarrow$ **B1**). In relatively large injuries, *tissue thrombokinase* (factor III), present on of nonvascular cells, is exposed to the blood, resulting in activation of VII. VII forms complexes with Ca^{2+} and phospholipids, thereby activating X (and IX).

Fibrin formation ($\rightarrow$ **B3**). After activation of X to Xa by endogenous and/or exogenous activation (the latter is faster), Xa activates V and conjugates with Va and Ca^{2+} on the surface of membranes. This complex, called *prothrombinase*, activates prothrombin (II) to **thrombin** (IIa). In the process, Ca^{2+} binds with phospholipids, and the N-terminal end of prothrombin splits off. The thrombin liberated in the process now activates (a) fibrinogen (I) to **fibrin** (Ia), (b) fibrin-stabilizing factor (XIII), and (c) V, VIII and XI (positive feedback). The single (monomeric) fibrin threads form a soluble meshwork (*fibrin$_S$; "s" for soluble) which XIIIa ultimately stabilizes to insoluble fibrin (fibrin$_i$)*. XIIIa is a transamidase that links the side chains of the fibrin threads via covalent bonds.

Fibrinolysis and Thromboprotection

To prevent excessive clotting and occlusion of major blood vessels (*thrombosis*) and *embolisms* due to clot migration, fibrin$_S$ is re-dissolved (fibrinolysis) and inhibitory factors are activated as soon as vessel repair is initiated.

Fibrinolysis is mediated by **plasmin** ($\rightarrow$ **C**). Various factors in blood (plasma kallikrein, factor XIIa), tissues (tissue plasminogen activator, tPA, endothelial etc.), and urine (urokinase) activate *plasminogen* to plasmin. Streptokinase, staphylokinase and tPA are used therapeutically to activate plasminogen. This is useful for dissolving a fresh thrombus located, e.g., in a coronary artery. Fibrin is split into fibrinopeptides which inhibit thrombin

formation and polymerization of fibrin to prevent further clot formation. *Alpha$_2$-antiplasmin* is an endogenous inhibitor of fibrinolysis. *Tranexamic acid* is administered therapeutically for the same purpose.

Thromboprotection. Antithrombin III, a serpin, is the most important thromboprotective plasma protein ($\rightarrow$ **D**).

It inactivates the protease activity of thrombin and factors IXa, Xa, XIa and XIIa by forming complexes with them. This is enhanced by **heparin** and heparin-like endothelial glucosaminoglycans. Heparin is produced naturally by mast cells and granulocytes, and synthetic heparin is injected for therapeutic purposes.

The binding of thrombin with endothelial **thrombomodulin** provides further thromboprotection. Only in this form does thrombin have anticoagulant effects ($\rightarrow$ **D**, negative feedback). Thrombomodulin activates protein C to Ca which, after binding to protein S, deactivates coagulation factors Va and VIIIa. The synthesis of proteins C and S is vitamin K-dependent. Other plasma proteins that inhibit thrombin are α_2-**macroglobulin** and α_1-**antitrypsin** ($\rightarrow$ **D**). Endothelial cells secrete **tissue thromboplastin inhibitor**, a substance that inhibits exogenous activation of coagulation, and **prostacyclin** (= prostaglandin I_2), which inhibits platelet adhesion to the normal endothelium.

Anticoagulants are administered for thromboprotection in patients at risk of blood clotting. Injected **heparin** has immediate action. Oral coumarin derivatives (phenprocoumon, warfarin, acenocoumarol) are **vitamin K antagonists** that work by inhibiting liver epoxide reductase, which is necessary for vitamin D recycling. Therefore, these drugs do not take effect until the serum concentration of vitamin K-dependent coagulation factors has decreased. **Cyclooxygenase inhibitors**, such as aspirin (acetylsalicylic acid), inhibit platelet aggregation by blocking thromboxane A_2 (TXA2) synthesis ($\rightarrow$ p. 269).

Hemorrhagic diatheses can have the following causes: a) Congenital deficiency of certain coagulation factors. Lack of VIII or IX, for example, leads to *hemophilia A or B*, respectively. b) Acquired deficiency of coagulation factors. The main causes are liver damage as well as vitamin K deficiency due to the destruction of vitamin K-producing intestinal flora or intestinal malabsorption. c) Increased consumption of coagulation factors, by *disseminated intravascular coagulation*. d) Platelet deficiency (thrombocytopenia) or platelet defect (thrombocytopathy). e) Certain vascular diseases, and f) excessive fibrinolysis.

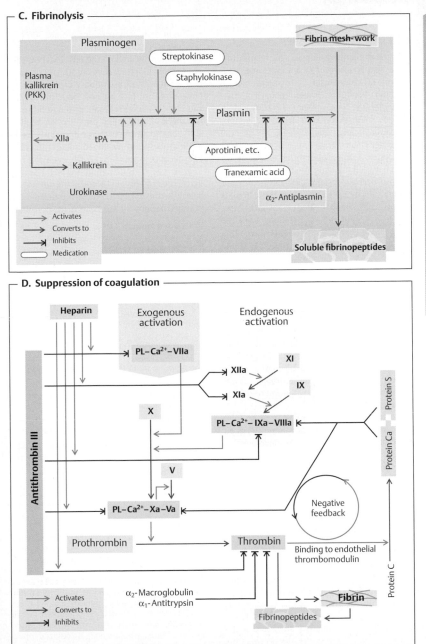

C. Fibrinolysis

Plasminogen

Streptokinase

Staphylokinase

Fibrin mesh-work

Plasma kallikrein (PKK)

Plasmin

XIIa ← tPA →

Kallikrein

Urokinase

Aprotinin, etc.

Tranexamic acid

α_2- Antiplasmin

→ Activates
→ Converts to
⇥ Inhibits
⬭ Medication

Soluble fibrinopeptides

D. Suppression of coagulation

Heparin

Exogenous activation

Endogenous activation

PL–Ca²⁺–VIIa

XIIa → XI

XIa → IX

X

PL–Ca²⁺– IXa–VIIIa

Protein S

Protein Ca

V

PL–Ca²⁺–Xa–Va

Negative feedback

Antithrombin III

Prothrombin → Thrombin

Binding to endothelial thrombomodulin

Protein C

α_2-Macroglobulin
α_1-Antitrypsin

Fibrin

Fibrinopeptides ←

→ Activates
→ Converts to
⇥ Inhibits

Plate 4.9 Fibrinolysis and Thromboprotection

5　Respiration

Lung Function, Respiration

The lung is mainly responsible for respiration, but it also has *metabolic functions*, e.g. conversion of angiotensin I to angiotensin II ($\rightarrow$ p. 184). In addition, the pulmonary circulation *buffers the blood volume* ($\rightarrow$ p. 204) and *filters* out small blood clots from the venous circulation before they obstruct the arterial circulation (heart, brain!).

External **respiration** is the **exchange of gases** between the body and the environment. (*Internal* or *tissue respiration* involves nutrient oxidation, $\rightarrow$ p. 228). **Convection (bulk flow)** is the means by which the body transports gases over long distances ($\rightarrow$ p. 24) along with the flow of air and blood. Both flows are driven by a pressure difference. **Diffusion** is used to transport gases over short distances of a few μm—e.g., through cell membranes and other physiological barriers ($\rightarrow$ p. 20ff.). The exchange of gas between the atmosphere and alveoli is called **ventilation**. Oxygen (O_2) in the inspired air is convected to the alveoli before diffusing across the alveolar membrane into the bloodstream. It is then transported via the bloodstream to the tissues, where it diffuses from the blood into the cells and, ultimately, to the intracellular mitochondria. *Carbon dioxide* (CO_2) produced in the mitochondria returns to the lung by the same route.

The **total ventilation per unit time**, $\dot{V}_T$ (also called *minute volume*) is the volume (V) of air inspired or expired per time. As the expiratory volume is usually measured, it is also abbreviated $\dot{V}_E$. (The dot means "per unit time"). At rest, the body maintains a $\dot{V}_E$ of about *8 L/min*, with a corresponding **oxygen consumption rate** ($\dot{V}_{O_2}$) of about 0.3 L/min and a **CO_2 elimination rate** ($\dot{V}_{CO_2}$) of about 0.25 L/min. Thus, about 26 L of air have to be inspired and expired to supply 1 L of O_2 (*respiratory equivalent* = 26). The **tidal volume** (V_T) is the volume of air moved in and out during one respiratory cycle. $\dot{V}_E$ is the product of V_T (ca. 0.5 L at rest) and **respiration rate** f (about 16/min at rest) (see p. 74 for values during physical work). Only around 5.6 L/min (at f = 16 min^{-1}) of the $\dot{V}_E$ of 8 L/min reaches the alveoli; this is known as **alveolar ventilation** ($\dot{V}_A$). The rest fills airways space not contributing to gas exchange (**dead space ventilation,** $\dot{V}_D$; $\rightarrow$ pp. 114 and 120).

The human body contains around 300 million **alveoli**, or thin-walled air sacs (ca. 0.3 mm in diameter) located on the terminal branches of the bronchial tree. They are surrounded by a dense network of **pulmonary capillaries** and have a total surface area of about 100 m^2. Because of this and the small air/blood diffusion distances of only a few μm ($\rightarrow$ p. 22, Eq. 1.7), sufficient quantities of O_2 can **diffuse** across the alveolar wall into the blood and CO_2 towards the alveolar space ($\rightarrow$ p. 120ff.), even at a tenfold increased O_2 demand ($\rightarrow$ p. 74). The oxygen-deficient "venous" blood of the pulmonary artery is thus oxygenated ("arterialized") and pumped back by the left heart to the periphery.

The **cardiac output (CO)** is the volume of blood pumped through the pulmonary and systemic circulation per unit time (5–6 L/min at rest). CO times the *arterial–venous O_2 difference (avDO_2)*—i.e., the difference between the arterial O_2 fraction in the aorta and in mixed venous blood of the right atrium (ca. 0.05 L of O_2 per L of blood)—gives the *O_2 volume transported per unit time from the lungs to the periphery*. At rest, it amounts to ($6 \times 0.05 =$) 0.3 L/min, a value matching that of $\dot{V}_{O_2}$ (see above). Inversely, if $\dot{V}_{O_2}$ and avDO_2 have been measured, CO can be calculated (**Fick's principle**):

$$CO = \dot{V}_{O_2}/avD_{O_2} \qquad [5.1]$$

The *stroke volume* **(SV)** is obtained by dividing CO by the heart rate (pulse rate).

According to *Dalton's law*, the total pressure (P_{total}) of a **mixture of gases** is the sum of the **partial pressures (P)** of the individual gases. The volume **fraction** (F, in L/L; $\rightarrow$ p. 376), of the individual gas relative to the total volume times P_{total} gives the partial pressure—in the case of O_2, for example, $P_{O_2} = F_{O_2} \times P_{total}$. The atmospheric partial pressures in *dry* ambient air at sea level (P_{total} = 101.3 kPa = 760 mmHg) are: F_{O_2} = 0.209, F_{CO_2} = 0.0004, and F_{N_2} + noble gases = 0.79 ($\rightarrow$ **A**, top right).

If the mixture of gases is "wet", the **partial pressure of water**, P_{H_2O} has to be subtracted from P_{total} (usually = atmospheric pressure). The other partial pressures will then be lower, since Px = Fx ($P_{total} - P_{H_2O}$). When passing through the respiratory tract (37 °C), inspired air is fully saturated with water. As a result, P_{H_2O} rises to 6.27 kPa (47 mmHg), and P_{O_2} drops 1.32 kPa lower than the dry atmospheric air ($\rightarrow$ p. 112). The partial pressures in the alveoli, arteries, veins (mixed venous blood), tissues, and expiratory air (all "wet") are listed in **A**.

Plate 5.1 Lung Function, Respiration

A. Gas transport

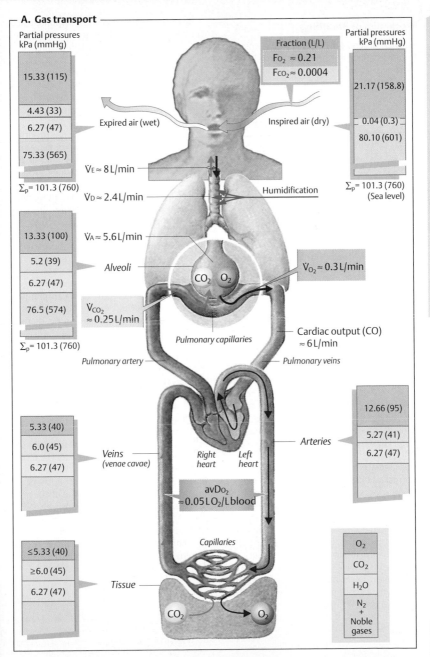

Partial pressures
kPa (mmHg)

15.33 (115)
4.43 (33)
6.27 (47)
75.33 (565)

Σ_p = 101.3 (760)

Expired air (wet)

Fraction (L/L)
$F_{O_2} \approx 0.21$
$F_{CO_2} \approx 0.0004$

Inspired air (dry)

Partial pressures
kPa (mmHg)

21.17 (158.8)
0.04 (0.3)
80.10 (601)

Σ_p = 101.3 (760)
(Sea level)

$\dot{V}_E \approx 8$ L/min

Humidification

$\dot{V}_D \approx 2.4$ L/min

$\dot{V}_A \approx 5.6$ L/min

13.33 (100)
5.2 (39)
6.27 (47)
76.5 (574)

Σ_p = 101.3 (760)

Alveoli

$\dot{V}_{O_2} \approx 0.3$ L/min

$\dot{V}_{CO_2} \approx 0.25$ L/min

CO_2 O_2

Pulmonary capillaries

Cardiac output (CO)
≈ 6 L/min

Pulmonary artery

Pulmonary veins

12.66 (95)
5.27 (41)
6.27 (47)

Arteries

5.33 (40)
6.0 (45)
6.27 (47)

Veins
(venae cavae)

Right
heart

Left
heart

avD_{O_2}
≈ 0.05 L O_2/L blood

$\leq$5.33 (40)
$\geq$6.0 (45)
6.27 (47)

Capillaries

Tissue

CO_2

CO_2 O_2

O_2
CO_2
H_2O
N_2
+
Noble
gases

Mechanics of Breathing

Pressure differences between the alveoli and the environment are the *driving "forces"* for the exchange of gases that occurs during ventilation. **Alveolar pressure** (P_A = intrapulmonary pressure; $\rightarrow$ **B**) must be lower than the barometric pressure (P_B) during **inspiration** (breathing in), and higher during **expiration** (breathing out). If P_B *is defined as zero*, the alveolar pressure is negative during inspiration and positive during expiration ($\rightarrow$ **B**). These pressure differences are created through coordinated movement of the diaphragm and chest (thorax), resulting in an increase in lung volume (V_{pulm}) during inspiration and a decrease during expiration ($\rightarrow$ **A1,2**).

The **inspiratory muscles** consist of the *diaphragm*, *scalene muscles*, and external intercostal muscles. Their contraction lowers (flattens) the diaphragm and raises and expands the chest, thus expanding the lungs. Inspiration is therefore *active*. The external intercostal muscles and accessory respiratory muscles are activated for deep breathing. During **expiration**, the diaphragm and other inspiratory muscles *relax*, thereby raising the diaphragm and lowering and reducing the volume of the chest and lungs. Since this action occurs primarily due to the intrinsic elastic recoil of the lungs ($\rightarrow$ p. 116), expiration is *passive* at rest. In deeper breathing, active mechanisms can also play a role in expiration: the *internal intercostal muscles* contract, and the diaphragm is pushed upward by *abdominal pressure* created by the muscles of the abdominal wall.

Two adjacent ribs are bound by internal and external intercostal muscle. Counteractivity of the muscles is due to variable leverage of the upper and lower rib ($\rightarrow$ **A3**). The distance separating the insertion of the external intercostal muscle on the upper rib (Y) from the axis of rotation of the upper rib (X) is smaller than the distance separating the insertion of the muscles on the lower rib (Z') and the axis of rotation of the lower rib (X'). Therefore, X'–Z' is longer and a more powerful lever than X–Y. The chest generally rises when the external intercostal muscles contract, and lowers when the opposing internal intercostal muscles contract.

To exploit the motion of the diaphragm and chest for ventilation, the lungs must be able to follow this motion without being completely attached to the diaphragm and chest. This is achieved with the aid of the **pleura**, a thin fluid-covered sheet of cells that invests each lung (*visceral pleura*), thereby separating it from the adjacent organs, which are covered by the pleura as well (*parietal pleura*).

In its natural state, the lung tends to shrink due to its *intrinsic elasticity* and *alveolar surface tension* ($\rightarrow$ p. 118). Since the fluid in the pleural space cannot expand, the lung sticks to the inner surface of the chest, resulting in suction (which still allows tangential movement of the two pleural sheets). **Pleural pressure (P_{pl})** is then negative with respect to atmospheric pressure. P_{pl}, also called *intrapleural* (P_{ip}) or *intrathoracic* pressure, can be measured during breathing (dynamically) using an esophageal probe ($\approx P_{pl}$). The intensity of suction (negative pressure) increases when the chest expands during inspiration, and decreases during expiration ($\rightarrow$ **B**). P_{pl} usually does not become positive unless there is very forceful expiration requiring the use of expiratory muscles. The difference between the alveolar and the pleural pressure ($P_A - P_{pl}$) is called **transpulmonary pressure** ($\rightarrow$ p. 114).

Characterization of breathing activity. The terms *hyperpnea* and *hypopnea* are used to describe abnormal increases or decreases in the depth and rate of respiratory movements. *Tachypnea* (too fast), *bradypnea* (too slow), and *apnea* (cessation of breathing) describe abnormal changes in the respiratory rate. The terms *hyperventilation* and *hypoventilation* imply that the volume of exhaled CO_2 is larger or smaller, respectively, than the rate of CO_2 production, and the arterial partial pressure of CO_2 (Pa_{CO_2}) decreases or rises accordingly ($\rightarrow$ p. 142). *Dyspnea* is used to describe difficult or labored breathing, and *orthopnea* occurs when breathing is difficult except in an upright position.

Plate 5.2 Mechanics of Breathing

A. Respiratory muscles

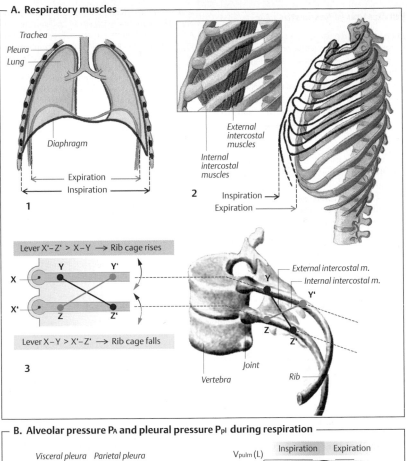

Trachea
Pleura
Lung
Diaphragm

Expiration
Inspiration

1

External intercostal muscles
Internal intercostal muscles

2 Inspiration →
Expiration →

Lever X'–Z' > X–Y ⟶ Rib cage rises

X
Y   Y'

X'
Z   Z'

Lever X–Y > X'–Z' ⟶ Rib cage falls

3

External intercostal m.
Internal intercostal m.
Y
Y'
Z
Z'

Joint
Vertebra
Rib

B. Alveolar pressure P_A and pleural pressure P_{pl} during respiration

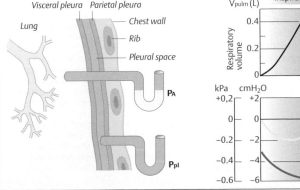

Visceral pleura   Parietal pleura
Lung
Chest wall
Rib
Pleural space

P_A

P_{pl}

V_{pulm} (L)   Inspiration | Expiration

Respiratory volume

0.4
0.2
0

kPa   cmH$_2$O
+0,2   +2
0   0
–0.2   –2
–0.4   –4
–0.6   –6

P_A

P_{pl}

Purification of Respiratory Air

Inhaled foreign particles are trapped by **mucus** in the nose, throat, trachea, and bronchial tree. The entrapped particles are engulfed by *macrophages* and are driven back to the trachea by the *cilia* lining the bronchial epithelium. **Cilial escalator:** The cilia move at a rate of $5–10\,s^{-1}$ and propel the mucus towards the mouth at a rate of 1 cm/min on a film of fluid secreted by the epithelium. Heavy smoking, mucoviscidosis genetic defects can impair cilial transport. A volume of 10–100 mL of mucus is produced each day, depending on the type and frequency of local irritation (e.g., smoke inhalation) and vagal stimulation. Mucus is usually swallowed, and the fluid fraction is absorbed in the gastrointestinal tract.

Artificial Respiration

Mouth-to-mouth resuscitation is an emergency measure performed when someone stops breathing. The patient is placed flat on the back. While pinching the patient's nostrils shut, the aid-giver places his or her mouth on the patient's mouth and blows forcefully into the patient's lungs ($\rightarrow$ **A3**). This raises the alveolar pressure ($\rightarrow$ p. 108) in the patient's lungs relative to the atmospheric pressure outside the chest and causes the lungs and chest to expand (inspiration). The rescuer then removes his or her mouth to allow the patient to exhale. Expulsion of the air blown into the lungs (expiration) occurs due to the intrinsic elastic recoil of the lungs and chest ($\rightarrow$ p. 109 A2). This process can be accelerated by pressing down on the chest. The rescuer should ventilate the patient at a rate of about $16\,min^{-1}$. The expiratory O_2 fraction ($\rightarrow$ p. 107 A) of the the rescuer is high enough to adequately oxygenate the patient's blood. The color change in the patient's skin from blue (cyanosis; $\rightarrow$ p. 130) to pink indicates that a resuscitation attempt was successful.

Mechanical ventilation. Mechanical *intermittent positive pressure ventilation* (**IPPV**) works on the same principle. This technique is used when the respiratory muscles are paralyzed due to disease, anesthesia, etc. The pump of the respirator drives air into the patient's lung during inspiration ($\rightarrow$ **A1**). The external inspiratory and expiratory pathways are separated by a valve (close to the patient's mouth as possible) to prevent enlargement of dead space ($\rightarrow$ p. 114). Ventilation frequency, tidal volume, inspiratory flow, as well as duration of inspiration and expiration can be preselected at the respirator. The drawback of this type of ventilation is that venous return to the heart is impaired to some extent ($\rightarrow$ p. 204). Today, the standard technique of mechanical respiration is *continuous positive pressure ventilation* (**CPPV**). In contrast to IPPV, the *endexpiratory pressure* is kept *positive* (PEEP) in CPPV. In any case, all ventilated patients should be continuously monitored (expiratory gas fraction; blood gas composition, etc.).

The **iron lung** (Drinker respirator) makes use of *negative-pressure respiration* ($\rightarrow$ **A2**). The patient's body is enclosed from the neck down in a metal tank. To achieve inhalation, pressure in the tank is decreased to a level below ambient pressure and, thus, below alveolar pressure. This pressure difference causes the chest to expand (inspiratory phase), and the cessation of negative pressure in the tank allows the patient to breathe out (expiratory phase). This type of respirator is used to ventilate patients who require long-term mechanical ventilation due to paralytic diseases, such as polio.

Pneumothorax

Pneumothorax occurs when air enters the pleural space and P_{pl} falls to zero ($\rightarrow$ p. 108), which can lead to collapse of the affected lung due to elastic recoil and respiratory failure ($\rightarrow$ **B**). The contralateral lung is also impaired because a portion of the inspired air travels back and forth between the healthy and collapsed lung and is not available for gas exchange. *Closed pneumothorax*, i.e., the leakage of air from the alveolar space into the pleural space, can occur spontaneously (e.g., lung rupture due to bullous emphysema) or due to lung injury (e.g., during mechanical ventilation = barotrauma; $\rightarrow$ p. 134). **Open pneumothorax** ($\rightarrow$ **B2**) can be caused by an open chest wound or blunt chest trauma (e.g., penetration of the pleura by a broken rib). **Valvular pneumothorax** ($\rightarrow$ **B3**) is a life-threatening form of pneumothorax that occurs when air enters the pleural space with every breath and can no longer be expelled. A flap of acts like a valve. Positive pressure develops in the pleural space on the affected side, as well as in the rest of the thoracic cavity. Since the tidal volume increases due to hypoxia, high pressure levels (4 kPa = 30 mmHg) quickly develop. This leads to increasing impairment of cardiac filling and compression of the healthy contralateral lung. Treatment of valvular pneumothorax consists of slow drainage of excess pressure and measures to prevent further valvular action.

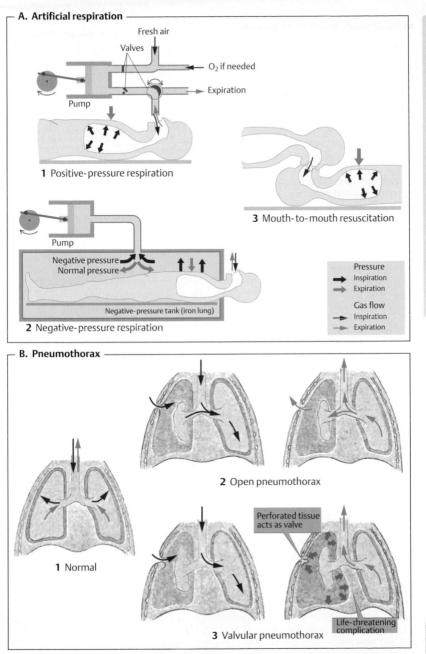

Plate 5.3 Artificial Respiration, Pneumothorax

A. Artificial respiration

Fresh air

Valves

O₂ if needed

Expiration

Pump

1 Positive-pressure respiration

3 Mouth-to-mouth resuscitation

Pump

Negative pressure

Normal pressure

Negative-pressure tank (iron lung)

2 Negative-pressure respiration

Pressure
→ Inspiration
→ Expiration

Gas flow
→ Inspiration
→ Expiration

B. Pneumothorax

2 Open pneumothorax

1 Normal

Perforated tissue acts as valve

Life-threatening complication

3 Valvular pneumothorax

Lung Volumes and their Measurement

At the end of normal quiet expiration, the lung–chest system returns to its intrinsic **resting position**. About 0.5 L of air is taken in with each breath during normal quiet respiration; this is called the **resting tidal volume** (VT). Inspiration can be increased by another 3 L or so on forced (maximum) inspiration; this is called the **inspiratory reserve volume** (IRV). Likewise, expiration can be increased by about 1.7 L more on forced (maximum) expiration. This is called the **expiratory reserve volume** (ERV). These reserve volumes are used during strenuous physical exercise ($\rightarrow$ p. 74) and in other situations where normal tidal volumes are insufficient. Even after forced expiration, about 1.3 L of air remains in the lungs; this is called the **residual volume** (RV). Lung capacities are sums of the individual lung volumes. The **vital capacity** (VC) is the maximum volume of air that can be moved in and out in a single breath. Therefore, VC = VT + IRV + ERV. The average 20-year-old male with a height of 1.80 m has a VC of about 5.3 L. Vital capacity decreases and residual volume increases with age (1.5 ⇒ 3 L). The **total lung capacity** is the sum of VC and RV—normally 6 to 7 L. The **functional residual capacity** is the sum of ERV and RV ($\rightarrow$ **A** and p. 114). The *inspiratory* capacity is the sum of VT and IRV. All numerical values of these volumes apply under body temperature–pressure saturation (BTPS) conditions (see below).

Spirometry. These lung volumes and capacities (except FRC, RV) can be measured by routine spirometry. The *spirometer* ($\rightarrow$ **A**) consists usually of a water-filled tank with a bell-shaped floating device. A tube connects the air space within the spirometer ($\rightarrow$ **A**) with the airways of the test subject. A counterweight is placed on the bell. The position of the bell indicates how much air is in the spirometer and is calibrated in volume units (L$_{ATPS}$; see below). The bell on the spirometer rises when the test subject blows into the device (expiration), and falls during inspiration ($\rightarrow$ **A**).

If the spirometer is equipped with a recording device (*spirograph*), it can be also used for graphic measurement of the total ventilation per unit time (V̇E; $\rightarrow$ pp. 106 and 118),

compliance ($\rightarrow$ p. 116), O_2 consumption (V̇O_2), and in dynamic lung function tests ($\rightarrow$ p. 118).

Range of normal variation. Lung volumes and capacities vary greatly according to age, height, physical constitution, sex, and degree of physical fitness. The range of normal variation of VC, for example, is 2.5 to 7 L. Empirical formulas were therefore developed to create normative values for better interpretation of lung function tests. For instance, the following formulas are used to calculate the range of normal values for VC in Caucasians:

Men: VC = 5.2 h – 0.022 a – 3.6 ($\pm$ 0.58)
Women: VC = 5.2 h – 0.018 a – 4.36 ($\pm$ 0.42),

where h = height (in meters) and a = age (in years); the standard deviation is given in parentheses. Because of the broad range of normal variation, patients with mild pulmonary disease may go undetected. Patients with lung disease should ideally be monitored by recording baseline values and observing *changes* over the course of time.

Conversion of respiratory volumes. The volume, V, of a gas (in L or m^3; 1 m^3 = 1000 L) can be obtained from the amount, M, of the gas (in mol), absolute temperature, T (in K), and total pressure, P (in Pa), using the *ideal gas equation*:

$$V = M \cdot R \cdot T/P, \qquad [5.2]$$

where P is barometric pressure (PB) minus water partial pressure (P$_{H_2O}$; $\rightarrow$ p. 106) and R is the universal gas constant = 8.31 J $\cdot$ K^{-1} $\cdot$ mol^{-1}.

Volume conditions

STPD:	Standard temperature pressure dry (273 K, 101 kPa, P$_{H_2O}$ = 0)
ATPS:	Ambient temperature pressure H_2O-saturated (T$_{amb}$, PB, P$_{H_2O}$ at T$_{Amb}$)
BTPS:	Body temperature pressure-saturated (310 K, PB, P$_{H_2O}$ = 6.25 kPa)

It follows that:

V_{STPD} = M $\cdot$ R $\cdot$ 273/101 000 [m^3]
V_{ATPS} = M $\cdot$ R $\cdot$ T$_{Amb}$/(PB – P$_{H_2O}$) [m^3]
V_{BTPS} = M $\cdot$ R $\cdot$ 310/(PB – 6250) [m^3].

Conversion factors are derived from the respective quotients (M $\cdot$ R is a reducing factor). Example: V_{BTPS}/V_{STPD} = 1.17. If V_{ATPS} is measured by spirometry at room temperature (T$_{Amb}$ = 20 °C; P$_{H_2O_{sat}}$ = 2.3 kPa) and PB = 101 kPa, V_{BTPS} ≈ 1.1 V_{ATPS} and V_{STPD} ≈ 0.9 V_{ATPS}.

A. Lung volumes and their measurement

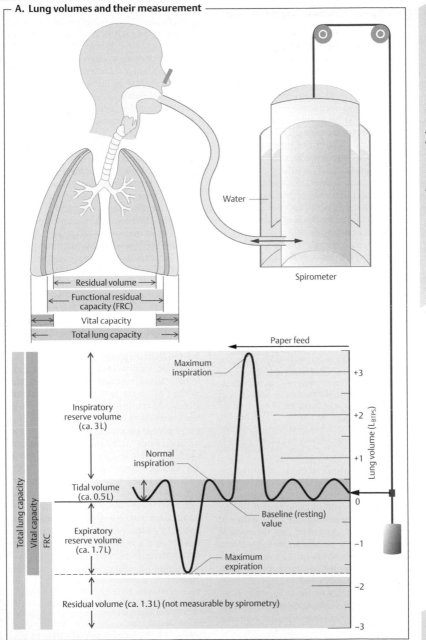

Water

Spirometer

Residual volume

Functional residual capacity (FRC)

Vital capacity

Total lung capacity

Paper feed

Maximum inspiration

Lung volume (L_{BTPS})

+3

+2

+1

Inspiratory reserve volume (ca. 3 L)

Normal inspiration

Tidal volume (ca. 0.5 L)

0

Baseline (resting) value

Expiratory reserve volume (ca. 1.7 L)

Maximum expiration

−1

−2

Residual volume (ca. 1.3 L) (not measurable by spirometry)

−3

Total lung capacity

Vital capacity

FRC

Plate 5.4 Lung Volumes and their Measurement

113

Dead Space, Residual Volume, Airway Resistance

The exchange of gases in the respiratory tract occurs in the alveoli. Only a portion of the tidal volume (V_T) reaches the alveoli; this is known as the **alveolar part** (V_A). The rest goes to dead space (not involved in gas exchange) and is therefore called **dead space** volume (V_D). The oral, nasal, and pharyngeal cavities plus the trachea and bronchi are jointly known as *physiological dead space* or *conducting zone of the airways*. The physiological dead space (ca. 0.15 L) is approximately equal to the *functional* dead space, which becomes larger than physiological dead space when the exchange of gases fails to take place in a portion of the alveoli ($\rightarrow$ p. 120). The **functions of dead space** are to conduct incoming air to the alveoli and to purify ($\rightarrow$ p. 110), humidify, and warm inspired ambient air. Dead space is also an element of the vocal organ ($\rightarrow$ p. 370).

The **Bohr equation** ($\rightarrow$ **A**) can be used to **estimate** the dead space.

Derivation: The expired tidal volume V_T is equal to the sum of its alveolar part V_A plus dead space V_D ($\rightarrow$ **A, top**). Each of these three variables has a characteristic CO_2 fraction ($\rightarrow$ p. 376): F_{ECO_2} in V_T, F_{ACO_2} in V_A, and F_{ICO_2} in V_D. F_{ICO_2} is extremely small and therefore negligible. The product of each of the three volumes and its corresponding CO_2 fraction gives the volume of CO_2 for each. The CO_2 volume in the expired air ($V_T \cdot F_{ECO_2}$) equals the sum of the CO_2 volumes in its two components, i.e. in V_A and V_D ($\rightarrow$ **A**).

Thus, three values must be known to **determine the dead space**: V_T, F_{ECO_2} and F_{ACO_2}. V_T can be measured using a spirometer, and F_{ECO_2} and F_{ACO_2} can be measured using a Bunte glass burette or an infrared absorption spectrometer. F_{ACO_2} is present in the last expired portion of V_T—i.e., in *alveolar gas*. This value can be measured using a Rahn valve or similar device.

The **functional residual capacity** (FRC) is the amount of air remaining in the lungs at the end of normal quiet expiration, and the **residual volume** (RV) is the amount present after forced maximum expiration ($\rightarrow$ p. 112). About 0.35 L of air (V_A) reaches the alveolar space with each breath during normal quiet respiration. Therefore, only about 12% of the 3 L total FRC is renewed at rest. The composition of gases in the alveolar space therefore remains relatively constant.

Measurement of FRC and RV cannot be performed by spirometry. This must be done using indirect techniques such as **helium dilution** ($\rightarrow$ **B**). Helium (He) is a poorly soluble inert gas. The test subject is instructed to repeatedly inhale and exhale a known volume (V_{Sp}) of a helium-containing gas mixture (e.g., $F_{He_0} = 0.1$) out of and into a spirometer. The helium distributes evenly in the lungs (V_L) and spirometer ($\rightarrow$ **B**) and is thereby diluted ($F_{He_x} < F_{He_0}$). Since the total helium volume does not change, the known initial helium volume ($V_{Sp} \cdot F_{He_0}$) is equal to the final helium volume ($V_{Sp} + V_L$) $\cdot F_{He_x}$. V_L can be determined once F_{He_x} in the spirometer has been measured at the end of the test ($\rightarrow$ **B**). V_L will be equivalent to RV if the test was started after a forced expiration, and will be equivalent to FRC if the test was started after normal expiration, i.e. from the resting position of lung and chest. The helium dilution method measures gases in *ventilated* airways only.

Body plethysmography can also detect gases in encapsulated spaces (e.g., cysts) in the lung. The test subject is placed in an airtight chamber and instructed to breathe through a *pneumotachygraph* (instrument for recording the flow rate of respired air). At the same time, respiration-dependent changes in air pressure in the subject's mouth and in the chamber are continuously recorded. FRC and RV can be derived from these measurements.

Such measurements can also be used to determine **airway resistance**, R_L, which is defined as the *driving pressure gradient* between the alveoli and the atmosphere *divided by the air flow per unit time*. Airway resistance is very low under normal conditions, especially during inspiration when (a) the lungs become more expanded (*lateral traction* of the airways), and (b) the *transpulmonary pressure* (P_A-P_{pl}) *rises* ($\rightarrow$ p. 108). P_A-P_{pl} represents the transmural pressure of the airways and widens them more and more as it increases. Airway resistance may become too high when the airway is narrowed by mucus—e.g., in chronic obstructive pulmonary disease, or when its smooth muscle contracts, e.g. in asthma ($\rightarrow$ p.118).

The *residual volume* (RV) *fraction of the total lung capacity* (TLC) is clinically significant ($\rightarrow$ p. 112). This fraction normally is no more than 0.25 in healthy subjects and somewhat higher in old age. It can rise to 0.55 and higher when pathological enlargement of the alveoli has occurred due, for example, to *emphysema*. The RV/TLC fraction is therefore a rough measure of the severity of such diseases.

Plate 5.5   Dead Space, Residual Volume, Airway Resist.

A. Measurement of dead space

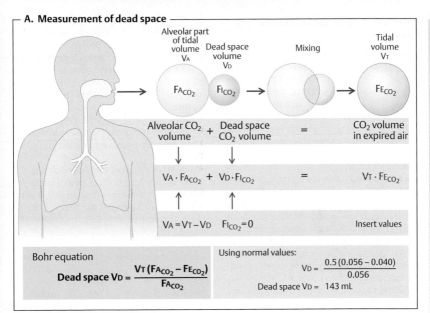

Alveolar part of tidal volume V_A

Dead space volume V_D

Mixing

Tidal volume V_T

FA_{CO_2} FI_{CO_2} FE_{CO_2}

Alveolar CO_2 volume	+	Dead space CO_2 volume	=	CO_2 volume in expired air

$$V_A \cdot FA_{CO_2} + V_D \cdot FI_{CO_2} = V_T \cdot FE_{CO_2}$$

$V_A = V_T - V_D$ $FI_{CO_2} = 0$ Insert values

Bohr equation

$$\text{Dead space } V_D = \frac{V_T (FA_{CO_2} - FE_{CO_2})}{FA_{CO_2}}$$

Using normal values:

$$V_D = \frac{0.5 (0.056 - 0.040)}{0.056}$$

Dead space $V_D = 143$ mL

B. Measurement of residual volume and functional residual capacity

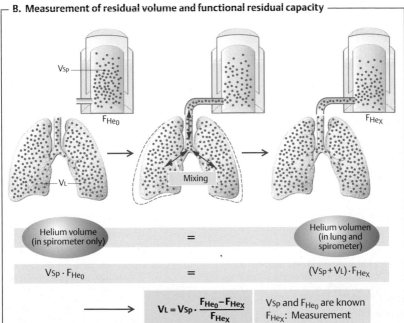

V_{Sp}

F_{He_0} F_{He_X}

Mixing

V_L

Helium volume (in spirometer only)	=	Helium volumen (in lung and spirometer)

$$V_{Sp} \cdot F_{He_0} = (V_{Sp} + V_L) \cdot F_{He_X}$$

$$V_L = V_{Sp} \cdot \frac{F_{He_0} - F_{He_X}}{F_{He_X}}$$

V_{Sp} and F_{He_0} are known
F_{He_X}: Measurement

Pressure–Volume Curve, Respiratory Work

Resting position (RP) is the position to which the lung–chest system returns at the end of normal quiet expiration; this lung volume equals the functional residual capacity (FRC, → p. 114). Its value is set at zero (V_{pulm} = 0) in **A–C**. RP (→ **A1**) is a stable central position characterized by the reciprocal cancellation of two passive forces: chest expansion force (CEF) and lung contraction force (LCF). As we breathe in and out, the lung – chest system makes excursions from resting position; thus, LCF > CEF during inspiration, and CEF > LCF during expiration. The difference between LCF and CEF, i.e. the net force (→ blue arrows in **A2, 3, 5, 6**), is equal to the alveolar pressure (P_A → p. 108) if the airway is closed (e.g., by turning a stopcock, as in **A1–3, 5, 6**) after a known air volume has been inhaled (V_{pulm} > 0, → **A2**) from a spirometer or expelled into it (V_{pulm} < 0, → **A3**). (In the resting position, CEF = LCF, and P_A = 0). Therefore, the relationship between V_{pulm} and P_A in the lung–chest system can be determined as illustrated by the **static resting pressure–volume (PV) curve** (→ **blue curve** in **A–C**) ("static" = measured while holding the breath; "resting" = with the respiratory muscles relaxed).

(Compression and expansion of V_{pulm} by a positive or negative P_A during measurement has to be taken into account; → **A**, dark-gray areas).

The **slope** of the static resting PV curve, $\Delta V_{pulm}/\Delta P_A$, represents the (static) **compliance** *of the lung–chest system* (→ **B**). The steepest part of the curve (range of greatest compliance; ca. 1 L/kPa in an adult) lies between RP and V_{pulm} = 1 L. This is in the normal respiratory range. The curve loses its steepness, i.e. compliance decreases, in old age or in the presence of lung disease. As a result, greater effort is needed to breathe the same tidal volume.

The above statements apply to lung-and-chest compliance. It is also possible to calculate compliance for the *chest wall alone* ($\Delta V_A/\Delta P_{pl}$ = 2 L/kPa) or for the *lung alone* ($\Delta V/\Delta[P_A - P_{pl}]$ = 2 L/kPa) when the pleural pressure (P_{pl}) is known (→ p. 108).

PV relationships can also be plotted during maximum expiratory and inspiratory effort to determine the **peak expiratory** and **inspiratory pressures** (→ **A**, red and green curves). Only a very small pressure can be generated from a position of near-maximum expiration ($V_{pulm} \ll$ 0; → **A7**) compared to a peak pressure of about 15 kPa ($\approx$ 110 mmHg) at $V_{pulm} \gg$ 0 (Valsalva's maneuver; → **A5**). Likewise, the greatest negative pressure (suction) (ca. – 10 kPa = 75 mmHg) can be generated from a position of maximum expiration (Müller's maneuver; → **A6**), but not from an inspiratory position (→ **A4**).

A **dynamic PV curve** is obtained *during* respiration (→ **C**). The result is a loop consisting of the opposing inspiratory (red) and expiratory (green) curves transected by the resting curve (blue) because airway flow resistance (R_L) must be overcome (mainly in the upper and middle airways) while inhaling in the one direction and exhaling in the other. The *driving pressure gradients (ΔP)* also oppose each other (inspiratory P_A < 0; expiratory P_A > 0; → p. 109 B). As in Ohm's law, $\Delta P = R_L \cdot$ respiratory flow rate ($\dot{V}$). Therefore, ΔP must increase if the bronchial tubes narrow and/or if the respiratory flow rate increases (→ **C**).

In **asthma**, the airway radius (r) decreases and a very high ΔP is needed for normal ventilation ($R_L \approx 1/r^4$!). During expiration, a high ΔP decreases the *transpulmonary pressure* (= $P_A - P_{pl}$) and thereby squeezes the airways ($R_L \uparrow$). The high R_L results in a pressure decrease along the expiratory airway ($P_{airway} \downarrow$) until $P_{airway} - P_{pl}$ < 0. At this point, the airway will collapse. This is called **dynamic airway compression**, which often results in a life-threatening vicious cycle: r↓ ⇨ $\Delta P \uparrow$ ⇨ r↓↓ ⇨ $\Delta P \uparrow\uparrow$ ....

Respiratory work. The colored areas within the loop (A_{Rinsp} and A_{Rexp}; → **C**) represent the inspiratory and expiratory PV work (→ p. 374) exerted to overcome flow resistance. The cross-hatched area (→ **C**) is the work required to overcome the intrinsic elastic force of the lungs and chest (A_{elast}). *Inspiratory work* is defined as A_{Rinsp} + A_{elast}. The inspiratory muscles (→ p. 108) must overcome the elastic force, whereas the same elastic force provides the (passive) driving force for expiration at rest (sign reverses for A_{elast}). Thus, *expiratory work* is A_{Rexp} – A_{elast}. Expiration can also require muscle energy if A_{Rexp} becomes larger than A_{elast}—e.g., during forced respiration or if R_L is elevated.

Plate 5.6   Pressure–Volume Curve, Respiratory Work

A. Pressure-volume curve of the lung-chest system

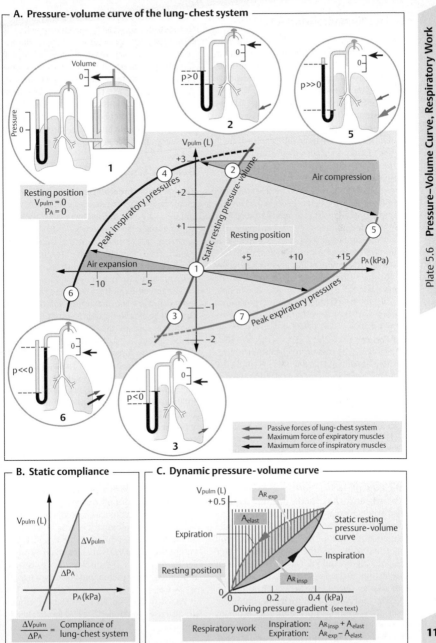

Resting position
$V_{pulm} = 0$
$P_A = 0$

Peak inspiratory pressures

Static resting pressure-volume

Air compression

Resting position

Air expansion

Peak expiratory pressures

Passive forces of lung-chest system
Maximum force of expiratory muscles
Maximum force of inspiratory muscles

B. Static compliance

V_{pulm} (L)

ΔV_{pulm}

ΔP_A

P_A (kPa)

$$\frac{\Delta V_{pulm}}{\Delta P_A} = \text{Compliance of lung-chest system}$$

C. Dynamic pressure-volume curve

V_{pulm} (L)
+0.5

$A_{R\,exp}$

A_{elast}

Expiration

Static resting pressure-volume curve

Inspiration

Resting position

$A_{R\,insp}$

0   0.2   0.4 (kPa)
Driving pressure gradient (see text)

Respiratory work   Inspiration: $A_{R\,insp} + A_{elast}$
Expiration: $A_{R\,exp} - A_{elast}$

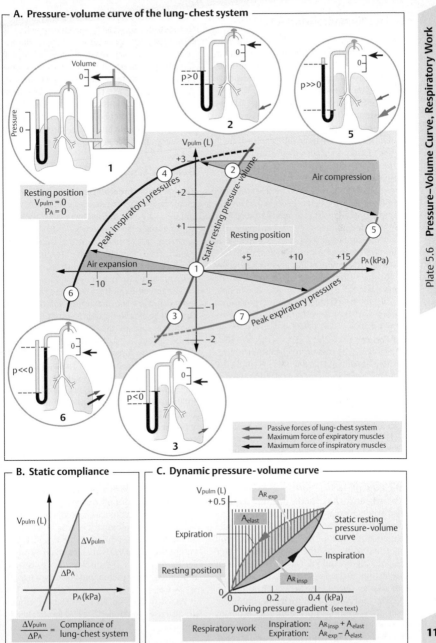

117

Surface Tension, Surfactant

Surface tension is the main factor that determines the *compliance* of the lung-chest system (→ p. 116) and develops at gas-liquid interfaces or, in the case of the lungs, on the *gas exchange surface of the alveoli* (ca. 100 m²).

The effectiveness of these forces can be demonstrated by filling an isolated and completely collapsed lung with (a) air or (b) liquid. In example (a), the lung exerts a much higher resistance, especially at the beginning of the filling phase. This represents the **opening pressure**, which raises the alveolar pressure (P_A) to about 2 kPa or 15 mmHg when the total lung capacity is reached (→ p. 113 A). In example (b), the resistance and therefore P_A is only one-fourth as large. Accordingly, the larger pressure requirement in example (a) is required to overcome surface tension.

If a *gas bubble* with radius r is surrounded by liquid, the surface tension γ (N · m^{-1}) of the liquid raises the pressure inside the bubble relative to the outside pressure (transmural pressure $\Delta P > 0$). According *to Laplace's law* (cf. p. 188):

$$\Delta P = 2\gamma/r \text{ (Pa)}. \qquad [5.3]$$

Since γ normally remains constant for the respective liquid (e.g., plasma: 10^{-3} N · m^{-1}), ΔP becomes larger and larger as r decreases.

Soap bubble model. If a flat soap bubble is positioned on the opening of a cylinder, r will be relatively large (→ **A1**) and ΔP small. (Since two air-liquid interfaces have to be considered in this case, Eq. 5.3 yields $\Delta P = 4\gamma/r$). For the bubble volume to expand, r must initially decrease and ΔP must increase (→ **A2**). Hence, a relatively high "opening pressure" is required. As the bubble further expands, r increases again (→ **A3**) and the pressure requirement/volume expansion ratio decreases. The alveoli work in a similar fashion. This model demonstrates that, in the case of two alveoli connected with each other (→ **A4**), the smaller one (ΔP_2 high) would normally become even smaller while the larger one (ΔP_1 low) becomes larger due to pressure equalization.

Surfactant (**surf**ace-**act**ive **age**nt) lining the inner alveolar surface prevents this problem by lowering γ in smaller alveoli more potently than in larger alveoli. Surfactant is a mixture of proteins and phospholipids (chiefly dipalmitoyl lecithin) secreted by alveolar type II cells. *Respiratory distress syndrome of the newborn*, a serious pulmonary gas exchange disorder, is caused by failure of the immature lung to produce sufficient quantities of surfactant. Lung damage related to O_2 toxicity (→ p. 136) is also partly due to oxidative destruction of surfactant, leading to reduced compliance. This can ultimately result in alveolar collapse (*atelectasis*) and pulmonary edema.

Dynamic Lung Function Tests

The **maximum breathing capacity** (MBC) is the greatest volume of gas that can be breathed (for 10 s) by voluntarily increasing the tidal volume and respiratory rate (→ **B**). The MBC normally ranges from 120 to 170 L/min. This capacity can be useful for monitoring diseases affecting the respiratory muscles, e.g., myasthenia gravis.

The **forced expiratory volume** (FEV or Tiffeneau test) is the maximum volume of gas that can be expelled from the lungs. In clinical medicine, FEV in the first second (FEV$_1$) is routinely measured. When its absolute value is related to the *forced vital capacity* (FVC), the *relative FEV$_1$* (normally > 0.7) is obtained. (FVC is the maximum volume of gas that can be expelled from the lungs as quickly and as forcefully as possible from a position of full inspiration; → **C**). It is often slightly lower than the vital capacity VC (→ p. 112). **Maximum expiratory flow**, which is measured using a pneumotachygraph during FVC measurement, is around 10 L/s.

Dynamic lung function tests are useful for distinguishing *restrictive lung disease* (RLD) from *obstructive lung disease* (OLD). RLD is characterized by a functional reduction of lung volume, as in pulmonary edema, pneumonia and impaired lung inflation due to spinal curvature, whereas OLD is characterized by physical narrowing of the airways, as in asthma, bronchitis, emphysema, and vocal cord paralysis (→ **C2**).

As with VC (→ p. 112), empirical formulas are also used to standardize FVC for age, height and sex.

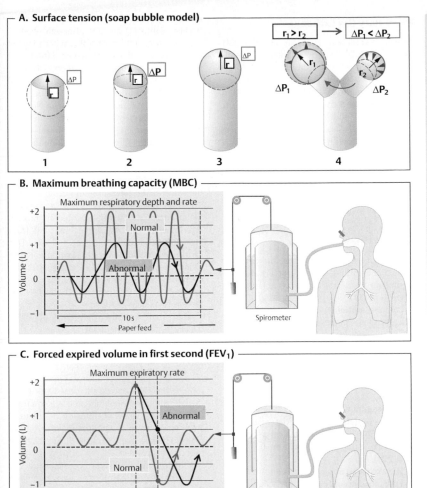

A. Surface tension (soap bubble model)

$$r_1 > r_2 \rightarrow \Delta P_1 < \Delta P_2$$

1 **2** **3** **4**

B. Maximum breathing capacity (MBC)

Maximum respiratory depth and rate

Normal

Abnormal

Volume (L)

10 s

Paper feed

Spirometer

C. Forced expired volume in first second (FEV$_1$)

Maximum expiratory rate

Abnormal

Normal

Volume (L)

1 s

Paper feed

1 Measurement

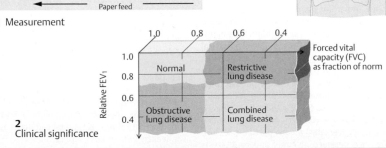

1.0   0.8   0.6   0.4

Forced vital capacity (FVC) as fraction of norm

Relative FEV$_1$

Normal

Restrictive lung disease

Obstructive lung disease

Combined lung disease

2 Clinical significance

Plate 5.7 Surface Tension, Lung Function Tests

Pulmonary Gas Exchange

Alveolar ventilation. Only the alveolar part (V_A) of the tidal volume (V_T) reaches the alveoli. The rest goes to dead space (V_D). It follows that $V_A = V_T - V_D$ (L) ($\rightarrow$ p. 114). Multiplying these volumes by the *respiratory rate* (**f** in min^{-1}) results in the respective ventilation, i.e., $\dot{V}_A$, $\dot{V}_E$ (or $\dot{V}_T$), and $\dot{V}_D$. Thus, $\dot{V}_A = \dot{V}_E - \dot{V}_D$ (L · min^{-1}). Since V_D is anatomically determined, $\dot{V}_D$ (= $V_D \cdot f$) rises with f. If, at a given total ventilation ($\dot{V}_E = V_T \cdot f$), the breathing becomes more frequent (f $\uparrow$) yet more shallow ($V_T \downarrow$), $\dot{V}_A$ will decrease because $\dot{V}_D$ increases.

Example: At a $\dot{V}_E$ of 8 L · min^{-1}, a V_D of 0.15 L and a normal respiratory rate f of 16 min^{-1} $\dot{V}_A = 5.6$ L · min^{-1} or 70% of $\dot{V}_E$. When f is doubled and V_T drops to one-half, $\dot{V}_A$ drops to 3.2 L · min^{-1} or 40% of V_T, although $\dot{V}_E$ (8 L · min^{-1}) remains unchanged.

Alveolar gas exchange can therefore decrease due to flat breathing and panting (e.g., due to a painful rib fracture) or artificial enlargement of V_D ($\rightarrow$ p. 134).

O_2 consumption ($\dot{V}_{O_2}$) is calculated as the *difference between* the *inspired* O_2 volume/time (= $\dot{V}_E \cdot F_{I_{O_2}}$, and the *expired* O_2 volume/time (= $\dot{V}_E \cdot F_{E_{O_2}}$. Therefore, $\dot{V}_{O_2} = \dot{V}_E$ (**$F_{I_{O_2}} - F_{E_{O_2}}$**). At rest, $\dot{V}_{O_2} \approx 8$ (0.21−0.17) = 0.32 L · min^{-1}.

The **eliminated CO_2 volume** ($\dot{V}_{CO_2}$) is calculated as $\dot{V}_T \cdot F_{E_{CO_2}}$ (≈ 0.26 L · min^{-1} at rest; $F_{I_{CO_2}} \approx 0$). $\dot{V}_{O_2}$ and $\dot{V}_{CO_2}$ increase about tenfold during strenuous physical work ($\rightarrow$ p. 74). The $\dot{V}_{CO_2}$ to $\dot{V}_{O_2}$ ratio is called the **respiratory quotient** (**RQ**), which depends on a person's nutritional state. RQ ranges from 0.7 to 1.0 ($\rightarrow$ p. 228).

The **exchange of gases** between the alveoli and the blood occurs by *diffusion*, as described by Fick's law of diffusion ($\rightarrow$ Eq. 1.7, p. 22,). The driving "force" for this diffusion is provided by the *partial pressure differences* between alveolar space and erythrocytes in pulmonary capillary blood ($\rightarrow$ **A**). The mean alveolar partial pressure of O_2 ($P_{A_{O_2}}$) is about 13.3 kPa (100 mmHg) and that of CO_2 ($P_{A_{CO_2}}$) is about 5.3 kPa (40 mmHg). The mean partial pressures in the "venous" blood of the pulmonary artery are approx. 5.3 kPa (40 mmHg) for O_2 ($P_{\bar{v}_{O_2}}$) and approx. 6.1 kPa (46 mmHg) for CO_2 ($P_{\bar{v}_{CO_2}}$). Hence, the mean **partial pressure difference** between alveolus and capillary is

about 8 kPa (60 mmHg) for O_2 and about 0.8 kPa (6 mmHg) for CO_2, although regional variation occurs ($\rightarrow$ p. 122). $P_{A_{O_2}}$ will rise when $P_{A_{CO_2}}$ falls (e.g., due to hyperventilation) and vice versa ($\rightarrow$ alveolar gas equation, p. 136).

O_2 diffuses about 1–2 μm from alveolus to bloodstream (**diffusion distance**). Under normal resting conditions, the blood in the pulmonary capillary is in contact with the alveolus for about 0.75 s. This **contact time** ($\rightarrow$ **A**) is long enough for the blood to equilibrate with the partial pressure of alveolar gases. The capillary blood is then *arterialized*. P_{O_2} and P_{CO_2} in **a**rterialized blood (Pa_{O_2} and Pa_{CO_2}) are about the same as the corresponding mean alveolar pressures ($P_{A_{O_2}}$ and $P_{A_{CO_2}}$). However, venous blood enters the arterialized blood through arteriovenous shunts in the lung and from bronchial and thebesian veins ($\rightarrow$ **B**). This *extra-alveolar shunt* as well as *ventilation–perfusion inequality* ($\rightarrow$ p. 122) make the Pa_{O_2} decrease from 13.3 kPa (after alveolar passage) to about 12.0 kPa (90 mmHg) in the aorta (Pa_{CO_2} increases slightly; $\rightarrow$ **A** and p. 107).

The small pressure difference of about 0.8 kPa is large enough for alveolar CO_2 exchange, since Krogh's diffusion coefficient K for CO_2 ($K_{CO_2} \approx 2.5 \cdot 10^{-16}$ $m^2 \cdot s^{-1} \cdot Pa^{-1}$ in tissue) is 23 times larger than that for O_2 ($\rightarrow$ p. 22). Thus, CO_2 diffuses much more rapidly than O_2. During physical work (high cardiac output), the contact time falls to a third of the resting value. If diffusion is impaired (see below), alveolar equilibration of O_2 partial pressure is less likely to occur during physical exercise than at rest.

Impairment of alveolar gas exchange can occur for several reasons: (a) when the blood flow rate along the alveolar capillaries decreases (e.g., due to *pulmonary infarction*; $\rightarrow$ **B2**), (b) if a *diffusion barrier exists* (e.g., due to a thickened alveolar wall, as in pulmonary edema; $\rightarrow$ **B3**), and (c) *if alveolar ventilation is reduced* (e.g., due to bronchial obstruction; $\rightarrow$ **B4**). Cases **B2** and **B3** lead to an *increase in functional dead space* ($\rightarrow$ p. 114); cases **B3** and **B4** lead to inadequate arterialization of the blood (*alveolar shunt*, i.e. non-arterialized blood mixing towards arterial blood). Gradual impairments of type **B2** and **B4** can occur even in healthy individuals ($\rightarrow$ p. 122).

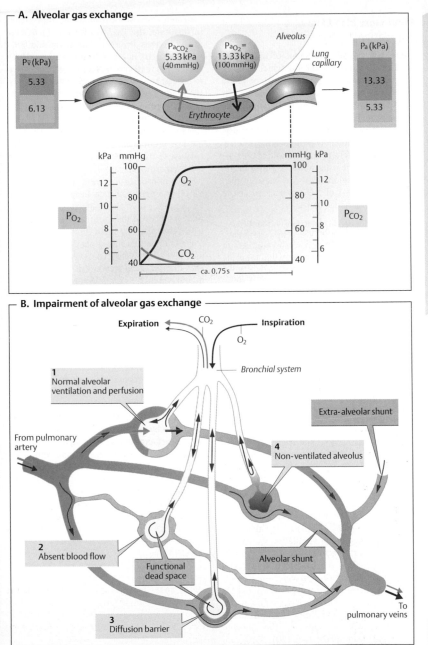

A. Alveolar gas exchange

$P\bar{v}$ (kPa)

5.33
6.13

Alveolus

$P_{aCO_2} =$ 5.33 kPa (40 mmHg)
$P_{aO_2} =$ 13.33 kPa (100 mmHg)

Lung capillary

Erythrocyte

P_a (kPa)

13.33
5.33

kPa mmHg
12 — 100
10 — 80
8 — 60
6 —
40

O_2

CO_2

mmHg kPa
100 — 12
80 — 10
60 — 8
— 6
40

P_{O_2}

P_{CO_2}

ca. 0.75 s

B. Impairment of alveolar gas exchange

Expiration ← CO_2 Inspiration →
O_2
Bronchial system

1 Normal alveolar ventilation and perfusion

From pulmonary artery

Extra-alveolar shunt

4 Non-ventilated alveolus

2 Absent blood flow

Functional dead space

Alveolar shunt

3 Diffusion barrier

To pulmonary veins

Plate 5.8 Pulmonary Gas Exchange

121

Pulmonary Blood Flow, Ventilation–Perfusion Ratio

Neglecting the slight amount of blood that reaches the lungs via the bronchial arteries, the mean **pulmonary perfusion ($\dot{Q}$)**, or blood flow to the lungs, is equal to the cardiac output (CO = 5–6 L/min). The **pulmonary arterial pressure** is about 25 mmHg in systole and 8 mmHg in diastole, with a mean ($\bar{P}$) of about 15 mmHg. $\bar{P}$ decreases to about 12 mmHg (P_{precap}) in the precapillary region (up to the origin of the pulmonary capillaries) and about 8 mmHg in the postcapillary region ($P_{postcap}$). These values apply to the areas of the lung located at the level of the pulmonary valve.

Uneven distribution of blood flow within the lung ($\rightarrow$ **A**). Due to the additive effect of hydrostatic pressure (up to 12 mmHg), P_{precap} increases in blood vessels below the pulmonary valves (near the base of the lung) when the chest is positioned upright. Near the apex of the lung, P_{precap} decreases in vessels above the pulmonary valve ($\rightarrow$ **A**, zone 1). Under these conditions, P_{precap} can even drop to subatmospheric levels, and the mean alveolar pressure (P_A) is atmospheric and can therefore cause extensive capillary compression ($P_A > P_{precap} > P_{postcap}$; $\rightarrow$ **A**). $\dot{Q}$ per unit of lung volume is therefore very small. In the central parts of the lung ($\rightarrow$ **A**, zone 2), luminal narrowing of capillaries can occur at their venous end, at least temporarily ($P_{precap} > P_A > P_{postcap}$), while the area near the base of the lung ($\rightarrow$ **A**, zone 3) is continuously supplied with blood ($P_{precap} > P_{postcap} > P_A$). $\dot{Q}$ per unit of lung volume therefore decreases from the apex of the lung to the base ($\rightarrow$ **A, B**, red line).

Uneven distribution of alveolar ventilation. Alveolar ventilation ($\dot{V}_A$) per unit of lung volume also increases from the apex to the base of the lungs due to the effects of gravity ($\rightarrow$ **B**, orange line), although not as much as $\dot{Q}$. Therefore, the **$\dot{V}_A/\dot{Q}$ ratio** decreases from the apex to the base of the lung ($\rightarrow$ **B**, green curve and top scale).

$\dot{V}_A/\dot{Q}$ imbalance. The mean $\dot{V}_A/\dot{Q}$ for the entire lung is 0.93 ($\rightarrow$ **C2**). This value is calculated from the mean alveolar ventilation $\dot{V}_A$ (ca. 5.6 L/min) and total perfusion $\dot{Q}$ (ca. 6 L/min), which is equal to the cardiac output (CO). Under extreme conditions in which one part of the lung is not ventilated at all, $\dot{V}_A/\dot{Q} = 0$ ($\rightarrow$ **C1**). In the other extreme in which blood flow is absent ($\dot{V}_A/\dot{Q}$ approaches infinity; $\rightarrow$ **C3**), fresh air conditions will prevail in the alveoli (functional dead space; $\rightarrow$ p. 120). $\dot{V}_A/\dot{Q}$ can vary tremendously—theoretically, from 0 to ∞. In this case, the P_{AO_2} will fluctuate between mixed venous $P\bar{v}_{O_2}$ and P_{IO_2} of (humidified) fresh air ($\rightarrow$ **D**). In a healthy upright lung, $\dot{V}_A/\dot{Q}$ decreases greatly (from 3.3 to 0.63) from apex to base at rest ($\rightarrow$ **B**, green line); P_{AO_2} (P_{ACO_2}) is therefore 17.6 (3.7) kPa in the "hyperventilated" lung apex, 13.3 (5.3) kPa in the normally ventilated central zone, and 11.9 (5.6) kPa in the hypoventilated lung base. These changes are less pronounced during physical exercise because $\dot{Q}$ also increases in zone 1 due to the corresponding increase in P_{precap}.

$\dot{V}_A/\dot{Q}$ imbalance *decreases the efficiency of the lungs for gas exchange*. In spite of the high P_{AO_2} at the apex of the lung (ca. 17.6 kPa; $\rightarrow$ **D**, right panel) and the fairly normal mean P_{AO_2} value, the relatively small $\dot{Q}$ fraction of zone 1 contributes little to the total $\dot{Q}$ of the pulmonary veins. In this case, $Pa_{O_2} < P_{AO_2}$ and an *alveolar–arterial O_2 difference* (AaD_{O_2}) exists (normally about 1.3 kPa). When a total arteriovenous shunt is present ($\dot{V}_A/\dot{Q} = 0$), even oxygen treatment will not help the patient, because it would not reach the pulmonary capillary bed ($\rightarrow$ **C1**).

Hypoxic vasoconstriction regulates alveolar perfusion and prevents the development of extreme $\dot{V}_A/\dot{Q}$ ratios. When the P_{AO_2} decreases sharply, receptors in the alveoli emit local signals that trigger constriction of the supplying blood vessels. This throttles shunts in poorly ventilated or non-ventilated regions of the lung, thereby routing larger quantities of blood for gas exchange to more productive regions.

$\dot{V}_A/\dot{Q}$ imbalance can cause severe complications in many lung diseases. In *shock lung*, for example, shunts can comprise 50% of $\dot{Q}$. Life-threatening lung failure can quickly develop if a concomitant pulmonary edema, alveolar diffusion barrier, or surfactant disorder exists ($\rightarrow$ p. 118).

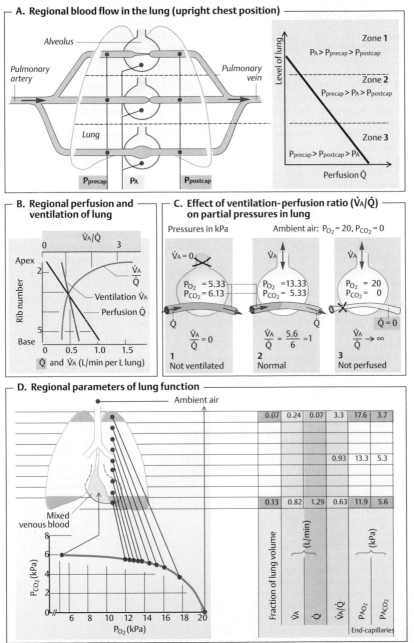

A. Regional blood flow in the lung (upright chest position)

Alveolus

Pulmonary artery

Pulmonary vein

Lung

P_{precap} P_A $P_{postcap}$

Zone 1
$P_A > P_{precap} > P_{postcap}$

Zone 2
$P_{precap} > P_A > P_{postcap}$

Zone 3
$P_{precap} > P_{postcap} > P_A$

Level of lung

Perfusion $\dot{Q}$

B. Regional perfusion and ventilation of lung

$\dot{V}_A/\dot{Q}$

Apex

Rib number

Base

$\dot{V}_A$ / $\dot{Q}$

Ventilation $\dot{V}_A$

Perfusion $\dot{Q}$

$\dot{Q}$ and $\dot{V}_A$ (L/min per L lung)

C. Effect of ventilation-perfusion ratio ($\dot{V}_A/\dot{Q}$) on partial pressures in lung

Pressures in kPa   Ambient air: $P_{O_2} = 20$, $P_{CO_2} = 0$

$\dot{V}_A = 0$

$P_{O_2} = 5.33$
$P_{CO_2} = 6.13$

$\dot{Q}$

$\dfrac{\dot{V}_A}{\dot{Q}} = 0$

1 Not ventilated

$\dot{V}_A$

$P_{O_2} = 13.33$
$P_{CO_2} = 5.33$

$\dot{Q}$

$\dfrac{\dot{V}_A}{\dot{Q}} = \dfrac{5.6}{6} \approx 1$

2 Normal

$\dot{V}_A$

$P_{O_2} = 20$
$P_{CO_2} = 0$

$\dot{Q} = 0$

$\dfrac{\dot{V}_A}{\dot{Q}} \rightarrow \infty$

3 Not perfused

D. Regional parameters of lung function

Ambient air

Mixed venous blood

P_{CO_2} (kPa)

P_{O_2} (kPa)

Fraction of lung volume	$\dot{V}_A$	$\dot{Q}$	$\dot{V}_A/\dot{Q}$	P_{AO_2}	P_{ACO_2}
	(L/min)			(kPa)	
0.07	0.24	0.07	3.3	17.6	3.7
			0.93	13.3	5.3
0.13	0.82	1.29	0.63	11.9	5.6

End-capillaries

(A, B, C, D after West et al.)

Plate 5.9   Pulmonary Blood Flow, $\dot{V}_A$–$\dot{Q}$ Ratio

CO₂ Transport in Blood

Carbon dioxide (CO₂) is the end-product of energy metabolism ($\to$ p. 228). CO_2 produced by cells of the body undergoes **physical dissolution** and diffuses into adjacent blood capillaries. A small portion of CO_2 in the blood remains dissolved, while the rest is **chemically bound** in form of HCO_3^- and carbamate residues of hemoglobin ($\to$ **A**, lower panel, blue arrows; $\to$ arteriovenous CO_2 difference given in the table). Circulating CO_2-loaded blood reaches the pulmonary capillaries via the right heart. CO_2 entering the pulmonary capillaries is released from the compounds ($\to$ **A**, red arrows), diffuses into the alveoli, and is expired into the atmosphere ($\to$ **A** and p. 106).

The enzyme **carbonic anhydrase** (carbonate dehydratase) catalyzes the reaction

$$HCO_3^- + H^+ \rightleftharpoons CO_2 + H_2O$$

in erythrocytes ($\to$ **A5, 7**). Because it accelerates the establishment of equilibrium, the short contact time (< 1 s) between red blood cells and alveolus or peripheral tissue is sufficient for the transformation $CO_2 \rightleftharpoons HCO_3^-$.

CO_2 diffusing from the **peripheral cells** ($\to$ **A**, bottom panel: "*Tissue*") increases P_{CO_2} (approx. 5.3 kPa = 40 mmHg in arterial blood) to a mean venous P_{CO_2} of about 6.3 kPa = 47 mmHg. It also increases the concentration of CO_2 dissolved in plasma. However, the major portion of the CO_2 diffuses into red blood cells, thereby increasing their content of dissolved CO_2. CO_2 (+ H_2O) within the cells is converted to **HCO_3^-** ($\to$ **A5, 2**) and **hemoglobin carbamate** ($\to$ **A3**). The HCO_3^- concentration in erythrocytes therefore becomes higher than in plasma. As a result, about three-quarters of the HCO_3^- ions exit the erythrocytes by way of an HCO_3^-/Cl^- antiporter. This **anion exchange** is also called *Hamburger shift* ($\to$ **A4**).

H⁺ ions are liberated when CO_2 in red cells circulating in the periphery is converted to HCO_3^- and hemoglobin (Hb) carbamate.

Bicarbonate formation:

$$CO_2 + H_2O \rightleftharpoons HCO_3^- + H^+, \tag{5.4}$$

Hemoglobin carbamate formation:

$$Hb-NH_2 + CO_2 \rightleftharpoons Hb-NH-COO^- + H^+. \tag{5.5}$$

Hemoglobin (Hb) is a key **buffer for H⁺** ions in the red cells ($\to$ **A6**; see also p. 140, "Non-bicarbonate buffers"). Since the removal of H⁺ ions

in reactions 5.4 and 5.5 prevents the rapid establishment of equilibrium, large quantities of CO_2 can be incorporated in HCO_3^- and Hb carbamate. Deoxygenated hemoglobin (Hb) can take up more H⁺ ions than oxygenated hemoglobin (Oxy-Hb) because Hb is a weaker acid ($\to$ **A**). This promotes CO_2 uptake in the peripheral circulation (*Haldane effect*) because of the simultaneous liberation of O_2 from erythrocytes, i.e. deoxygenation of Oxy-Hb to Hb.

In the **pulmonary capillaries**, these reactions proceed in the opposite direction ($\to$ **A**, top panel, red and black arrows). Since the P_{CO_2} in alveoli is lower than in venous blood, CO_2 diffuses into the alveoli, and reactions 5.4 and 5.5 proceed to the left. CO_2 is released from HCO_3^- and Hb carbamate whereby H⁺ ions (released from Hb) are bound in both reactions ($\to$ **A7, A8**), and the direction of HCO_3^-/Cl^- exchange reverses ($\to$ **A9**). Reoxygenation of Hb to Oxy-Hb in the lung promotes this process by increasing the supply of H⁺ ions (*Haldane effect*).

CO₂ distribution in blood (mmol/L blood, 1 mmol = 22.26 mL CO₂)

	Dis-solved CO₂	HCO₃⁻	Carba-mate	Total
Arterial blood:				
Plasma*	0.7	13.2	0.1	14.0
Erythrocytes**	0.5	6.5	1.1	8.1
Blood	1.2	19.7	1.2	22.1
Mixed venous blood:				
Plasma*	0.8	14.3	ca. 0.1	15.2
Erythrocytes**	0.6	7.2	1.4	9.2
Blood	1.4	21.5	1.5	24.4
Arteriovenous CO₂ difference in blood				
	0.2	1.8	0.3	2.3
Percentage of total arteriovenous difference				
	9%	78%	13%	100%

* Approx. 0.55 L plasma/L blood; ** ca. 0.45 L erythrocytes/L blood

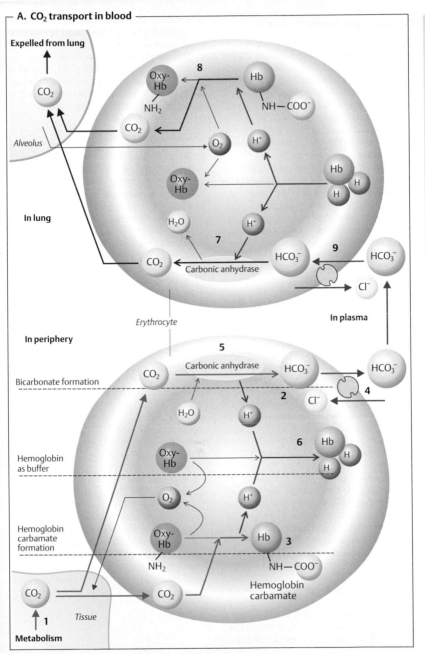

A. CO₂ transport in blood

Expelled from lung

In lung

In periphery

Bicarbonate formation

Hemoglobin as buffer

Hemoglobin carbamate formation

Metabolism

Plate 5.10 CO₂-Transport in Blood

CO_2 Binding in Blood, CO_2 in CSF

The **total carbon dioxide** concentration (= chemically bound "CO_2" + dissolved CO_2) of mixed venous blood is about 24–25 mmol/L; that of arterial blood is roughly 22–23 mmol/L. Nearly 90% of this is present as HCO_3^- ($\rightarrow$ **A**, right panel, and table on p. 124). The partial pressure of CO_2 (P_{CO_2}) is the chief factor that determines the CO_2 content of blood. The **CO_2 dissociation curve** illustrates how the total CO_2 concentration depends on P_{CO_2} ($\rightarrow$ **A**).

The concentration of **dissolved CO_2, [CO_2]**, in plasma is directly proportional to the P_{CO_2} in plasma and can be calculated as follows:

$$[CO_2] = \alpha_{CO_2} \cdot P_{CO_2} \text{ (mmol/L plasma}$$
$$\text{or mL/L plasma),} \qquad [5.6]$$

where α_{CO_2} is the (Bunsen) *solubility coefficient* for CO_2. At 37 °C,

$\alpha_{CO_2} = 0.225$ mmol $\cdot$ L^{-1} $\cdot$ kPa^{-1},

After converting the amount of CO_2 into volume CO_2 (mL = mmol $\cdot$ 22.26), this yields

$\alpha_{CO_2} = 5$ mL $\cdot$ L^{-1} $\cdot$ kPa^{-1}.

The *curve for dissolved* CO_2 is therefore linear ($\rightarrow$ **A**, green line).

Since the buffering and carbamate formation capacities of hemoglobin are limited, the relation between **bound "CO_2"** and P_{CO_2} is curvilinear. The *dissociation curve for total* CO_2 is calculated from the sum of dissolved and bound CO_2 ($\rightarrow$ **A**, red and violet lines).

CO_2 binding with hemoglobin depends on the degree of **oxygen saturation** (S_{O_2}) of hemoglobin. Blood completely saturated with O_2 is not able to bind as much CO_2 as O_2-free blood at equal P_{CO_2} levels ($\rightarrow$ **A**, red and violet lines). When venous blood in the lungs is loaded with O_2, the buffer capacity of hemoglobin and, consequently, the levels of chemical CO_2 binding decrease due to the Haldane effect ($\rightarrow$ p. 124). Venous blood is never completely void of O_2, but is always O_2-saturated to a certain degree, depending on the degree of O_2 extraction ($\rightarrow$ p. 130) of the organ in question. The S_{O_2} of mixed venous blood is about 0.75. The CO_2 dissociation curve for S_{O_2} = 0.75 therefore lies between those for S_{O_2} = 0.00 and 1.00 ($\rightarrow$ **A**, dotted line). In arterial blood, $P_{CO_2} \approx 5.33$ kPa and $S_{O_2} \approx 0.97$ ($\rightarrow$ **A**, point a). In mixed venous blood, $P_{CO_2} \approx 6.27$ kPa and $S_{O_2} \approx 0.75$ ($\rightarrow$ **A**, point $\triangledown$). The normal range of CO_2

dissociation is determined by connecting these two points by a line called *"physiologic CO_2 dissociation curve."*

The concentration ratio of HCO_3^- to dissolved CO_2 in plasma and red blood cells differs (about 20 : 1 and 12 : 1, respectively). This reflects the difference in the pH of plasma (7.4) and erythrocytes (ca. 7.2) ($\rightarrow$ p. 138ff.).

CO_2 in Cerebrospinal Fluid

Unlike HCO_3^- and H^+, CO_2 can cross the blood-cerebrospinal fluid (CSF) barrier with relative ease ($\rightarrow$ **B1** and p. 310). The P_{CO_2} in CSF therefore adapts quickly to **acute changes in the P_{CO_2} in blood**. CO_2-related (respiratory) pH changes in the body can be buffered by *non-bicarbonate buffers* (NBBs) only ($\rightarrow$ p. 144). Since the concentration of non-bicarbonate buffers in CSF is very low, an acute rise in P_{CO_2} (respiratory acidosis; $\rightarrow$ p. 144) leads to a relatively sharp decrease in the pH of CSF ($\rightarrow$ **B1**, pH $\downarrow\downarrow$). This decrease is registered by central chemosensors (or chemoreceptors) that adjust respiratory activity accordingly ($\rightarrow$ p. 132). (In this book, sensory receptors are called sensors in order to distinguish them from hormone and transmitter receptors.)

The concentration of non-bicarbonate buffers in blood (hemoglobin, plasma proteins) is high. When the CO_2 concentration increases, the liberated H^+ ions are therefore effectively buffered in the blood. The actual HCO_3^- concentration in blood then rises relatively slowly, to ultimately become higher than in the CSF. As a result, HCO_3^- diffuses (relatively slowly) into the CSF ($\rightarrow$ **B2**), resulting in a renewed increase in the pH of the CSF because the HCO_3^-/CO_2 ratio increases ($\rightarrow$ p. 140). This, in turn, leads to a reduction in respiratory activity (via central chemosensors), a process enhanced by renal compensation, i.e., a pH increase through HCO_3^- retention ($\rightarrow$ p. 144). By this mechanism, the body ultimately adapts to **chronic elevation in P_{CO_2}** — i.e., a chronically elevated P_{CO_2} will no longer represent a respiratory drive (cf. p. 132).

A. CO₂ dissociation curve

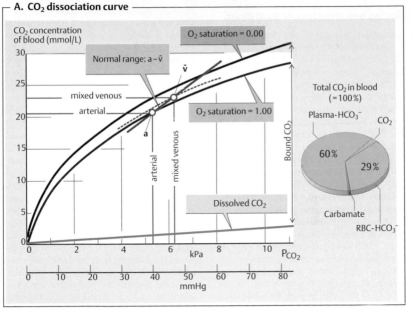

CO₂ concentration of blood (mmol/L)

O₂ saturation = 0.00

Normal range: a–v̄

ṽ

mixed venous

arterial

O₂ saturation = 1.00

a

Bound CO₂

Dissolved CO₂

arterial

mixed venous

kPa   P_CO₂

mmHg

Total CO₂ in blood (=100%)

Plasma-HCO₃⁻   CO₂

60%   29%

Carbamate

RBC-HCO₃⁻

B. Effect of CO₂ on pH of CSF

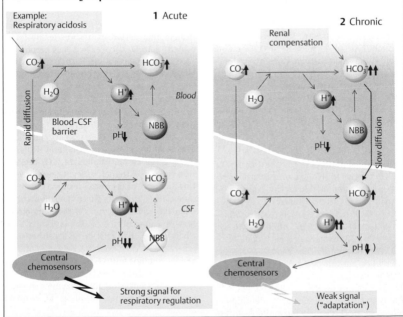

Example: Respiratory acidosis

1 Acute

CO₂↑   HCO₃⁻↑

H₂O   H⁺↑↑   NBB   *Blood*

pH↓

Rapid diffusion

Blood-CSF barrier

CO₂↑   HCO₃⁻

H₂O   H⁺↑↑   *CSF*

pH↓↓   NBB

Central chemosensors

Strong signal for respiratory regulation

2 Chronic

Renal compensation

CO₂↑   HCO₃⁻↑↑

H₂O   H⁺↑   NBB   *Blood*

pH↓

Slow diffusion

CO₂↑   HCO₃⁻↑

H₂O   H⁺↑↑

pH(↓)

Central chemosensors

Weak signal ("adaptation")

Plate 5.11 CO₂ Binding in Blood, CO₂ in CSF

Binding and Transport of O₂ in Blood

Hemoglobin (Hb) is the O_2-*carrying protein* of red blood cells (RBCs) (mol. mass: 64 500 Da). Hb is also involved in *CO_2 transport* and is an important blood *pH buffer* ($\rightarrow$ pp. 124 and 138ff.). Hb is a tetramer with 4 subunits (adults: 98%: $2\alpha + 2\beta$ = **HbA**; 2% $2\alpha + 2\delta$ = HbA₂), each with its own heme group. **Heme** consists of porphyrin and **Fe(II)**. Each of the four Fe(II) atoms (each linked with one histidine residue of Hb) binds reversibly with an O_2 molecule. This is referred to as **oxygenation** (not oxidation) of Hb to oxyhemoglobin (Oxy-Hb). The amount of O_2 which combines with Hb depends on the partial pressure of O_2 (P_{O_2}): **oxygen dissociation curve** ($\rightarrow$ **A**, red line). The curve has a sigmoid shape, because initially bound O_2 molecules change the conformation of the Hb tetramer (positive cooperativity) and thereby increase *hemoglobin-O_2 affinity*.

When fully saturated with O_2, 1 mol of tetrameric Hb combines with 4 mol O_2, i.e., 64 500 g of Hb combine with 4×22.4 L of O_2. Thus, 1 g Hb can theoretically transport 1.39 mL O_2, or 1.35 mL in vivo (Hüfner number). The total Hb concentration of the blood ([Hb]$_{total}$) is a mean 150 g/L ($\rightarrow$ p. 88), corresponding to a maximum O_2 concentration of 9.1 mmol/L or an O_2 fraction of 0.203 L O_2/L blood. This **oxygen-carrying capacity** is a function of [Hb]$_{total}$ ($\rightarrow$ **A**, yellow and purple curves as compared to the red curve).

The O_2 content of blood is virtually equivalent to the amount of O_2 bound by Hb since only 1.4% of O_2 in blood is dissolved at a P_{O_2} of 13.3 kPa ($\rightarrow$ **A**, orange line). The solubility coefficient (α_{O_2}), which is $10\,\mu mol \cdot [L\ of\ plasma]^{-1} \cdot kPa^{-1}$, is 22 times smaller than α_{CO_2} ($\rightarrow$ p. 126).

Oxygen saturation (S_{O_2}) is the fraction of Oxy-Hb relative to [Hb]$_{total}$, or the ratio of *actual O_2 concentration/ O_2-carrying capacity*. At normal P_{O_2} in arterial blood (e.g., Pa$_{O_2}$ = 12.6 kPa or 95 mmHg), S_{O_2} will reach a *saturation plateau* at approx. 0.97, while S_{O_2} will still amount to 0.73 in mixed venous blood (P$\bar{v}_{O_2}$ = 5.33 kPa or 40 mmHg). The venous S_{O_2} values in different organs can vary greatly ($\rightarrow$ p. 130).

O_2 dissociation is independent of total Hb if plotted as a function of S_{O_2} ($\rightarrow$ **B**). Changes in O_2 affinity to Hb can then be easily identified as **shifting of the O_2 dissociation curve**. A shift to

the right signifies an affinity decrease, and a shift to the left signifies an affinity increase, resulting in flattening and steepening, respectively, of the initial part of the curve. **Shifts to the left** are caused by increases in pH (with or without a P_{CO_2} decrease) and/or decreases in P_{CO_2}, temperature and 2,3-bisphosphoglycerate (BPG; normally 1 mol/mol Hb tetramer). **Shifts to the right** occur due to decreases in pH and/or increases in P_{CO_2}, temperature and 2,3-BPG ($\rightarrow$ **B**). The **half-saturation pressure** (**P₀.₅** or P_{50}) of O_2 ($\rightarrow$ **B**, dotted lines) is the P_{O_2} at which S_{O_2} is 0.5 or 50%. The $P_{0.5}$, which is normally 3.6 kPa or 27 mmHg, is a measure of shifting to the right ($P_{0.5} \uparrow$) or left ($P_{0.5} \downarrow$). Displacement of the O_2 dissociation curve due to changes in pH and P_{CO_2} is called the **Bohr effect**. A shift to the right means that, in the periphery (pH ↓, $P_{CO_2} \uparrow$), larger quantities of O_2 can be absorbed from the blood without decreasing the P_{O_2}, which is the driving force for O_2 diffusion ($\rightarrow$ **B**, broken lines). A higher affinity for O_2 is then re-established in the pulmonary capillaries (pH ↑, $P_{CO_2} \downarrow$). A shift to the left is useful when the Pa$_{O_2}$ is decreased (e.g., in altitude hypoxia), a situation where arterial S_{O_2} lies to the left of the S_{O_2} plateau.

Myoglobin is an Fe(II)-containing muscle protein that serves as a short-term storage molecule for O_2 ($\rightarrow$ p. 72). As it is monomeric (no positive cooperativity), its O_2 dissociation curve at low P_{O_2} is much steeper than that of HbA ($\rightarrow$ **C**). Since the O_2 dissociation curve of **fetal Hb** ($2\alpha + 2\gamma$ = HbF) is also steeper, S_{O_2} values of 45 to 70% can be reached in the fetal umbilical vein despite the low P_{O_2} (3–4 kPa or 22–30 mmHg) of maternal placental blood. This is sufficient, because the fetal [Hb]$_{total}$ is 180 g/L. The **carbon monoxide (CO) dissociation curve** is extremely steep. Therefore, even tiny amounts of CO in the respiratory air will dissociate O_2 from Hb. This can result in *carbon monoxide poisoning* ($\rightarrow$ **C**). **Methemoglobin**, Met-Hb (normally 1% of Hb), is formed from Hb by oxidation of Fe(II) to Fe(III) either spontaneously or via exogenous oxidants. Met-Hb cannot combine with O_2 ($\rightarrow$ **C**). *Methemoglobin reductase* reduces Fe(III) of Met-Hb back to Fe(II); deficiencies of this enzyme can cause methemoglobinemia, resulting in neonatal anoxia.

Plate 5.12 Binding and Transport of O_2 in Blood

A. O_2 dissociation curve: O_2-carrying capacity

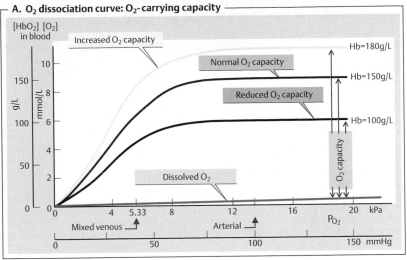

B. O_2 dissociation curve: O_2 saturation

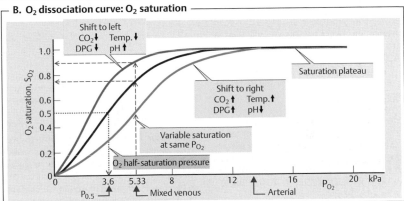

C. O_2 and carbon monoxide (CO) dissociation curves

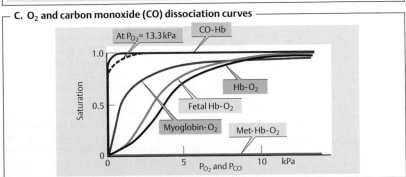

Internal (Tissue) Respiration, Hypoxia

O_2 **diffuses** from peripheral blood to adjacent tissues and CO_2 in the opposite direction ($\rightarrow$ pp. 20ff. and 106). Since CO_2 diffuses much faster ($\rightarrow$ p. 120), O_2 diffusion is the limiting factor. Sufficient O_2 delivery is ensured by a dense capillary network with a gas *exchange area* of about 1000 m². The *diffusion distance* ($\rightarrow$ R in **A**) is only 10–25 µm. The *driving force* for diffusion is the difference in partial pressures of oxygen (ΔP_{O_2}) in capillary blood and mitochondria, where the P_{O_2} must not fall below 0.1 kPa $\approx$ 1 mmHg. Since P_{O_2} decreases with distance parallel and perpendicular to the course of capillaries, the O_2 supply to cells at the venous end far away from the capillaries (large R) is lowest, as shown using *Krogh's cylinder model* ($\rightarrow$ **A1**). Since these cells are also the first to be affected by oxygen deficiency (*hypoxia*), this is sometimes called the "lethal corner" ($\rightarrow$ **A2**).

Using *Fick's principle* ($\rightarrow$ p. 106), **oxygen consumption** of a given organ, $\dot{V}_{O_2}$ (in L/min), is calculated as the difference between the **a**rterial supply ($\dot{Q} \cdot [O_2]a$) and non-utilized **v**enous O_2 volume/time ($\dot{Q} \cdot [O_2]v$), where $\dot{Q}$ is rate of blood flow in the organ (L/min) and $[O_2]$ is the oxygen fraction (L O_2/L blood):

$$\dot{V}_{O_2} = \dot{Q} \, ([O_2]_a - [O_2]_v) \qquad [5.7]$$

To meet **increased O_2 demands**, $\dot{Q}$ *can therefore be increased* by **vasodilatation** in the organ in question and/or by **raising the oxygen extraction (E_{O_2})**. E_{O_2} describes the O_2 consumption in the organ (= $\dot{Q} \,([O_2]a - [O_2]v)$; see Eq. 5.7) relative to the arterial O_2 supply ($\dot{Q} \cdot [O_2]a$). Since $\dot{Q}$ can be simplified,

$$E_{O_2} = ([O_2]_a - [O_2]_v)/ [O_2]_a \qquad [5.8]$$

E_{O_2} varies according to the type and function of the organ under resting conditions: skin 0.04 (4%), kidney 0.07; brain, liver and resting skeletal muscle ca. 0.3, myocardium 0.6. The E_{O_2} of muscle during strenuous exercise can rise to 0.9. Skeletal muscle can therefore meet **increased O_2 demands** by raising the E_{O_2} (0.3 $\Rightarrow$ 0.9), as can myocardial tissue to a much smaller extent ($\rightarrow$ p. 210).

Hypoxia. An abnormally reduced O_2 supply to tissue is classified as follows:

1. *Hypoxic hypoxia* ($\rightarrow$ **A2, B1**): an insufficient O_2 supply reaches the blood due, for example, to decreased atmospheric P_{O_2} at high altitudes ($\rightarrow$ p. 136), reduced alveolar ventilation, or impaired alveolar gas exchange.

2. *Anemic hypoxia* ($\rightarrow$ **B2**): reduced O_2-carrying capacity of blood ($\rightarrow$ p. 128), e.g., due to decreased total Hb in iron deficiency anemia ($\rightarrow$ p. 90).

3. *Stagnant* or *ischemic hypoxia* ($\rightarrow$ **B3**): insufficient O_2 reaches the tissue due to reduced blood flow ($\dot{Q} \downarrow$). The cause can be systemic (e.g., heart failure) or local (e.g., obstructed artery). The reduction of blood flow must be compensated for by a rise in E_{O_2} to maintain an adequate O_2 delivery (see Eq. 5.7). This is not the case in hypoxic and anemic hypoxia. The influx and efflux of substrates and metabolites is also impaired in stagnant hypoxia. Anaerobic glycolysis ($\rightarrow$ p. 72) is therefore of little help because neither the uptake of glucose nor the discharge of H^+ ions dissociated from lactic acid is possible.

4. Hypoxia can also occur when the *diffusion distance* is increased due to tissue thickening without a corresponding increase in the number of blood capillaries. This results in an insufficient blood supply to cells lying outside the O_2 supply radius (R) of the Krogh cylinder ($\rightarrow$ **A**).

5. *Histotoxic* or *cytotoxic hypoxia* occurs due to impaired utilization of O_2 by the tissues despite a sufficient supply of O_2 in the mitochondria, as observed in *cyanide poisoning*. Cyanide (HCN) blocks oxidative cellular metabolism by inhibiting cytochromoxidase.

Brain tissue is extremely susceptible to hypoxia, which can cause critical damage since dead nerve cells generally cannot be replaced. **Anoxia**, or a total lack of oxygen, can occur due to heart or respiratory failure. The cerebral survival time is thus the limiting factor for overall survival. Unconsciousness occurs after only 15 s of anoxia, and irreparable brain damage occurs if anoxia lasts for more than 3 min or so.

Cyanosis is a bluish discoloration of the skin, lips, nails, etc. due to excessive arterial deoxyhemoglobin (> 50 g/L). Cyanosis is a sign of hypoxia in individuals with normal or only moderately reduced total Hb levels. When total Hb is extremely low, O_2 deficiencies (anemic hypoxia) can be life-threatening, even in the absence of cyanosis. Cyanosis can occur in absence of significant hypoxia when the Hb level is elevated.

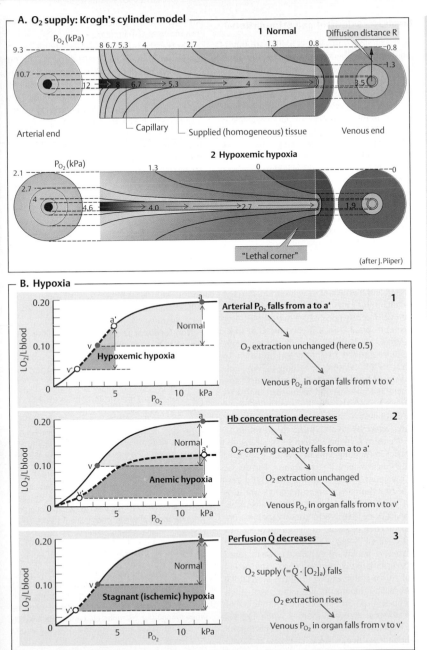

A. O₂ supply: Krogh's cylinder model

1 Normal

P_{O_2} (kPa)

Diffusion distance R

9.3 — 8 6.7 5.3 — 4 — 2.7 — 1.3 — 0.8 — 0.8
10.7 — — 1.3
12 → 8 → 6.7 → 5.3 → 4 → 3.5
Arterial end — Capillary — Supplied (homogeneous) tissue — Venous end

2 Hypoxemic hypoxia

P_{O_2} (kPa)

2.1 — 1.3 — 0 — 0
2.7 —
4 → 4.6 → 4.0 → 2.7 → 1.9

"Lethal corner"

(after J. Piiper)

B. Hypoxia

1 **Arterial P_{O_2} falls from a to a'**

↓

O_2 extraction unchanged (here 0.5)

↓

Venous P_{O_2} in organ falls from v to v'

LO_2/L blood — Normal — Hypoxemic hypoxia — P_{O_2} — kPa

2 **Hb concentration decreases**

↓

O_2-carrying capacity falls from a to a'

↓

O_2 extraction unchanged

↓

Venous P_{O_2} in organ falls from v to v'

LO_2/L blood — Normal — Anemic hypoxia — P_{O_2} — kPa

3 **Perfusion $\dot{Q}$ decreases**

↓

O_2 supply $(=\dot{Q} \cdot [O_2]_a)$ falls

↓

O_2 extraction rises

↓

Venous P_{O_2} in organ falls from v to v'

LO_2/L blood — Normal — Stagnant (ischemic) hypoxia — P_{O_2} — kPa

Plate 5.13 Internal (Tissue) Respiration, Hypoxia

131

Respiratory Control and Stimulation

The respiratory muscles ($\rightarrow$ p. 108) are innervated by nerve fibers extending from the cervical and thoracic medulla (C4–C8 and T1–T7). The most important **control centers** are located in the *medulla oblongata* and cervical *medulla* (C1–C2), where interactive *inspiratory and expiratory neurons* on different levels ($\rightarrow$ **A1,** red and green areas). The network of these spatially separate neuron groups form a **rhythm generator (respiratory "center")** where respiratory rhythm originates ($\rightarrow$ **A1**). The neuron groups are triggered alternately, resulting in rhythmic inspiration and expiration. They are activated in a tonic (non-rhythm-dependent) manner by the *formatio reticularis*, which receives signals from *respiratory stimulants* in the periphery and higher centers of the brain.

Respiratory sensors or receptors are involved in respiratory control circuits ($\rightarrow$ p. 4). *Central* and *peripheral chemosensors* on the medulla oblongata and in the arterial circulation continuously register gas partial pressures in cerebrospinal fluid (CSF) and blood, respectively, and *mechanosensors* in the chest wall respond to stretch of intercostal muscles to modulate the depth of breathing ($\rightarrow$ **A2**). *Pulmonary stretch sensors* in the tracheal and bronchial walls respond to marked increases in lung volume, thereby limiting the depth of respiration in humans (*Hering–Breuer reflex*). *Muscle spindles* ($\rightarrow$ p. 318) in the respiratory muscles also respond to changes in airway resistance in the lung and chest wall.

Chemical respiratory stimulants. The extent of involuntary ventilation is mainly determined by the partial pressures of O_2 and CO_2 and the pH of blood and CSF. Chemosensors respond to any changes in these variables. **Peripheral chemosensors** in the glomera aortica and carotica ($\rightarrow$ **A3**) register changes in the arterial P_{O_2}. If it falls, they stimulate an increase in ventilation via the vagus (X) and glossopharyngeal nerves (IX) until the arterial P_{O_2} rises again. This occurs, for example, at high altitudes ($\rightarrow$ p. 136). The impulse frequency of the sensors increases sharply when the P_{O_2} drops below 13 kPa or 97 mmHg (**peripheral ventilatory drive**). These changes are even stronger when P_{CO_2} and/or the H$^+$ concentration in blood also increase.

Central chemosensors in the medulla react to CO_2 and H$^+$ increases (= pH decrease) in the CSF ($\rightarrow$ **A4** and p. 126). Ventilation is then increased until P_{CO_2} and the H$^+$ concentration in blood and CSF decrease to normal values. This mostly **central respiratory drive** is very effective in responding to acute changes. An increase in arterial P_{CO_2} from, say, 5 to 9 kPa increases the total ventilation $\dot{V}_E$ by a factor of ten, as shown in the *CO_2 response curve* ($\rightarrow$ **A6**).

When a **chronic rise in P_{CO_2}** occurs, the previously increased central respiratory drive decreases ($\rightarrow$ p. 126). If O_2 supplied by artificial respiration tricks the peripheral chemosensors into believing that there is adequate ventilation, the residual peripheral respiratory drive will also be in jeopardy.

During **physical work** ($\rightarrow$ **A5**), the total ventilation increases due to (a) co-innervation of the respiratory centers (by collaterals of cortical efferent motor fibers) and (b) through impulses transmitted by proprioceptive fibers from the muscles.

Non-feedback sensors and stimulants also play an important role in modulating the basic rhythm of respiration. They include

◆ *Irritant sensors* in the bronchial mucosa, which quickly respond to lung volume decreases by increasing the respiratory rate (deflation reflex or Head's reflex), and to dust particles or irritating gases by triggering the cough reflex.

◆ *J sensors* of free C fiber endings on alveolar and bronchial walls; these are stimulated in pulmonary edema, triggering symptoms such as apnea and lowering the blood pressure.

◆ *Higher central nervous centers* such as the cortex, limbic system, hypothalamus or pons. They are involved in the expression of emotions like fear, pain and joy; in reflexes such as sneezing, coughing, yawning and swallowing; and in voluntary control of respiration while speaking, singing, etc.

◆ *Pressosensors* ($\rightarrow$ p. 214), which are responsible for increasing respiration when the blood pressure decreases.

◆ *Heat and cold sensors* in the skin and thermoregulatory center. Increases (fever) and decreases in body temperature lead to increased respiration.

◆ *Certain hormones* also help to regulate respiration. Progesterone, for example, increases respiration in the second half of the menstrual cycle and during pregnancy.

Plate 5.14

A. Respiratory control and stimulation

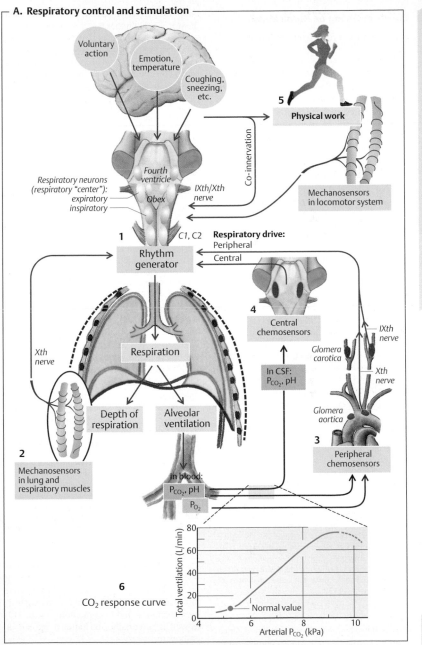

Voluntary action

Emotion, temperature

Coughing, sneezing, etc.

5 **Physical work**

Co-innervation

Respiratory neurons (respiratory "center"):
expiratory
inspiratory

Fourth ventricle

Obex

IXth/Xth nerve

Mechanosensors in locomotor system

1 C1, C2

Rhythm generator

Respiratory drive:
Peripheral
Central

4

Central chemosensors

In CSF: P_{CO_2}, pH

Xth nerve

Respiration

Glomera carotica

IXth nerve

Xth nerve

Depth of respiration

Alveolar ventilation

Glomera aortica

2

Mechanosensors in lung and respiratory muscles

in blood:
P_{CO_2}, pH
P_{O_2}

3

Peripheral chemosensors

6

CO_2 response curve

Total ventilation (L/min)

80
60
40
20
0

Normal value

4   6   8   10

Arterial P_{CO_2} (kPa)

Effects of Diving on Respiration

Diving creates a problem for respiration due to the lack of normal ambient air supply and to higher the outside pressures exerted on the body. The total pressure on the body underwater is equal to the water pressure (98 kPa or 735 mmHg for each 10 m of water) plus the atmospheric pressure at the water surface.

A **snorkel** can be used when diving just below the water surface ($\rightarrow$ **A**), but it increases dead space ($\rightarrow$ pp. 114 and 120), making it harder to breathe. The additional pressure load from the water on chest and abdomen must also be overcome with each breath.

The depth at which a snorkel can be used is limited 1) because an intolerable increase in dead space or airway resistance will occur when using an extremely long or narrow snorkel, and 2) because the water pressure at lower depths will prevent inhalation. The maximum suction produced on inspiration is about 11 kPa, equivalent to 112 cm H_2O (peak inspiratory pressure, $\rightarrow$ p. 116). Inspiration therefore is no longer possible at aquatic depths of about 112 cm or more due to the risk of hypoxic anoxia ($\rightarrow$ **A**).

Scuba diving equipment (scuba = **s**elf-**c**ontained **u**nderwater **b**reathing **a**pparatus) is needed to breathe at lower depths (up to about 70 m). The inspiratory air pressure (from pressurized air cylinders) is automatically adjusted to the water pressure, thereby permitting the diver to breathe with normal effort.

However, the additional water pressure increases the partial pressure of **nitrogen** P_{N_2} ($\rightarrow$ **B**), resulting in higher concentrations of dissolved N_2 in the blood. The pressure at a depth of 60 meters is about seven times higher than at the water surface. The pressure decreases as the diver returns to the water surface, but the additional N_2 does not remain dissolved. The diver must therefore ascend slowly, in gradual stages so that the excess N_2 can return to and be expelled from the lungs. Resurfacing too quickly would lead to the development of N_2 bubbles in tissue (pain!) and blood, where they can cause obstruction and embolism of small blood vessels. This is called **decompression sickness** or **caisson disease** ($\rightarrow$ **B**). Euphoria (N_2 narcosis?), also called *rapture of the deep*, can occur when diving at depths of over 40 to 60 meters. Oxygen toxicity can occur at depths of 75 m or more ($\rightarrow$ p. 136).

When **diving unassisted**, i.e., simply by holding one's breath, P_{CO_2} in the blood rises, since the CO_2 produced by the body is not exhaled. Once a certain P_{CO_2} has been reached, chemosensors ($\rightarrow$ p. 132) trigger a sensation of shortness of breath, signaling that it is time to resurface.

To delay the time to resurface, it is possible to lower the P_{CO_2} in blood by **hyperventilating before diving**. Experienced divers use this trick to stay under water longer. The course of alveolar partial pressures over time and the direction of alveolar gas exchange while diving (depth: 10 m; duration 40 s) is shown in **C**: Hyperventilating before a dive reduces the P_{CO_2} (solid green line) and slightly increases the P_{O_2} (red line) in the alveoli (and in blood). Diving at a depth of 10 m doubles the pressure on the chest and abdominal wall. As a result, the partial pressures of gases in the alveoli (P_{CO_2}, P_{O_2}, P_{N_2}) increase sharply. Increased quantities of O_2 and CO_2 therefore diffuse from the alveoli into the blood. once the P_{CO_2} in blood rises to a certain level, the body signals that it is time to resurface. If the diver resurfaces at this time, the P_{O_2} in the alveoli and blood drops rapidly (O_2 consumption + pressure decrease) and the alveolar O_2 exchange stops. Back at the water surface, the P_{O_2} reaches a level that is just tolerable. If the diver excessively hyperventilates before the dive, the signal to resurface will come too late, and the P_{O_2} will drop to zero (**anoxia**) before the person reaches the water surface, which can result in unconsciousness and drowning ($\rightarrow$ **C**, dotted lines).

Barotrauma. The increased pressure associated with diving leads to compression of air-filled organs, such as the lung and middle ear. Their gas volumes are compressed to $1/2$ their normal size at water depths of 10 m, and to $1/4$ at depths of 30 m.

The missing volume of air in the lungs is automatically replaced by the scuba, but not that of the middle ear. The middle ear and throat are connected by the Eustachian tube, which is open only at certain times (e.g., when swallowing) or not at all (e.g., in pharyngitis). If volume loss in the ear is not compensated for during a dive, the increasing water pressure in the outer auditory canal distends the eardrum, causing pain or even eardrum rupture. As a result, cold water can enter the middle ear and impair the organ of equilibrium, leading to nausea, dizziness, and disorientation. This can be prevented by pressing air from the lungs into the middle ear by holding the nose and blowing with the mouth closed.

The air in air-filled organs expand when the diver ascends to the water surface. Resurfacing too quickly, i.e., without expelling air at regular intervals, can lead to complications such as lung laceration and pneumothorax ($\rightarrow$ p. 110) as well as potentially fatal hemorrhage and air embolism.

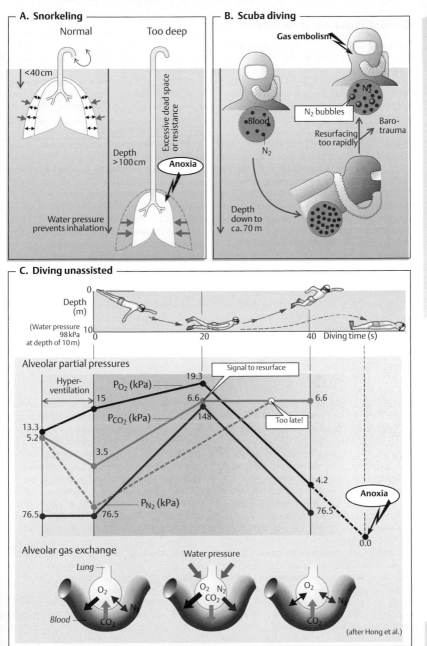

A. Snorkeling

Normal　　　　Too deep

<40 cm

Depth >100 cm

Excessive dead space or resistance

Anoxia

Water pressure prevents inhalation

B. Scuba diving

Gas embolism

N_2 bubbles

N_2

Blood

N_2

Resurfacing too rapidly

Baro-trauma

Depth down to ca. 70 m

C. Diving unassisted

Depth (m)

(Water pressure 98 kPa at depth of 10 m)

0

10

0　　　　20　　　　40　Diving time (s)

Alveolar partial pressures

Hyper-ventilation

Signal to resurface

P_{O_2} (kPa)　19.3

15

6.6

P_{CO_2} (kPa)　148

13.3

5.2

3.5

6.6

Too late!

P_{N_2} (kPa)

4.2

76.5　76.5

76.5

Anoxia

0.0

Alveolar gas exchange

Water pressure

Lung

O_2　　N_2

O_2　N_2
CO_2

O_2　N_2

Blood

CO_2

CO_2

(after Hong et al.)

Plate 5.15　**Effects of Diving on Respiration**

135

Effects of High Altitude on Respiration

At sea level, the average barometric pressure (P_B) ≈ 101 kPa (760 mmHg), the O_2 fraction in ambient air (F_{IO_2}) is 0.209, and the inspiratory partial pressure of O_2 (P_{IO_2}) ≈ 21 kPa (→ p. 106). However, P_B decreases with increasing altitude (h, in km):

$$P_B \text{ (at h)} = P_B \text{ (at sea level)} \cdot e^{-0.127 \cdot h} \qquad [5.9]$$

This results in a drop in P_{IO_2} (→ **A**, column 1), alveolar P_{O_2} (P_{AO_2}) and arterial P_{O_2} (P_{aO_2}). The P_{AO_2} at sea level is about 13 kPa (→ **A**, column 2). P_{AO_2} is an important measure of oxygen supply. If the P_{AO_2} falls below a critical level (ca. 4.7 kPa = 35 mmHg), hypoxia (→ p. 130) and impairment of cerebral function will occur. The *critical P_{AO_2}* would be reached at heights of about 4000 m above sea level during normal ventilation (→ **A**, dotted line in column 2). However, the low P_{aO_2} triggers chemosensors (→ p. 132) that stimulate an increase in total ventilation ($\dot{V}_E$); this is called **O_2 deficiency ventilation** (→ **A**, column 4). As a result, larger volumes of CO_2 are exhaled, and the P_{ACO_2} and P_{aCO_2} decrease (see below). As described by the *alveolar gas equation*,

$$P_{AO_2} = P_{IO_2} - \frac{P_{ACO_2}}{RQ} \qquad [5.10]$$

where RQ is the respiratory quotient (→ pp. 120 and 228), any fall in P_{ACO_2} will lead to a rise in the P_{AO_2}. O_2 deficiency ventilation stops the P_{AO_2} from becoming critical up to altitudes of about 7000 m (*altitude gain*, → **A**).

The maximal increase in ventilation (≈ 3 × resting rate) during acute O_2 deficiency is relatively small compared to the increase (≈ 10 times the resting rate) during strenuous physical exercise at normal altitudes (→ p. 74, C3) because increased ventilation at high altitudes reduces the P_{aCO_2} (= **hyperventilation**, → p. 108), resulting in the development of **respiratory alkalosis** (→ p. 144). Central chemosensors (→ p. 132) then emit signals to lower the respiratory drive, thereby counteracting the signals from O_2 chemosensors to increase the respiratory drive. As the mountain climber adapts, respiratory alkalosis is compensated for by increased renal excretion of HCO_3^- (→ p. 144). This helps return the pH of the blood toward normal, and the O_2 deficiency-related increase in respiratory drive can now prevail. Stimulation of O_2 chemosensors at high altitudes also leads to an increase in the heart rate and a corresponding *increase in cardiac output*, thereby increasing the O_2 supply to the tissues.

High altitude also stimulates **erythropoiesis** (→ p. 88ff.). Prolonged exposure to high altitudes increases the hematocrit levels, although this is limited by the corresponding rise in blood viscosity (→ pp. 92, 188).

Breathing oxygen from pressurized O_2 cylinders is necessary for survival at altitudes above 7000 m, where P_{IO_2} is almost as high as the barometric pressure P_B (→ **A**, column 3). The critical P_{AO_2} level now occurs at an altitude of about 12 km with normal ventilation, and at about 14 km with increased ventilation. Modern long-distance planes fly slightly below this altitude to ensure that the passengers can survive with an oxygen mask in case the cabin pressure drops unexpectedly.

Survival at altitudes above 14 km is not possible without pressurized chambers or pressurized suits like those used in space travel. Otherwise, the body fluids would begin to boil at altitudes of 20 km or so (→ **A**), where P_B is lower than water vapor pressure at body temperature (37 °C).

Oxygen Toxicity

Hyperoxia occurs when P_{IO_2} is above normal (> 22 kPa or 165 mmHg) due to an increased O_2 fraction (*oxygen therapy*) or to an overall pressure increase with a normal O_2 fraction (e.g. in diving, → p. 134). The degree of O_2 toxicity depends on the P_{IO_2} level (critical: ca. 40 kPa or 300 mmHg) and duration of hyperoxia. Lung dysfunction (→ p. 118, *surfactant deficiency*) occurs when a P_{IO_2} of about 70 kPa (525 mmHg) persists for several days or 200 kPa (1500 mmHg) for 3–6 hours. Lung dysfunction initially manifests as *coughing* and *painful breathing*. *Seizures* and *unconsciousness* occur at P_{IO_2} levels above 220 kPa (1650 mmHg), e.g., when diving at a depth of about 100 m using pressurized air.

Newborns will go blind if exposed to P_{IO_2} levels much greater than 40 kPa (300 mmHg) for long periods of time (e.g., in an incubator), because the vitreous body then opacifies.

A. Respiration at high altitudes (without acclimatization)

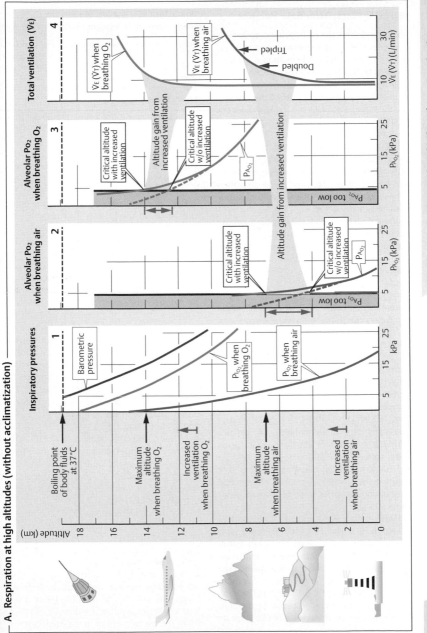

Plate 5.16 **Effects of High Altitude on Respiration**

137

6 Acid–Base Homeostasis

pH, pH Buffers, Acid–Base Balance

The **pH** indicates the hydrogen ion activity or the "effective" H⁺ concentration of a solution (H⁺ activity = $f_H \cdot$ [H⁺], where square brackets mean concentration; $\rightarrow$ p. 378), where

$$pH = -\log(f_H \cdot [H^+]) \qquad [6.1]$$

In healthy individuals, the **pH of the blood** is usually a mean pH **7.4** (see p. 142 for normal range), corresponding to H⁺ activity of about 40 nmol/L. The maintenance of a constant pH is important for human survival. Large deviations from the norm can have detrimental effects on metabolism, membrane permeability, and electrolyte distribution. Blood pH values below 7.0 and above 7.8 are not compatible with life.

Various **pH buffers** are responsible for maintaining the body at a constant pH ($\rightarrow$ p. 379). One important buffer for blood and other body fluids is the **bicarbonate/carbon dioxide (HCO_3^-/CO_2) buffer system**:

$$CO_2 + H_2O \rightleftharpoons HCO_3^- + H^+. \qquad [6.2]$$

The *pK_a value* ($\rightarrow$ p. 378f.) determines the prevailing *concentration ratio of the buffer base and buffer acid* ([HCO_3^-] and [CO_2], respectively in Eq. 6.2) at a given pH (*Henderson–Hasselbalch equation*; $\rightarrow$ **A**).

The primary **function of the CO_2/HCO_3^- buffer system** in blood is to buffer H⁺ ions. However, this system is especially important because the concentrations of the two buffer components can be modified largely independent of each other: [CO_2] by respiration and [HCO_3^-] by the liver and kidney ($\rightarrow$ **A**; see also p. 174). It is therefore classified as an **open buffer system** ($\rightarrow$ p. 140).

Hemoglobin in red blood cells (320 g Hb/L erythrocytes! $\rightarrow$ MCHC, p. 89 C), the second most important buffer in blood, is a **non-bicarbonate buffer**.

$$HbH \rightleftharpoons Hb^- + H^+ \qquad [6.3]$$
$$Oxy\text{-}HbH \rightleftharpoons Oxy\text{-}Hb^- + H^+ \qquad [6.4]$$

The relatively acidic oxyhemoglobin anion (Oxy-Hb⁻) combines with fewer H⁺ ions than deoxygenated Hb⁻, which is less acidic (see also p. 124). H⁺ ions are therefore liberated upon oxygenation of Hb to Oxy-Hb in the lung. Reaction 6.2 therefore proceeds to the left, thereby promoting the release of CO_2 from its bound forms ($\rightarrow$ p. 124). This, in turn, increases the pulmonary elimination of CO_2.

Other non-bicarbonate buffers of the blood include *plasma proteins* and inorganic *phosphate* ($H_2PO_4^- \rightleftharpoons H^+ + HPO_4^{2-}$) as well as organic phosphates (in red blood cells). Intracellular organic and inorganic substances in various tissues also function as buffers.

The **buffer capacity** is a measure of the buffering power of a buffer system (mol·L⁻¹ · [ΔpH]⁻¹). It corresponds to the number of added H⁺ or OH⁻ ions per unit volume that change the pH by one unit. The buffer capacity therefore corresponds to the slope of the titration curve for a given buffer ($\rightarrow$ p. 380, B). The buffer capacity is dependent on (a) the buffer concentration and (b) the pH. The farther the pH is from the pK_a of a buffer system, the smaller the buffer capacity ($\rightarrow$ p. 380). The buffer capacity of the blood is about 75 mmol·L⁻¹·(ΔpH)⁻¹ at pH 7.4 and constant P_{CO_2}. Since the buffer capacity is dependent on the prevailing P_{CO_2}, the **buffer base concentration** of the blood (normally about 48 mEq/L) is normally used as the measure of buffering power of the blood in clinical medicine ($\rightarrow$ pp. 142 and 146). The buffer base concentration is the sum of the concentrations of all buffer components that accept hydrogen ions, i.e., HCO_3^-, Hb⁻, Oxy-Hb⁻, diphosphoglycerate anions, plasma protein anions, HPO_4^{2-}, etc.

Changes in the pH of the blood are chiefly due to changes in the following factors ($\rightarrow$ **A** and p. 142ff.):

◆ *H⁺ ions*: Direct uptake in foodstuffs (e.g., vinegar) or by metabolism, or removal from the blood (e.g., by the kidney; $\rightarrow$ p. 174ff.).

◆ *OH⁻ ions*: Uptake in foodstuffs containing (basic) salts of weak acids, especially in primarily vegetarian diet.

◆ *CO_2*: Its concentration, [CO_2], can change due to alterations in metabolic production or pulmonary elimination of CO_2. A drop in [CO_2] leads to a rise in pH and vice versa ($\rightarrow$ **A**: [CO_2] is the denominator in the equation).

◆ *HCO_3^-*: It can be eliminated directly from the blood by the kidney or gut (in diarrhea) ($\rightarrow$ pp. 176, 142). A rise or fall in [HCO_3^-] will lead to a corresponding rise or fall in pH ($\rightarrow$ **A**: [HCO_3^-] is the numerator in the equation).

Plate 6.1 pH, pH Buffers, Acid–Base Balance

A. Factors that affect the blood pH

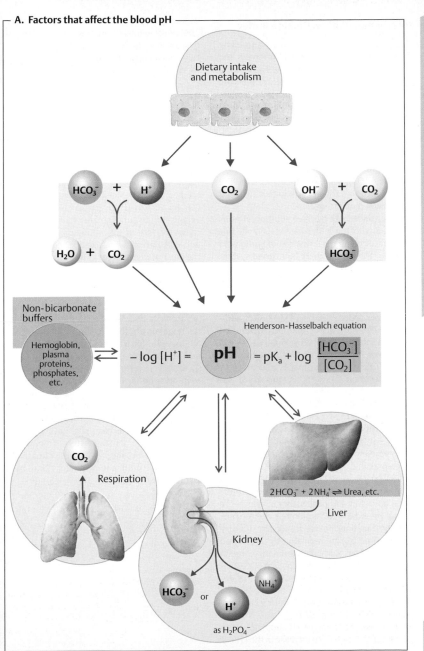

Dietary intake and metabolism

HCO_3^- + H^+ CO_2 OH^- + CO_2

H_2O + CO_2 HCO_3^-

Non-bicarbonate buffers

Hemoglobin, plasma proteins, phosphates, etc.

Henderson-Hasselbalch equation

$$-\log[H^+] = \mathbf{pH} = pK_a + \log\frac{[HCO_3^-]}{[CO_2]}$$

CO_2

Respiration

$2\,HCO_3^- + 2\,NH_4^+ \rightleftharpoons$ Urea, etc.

Liver

Kidney

HCO_3^- or H^+ NH_4^+

as $H_2PO_4^-$

Bicarbonate/Carbon Dioxide Buffer

The pH of any buffer system is determined by the concentration *ratio* of the buffer pairs and the pK_a of the system ($\rightarrow$ p. 378). The pH of a bicarbonate solution is the concentration ratio of bicarbonate and dissolved carbon dioxide ([**HCO$_3^-$**]/[**CO$_2$**]), as defined in the Henderson–Hasselbalch equation ($\rightarrow$ **A1**). Given [HCO$_3^-$] = 24 mmol/L and [CO$_2$] = 1.2 mmol/l, [HCO$_3^-$]/[CO$_2$] = 24/1.2 = 20. Given log20 = 1.3 and pK_a = 6.1, a pH of 7.4 is derived when these values are set into the equation ($\rightarrow$ **A2**). If [HCO$_3^-$] drops to 10 and [CO$_2$] decreases to 0.5 mmol/L, the *ratio* of the two variables will not change, and the pH will remain constant.

When added to a buffered solution, H$^+$ ions combine with the buffer base (HCO$_3^-$ in this case), resulting in the formation of buffer acid (HCO$_3^-$ + H$^+$ $\rightarrow$ CO$_2$ + H$_2$O). In a **closed system** from which CO$_2$ cannot escape ($\rightarrow$ **A3**), the amount of buffer acid formed (CO$_2$) equals the amount of buffer base consumed (HCO$_3^-$). The inverse holds true for the addition of hydroxide ions (OH$^-$ + CO$_2$ $\rightarrow$ HCO$_3^-$). After **addition of** 2 mmol/L of **H$^+$**, the aforementioned baseline ratio [HCO$_3^-$]/[CO$_2$] of 24/1.2 ($\rightarrow$ **A2**) changes to 22/3.2, making the pH fall to 6.93 ($\rightarrow$ **A3**). Thus, the buffer capacity of the HCO$_3^-$/CO$_2$ buffer at pH 7.4 is very low in a *closed system*, for which the pK_a of 6.1 is too far from the target pH of 7.4 ($\rightarrow$ pp. 138, 378 ff).

If, however, the additionally produced CO$_2$ is eliminated from the system (**open system**; $\rightarrow$ **A4**), only the [HCO$_3^-$] will change when the same amount of **H$^+$ is added** (2 mmol/L). The corresponding decrease in the [HCO$_3^-$]/[CO$_2$] ratio (22/1.2) and pH (7.36) is much less than in a closed system. In the body, bicarbonate buffering occurs in an open system in which the partial pressure (P_{CO_2}) and hence the concentration of carbon dioxide in plasma ([CO$_2$] = $\alpha \cdot P_{CO_2}$; $\rightarrow$ p. 126) are regulated by respiration ($\rightarrow$ **B**). The lungs normally eliminate as much CO$_2$ as produced by metabolism (15 000–20 000 mmol/day), while the alveolar P_{CO_2} remains constant ($\rightarrow$ p. 120 ff.). Since the plasma P_{CO_2} adapts to the alveolar P_{CO_2} during each respiratory cycle, the arterial P_{CO_2} (Pa$_{CO_2}$) also remains constant. An increased supply of H$^+$ in the periphery leads to an increase in the P_{CO_2} of

venous blood (H$^+$ + HCO$_3^-$ $\rightarrow$ CO$_2$ + H$_2$O) ($\rightarrow$ **B1**). The lungs eliminate the additional CO$_2$ so quickly that the arterial P_{CO_2} remains practically unchanged despite the addition of H$^+$ (open system!).

The following example demonstrates the quantitatively small impact of increased **pulmonary CO$_2$ elimination**. A two-fold increase in the amount of H$^+$ ions produced within the body on a given day (normally 60 mmol/day) will result in the added production of 60 mmol more of CO$_2$ per day (disregarding non-bicarbonate buffers). This corresponds to only about 0.3% of the normal daily CO$_2$ elimination rate.

An increased supply of OH$^-$ ions in the periphery has basically similar effects. Since OH$^-$ + CO$_2$ $\rightarrow$ HCO$_3^-$, [HCO$_3^-$] increases and the venous P_{CO_2} becomes smaller than normal. Because the rate of CO$_2$ elimination is also reduced, the arterial P_{CO_2} also does not change in the illustrated example ($\rightarrow$ **B2**).

At a pH of 7.4, the open HCO$_3^-$/CO$_2$ buffer system makes up about two-thirds of the buffer capacity of the blood when the P_{CO_2} remains constant at 5.33 kPa ($\rightarrow$ p. 138). Mainly intracellular non-bicarbonate buffers provide the remaining buffer capacity.

Since **non-bicarbonate buffers** (NBBs) function in *closed systems*, their total concentration ([NBB base] + [NBB acid]) remains constant, even after buffering. The total concentration changes in response to changes in the hemoglobin concentration, however, since hemoglobin is the main constituent of NBBs ($\rightarrow$ pp. 138, 146). NBBs supplement the HCO$_3^-$/CO$_2$ buffer in non-respiratory (metabolic) acid–base disturbances ($\rightarrow$ p. 142), but are the *only* effective buffers in respiratory acid–base disturbances ($\rightarrow$ p. 144).

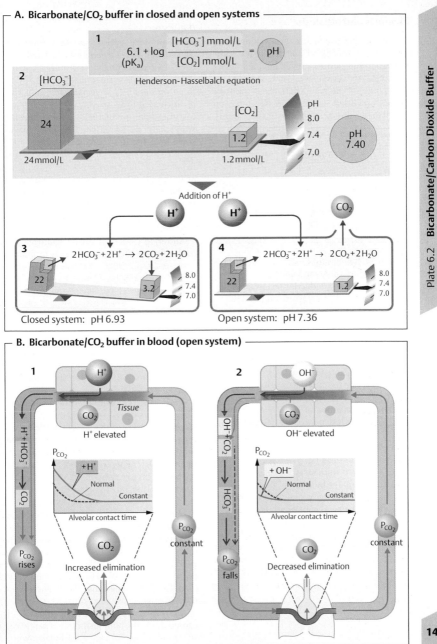

A. Bicarbonate/CO₂ buffer in closed and open systems

1

$$6.1 + \log \frac{[HCO_3^-] \text{ mmol/L}}{[CO_2] \text{ mmol/L}} = pH$$
(pK_a)

Henderson-Hasselbalch equation

2 $[HCO_3^-]$

24

24 mmol/L

$[CO_2]$

1.2

1.2 mmol/L

pH
8.0
7.4
7.0

pH
7.40

Addition of H⁺

H⁺   H⁺ CO_2

3 $2HCO_3^- + 2H^+ \rightarrow 2CO_2 + 2H_2O$

22   3.2

8.0
7.4
7.0

4 $2HCO_3^- + 2H^+ \rightarrow 2CO_2 + 2H_2O$

22   1.2

8.0
7.4
7.0

Closed system: pH 6.93   Open system: pH 7.36

B. Bicarbonate/CO₂ buffer in blood (open system)

1

H⁺

CO_2 *Tissue*

H⁺ elevated

$H^+ + HCO_3^-$

CO_2

P_{CO_2} rises

P_{CO_2}
+H⁺
Normal
Constant
Alveolar contact time

CO_2
Increased elimination

P_{CO_2} constant

2

OH⁻

CO_2

OH⁻ elevated

$OH^- + CO_2$

HCO_3^-

P_{CO_2} falls

P_{CO_2}
+ OH⁻
Normal
Constant
Alveolar contact time

CO_2
Decreased elimination

P_{CO_2} constant

Plate 6.2 Bicarbonate/Carbon Dioxide Buffer

Acidosis and Alkalosis

The main objective of acid–base regulation is to keep the pH of blood, and thus of the body, constant. The normal ranges for parameters relevant to acid–base homeostasis, as measured in plasma (arterialized capillary blood) are listed in the table (see table on p. 124 for erythrocyte P_{CO_2} and $[HCO_3^-]$ values).

Normal range of acid–base parameters in plasma

	Women	Men
$[H^+]$ (nmol/L)	39.8 ± 1.4	40.7 ± 1.4
pH	7.40 ± 0.015	7.39 ± 0.015
P_{CO_2} (kPa)	5.07 ± 0.3	5.47 ± 0.3
(mmHg)	38.9 ± 2.3	41.0 ± 2.3
$[HCO_3^-]$ (mmol/L)	24 ± 2.5	24 ± 2.5

Acid–base homeostasis exists when the following balances are maintained:

1. (H^+ addition or production) – (HCO_3^- addition or production) = (H^+ excretion) – (HCO_3^- excretion) ≈ 60 mmol/day (diet-dependent).
2. (CO_2 production) = (CO_2 excretion) ≈ 15 000–20 000 mmol/day.

H^+ production (HCl, H_2SO_4, lactic acid, H_3PO_4, etc.) and adequate renal H^+ excretion (→ p. 174ff.) are the main factors that influence the first balance. A vegetarian diet can lead to a considerable addition of HCO_3^- (metabolism: $OH^- + CO_2 \rightarrow HCO_3^-$; → p. 138). HCO_3^- is excreted in the urine to compensate for the added supply (the urine of vegetarians therefore tends to be alkaline).

Acid–base disturbances. *Alkalosis* occurs when the pH of the blood rises above the normal range (see table), and *acidosis* occurs when it falls below the lower limits of normal. *Respiratory* acid–base disturbances occur due to primary changes in P_{CO_2} (→ p. 144), whereas non-*respiratory (metabolic)* disturbances occur due to a primary change in $[HCO_3^-]$. Acid–base disturbances can be partially or almost completely compensated.

Nonrespiratory (Metabolic) Acid–Base Disturbances

Nonrespiratory acidosis is most commonly caused by (1) *renal failure* or isolated renal tubular H^+ secretion defect resulting in inability to eliminate normal quantities of H^+ ions (*renal acidosis*); (2) *hyperkalemia* (→ p. 180); (3) increased β-hydroxybutyric acid and acetoacetic acid production (*diabetes mellitus, starvation*); (4) increased anaerobic conversion of glucose to lactic acid (→ lactate$^-$ + H^+), e.g., due to *strenuous physical work* (→ p. 74) or hypoxia; (5) increased metabolic production of HCl and H_2SO_4 in individuals with a *high intake of dietary proteins*; and (6) *loss of HCO_3^-* through renal excretion (proximal renal tubular acidosis, use of carbonic anhydrase inhibitors) or *diarrhea*.

Buffering (→ **A1**) of excess hydrogen ions occurs in the first stage of non-respiratory acidosis (every HCO_3^- lost results in an H^+ gained). Two-thirds and one-third of the buffering is achieved by HCO_3^- and non-bicarbonate buffer bases (NBB$^-$), respectively, and the CO_2 arising from HCO_3^- buffering is eliminated from the body by the lungs (open system; → p. 140). The standard bicarbonate concentration $[HCO_3^-]_{St}$, the actual bicarbonate concentration $[HCO_3^-]_{Act}$ and the buffer base concentration [BB] *decrease* (negative base excess; → p. 146).

Respiratory compensation of non-respiratory acidosis (→ **A2**) occurs in the second stage. The total ventilation rises in response to the reduced pH levels (via central chemosensors), leading to a decrease in the alveolar and arterial P_{CO_2} (hyperventilation; → **A2a**). This not only helps to return the $[HCO_3^-]/[CO_2]$ ratio towards normal (20:1), but also converts NBB-H back to NBB$^-$ (due to the increasing pH) (→ **A2b**). The latter process also requires HCO_3^- and, thus, further compensatory pulmonary elimination of CO_2 (→ **A2c**). If the cause of acidosis persists, respiratory compensation will eventually become insufficient, and increased renal excretion of H^+ ions will occur (→ p. 174ff.), provided that the acidosis is not of renal origin (see above, cause 1).

Nonrespiratory (metabolic) alkalosis is caused by (1) the administration of bases (e.g., HCO_3^- *infusion*); (2) increased breakdown of

▶

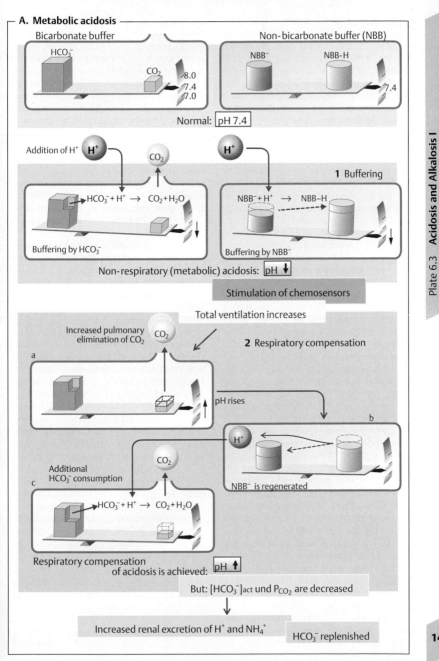

A. Metabolic acidosis

Bicarbonate buffer

HCO_3^- CO_2 8.0 / 7.4 / 7.0

Non-bicarbonate buffer (NBB)

NBB^- NBB-H 7.4

Normal: pH 7.4

Addition of H^+ H^+ CO_2 H^+

1 Buffering

$HCO_3^- + H^+ \rightarrow CO_2 + H_2O$

Buffering by HCO_3^-

$NBB^- + H^+ \rightarrow NBB-H$

Buffering by NBB^-

Non-respiratory (metabolic) acidosis: pH ↓

Stimulation of chemosensors

Total ventilation increases

Increased pulmonary elimination of CO_2 CO_2

2 Respiratory compensation

a

pH rises

b

H^+

NBB⁻ is regenerated

Additional HCO_3^- consumption CO_2

c

$HCO_3^- + H^+ \rightarrow CO_2 + H_2O$

Respiratory compensation of acidosis is achieved: pH ↑

But: $[HCO_3^-]_{act}$ und P_{CO_2} are decreased

Increased renal excretion of H^+ and NH_4^+ HCO_3^- replenished

Plate 6.3   Acidosis and Alkalosis I

organic anions (e.g., lactate⁻, α-ketoglutarate²⁻); (3) loss of H⁺ ions due to *vomiting* (→ p. 238) or *hypokalemia;* and (4) *volume depletion*. **Buffering** in metabolic alkalosis is similar to that of non-respiratory acidosis (rise in [HCO_3^-]$_{St}$, positive base excess). Nonetheless, the capacity for **respiratory compensation** through hypoventilation is very limited because of the resulting O_2 deficit.

Respiratory Acid–Base Disturbances

Respiratory alkalosis (→ **B**) occurs when the lungs eliminate more CO_2 than is produced by metabolism (*hyperventilation*), resulting in a decrease in plasma P_{CO_2} (hypocapnia). Inversely, respiratory acidosis occurs (→ **B**) when less CO_2 is eliminated than produced (*hypoventilation*), resulting in an increase in plasma P_{CO_2} (hypercapnia). Whereas bicarbonate and non-bicarbonate buffer bases (NBB⁻) jointly buffer the pH decrease in metabolic acidosis (→ p. 142), the two buffer systems behave very differently in respiratory alkalosis (→ **B1**). In the latter case, the HCO_3^-/CO_2 system is not effective because the change in P_{CO_2} is the primary *cause*, not the result of respiratory alkalosis.

Respiratory acidosis can occur as the result of lung tissue damage (e.g., tuberculosis), impairment of alveolar gas exchange (e.g., pulmonary edema), paralysis of respiratory muscles (e.g., polio), insufficient respiratory drive (e.g., narcotic overdose), reduced chest motility (e.g., extreme spinal curvature), and many other conditions. The resulting increase in plasma CO_2 ([CO_2] = $\alpha \cdot P_{CO_2}$) is followed by increased HCO_3^- and H⁺ production (→ **B1**, left panel). The H⁺ ions are buffered by NBB bases (NBB⁻ + H⁺ → NBB-H; → **B1**, right panel) while [HCO_3^-]$_{Act}$ increases. Unlike non-respiratory acidosis, [HCO_3^-]$_{St}$ remains unchanged (at least initially since it is defined for normal P_{CO_2}; → p. 146) and [BB] remains unchanged because the *[NBB⁻] decrease equals the [HCO_3^-]$_{Act}$ increase*. Since the percentage increase in [HCO_3^-]$_{Act}$ is much lower than the rise in [CO_2], the [HCO_3^-]/[CO_2] ratio and pH are lower than normal (acidosis).

If the increased P_{CO_2} persists, **renal compensation** (→ **B2**) of the respiratory disturbance will occur. The kidneys begin to excrete increased quantities of H⁺ in form of titratable acidity (→ p. 174f.) or NH₄⁺ as well and, after a latency period of 1 to 2 days. Each NH₄⁺ ion excreted results in the sparing of one HCO_3^- ion in the liver, and each H⁺ ion excreted results in the tubular cellular release of one HCO_3^- ion into the blood (→ p. 174ff.). This process continues until the pH has been reasonably normalized despite the P_{CO_2} increase. A portion of the HCO_3^- is used to buffer the H⁺ ions liberated during the reaction NBB-H → NBB⁻ + H⁺ (→ **B2, right** panel). Because of the relatively long latency for renal compensation, the drop in pH is more pronounced in acute respiratory acidosis than in chronic respiratory acidosis. In the chronic form, [HCO_3^-]$_{Act}$ can rise by about 1 mmol per 1.34 kPa (10 mmHg) increase in P_{CO_2}.

Respiratory alkalosis is usually *caused* by hyperventilation due to anxiety or high altitude (oxygen deficit ventilation; → p. 136), resulting in a fall in plasma P_{CO_2}. This leads to a slight decrease in [HCO_3^-]$_{Act}$ since a small portion of the HCO_3^- is converted to CO_2 (H⁺ + HCO_3^- → CO_2 + H_2O); the HCO_3^- required for this reaction is supplied by H⁺ ions from NBB's (**buffering**: NBB-H → NBB⁻ + H⁺). This is also the reason for the additional drop in [HCO_3^-]$_{Act}$ when respiratory compensation of non-respiratory acidosis occurs (→ p. 143 **A**, **bottom** panel, and p. 146). Further reduction of [HCO_3^-]$_{Act}$ is required for adequate pH normalization (compensation). This is achieved through reduced renal tubular secretion of H⁺. As a consequence, increased renal excretion of HCO_3^- will occur (renal compensation).

In acute respiratory acidosis or alkalosis, CO_2 diffuses more rapidly than HCO_3^- and H⁺ from the blood into the cerebrospinal fluid (CSF). The low NBB concentrations there causes relatively strong fluctuations in the pH of the CSF (→ p. 126), providing an adequate stimulus for central chemosensors (→ p. 132).

B. Respiratory acidosis

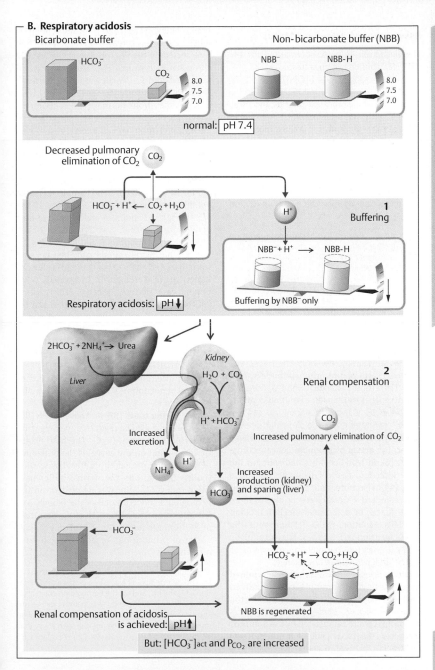

Bicarbonate buffer

HCO_3^- CO_2 8.0 7.5 7.0

Non-bicarbonate buffer (NBB)

NBB$^-$ NBB-H   8.0 7.5 7.0

normal: pH 7.4

Decreased pulmonary elimination of CO_2 CO_2

$HCO_3^- + H^+ \leftarrow CO_2 + H_2O$

H^+ **1** Buffering

$NBB^- + H^+ \longrightarrow NBB\text{-}H$

Buffering by NBB$^-$ only

Respiratory acidosis: pH ↓

$2HCO_3^- + 2NH_4^+ \rightarrow$ Urea

Liver

Kidney $H_2O + CO_2$

2 Renal compensation

$H^+ + HCO_3^-$

Increased excretion NH_4^+ H^+

CO_2

Increased pulmonary elimination of CO_2

Increased production (kidney) and sparing (liver)

HCO_3^-

HCO_3^-

$HCO_3^- + H^+ \rightarrow CO_2 + H_2O$

Renal compensation of acidosis is achieved: pH ↑

NBB is regenerated

But: $[HCO_3^-]_{act}$ and P_{CO_2} are increased

Plate 6.4   **Acidosis and Alkalosis II**

145

Assessment of Acid–Base Status

The *Henderson–Hasselbalch equation* for the HCO_3^-/CO_2 buffer system states:

$$pH = pK_a + \log ([HCO_3^-]/[CO_2]). \qquad [6.5]$$

Since $[CO_2] = \alpha \cdot P_{CO_2}$ ($\rightarrow$ p. 126), Equation 6.5 contains two constants (pK_a and α) and three variables (pH, $[HCO_3^-]$, and P_{CO_2}). At 37 °C in plasma, $pK_a = 6.1$ and $\alpha = 0.225$ mmol $\cdot$ L^{-1} $\cdot$ kPa^{-1} (cf. p. 126). When one of the variables remains constant (e.g., $[HCO_3^-]$), the other two (e.g., P_{CO_2} and pH) are interdependent. In a graphic representation, this dependency is reflected as a straight line when the logarithm of P_{CO_2} is plotted against the pH ($\rightarrow$ **A–C** and p. 382).

When the P_{CO_2} varies in a **bicarbonate solution** (without other buffers), the pH changes but $[HCO_3^-]$ remains constant ($\rightarrow$ **A**, solid line). One can also plot the lines for different HCO_3^- concentrations, all of which are parallel ($\rightarrow$ **A**, **B**, dotted orange lines). Figures **A** through **C** use a scale that ensures that the bicarbonate lines intersect the coordinates at 45° angles. The **Siggaard–Andersen nomogram** ($\rightarrow$ **C**) does not use the lines, but only the points of intersection of the $[HCO_3^-]$ lines with the normal P_{CO_2} of 5.33 kPa (40 mmHg).

The **blood** contains not only the HCO_3^-/CO_2 buffer but also **non-bicarbonate buffers, NBB** ($\rightarrow$ p. 138). Thus, a change in the P_{CO_2} does not alter the pH as much as in a solution containing the HCO_3^-/CO_2 buffer alone ($\rightarrow$ p. 144). In the P_{CO_2}/pH nomogram, the slope is therefore steeper than 45° ($\rightarrow$ **B**, green and red lines). Hence, the **actual bicarbonate** concentration, **$[HCO_3^-]_{Act}$**, in blood changes and shifts in the same direction as the P_{CO_2} if the pH varies ($\rightarrow$ p. 144). Therefore, both the $[HCO_3^-]_{Act}$ and the **standard bicarbonate** concentration, **$[HCO_3^-]_{St}$**, can be determined in clinical blood tests. By definition, $[HCO_3^-]_{St}$ represents the *$[HCO_3^-]$ at a normal P_{CO_2}* of 5.33 kPa (40 mmHg). $[HCO_3^-]_{St}$ therefore permits an assessment of $[HCO_3^-]$ independent of P_{CO_2} changes.

$[HCO_3^-]_{St}$ and **$[HCO_3^-]_{Act}$** are **determined** using measured P_{CO_2} and pH values obtained with a blood gas analyzer. When plotted on the *Siggaard–Andersen nomogram*, $[HCO_3^-]_{St}$ is read from the line as indicated by the points of intersect of the $[HCO_3^-]$ line ($\rightarrow$ **B**, orange lines) and the P_{CO_2}/pH line (**B** and **C**, green and red

lines) *at normal P_{CO_2}* = 5.33 ($\rightarrow$ **B** and **C**, points *D* and *d*). $[HCO_3^-]_{Act}$ is read from the $[HCO_3^-]$ line intersected by the P_{CO_2}/pH line at the level of the *actually measured P_{CO_2}*. Since the normal and measured P_{CO_2} values agree in normals, their $[HCO_3^-]_{Act}$ is usually equal to $[HCO_3^-]_{St}$. If P_{CO_2} deviates from normal ($\rightarrow$ **B**, **C**, point *c*), $[HCO_3^-]_{Act}$ is read at point *e* on the HCO_3^- line ($\rightarrow$ **B**, **C**, interrupted 45° line) on which the actually measured P_{CO_2} lies ($\rightarrow$ **B**, **C**, point *c*).

Blood P_{CO_2} and pH measurement. When using the equilibration method (*Astrup method*), three pH measurements are taken: (1) in the unchanged blood sample; (2) after equilibration with a high P_{CO_2} (e.g., 10 kPa [75 mmHg]; $\rightarrow$ **C**, points *A* and *a*), and (3) after equilibration with a low P_{CO_2} (e.g., 2.7 kPa [20 mmHg]; $\rightarrow$ **C**, points *B* and *b*). The P_{CO_2} of the original blood sample can then be read from lines *A–B* and *a–b* using the pH value obtained in measurement 1. In normals ($\rightarrow$ **C**, upper case letters, green), $[HCO_3^-]_{Act} = [HCO_3^-]_{St} = 24$ mmol/L ($\rightarrow$ **C**, points *E* and *D*). Example 2 ($\rightarrow$ **C**, lower case letters, red) shows an acid–base disturbance: The pH is too low (7.2) and $[HCO_3^-]_{St}$ ($\rightarrow$ **C**, point *d*) has dropped to 13 mmol/L (metabolic acidosis). This has been partially compensated ($\rightarrow$ p. 142) by a reduction in P_{CO_2} to 4 kPa, which led to a consequent reduction in $[HCO_3^-]_{Act}$ to 11 mmol/L ($\rightarrow$ **C**, point *e*).

Total **buffer bases** (BB) and **base excess** (BE) ($\rightarrow$ p. 142) can also be read from the Siggaard–Andersen nomogram ($\rightarrow$ **C**). The base excess (points F and f on the curve) is the difference between the measured buffer base value (points *G* or *g*) and the normal buffer base value (point *G*). Point *G* is dependent on the hemoglobin concentration of the blood ($\rightarrow$ **C**; [Hb]/BB comparison). Like $[HCO_3^-]_{St}$, deviation of BB from the norm (0 ± 2.5 mEq/L) is diagnostic of primary non-respiratory acid–base disturbances.

The P_{CO_2}/pH line of the blood sample in plate **C** can also be determined if (1) the P_{CO_2} (without equilibration), (2) the pH, and (3) the hemoglobin concentration are known. One point ($\rightarrow$ **C**, point *c*) on the unknown line can be drawn using (1) and (2). The line must be drawn through the point in such a way that BB (point *g*) – BB$_{normal}$ (dependent on Hb value) = BE (point *f*).

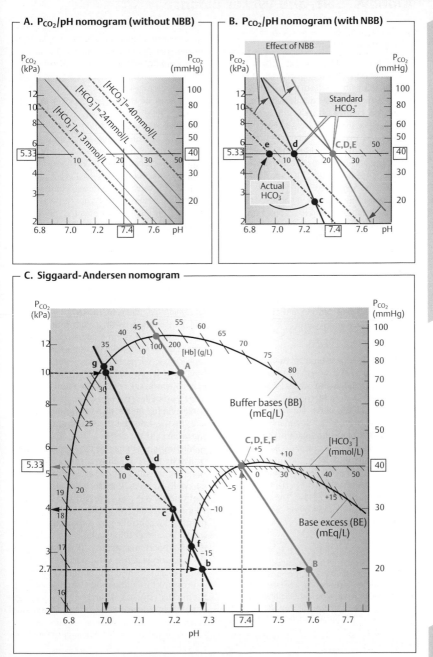

A. P$_{CO_2}$/pH nomogram (without NBB)

P$_{CO_2}$ (kPa) — P$_{CO_2}$ (mmHg)

[HCO$_3^-$] = 40 mmol/L
[HCO$_3^-$] = 24 mmol/L
[HCO$_3^-$] = 13 mmol/L

B. P$_{CO_2}$/pH nomogram (with NBB)

Effect of NBB

P$_{CO_2}$ (kPa) — P$_{CO_2}$ (mmHg)

Standard HCO$_3^-$

C,D,E

Actual HCO$_3^-$

C. Siggaard-Andersen nomogram

P$_{CO_2}$ (kPa) — P$_{CO_2}$ (mmHg)

[Hb] (g/L)

Buffer bases (BB) (mEq/L)

C,D,E,F

[HCO$_3^-$] (mmol/L)

Base excess (BE) (mEq/L)

pH

Plate 6.5 Assessment of Acid–Base Status

Kidney Structure and Function

Three fundamental mechanisms characterize kidney function: (1) large quantities of water and solutes are **filtered** from the blood. (2) This *primary urine* enters the tubule, where most of it is **reabsorbed**, i.e., it exits the tubule and passes back into the blood. (3) Certain substances (e.g., toxins) are not only not reabsorbed but actively **secreted** into the tubule lumen. The non-reabsorbed residual filtrate is **excreted** together with the secreted substances in the *final urine*.

Functions: The kidneys (1) adjust salt and water excretion to maintain a constant *extracellular fluid volume* and *osmolality*; (2) they help to maintain *acid-base homeostasis*; (3) they *eliminate end-products* of metabolism and foreign substances while (4) preserving useful compounds (e.g., glucose) by reabsorption; (5) the produce *hormones* (e.g., erythropoietin) and hormone activators (renin), and (6) have *metabolic functions* (protein and peptide catabolism, gluconeogenesis, etc.).

Nephron Structure

Each kidney contains about 10^6 **nephrons**, each consisting of the malpighian body and the tubule. The **malpighian body** is located in the renal cortex ($\rightarrow$ **A**) and consists of a tuft of capillaries (**glomerulus**) surrounded by a double-walled capsule (**Bowman's capsule**). The primary urine accumulates in the capsular space between its two layers ($\rightarrow$ **B**). Blood enters the glomerulus by an afferent arteriole (*vas afferens*) and exits via an efferent arteriole (*vas efferens*) from which the peritubular capillary network arises ($\rightarrow$ p. 150). The **glomerular filter** ($\rightarrow$ **B**) separates the blood side from the Bowman's capsular space.

The **glomerular filter** comprises the fenestrated endothelium of the glomerular capillaries (50–100 nm pore size) followed by the basal membrane as the second layer and the visceral membrane of Bowman's capsule on the urine side. The latter is formed by *podocytes* with numerous interdigitating footlike processes (*pedicels*). The slit-like spaces between them are covered by the *slit membrane*, the *pores* of which are about 5 nm in diameter. They are shaped by the protein *nephrine*, which is anchored to the cytoskeleton of the podocytes.

♦ The **proximal tubule** ($\rightarrow$ **A**, dark green) is the longest part of a nephron (ca. 10 mm). Its twisted initial segment (*proximal convoluted tubule, PCT*; $\rightarrow$ **A3**) merges into a straight part, PST (*pars recta*; $\rightarrow$ **A4**).

♦ The **loop of Henle** consists of a *thick descending limb* that extends into the renal medulla ($\rightarrow$ **A4** = PST), a *thin descending limb* ($\rightarrow$ **A5**), a *thin ascending limb* (only in juxtamedullary nephrons which have long loops), and a *thick ascending limb, TAL* ($\rightarrow$ **A6**). It contains the **macula densa** ($\rightarrow$ p. 184), a group of specialized cells that closely communicate with the glomerulus of the respective nephron. Only about 20% of all Henle's loops (those of the deep juxtamedullary nephrons) are long enough to penetrate into the inner medulla. Cortical nephrons have shorter loops ($\rightarrow$ **A** and p. 150).

♦ The **distal tubule** ($\rightarrow$ **A**, grayish green) has an initially straight part (= TAL of Henle's loop; $\rightarrow$ **A6**) that merges with a convoluted part (*distal convoluted tubule, DCT*; $\rightarrow$ **A7**).

The DCT merges with a connecting tubule ($\rightarrow$ **A8**). Many of them lead into a **collecting duct, CD** ($\rightarrow$ **A9**) which extends through the renal cortex (cortical CD) and medulla (medullary CD). At the *renal papilla* the collecting ducts opens in the *renal pelvis*. From there, the urine (propelled by peristaltic contractions) passes via the *ureter* into the *urinary bladder* and, finally, into the *urethra*, through which the urine exits the body.

Micturition. Voiding of the bladder is controlled by reflexes. Filling of the bladder activates the smooth detrusor muscle of the bladder wall via stretch sensors and parasympathetic neurons (S_2–S_4, $\rightarrow$ p. 78 ff.). At low filling volumes, the wall relaxes via sympathetic neurons (L_1–L_2) controlled by supraspinal centers (pons). At higher filling volumes (> 0.3 L), the threshold pressure (about 1 kPa) that triggers the *micturition reflex* via a positive feedback loop is reached: The detrusor muscle contracts $\rightarrow$ pressure $\uparrow$ $\rightarrow$ contraction $\uparrow\uparrow$ and so on until the *internal* (smooth m.) and *external sphincter* (striated m.) open so the urine can exit the body.

Plate 7.1 Kidney Structure and Function

A. Anatomy of the kidney (schematic diagram)

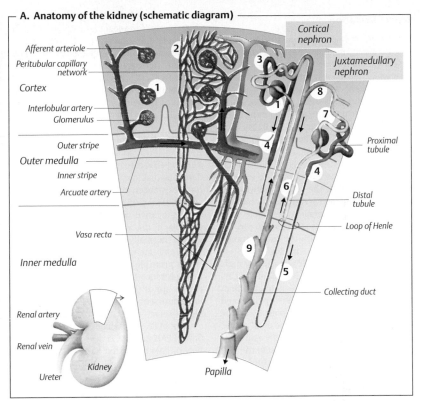

Cortical nephron

Juxtamedullary nephron

Afferent arteriole

Peritubular capillary network

Cortex

Interlobular artery

Glomerulus

Outer stripe

Outer medulla

Inner stripe

Arcuate artery

Vasa recta

Inner medulla

Renal artery

Renal vein

Kidney

Ureter

Proximal tubule

Distal tubule

Loop of Henle

Collecting duct

Papilla

B. Glomerulus and Bowman's capsule

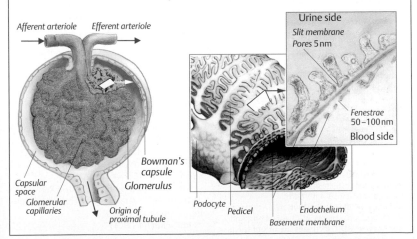

Afferent arteriole

Efferent arteriole

Urine side

Slit membrane
Pores 5 nm

Fenestrae
50–100 nm

Blood side

Bowman's capsule

Glomerulus

Capsular space

Glomerular capillaries

Origin of proximal tubule

Podocyte

Pedicel

Endothelium

Basement membrane

Renal Circulation

The *arcuate arteries* ($\rightarrow$ **A1**) pass between the renal cortex and medulla. They branch towards the cortex into the *interlobular arteries* ($\rightarrow$ **A2**) from which the *afferent arterioles* (or *vasa afferentia*) arise ($\rightarrow$ **A3**). Unlike other organs, the kidney has *two successive capillary networks* that are connected with each other by an *efferent arteriole* (or *vas efferens*) ($\rightarrow$ **A, B**). *Pressure* in the first network of **glomerular capillaries** ($\rightarrow$ p. 148) is a relatively *high* ($\rightarrow$ **B** and p. 152) and is regulated by adjusting the width of interlobular artery, the afferent and/or efferent arterioles ($\rightarrow$ **A 3,4**). The second network of **peritubular capillaries** ($\rightarrow$ **A**) winds around the cortical tubules. It supplies the tubule cells with blood, but it also contributes the to exchange of substances with the tubule lumen (reabsorption, secretion; $\rightarrow$ p. 154 ff.).

The **renal blood flow** (RBF) is relatively high, ca. 1.2 L/min, equivalent to 20–25% of the cardiac output. This is required to maintain a high glomerular filtration rate (GFR; $\rightarrow$ p. 152) and results in a very low arteriovenous O_2 difference (ca. 15 mL/L of blood). In the *renal cortex*, O_2 **is consumed** (ca. 18 mL/min) for oxidative **metabolism** of fatty acids, etc. Most of the ATP produced in the process is used to fuel active transport. In the renal medulla, metabolism is mainly anaerobic ($\rightarrow$ p. 72).

Around 90% of the renal blood supply goes to the cortex. Per gram of tissue, approximately 5, 1.75 and 0.5 mL/min of blood pass through the cortex, external medulla, and internal medulla, respectively. The latter value is still higher than in most organs ($\rightarrow$ p. 213 A).

The kidney contains **two types of nephrons** that differ with respect to the features of their second network of capillaries ($\rightarrow$ **A**).

◆ **Cortical nephrons** are supplied by peritubular capillaries (see above) and have short loops of Henle.

◆ **Juxtamedullary nephrons** are located near the cortex-medulla junction. Their efferent arterioles give rise to relatively long (≤ 40 mm), straight arterioles (**vasa recta**) that descend into the renal medulla. The vasa recta supply the renal medulla and can accompany long loops of Henle of juxtamedullary nephrons as far as the tip of the renal papilla ($\rightarrow$ p. 148). Their hairpin shape is important for the concentration of urine ($\rightarrow$ p. 164 ff.).

Any **change in blood distribution** to these two types of nephrons affects NaCl excretion. Antidiuretic hormone (ADH) increases the GFR of the juxtamedullary nephrons.

Due to **autoregulation of renal blood flow**, only slight changes in renal plasma flow (RPF) and glomerular filtration rate (GFR) occur (even in a denervated kidney) when the systemic blood pressure fluctuates between 80 and about 180 mmHg ($\rightarrow$ **C**). Resistance in the *interlobular arteries* and *afferent arterioles* located upstream to the cortical glomeruli is automatically adjusted when the mean blood pressure changes ($\rightarrow$ **B, C**). If the blood pressure falls below about 80 mmHg, however, renal circulation and filtration will ultimately fail ($\rightarrow$ **C**). RBF and GFR can also be regulated independently by making isolated changes in the (serial) resistances of the afferent and efferent arterioles ($\rightarrow$ p. 152).

Non-invasive **determination of RBF** is possible if the **renal plasma flow** (**RPF**) is known (normally about **0.6 L/min**). RPF is obtained by measuring the amount balance (Fick's principle) of an intravenously injected test substance (e.g., *p*-aminohippurate, **PAH**) that is almost completely eliminated in the urine during one renal pass (PAH is filtered and highly secreted, $\rightarrow$ p. 156 ff.). The eliminated amount of PAH is calculated as the arterial inflow of PAH into the kidney minus the venous flow of PAH out of the kidney per unit time. Since

Amount/time =
(volume/time) $\cdot$ concentration $\qquad$ [7.1]

$$(RPF \cdot Pa_{PAH}) - (RPF \cdot Prv_{PAH}) = \dot{V}_U \cdot U_{PAH} \quad [7.2]$$

or

$$RPF = \dot{V}_U \cdot U_{PAH}/(Pa_{PAH} - Prv_{PAH}). \quad [7.3]$$

where Pa_{PAH} is the arterial PAH conc., Prv_{PAH} is the renal venous PAH conc., U_{PAH} is the urinary PAH conc., and $\dot{V}_U$ is the urine output/time. Prv_{PAH} makes up only about 10% of the Pa_{PAH} and normally is not measured directly, but is estimated by dividing **PAH clearance** ($= \dot{V}_U \cdot U_{PAH}/Pa_{PAH}$; $\rightarrow$ p. 152) by a factor of 0.9. Therefore,

$$\mathbf{RPF = \dot{V}_U \cdot U_{PAH}/(0{,}9 \cdot Pa_{PAH}).} \quad [7.4]$$

This equation is only valid when the Pa_{PAH} is not too high. Otherwise, PAH secretion will be saturated and PAH clearance will be much smaller than RPF ($\rightarrow$ p. 161 A).

RBF is derived by inserting the known hematocrit (HCT) value ($\rightarrow$ p. 88) into the following equation:

$$\mathbf{RBF = RPF/(1 - HCT)} \quad [7.5]$$

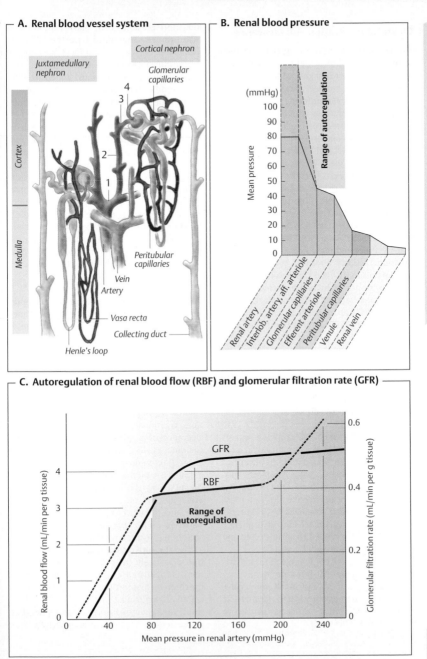

A. Renal blood vessel system

Juxtamedullary nephron

Cortical nephron

Glomerular capillaries

Cortex

Medulla

4
3
2
1

Peritubular capillaries

Vein

Artery

Vasa recta

Collecting duct

Henle's loop

B. Renal blood pressure

(mmHg)

Mean pressure

Range of autoregulation

100
90
80
70
60
50
40
30
20
10
0

Renal artery

Interlob. artery, aff. arteriole

Glomerular capillaries

Efferent arteriole

Peritubular capillaries

Venule

Renal vein

C. Autoregulation of renal blood flow (RBF) and glomerular filtration rate (GFR)

GFR

RBF

Range of autoregulation

Renal blood flow (mL/min per g tissue)

Glomerular filtration rate (mL/min per g tissue)

Mean pressure in renal artery (mmHg)

Glomerular Filtration and Clearance

The **glomerular filtration rate (GFR)** is the total volume of fluid filtered by the glomeruli per unit time. It is normally about **120 mL/min** per 1.73 m^2 of body surface area, equivalent to around 180 L/day. Accordingly, the volume of exchangeable extracellular fluid of the whole body (ca. 17 L) enters the renal tubules about 10 times a day. About 99% of the GFR returns to the extracellular compartment by tubular reabsorption. The mean *fractional excretion of H$_2$O* is therefore about 1% of the GFR, and absolute H$_2$O excretion (= **urine output/time** = $\dot{V}_U$) is about 1 to 2 L per day. (The filtration of dissolved substances is described on p. 154).

The GFR makes up about 20% of renal plasma flow, RPF ($\rightarrow$ p. 150). The **filtration fraction** (FF) is defined as the ratio of GFR/RPF. The filtration fraction is increased by *atriopeptin*, a peptide hormone that increases efferent arteriolar resistance (R_e) while lowering afferent arteriolar resistance (R_a). This raises the effective filtration pressure in the glomerular capillaries without significantly changing the overall resistance in the renal circulation.

The effective filtration pressure (P_{eff}) is the driving "force" for filtration. P_{eff} is the glomerular capillary pressure ($P_{cap} \approx 48$ mmHg) minus the pressure in Bowman's capsule ($P_{Bow} \approx 13$ mmHg) and the oncotic pressure in plasma (π_{cap} = 25 to 35 mmHg):

$$P_{eff} = P_{cap} - P_{Bow} - \pi_{cap} \qquad [7.6]$$

P_{eff} at the arterial end of the capillaries equals 48−13−25 = 10 mmHg. Because of the high filtration fraction, the plasma protein concentration and, therefore, π_{cap} values along the glomerular capillaries increase ($\rightarrow$ p. 378) and P_{eff} decreases. (The mean effective filtration pressure, $\overline{P}_{eff}$, is therefore used in Eq. 7.7.) Thus, filtration ceases (near distal end of capillary) when π_{cap} rises to about 35 mmHg, decreasing P_{eff} to zero (*filtration equilibrium*).

GFR is the product of $\overline{P}_{eff}$ (mean for all glomeruli), the glomerular filtration area A (dependent on the number of intact glomeruli), and the water permeability k of the glomerular filter. The *ultrafiltration coefficient K_f* is used to represent A · k. This yields

$$GFR = \overline{P}_{eff} \cdot K_f. \qquad [7.7]$$

Indicators present in the plasma are used to **measure GFR.** They must have the following properties:

— They must be freely filterable
— Their filtered amount must not change due to resorption or secretion in the tubule
— They must not be metabolized in the kidney
— They must not alter renal function

Inulin, which must be infused intravenously, fulfills these requirements. **Endogenous creatinine** (normally present in blood) can also be used with certain limitations.

The amount of indicator filtered over time ($\rightarrow$ **A**) is calculated as the plasma concentration of the indicator (P_{In}, in g/L or mol/L) times the GFR in L/min. The same amount of indicator/time appears in the urine (conditions 2 and 3; see above) and is calculated as $\dot{V}_U$ (in L/min), times the indicator conc. in urine (U_{In}, in g/L or mol/L, resp.), i.e. $P_{In} \cdot GFR = \dot{V}_U \cdot U_{In}$, or:

$$GFR = \frac{\dot{V}_U \cdot U_{In}}{P_{In}} \text{ [L/min] } (\rightarrow \textbf{A}). \qquad [7.8]$$

The expression on the right of Eq. 7.8 represents **clearance**, regardless of which substance is being investigated. Therefore, the *inulin or creatinine clearance represents the GFR.* (Although the plasma concentration of creatinine, P_{cr}, rises as the GFR falls, P_{cr} alone is a quite unreliable measure of GFR.)

Clearance can also be regarded as the completely indicator-free (or cleared) plasma volume flowing through the kidney per unit time. **Fractional excretion (FE)** is the ratio of clearance of a given substance X to inulin clearance (C_X/C_{In}) and defines which fraction of the filtered quantity of X was excreted (cf. p. 154). FE < 1 if the substance is removed from the tubule by reabsorption (e.g. Na$^+$, Cl$^-$, amino acids, glucose, etc.; $\rightarrow$ **B1**), and FE > 1 if the substance is subject to filtration plus tubular secretion ($\rightarrow$ **B2**). For PAH ($\rightarrow$ p. 150), tubular secretion is so effective that $FE_{PAH} \approx 5$ (500%).

The **absolute rate of reabsorption or secretion** of a freely filterable substance X (mol/min) is calculated as the difference between the filtered amount/time (GFR · P_X) and the excreted amount/time ($\dot{V}_U \cdot U_X$), where a positive result means net reabsorption and a negative net secretion. (For inulin, the result would be zero.)

A. Inulin clearance = glomerular filtration rate (GFR)

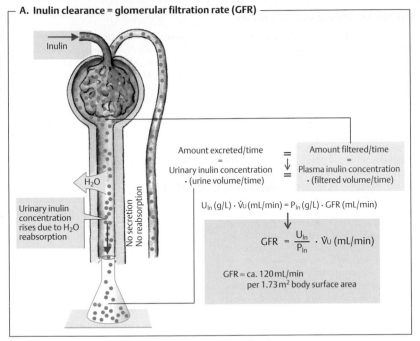

Inulin

H₂O

No secretion
No reabsorption

Urinary inulin concentration rises due to H₂O reabsorption

Amount excreted/time
=
Urinary inulin concentration · (urine volume/time)

↓↓

Amount filtered/time
=
Plasma inulin concentration · (filtered volume/time)

$U_{In}\,(g/L) \cdot \dot{V}_U\,(mL/min) = P_{In}\,(g/L) \cdot GFR\,(mL/min)$

$$GFR = \frac{U_{In}}{P_{In}} \cdot \dot{V}_U \ (mL/min)$$

GFR ≈ ca. 120 mL/min
per 1.73 m² body surface area

Plate 7.3   Glomerular Filtration and Clearance

B. Clearance levels (1) lower or (2) higher than inulin clearance

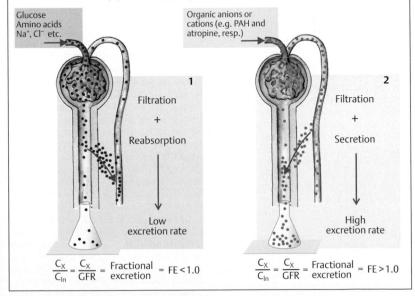

Glucose
Amino acids
Na⁺, Cl⁻ etc.

Organic anions or
cations (e.g. PAH and
atropine, resp.)

1

Filtration
+
Reabsorption

Low
excretion rate

$$\frac{C_X}{C_{In}} = \frac{C_X}{GFR} = \text{Fractional excretion} = FE < 1.0$$

2

Filtration
+
Secretion

High
excretion rate

$$\frac{C_X}{C_{In}} = \frac{C_X}{GFR} = \text{Fractional excretion} = FE > 1.0$$

Transport Processes at the Nephron

Filtration of solutes. The glomerular filtrate also contains small dissolved molecules of plasma (*ultrafiltrate*) ($\rightarrow$ p. 152). The *glomerular sieving coefficient GSC* of a substance (= concentration in filtrate/concentration in plasma water) is a measure of the permeability of the glomerular filter for this substance ($\rightarrow$ p. 148). Molecules with a radius of r < 1.8 nm (molecular mass < ca. 10 000 Da) can freely pass through the filter (GSC $\approx$ 1.0), while those with a radius of r > 4.4 nm (molecular mass > 80 000 DA, e.g., globulins) normally cannot pass through it (GSC = 0). Only a portion of molecules where 1.8 nm < r < 4.4 nm applies are able to pass through the filter (GSC ranges between 1 and 0). Negatively charged particles (e.g., albumin: r = 3.4 nm; GSC $\approx$ 0.0003) are less permeable than neutral substances of equal radius because negative charges on the wall of the glomerular filter repel the ions. When small molecules are bound to plasma proteins (**protein binding**), the bound fraction is practically non-filterable ($\rightarrow$ p. 24).

Molecules entrapped in the glomerular filter are believed to be eliminated by phagocytic mesangial macrophages ($\rightarrow$ p. 94ff.) and glomerular podocytes.

Tubular epithelium. The epithelial cells lining the renal tubule and collecting duct are *polar cells*. As such, their luminal (or apical) membrane on the urine side differs significantly from that of the basolateral membrane on the blood side. The luminal membrane of the proximal tubule has a high *brush border* consisting of *microvilli* that greatly increase the surface area (especially in the convoluted proximal tubule). The basolateral membrane of this tubule segment has deep folds (*basal labyrinth*) that are in close contact with the intracellular mitochondria ($\rightarrow$ p. 9 B), which produce the ATP needed for Na^+-K^+-ATPase ($\rightarrow$ p. 26) located in the basolateral membrane (of all epithelial cells). The large surface areas (about 100 m²) of the proximal tubule cells of both kidneys are needed to reabsorb the lion's share of filtered solutes within the contact time of a couple of seconds. Postproximal tubule cells do not need a brush border since the amount of substances reabsorbed decreases sharply from the proximal to the distal segments of the tubules.

Whereas permeability of the two membranes in series is decisive for **transcellular transport** (reabsorption, secretion), the tightness of *tight junctions* ($\rightarrow$ p. 18) determines the paracellular permeability of the epithelium for water and solutes that cross the epithelium by **paracellular transport**. The tight junctions in the proximal tubule are relatively permeable to water and small ions which, together with the large surface area of the cell membranes, makes the epithelium well equipped for para- and transcellular mass transport ($\rightarrow$ **D**, column 2). The thin limbs of Henle's loop are relatively "leaky", while the thick ascending limb and the rest of the tubule and collecting duct are "moderately tight" epithelia. The tighter epithelia can develop much higher *transepithelial chemical and electrical gradients* than "leaky" epithelia.

Measurement of reabsorption, secretion and excretion. Whether and to which degree a substance filtered by the glomerulus is reabsorbed or secreted at the tubule and collecting duct cannot be determined based on its urinary concentration alone as concentrations rise as due to the reabsorption of water ($\rightarrow$ p. 164). The urinary/plasma inulin (or creatinine) concentration ratio, U_{in}/P_{in} *is a measure of the degree of water reabsorption.* These substances can be used as indicators because they are neither reabsorbed nor secreted ($\rightarrow$ p. 152). Thus, changes in indicator concentration along the length of the tubule occur due to the **H_2O reabsorption** alone ($\rightarrow$ **A**). If U_{in}/P_{in} = 200, the inulin concentration in the final urine is 200 times higher than in the original filtrate. This implies that *fractional excretion of H_2O (FE_{H_2O})* is 1/200 or 0.005 or 0.5% of the GFR. Determination of the concentration of a (freely filterable and perhaps additionally secreted) substance X in the same plasma and urine samples for which U_{in}/P_{in} was measured will yield U_X/P_X. Considering U_{in}/P_{in}, the fractional excretion of X, FE_X can be calculated as follows ($\rightarrow$ **A** and **D**, in % in column 5):

$$FE_X = (U_X/P_X)/(U_{in}/P_{in}) \qquad [7.9]$$

Eq. 7.9 can also be derived from C_X/C_{in} ($\rightarrow$ p. 152) when simplified for $\dot{V}_U$. The **fractional reabsorption** of X (**FR_X**) is calculated as

$$FR_X = 1 - FE_X \qquad [7.10]$$

▶

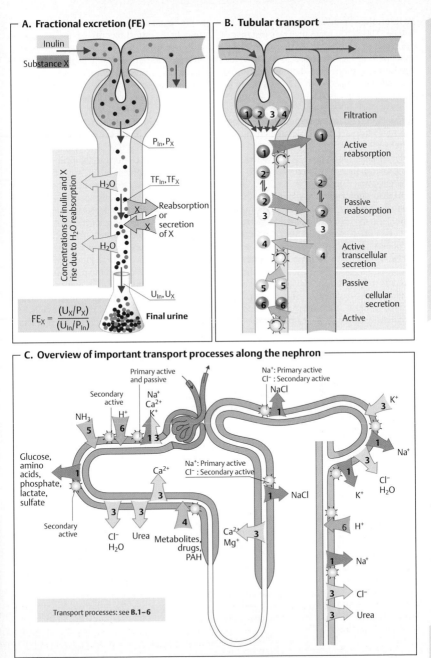

A. Fractional excretion (FE)

Inulin

Substance X

P_{In}, P_X

Concentrations of inulin and X rise due to H_2O reabsorption

H_2O

TF_{In}, TF_X

Reabsorption or secretion of X

H_2O

U_{In}, U_X

Final urine

$$FE_X = \frac{(U_X/P_X)}{(U_{In}/P_{In})}$$

B. Tubular transport

Filtration

Active reabsorption

Passive reabsorption

Active transcellular secretion

Passive cellular secretion

Active

C. Overview of important transport processes along the nephron

Primary active and passive

Na^+
Ca^{2+}
K^+

Secondary active

NH_3

H^+

Na^+: Primary active
Cl^-: Secondary active
NaCl

K^+

Na^+

Cl^-
H_2O

Glucose, amino acids, phosphate, lactate, sulfate

Secondary active

Cl^-
H_2O

Urea

Ca^{2+}

Metabolites, drugs, PAH

Na^+: Primary active
Cl^-: Secondary active

NaCl

Ca^{2+}
Mg^+

K^+

H^+

Na^+

Cl^-

Urea

Transport processes: see **B.1–6**

Plate 7.4 Transport Processes at the Nephron I

Reabsorption in different segments of the tubule. The concentration of a substance X (TF_X) and inulin (TF_{in}) in tubular fluid can be measured via micropuncture ($\to$ **A**). The values can be used to calculate the non-reabsorbed fraction (*fractional delivery*, FD) of a freely filtered substance X as follows:

$$FD = (TF_X/P_X)/(TF_{in}/P_{in}),$$

where P_X and P_{in} are the respective concentrations in plasma (more precisely: in plasma water).

Fractional reabsorption (FR) up to the sampling site can then be derived from $1 - FD$ ($\to$ **D**, columns 2 and 3, in %).

Reabsorption and secretion of various substances (see pp. 16–30, transport mechanisms). Apart from H_2O, many inorganic ions (e.g., Na^+, Cl^-, K^+, Ca^{2+}, and Mg^{2+}) and organic substances (e.g., HCO_3^-, D-glucose, L-amino acids, urate, lactate, vitamin C, peptides and proteins; $\to$ **C, D**, p. 158ff.) are also subject to tubular **reabsorption** ($\to$ **B1–3**). Endogenous products of metabolism (e.g., urate, glucuronides, hippurates, and sulfates) and foreign substances (e.g., penicillin, diuretics, and PAH; $\to$ p. 150) enter the tubular urine by way of **transcellular secretion** ($\to$ **B4, C**). Many substances, such as ammonia (NH_3) and H^+ are first produced by tubule cells before they enter the tubule by **cellular secretion**. NH_3 enters the tubule lumen by passive transport ($\to$ **B5**), while H^+ ions are secreted by active transport ($\to$ **B6** and p. 174ff.).

Na^+/K^+ transport by **Na^+-K^+-ATPase** ($\to$ p. 26) in the basolateral membrane of the tubule and collecting duct serves as the **"motor"** for most of these transport processes. By *primary active transport* (fueled directly by ATP consumption), Na^+-K^+-ATPase pumps Na^+ out of the cell into the blood while pumping K^+ in the opposite direction (subscript "i" = intracellular and "o" = extracellular). This creates two driving "forces" essential for the transport of numerous substances (including Na^+ and K^+): first, a **chemical Na^+ gradient** ($[Na^+]o > [Na^+]i$) and (because $[K^+]i > [K^+]o$), second, a **membrane potential** (inside the cell is negative relative to the outside) which represents an **electrical gradient** and can drive ion transport ($\to$ pp. 32ff. and 44).

Transcellular transport implies that two membranes must be crossed, usually by two different mechanisms. If a given substance (D-glucose, PAH, etc.) is actively transported across an epithelial barrier (i.e., against an electrochemical gradient; $\to$ see p. 26ff.), at least one of the two serial membrane transport steps must also be active.

Interaction of transporters. Active and passive transport processes are usually closely interrelated. The active absorption of a solute such as Na^+ or D-glucose, for example, results in the development of an *osmotic gradient* ($\to$ p. 24), leading to the passive absorption of water. When water is absorbed, certain solutes are carried along with it (*solvent drag*; $\to$ p. 24), while other substrates within the tubule become more *concentrated*. The latter solutes (e.g., Cl^- and urea) then return to the blood along their concentration gradients by passive reabsorption. Electrogenic ion transport and ion-coupled transport ($\to$ p. 28) can depolarize or hyperpolarize only the luminal or only the basolateral membrane of the tubule cells. This causes a *transepithelial potential* which serves as the driving "force" for paracellular ion transport in some cases.

Since non-ionized forms of weak electrolytes are more lipid-soluble than ionized forms, they are better able to penetrate the membrane (**non-ionic diffusion**; $\to$ **B2**). Thus, the pH of the urine has a greater influence on passive reabsorption by non-ionic diffusion. *Molecular size* also influences diffusion: the smaller a molecule, the larger its diffusion coefficient ($\to$ p. 20ff.).

D. Reabsorption, secretion and fractional excretion

Plate 7.5   Transport Processes at the Nephron II

Substance	1 Concentration in plasma water (P) [mmol/l]	Fractional reabsorption (FR) [%]			5 Fractional excretion (FE) [% of filtered amount]	6
		2 % in proximal tubule (TF/P)	3 % in loop of Henle (TF/P)	4 Total %		P = Plasma concentration TF = Concentration in tubular urine Effects: ↑ raises FE ↓ lowers FE
H_2O	---	65%	10%	93%–99.5%	0.5%–7%	ADH: ↓
Na^+	153	65% (1.0)	25% (0.4)	95%–99.5%	0.5%–5%	Aldosterone: ↓ ADH: ↓ ANP: ↑
K^+	4.6	65% (1.0)	10%–20%	Secretion possible	2%–150%	Aldosterone: ↑
Ca^{2+}	Free: 1.6	60% (1.1)	30%	95%–99%	1%–5%	PTH: ↓ Acidosis: ↑
Mg^{2+}	Free: 0.6	15% (2.5)	ca. 70%	80%–95%	5%–20%	P rise: ↑
Cl^-	112	55% (1.3)	ca. 20%	95%–99.5%	0.5%–5%	---
HCO_3^-	24	93% (0.2)		98%–99%	1%–2%	Alkalosis: ↑
Phosphate	2.2	65% (1.0)	15%	80%–97%	3%–20%	P rise: ↑ PTH: ↑ Ca^{2+} falls: ↑ Acidosis: ↑
D-Glucose	5	96% (0.1)	4%	≈100%	≈0%	Sharp P rise: ↑
Urea	5	50% (1.4)	Secretion	ca. 60%	ca. 40%	Diuresis: ↑
Creatinine	0.1	0% (2.9)	0%	0%	100%	---
PAH (I.V.)	C_{test}	Secretion	Secretion	Secretion	≈500%	Sharp P rise: ↓

Reabsorption of Organic Substances

The **filtered load of a substance** is the product of its plasma concentration and GFR. Since the GFR is high (ca. 180 L/day), enormous quantities of substances enter the primary urine each day (e.g., 160 g/day of D-glucose).

Fractional excretion (FE, → p. 154) of **D-glucose** is very low (FE ≈ 0.4%). This virtually complete reabsorption is achieved by secondary active transport (*Na$^+$-glucose symport*) at the luminal cell membrane (→ **B** and p. 29 B1). About 95% of this activity occurs in the proximal tubule. If the plasma glucose conc. exceeds 10–15 mmol/L, as in diabetes mellitus (normally 5 mmol/L), *glucosuria* develops, and urinary glucose conc. rises (→ **A**). Glucose reabsorption therefore exhibits *saturation kinetics* (Michaelis-Menten kinetics; → p. 28). The above example illustrates *prerenal glucosuria*. *Renal glucosuria* can also occur when one of the tubular glucose carriers is defective.

Low-affinity **carriers** in the luminal cell membrane of the pars convoluta (sodium-glucose transporter type 2 = SGLT2) and high-affinity carriers (SGLT1) in the pars recta are responsible for D-glucose reabsorption. The co-transport of D-glucose and Na$^+$ occurs in each case, namely at a ratio of 1 : 1 with SGLT2 and 1 : 2 with SGLT1. The energy required for this form of *secondary active* glucose transport is supplied by the electrochemical Na$^+$ gradient directed towards the cell interior. Because of the co-transport of two Na$^+$ ions, the gradient for SGLT1 is twice as large as that for SGLT2. A *uniporter* (GLUT2 = glucose transporter type 2) on the blood side facilitates the passive transport of accumulated intracellular glucose out of the cell (*facilitated diffusion*, → p. 22). **D-galactose** also makes use of the SGLT1 carrier, while **D-fructose** is passively absorbed by tubule cells (GLUT5).

The plasma contains over 25 **amino acids**, and about 70 g of amino acids are filtered each day. Like D-glucose, most L-amino acids are reabsorbed at proximal tubule cells by Na$^+$-coupled secondary active transport (→ **B** and p. 29 B3). At least 7 different amino acid transporters are in the proximal tubule, and the specificities of some overlap. J_{max} and K_M (→ p. 28) and, therefore, saturability and reabsorption capacities vary according to the type of amino acid and carrier involved. Fractional excretion of most amino acids ≈ to 1% (ranging from 0.1% for L-valine to 6% for L-histidine).

Increased urinary excretion of amino acids (**hyperaminoaciduria**) can occur. *Prerenal* hyperaminoaciduria occurs when plasma amino acid concentrations are elevated (and reabsorption becomes saturated, as in **A**), whereas *renal* hyperaminoaciduria occurs due to deficient transport. Such a dysfunction may be *specific* (e.g., in cystinuria, where only L-cystine, L-arginine and L-lysine are hyperexcreted) or *unspecific* (e.g., in Fanconi's syndrome, where not only amino acids but also glucose, phosphate, bicarbonate etc. are hyperexcreted).

Certain substances (lactate, sulfate, phosphate, dicarboxylates, etc.) are also reabsorbed at the proximal tubule by way of Na$^+$ symport, whereas urea is subject to passive back diffusion (→ p. 166). **Urate** and **oxalate** are both reabsorbed and secreted (→ p. 160), with the predominant process being reabsorption for urate (FE ≈ 0.1) and secretion for oxalate (FE > 1). If the urinary conc. of these poorly soluble substances rises above normal, they will start to precipitate (increasing the risk of **urinary calculus formation**). Likewise, the excessive urinary excretion of **cystine** can lead to cystine calculi.

Oligopeptides such as glutathione and angiotensin II are broken down so quickly by *luminal peptidases* in the brush border that they can be reabsorbed as free amino acids (→ **C1**). **Dipeptides** resistant to luminal hydrolysis (e.g., carnosine) must be reabsorbed as intact molecules. A symport carrier (*PepT2*) driven by the inwardly directed H$^+$ gradient (→ p. 174) transports the molecules into the cells *tertiary active* H$^+$ symport; → p. 26, 29 B4). The dipeptides are then hydrolyzed within the cell (→ **C2**). The PepT2 carrier is also used by certain drugs and toxins.

Proteins. Although *albumin* has a low sieving coefficient of 0.0003 (→ p. 154, 2400 mg/day are filtered at a plasma conc. of 45 g/L (180 L/day · 45 g/L · 0.0003 = 2400 mg/day). Only 2 to 35 mg of albumin are excreted each day (FE ≈ 1%). In the proximal tubule, albumin, lysozyme, α_1-microglobulin, β_2-microglobulin and other proteins are reabsorbed by *receptor-mediated endocytosis* (→ p. 28) and are "digested" by lysosomes (→ **D**). Since this type of reabsorption is nearly saturated at normal filtered loads of proteins, an elevated plasma protein conc. or increased protein sieving coefficient will lead to *proteinuria*.

25-OH-cholecalciferol, which is bound to **DBP** (vitamin D-binding protein) in plasma and glomerular filtrate, is reabsorbed in combination with DBP by receptor-mediated endocytosis (→ p. 292).

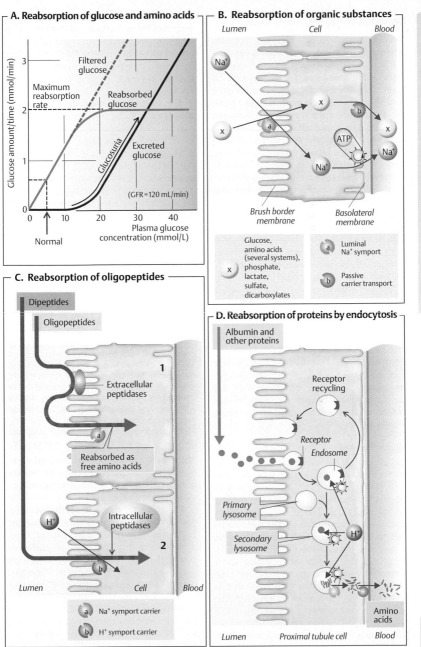

A. Reabsorption of glucose and amino acids

Glucose amount/time (mmol/min)

Filtered glucose

Maximum reabsorption rate

Reabsorbed glucose

Glucosuria

Excreted glucose

(GFR = 120 mL/min)

Plasma glucose concentration (mmol/L)

Normal

B. Reabsorption of organic substances

Lumen   Cell   Blood

Na^+

x

x

b

x

a

ATP

Na^+ Na^+

Brush border membrane

Basolateral membrane

Glucose, amino acids (several systems), phosphate, lactate, sulfate, dicarboxylates — x

a — Luminal Na^+ symport

b — Passive carrier transport

C. Reabsorption of oligopeptides

Dipeptides

Oligopeptides

Extracellular peptidases

1

a

Reabsorbed as free amino acids

H^+

Intracellular peptidases

2

b

Lumen   Cell   Blood

a — Na^+ symport carrier

b — H^+ symport carrier

D. Reabsorption of proteins by endocytosis

Albumin and other proteins

Receptor recycling

Receptor

Endosome

Primary lysosome

Secondary lysosome

H^+

Amino acids

Lumen   Proximal tubule cell   Blood

Plate 7.6 Reabsorption of Organic Substances

Excretion of Organic Substances

Food provides necessary nutrients, but also contains inert and harmful substances. The body can usually sort out these substances already at the time of **intake**, either based on their smell or taste or, if already eaten, with the help of specific digestive enzymes and intestinal absorptive mechanisms (e.g., D-glucose and L-amino acids are absorbed, but L-glucose and D-amino acids are not). Similar distinctions are made in **hepatic excretion** ($\Rightarrow$ bile $\Rightarrow$ stools): useful bile salts are almost completely reabsorbed from the gut by way of specific carriers, while waste products such as bilirubin are mainly eliminated in the feces. Likewise, the **kidney** reabsorbs hardly any useless or harmful substances (including end-products such as creatinine). Valuable substances (e.g., D-glucose and L-amino acids), on the other hand, are reabsorbed via specific transporters and thus spared from excretion ($\rightarrow$ p. 158).

The liver and kidney are also able to modify endogenous waste products and foreign compounds (xenobiotics) so that they are **"detoxified"** if toxic and made ready for rapid elimination. In unchanged form or after the enzymatic addition of an OH or COOH group, the substances then combine with glucuronic acid, sulfate, acetate or glutathione to form **conjugates**. The conjugated substances are then secreted into the bile and proximal tubule lumen (with or without further metabolic processing).

Tubular Secretion

The proximal tubule utilizes **active transport mechanisms** to secrete numerous waste products and xenobiotics. This is done by way of carriers for organic anions (OA^-) and organic cations (OC^+). The secretion of these substances makes it possible to raise their clearance level above that of inulin and, therefore, to raise their fraction excretion (FE) above 1.0 = 100% ($\rightarrow$ p. 152) in order to eliminate them more effectively ($\rightarrow$ **A**; compare red and blue curves). Secretion is carrier-mediated and is therefore subject to saturation kinetics. Unlike reabsorbed substances such as D-glucose ($\rightarrow$ p. 159 A), the fractional excretion (FE) of organic anions and cations decreases when their plasma concentrations rise ($\rightarrow$ **A**; PAH secre-

tion curve reaches plateau, and slope of PAH excretion curve decreases). Some organic anions (e.g., urate and oxalate) and cations (e.g., choline) are both secreted and reabsorbed (bidirectional transport), which results in net reabsorption (urate, choline) or net secretion (oxalate).

The secreted **organic anions** (OA^-) include indicators such as PAH (p-aminohippurate; $\rightarrow$ p. 150) and phenol red; endogenous substances such as oxalate, urate, hippurate; drugs such as penicillin G, barbiturates, and numerous diuretics ($\rightarrow$ p. 172); and conjugated substances (see above) containing glucuronate, sulfate or glutathione. Because of its high affinity for the transport system, *probenecid* is a potent inhibitor of OA^- secretion.

The *active step of* OA^- *secretion* ($\rightarrow$ **B**) occurs across the *basolateral membrane* of proximal tubule cells and accumulates organic anions in the cell whereby the inside-negative membrane potential has to be overcome. The membrane has a broad specificity carrier (OAT1 = *organic anion transporter* type 1) that transports OA^- from the blood into the tubule cells in exchange for a dicarboxylate, such as succinate^{2-} or α-ketoglutarate^{2-}; $\rightarrow$ **B1**). The latter substance arises from the glutamine metabolism of the cell ($\rightarrow$ p. 177 D2); the human Na^+-dicarboxylate transporter hNADC-1 also conveys dicarboxylates (in combination with 3 Na^+) into the cell by secondary active transport ($\rightarrow$ **B2**). The transport of OA^- is therefore called *tertiary active transport*. The efflux of OA^- into the lumen is passive (facilitated diffusion; $\rightarrow$ **B3**). An ATP-dependent *conjugate pump* (MRP2 = multi-drug resistance protein type 2) in the luminal membrane is also used for secretion of amphiphilic conjugates, such as glutathione-linked lipophilic toxins ($\rightarrow$ **B4**).

The **organic cations** (OC^+) secreted include endogenous substances (epinephrine, choline, histamine, serotonin, etc.) and drugs (atropine, quinidine, morphine, etc.).

The *active step* of OC^+ **secretion** occurs across the *luminal* membrane of proximal tubule cells (luminal accumulation occurs after overcoming the negative membrane potential inside the cell). The membrane contains (a) direct ATP-driven carriers for organic cations (mdr1; primary active OC^+ transport; $\rightarrow$ **C1**) and (b) a multispecific OC^+/H^+ antiporter (tertiary active transport; $\rightarrow$ **C2**). The OC^+ diffuse passively from the blood into the cell by way of a polyspecific organic cation transporter (OCT; $\rightarrow$ **C3**).

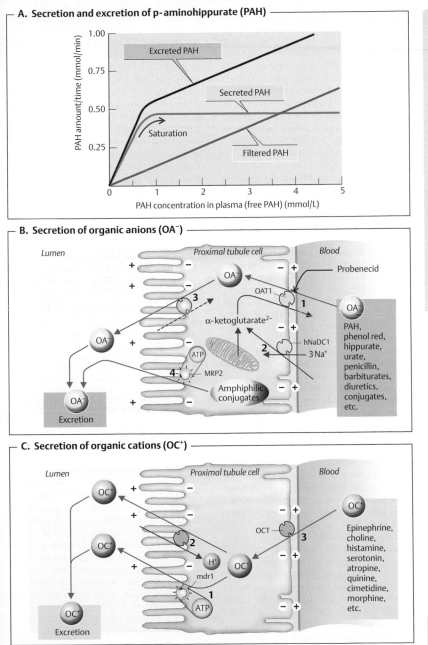

A. Secretion and excretion of p-aminohippurate (PAH)

Excreted PAH

Secreted PAH

Saturation

Filtered PAH

PAH amount/time (mmol/min)

1.00
0.75
0.50
0.25

0   1   2   3   4   5

PAH concentration in plasma (free PAH) (mmol/L)

B. Secretion of organic anions (OA⁻)

Lumen

Proximal tubule cell

Blood

OA⁻

OAT1

Probenecid

α-ketoglutarate²⁻

hNaDC1

3 Na⁺

ATP

MRP2

Amphiphilic conjugates

OA⁻

Excretion

OA⁻

PAH, phenol red, hippurate, urate, penicillin, barbiturates, diuretics, conjugates, etc.

C. Secretion of organic cations (OC⁺)

Lumen

Proximal tubule cell

Blood

OC⁺

OCT

OC⁺

H⁺

mdr1

ATP

OC⁺

Excretion

OC⁺

Epinephrine, choline, histamine, serotonin, atropine, quinine, cimetidine, morphine, etc.

Plate 7.7 Excretion of Organic Substances

Reabsorption of Na⁺ and Cl⁻

About 99% of the *filtered load of Na⁺* is reabsorbed (ca. 27 000 mmol/day), i.e., the fractional excretion of Na⁺ (FE_{Na}) is about 1%. The precise value of FE_{Na} needed (range 0.5 to 5%) is regulated by aldosterone, atriopeptin (ANF) and other hormones ($\rightarrow$ p. 170).

Sites of Na⁺ reabsorption. The reabsorption of Na⁺ occurs in all parts of the renal tubule and collecting duct. About 65% of the filtered Na⁺ is reabsorbed in the proximal tubule, while the luminal Na⁺ conc. remains constant ($\rightarrow$ p. 166). Another 25% is reabsorbed in the loop of Henle, where luminal Na⁺ conc. drops sharply; $\rightarrow$ p. 157 D, columns 2 and 3). The distal convoluted tubule and collecting duct also reabsorb Na⁺. The latter serves as the site of hormonal *fine adjustment* of Na⁺ excretion.

Mechanisms of Na⁺ reabsorption. Na⁺-K⁺-ATPase pumps Na⁺ ions out of the cell while conveying K⁺ ions into the cell ($\rightarrow$ **A** and p. 156), thereby producing a chemical Na⁺ gradient ($\rightarrow$ **A2**). Back diffusion of K⁺ ($\rightarrow$ **A3**) also leads to the development of a membrane potential ($\rightarrow$ **A4**). Both combined result in a high electrochemical Na⁺ gradient that provides the driving "force" for passive Na⁺ influx, the features of which vary in the individual nephron segments ($\rightarrow$ **B**).

◆ In the **proximal tubule**, Na⁺ ions diffuse passively from the tubule lumen into the cells via (a) the electroneutral Na⁺/H⁺ exchanger type 3 (NHE3), an *Na⁺/H⁺-antiport* carrier for electroneutral exchange of Na⁺ for H⁺ ($\rightarrow$ **B1**, p. 29 B4 and p. 174) and (b) various *Na⁺ symport carriers* for reabsorption of D-glucose etc. ($\rightarrow$ **B1** and p. 158). Since most of these symport carriers are electrogenic, the luminal cell membrane is depolarized, and an early proximal *lumen-negative transepithelial potential* (LNTP) develops.

◆ In the thick ascending limb (**TAL**) of the **loop of Henle** ($\rightarrow$ **B6**), Na⁺ is reabsorbed via the bumetanide-sensitive co-transporter BSC, a *Na⁺-K⁺-2 Cl⁻ symport carrier* ($\rightarrow$ p. 172). Although BSC is primarily electroneutral, the absorbed K⁺ recirculate back to the lumen through K⁺ channels. This hyperpolarizes the luminal membrane, resulting in the development of a lumen-positive transepithelial potential (LPTP).

◆ In the **distal convoluted tubule, DCT** ($\rightarrow$ **B8**), Na⁺ is reabsorbed via the thiazide-sensitive co-transporter TSC, an electroneutral Na⁺-Cl⁻ symport carrier ($\rightarrow$ p. 172).

◆ In principal cells of the connecting tubule and collecting duct ($\rightarrow$ **B9**), Na⁺ exits the lumen via *Na⁺ channels* activated by aldosterone and antidiuretic hormone (ADH) and inhibited by prostaglandins and ANF ($\rightarrow$ p. 170).

Since these four passive Na⁺ transport steps in the luminal membrane are serially connected to active Na⁺ transport in the basolateral membrane (Na⁺-K⁺-ATPase), the associated **transepithelial Na⁺ reabsorption** is also *active*. This makes up about $^1/_3$ of the Na⁺ reabsorption in the proximal tubule, and 1 ATP molecule is consumed for each 3 Na⁺ ions absorbed ($\rightarrow$ p. 26). The other $^2/_3$ of proximal sodium reabsorption is passive and **paracellular**.

Two **driving "forces"** are responsible for this: (1) the LPTP in the mid and late proximal tubule ($\rightarrow$ **B5**) and in the loop of Henle ($\rightarrow$ **B7**) drives Na⁺ and other cations onto the blood side of the epithelium. (2) **Solvent drag** ($\rightarrow$ p. 24): When water is reabsorbed, solutes for which the reflection coefficient < 1 (including Na⁺) are "dragged along" due to friction forces (like a piece of wood drifts with flowing water). Since driving forces (1) and (2) are indirect products of Na⁺-K⁺-ATPase, the **energy balance** rises to about 9 Na⁺ per ATP molecule in the proximal tubule (and to about 5 Na⁺ per ATP molecule in the rest of the kidney).

On the basolateral side, Na⁺ ions exit the proximal tubule cell via Na⁺-K⁺-ATPase and an Na⁺-3 HCO₃⁻ symport carrier ($\rightarrow$ p. 174). In the latter case, Na⁺ exits the cell via tertiary active transport as secondary active secretion of H⁺ (on the opposite cell side) results in intracellular accumulation of HCO₃⁻.

The **fractional Cl⁻ excretion** (**FE**$_{Cl}$) ranges from 0.5% to 5%. About 50% of all **Cl⁻ reabsorption** occurs in the *proximal tubule*. The early proximal LNTP drives Cl⁻ through paracellular spaces out of the lumen ($\rightarrow$ **B3**). The reabsorption of Cl⁻ lags behind that of Na⁺ and H₂O, so the luminal Cl⁻ conc. rises. As a result, Cl⁻ starts to diffuse down its chemical gradient paracellularly along the mid and late proximal tubule ($\rightarrow$ **B4**), thereby producing a LPTP (reversal of potential, $\rightarrow$ **B5**). At the TAL and the DCT, Cl⁻ enters the cells by secondary active transport and exits passively through ADH-activated basolateral Cl⁻ channels ($\rightarrow$ **B6, 8**).

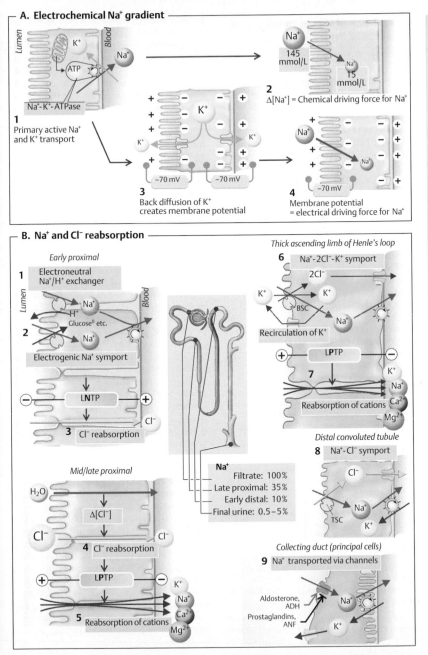

A. Electrochemical Na⁺ gradient

Lumen | *Blood*

K⁺
ATP
Na⁺
Na⁺-K⁺-ATPase

1
Primary active Na⁺
and K⁺ transport

Na⁺
145
mmol/L
Na⁺
15
mmol/L

2
Δ[Na⁺] = Chemical driving force for Na⁺

K⁺
K⁺ K⁺
−70 mV   −70 mV

3
Back diffusion of K⁺
creates membrane potential

Na⁺
Na⁺
−70 mV

4
Membrane potential
= electrical driving force for Na⁺

B. Na⁺ and Cl⁻ reabsorption

Early proximal

1 Electroneutral
Na⁺/H⁺ exchanger

Lumen | *Blood*

Na⁺
H⁺
Glucose⁰ etc.
2
Na⁺
Electrogenic Na⁺ symport

LNTP

3 Cl⁻ reabsorption

Cl⁻

Mid/late proximal

H₂O
Δ[Cl⁻]
Cl⁻   Cl⁻
4 Cl⁻ reabsorption

LPTP
K⁺
Na⁺
5 Reabsorption of cations
Ca²⁺
Mg²⁺

Na⁺
Filtrate: 100%
Late proximal: 35%
Early distal: 10%
Final urine: 0.5–5%

Thick ascending limb of Henle's loop

6 Na⁺-2Cl⁻-K⁺ symport

2Cl⁻
K⁺   K⁺
BSC
Na⁺
Recirculation of K⁺

⊕   LPTP   ⊖

7
K⁺
Na⁺
Reabsorption of cations
Ca²⁺
Mg²⁺

Distal convoluted tubule

8 Na⁺-Cl⁻ symport

Cl⁻
Na⁺
TSC
K⁺

Collecting duct (principal cells)

9 Na⁺ transported via channels

Na⁺
Aldosterone,
ADH
Prostaglandins,
ANF
K⁺

Plate 7.8 Reabsorption of Na⁺ and Cl⁻

Reabsorption of Water, Formation of Concentrated Urine

The glomeruli filter around 180 L of plasma water each day (= GFR; → p. 152). By comparison, the normal **urine output** ($\dot{V}u$) is relatively small (**0.5 to 2 L/day**). Normal fluctuations are called *antidiuresis* (low $\dot{V}u$) and *diuresis* (high $\dot{V}u$; → p. 172). Urine output above the range of normal is called *polyuria*. Below normal output is defined as *oliguria* (< 0.5 L/day) or *anuria* (< 0.1 L/day). The **osmolality** (→ p. 377) of plasma and glomerular filtrate is about 290 mOsm/kg H_2O (= P_{osm}); that of the final urine (U_{osm}) ranges from 50 (hypotonic urine in extreme water diuresis) to about 1200 mOsm/kg H_2O (hypertonic urine in maximally concentrated urine). Since *water diuresis* permits the excretion of large volumes of H_2O without the simultaneous loss of NaCl and other solutes, this is known as "free water excretion," or "free water clearance" (C_{H_2O}). This allows the kidney to normalize decreases in plasma osmolality, for example (→ p. 170). The C_{H_2O} represents to the volume of water that could be theoretically extracted in order for the urine to reach the same osmolality as the plasma:

$$C_{H_2O} = \dot{V}u \, (1 - [U_{osm}/P_{osm}]). \qquad [7.11]$$

Countercurrent Systems

A **simple exchange** system (→ **A1**) can consist of two tubes in which *parallel streams* of water flow, one cold (0 °C) and one hot (100 °C). Due to the exchange of heat between them, the water leaving the ends of both tubes will be about 50 °C, that is, the initially steep temperature gradient of 100 °C will be offset.

In **countercurrent exchange** of heat (→ **A2**), the fluid within the tubes flows in *opposite directions*. Since a temperature gradient is present in all parts of the tube, heat is exchanged along the entire length. Molecules can also be exchanged, provided the wall of the tube is permeable to them and that a concentration gradient exists for the substance.

If the countercurrent exchange of heat occurs in a **hairpin-shaped loop**, the bend of which is in contact with an environment with a temperature different from that inside the tube (ice, → **A3**), the fluid exiting the loop will be only slightly colder than that entering it, because heat always passes from the warmer limb of the loop to the colder limb.

Countercurrent exchange of **water** in the **vasa recta** of the *renal medulla* (→ **A6** and p. 150) occurs if the medulla becomes increasingly hypertonic towards the papillae (see below) and if the vasa recta are permeable to water. Part of the water diffuses by osmosis from the descending vasa recta to the ascending ones, thereby "bypassing" the inner medulla (→ **A4**). Due to the extraction of water, the concentration of all other blood components increases as the blood approaches the papilla. The plasma osmolality in the vasa recta is therefore continuously adjusted to the osmolality of the surrounding interstitium, which rises towards the papilla. The hematocrit in the vasa recta also rises. Conversely, substances entering the blood in the renal medulla diffuse from the ascending to the descending vasa recta, provided the walls of both vessels are permeable to them (e.g., urea; → **C**). The countercurrent exchange in the vasa recta permits the necessary supply of blood to the renal medulla without significantly altering the high osmolality of the renal medulla and hence impairing the urine concentration capacity of the kidney.

In a **countercurrent multiplier** such as the **loop of Henle**, a concentration gradient between the two limbs is maintained by the expenditure of energy (→ **A5**). The countercurrent flow amplifies the relatively small gradient at all points between the limbs (*local gradient* of about 200 mOsm/kg H_2O) to a relatively large gradient along the limb of the loop (about 1000 mOsm/kg H_2O). The longer loop and the higher the one-step gradient, the steeper the multiplied gradient. In addition, it is inversely proportional to (the square of) the flow rate in the loop.

Reabsorption of Water

Approximately 65% of the GFR is reabsorbed at the **proximal convoluted tubule, PCT** (→ **B** and p. 157 D). The driving "force" for this is the reabsorption of solutes, especially Na^+ and Cl^-. This slightly dilutes the urine in the tubule, but H_2O immediately follows this small osmotic gradient because the PCT is "leaky" (→ p. 154). The reabsorption of water can occur by a *paracellular* route (through leaky tight junctions) or *transcellular* route, i.e., through *water channels* (aquaporin type 1 = *AQP1*) in the two cell membranes. The urine in PCT therefore remains (virtually) isotonic. *Oncotic pressure* (→ p. 378) in the peritubular capillaries pro-

▶

Plate 7.9 **Water Reabsorption, Concentration of Urine I**

A. Countercurrent systems

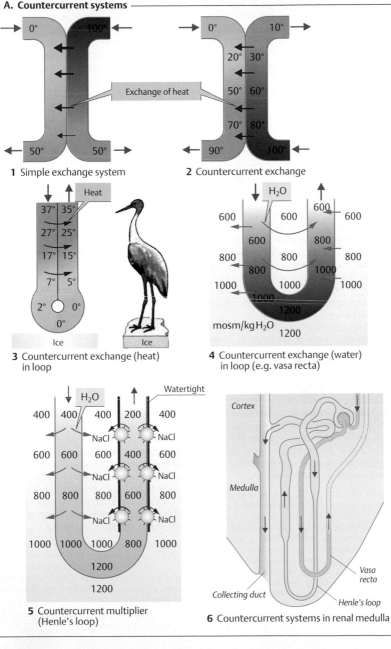

1 Simple exchange system

2 Countercurrent exchange

Exchange of heat

Heat

3 Countercurrent exchange (heat) in loop

Ice

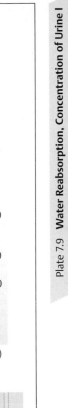

Ice

4 Countercurrent exchange (water) in loop (e.g. vasa recta)

H_2O

mosm/kg H_2O

H_2O

Watertight

NaCl

5 Countercurrent multiplier (Henle's loop)

Cortex

Medulla

Collecting duct

Vasa recta

Henle's loop

6 Countercurrent systems in renal medulla

vides an additional driving force for water reabsorption. The more water filtered at the glomerulus, the higher this oncotic pressure. Thus, the reabsorption of water at the proximal tubule is, to a certain extent, adjusted in accordance with the GFR (**glomerulotubular balance**).

Because the *descending limb* of the **loop of Henle** has aquaporins (AQP1) that make it permeable to water, the urine in it is largely in osmotic balance with the hypertonic interstitium, the content of which becomes increasingly hypertonic as it approaches the papillae ($\rightarrow$ **A5**). The urine therefore becomes increasingly concentrated as it flows in this direction. In the *thin* descending limb, which is only sparingly permeable to salt, this increases the conc. of Na^+ and Cl^-. Most water drawn into the interstitium is carried off by the vasa recta ($\rightarrow$ **B**). Since the *thin* and *thick ascending limbs* of the loop of Henle are largely impermeable to water, Na^+ and Cl^- passively diffuses (thin limb) and is actively transported (thick limb) out into the interstitium ($\rightarrow$ **B**). Since water cannot escape, the urine leaving the loop of Henle is *hypotonic*.

Active reabsorption of Na^+ and Cl^- from the thick ascending limb of the loop of Henle (TAL; $\rightarrow$ p. 162) creates a **local gradient** (ca. $200\,mOsm/kg\ H_2O$; $\rightarrow$ **A5**) at all points between the TAL on the one side and the descending limb and the medullary interstitium on the other. Since the high osmolality of fluid in the medullary interstice is the reason why water is extracted from the collecting duct (see below), active NaCl transport is the ATP-consuming *"motor" for the kidney's urine-concentrating mechanism* and is up-regulated by sustained stimulation of ADH secretion.

Along the course of the **distal convoluted tubule** and, at the latest, at the **connecting tubule**, which contains aquaporins and ADH receptors of type V_2 (explained below), the fluid in the tubule will again become isotonic (in osmotic balance with the isotonic interstice of the renal cortex) if *ADH* is present ($\rightarrow$ p. 168), i.e., when *antidiuresis* occurs. Although Na^+ and Cl^- are still reabsorbed here ($\rightarrow$ p. 162), the osmolality does not change significantly because H_2O is reabsorbed (ca. 5% of the GFR) into the interstitial space due to osmotic forces and *urea* increasingly determines the osmolality of the tubular fluid.

Final adjustment of the excreted urine volume occurs in the **collecting duct**. In the presence of antidiuretic hormone, **ADH** (which binds to basolateral V_2 receptors, named after vasopressin, the synonym for ADH), *aquaporins* (**AQP2**) in the (otherwise water-impermeable) luminal membrane of principal cells are used to extract enough water from the urine passing through the increasingly hypertonic renal medulla. Thereby, the U_{osm} rises about four times higher than the P_{osm} ($U_{osm}/P_{osm} \approx 4$), corresponding to *maximum antidiuresis*. The *absence of ADH* results in *water diuresis*, where U_{osm}/P_{osm} can drop to < 0.3. The U_{osm} can even fall below the osmolality at the end of TAL, since reabsorption of Na^+ and Cl^- is continued in the distal convoluted tubule and collecting duct ($\rightarrow$ p. 162) but water can hardly follow.

Urea also plays an important role in the formation of concentrated urine. A protein-rich diet leads to increased urea production, thus increased the urine-concentrating capacity of the kidney. About 50% of the filtered urea leaves the proximal tubule by diffusion ($\rightarrow$ **C**). Since the ascending limb of the loop of Henle, the distal convoluted tubule, and the cortical and outer medullary sections of the collecting duct are only sparingly permeable to urea, its conc. increases downstream in these parts of the nephron ($\rightarrow$ **C**). ADH can (via V_2 receptors) introduce *urea carriers* (urea transporter type 1, UT1) in the luminal membrane, thereby making the inner medullary collecting duct permeable to urea. Urea now diffuses back into the interstitium (where urea is responsible for half of the high osmolality there) via UT1 and is then transported by UT2 carriers back into the descending limb of the loop of Henle, comprising the **recirculation of urea** ($\rightarrow$ **C**). The non-reabsorbed fraction of urea is excreted: $FE_{urea} \approx$ 40%. Urea excretion increases in water diuresis and decreases in antidiuresis, presumably due to up-regulation of the UT2 carrier.

Urine concentration disorders primarily occur due to (a) excessive medullary blood flow (washing out Na^+, Cl^- and urea); (b) osmotic diuresis; (c) loop diuretics ($\rightarrow$ p. 172); (d) deficient secretion or effectiveness of ADH, as seen in *central* or *peripheral diabetes insipidus*, respectively.

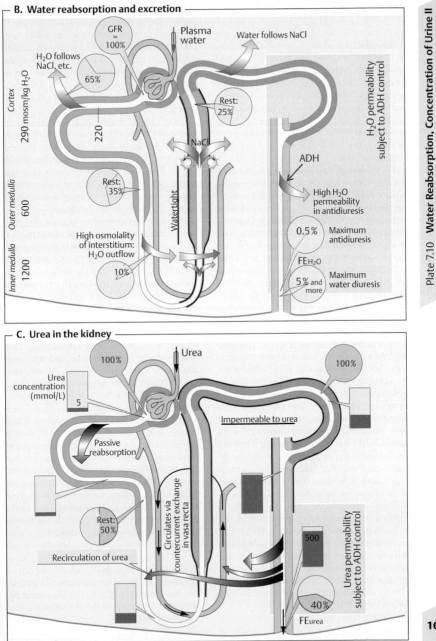

B. Water reabsorption and excretion

Plasma water

GFR = 100%

Water follows NaCl

H_2O follows NaCl, etc.

65%

Cortex
290 mosm/kg H_2O

220

Rest: 25%

H_2O permeability subject to ADH control

NaCl

Watertight

Rest: 35%

Outer medulla
600

ADH

High H_2O permeability in antidiuresis

High osmolality of interstitium: H_2O outflow

Inner medulla
1200

10%

0.5% Maximum antidiuresis

FE_{H_2O}

5% and more Maximum water diuresis

Plate 7.10 **Water Reabsorption, Concentration of Urine II**

C. Urea in the kidney

Urea

100%

100%

Urea concentration (mmol/L)

5

Impermeable to urea

Passive reabsorption

Circulates via countercurrent exchange in vasa recta

Rest: 50%

Recirculation of urea

Urea permeability subject to ADH control

500

40%

FE_{urea}

Body Fluid Homeostasis

Life cannot exist without water. Water is the initial and final product of countless biochemical reactions. It serves as a solvent, transport vehicle, heat buffer, and coolant, and has a variety of other functions. Water is present in cells as *intracellular fluid*, and surrounds them as *extracellular fluid*. It provides a constant environment (*internal milieu*) for cells of the body, similar to that of the primordial sea surrounding the first unicellular organisms (→ p. 2).

The volume of fluid circulating in the body remains relatively constant when the **water balance** (→ **A**) is properly regulated. The average **fluid intake** of ca. 2.5 L per day is supplied by *beverages, solid foods,* and *metabolic oxidation* (→ p. 229 C). The fluid intake must be high enough to counteract **water losses** due to *urination, respiration, perspiration,* and *defecation* (→ p. 265 C). The mean daily H_2O turnover is 2.5 L/70 kg (1/30 th the body weight [BW]) in adults and 0.7 liters/7 kg (1/10th the BW) in infants. The water balance of infants is therefore more susceptible to disturbance.

Significant **rises in the H_2O turnover** can occur, but must be adequately compensated for if the body is to function properly (regulation, → p. 170). Respiratory H_2O losses occur, for example, due to *hyperventilation* at high altitudes (→ pp. 106 and 136), and *perspiration losses* (→ p. 222) occur due to exertion at high temperatures (e.g., hiking in the sun or hot work environment as in an ironworks). Both can lead to the loss of several liters of water per hour, which must be compensated for by increasing the intake of fluids (and salt) accordingly. Conversely, an increased intake of fluids will lead to an increased volume of urine being excreted (→ p. 170).

Water loss (hypovolemia) results in the stimulation of **thirst**, a sensation controlled by the so-called *thirst center in the hypothalamus.* Thirst is triggered by significant rises in the osmolality of body fluids and angiotensin II concentration of cerebrospinal fluid (→ p. 170).

Body water content. The fraction of total body water (TBW) to body weight (BW = 1.0) ranges from **0.46** (46%) **to 0.75** depending on a person's age and sex (→ **B**). The TBW content in *infants* is 0.75 compared to only 0.64 (0.53) in young men (women) and 0.53 (0.46) in elderly men (women). Gender-related differences (and interindividual differences) are mainly due to differences in a person's total *body fat content.* The average fraction of water in most body tissues (in young adults) is 0.73 compared to a fraction of only about 0.2 in fat (→ **B**).

Fluid compartments. In a person with an average TBW of ca. 0.6, about 3/5 (0.35 BW) of the TBW is intracellular fluid (**ICF**), and the other 2/5 (0.25 BW) is extracellular. Extracellular fluid (**ECF**) is located between cells (*interstice,* 0.19), in blood (*plasma water* (0.045) and in "transcellular" compartments (0.015) such as the CSF and intestinal lumen (→ **C**). The *protein concentration* of the plasma is significantly different than that of the rest of the ECF. Moreover, there are fundamental differences in the *ionic composition* of the ECF and the ICF (→ p. 93 B). Since most of the body's supply of Na^+ ions are located in extracellular compartments, the total Na^+ content of the body determines its ECF volume (→ p. 170).

Measurement of fluid compartments. In clinical medicine, the body's fluid compartments are usually measured by *indicator dilution techniques.* Provided the indicator substance, S, injected into the bloodstream spreads to the target compartment only (→ **C**), its volume V can be calculated from:

$$V[L] = \text{injected amount of}$$
$$S \text{ [mol]}/C_S \text{ [mol/L]} \qquad [7.12]$$

where C_S is the concentration of S after it spreads throughout the target compartment (measured in collected blood specimens). The ECF volume is generally measured using *inulin* as the indicator (does not enter cells), and the TBW volume is determined using *antipyrine.* The ICF volume is approximately equal to the antipyrine distribution volume minus the inulin distribution volume. *Evans blue,* a substance entirely bound by plasma proteins, can be used to measure the plasma volume. Once this value is known, the blood volume can be calculated as the plasma volume divided by [1 – hematocrit] (→ p. 88), and the interstitial volume is calculated as the ECF volume minus the plasma volume.

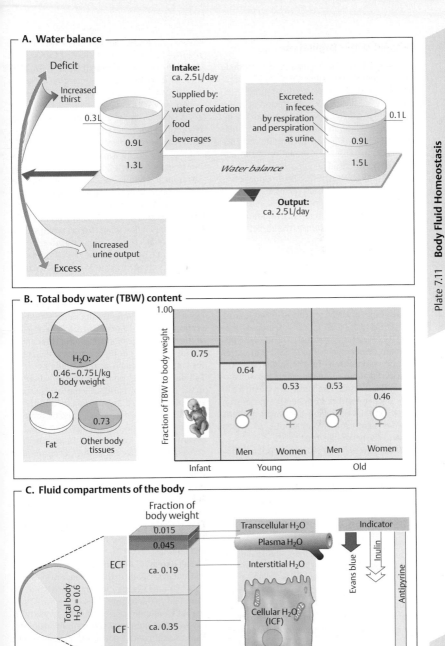

A. Water balance

Deficit

Increased thirst

Intake:
ca. 2.5 L/day

Supplied by:
water of oxidation
food
beverages

0.3 L
0.9 L
1.3 L

Water balance

Excreted:
in feces
by respiration
and perspiration
as urine

0.1 L
0.9 L
1.5 L

Output:
ca. 2.5 L/day

Increased urine output

Excess

B. Total body water (TBW) content

H₂O:
0.46–0.75 L/kg
body weight

0.2

0.73

Fat

Other body tissues

Fraction of TBW to body weight

1.00

0.75

0.64

0.53 0.53

0.46

♂ ♀ ♂ ♀

Men Women Men Women

Infant Young Old

C. Fluid compartments of the body

Fraction of body weight

0.015 — Transcellular H₂O

0.045 — Plasma H₂O

ECF ca. 0.19 — Interstitial H₂O

ICF ca. 0.35 — Cellular H₂O (ICF)

Total body H₂O = 0.6

Indicator

Evans blue

Inulin

Antipyrine

Plate 7.11 Body Fluid Homeostasis

Salt and Water Regulation

Osmoregulation. The osmolality of most body fluids is about 290 mOsm/kg H_2O. Any increase in the osmolality of extracellular fluid (ECF) due, for example, to NaCl absorption or water loss, results in an outflow of water from the intracellular space, because the intracellular fluid (ICF) and ECF are in osmotic balance ($\rightarrow$ p. 173; B2, B6). The osmolality of the ECF must be tightly regulated to protect cells from large volume fluctuations. Osmoregulation is controlled by *osmosensors (or osmoreceptors)* found mainly in the hypothalamus, *hormones* (e.g., antidiuretic hormone = ADH = adiuretin = vasopressin) and the *kidney*, the target organ of ADH ($\rightarrow$ p. 166).

Water deficit ($\rightarrow$ **A1**). Net water losses (hypovolemia) due, for example, to sweating, urination or respiration, make the ECF hypertonic. Osmolality rises of 1% or more (≥ 3 mOsm/kg H_2O) are sufficient to stimulate the secretion of **ADH** from the posterior lobe of the pituitary ($\rightarrow$ p. 280). ADH decreases urinary H_2O excretion ($\rightarrow$ p. 166). The likewise hypertonic cerebrospinal fluid (CSF) stimulates central osmosensors in the hypothalamus, which trigger *hyperosmotic thirst*. The perception of **thirst** results in an urge to replenish the body's water reserves. Peripheral osmosensors in the portal vein region and vagal afferent neurons warn the hypothalamus of water shifts in the gastrointestinal tract.

Water excess ($\rightarrow$ **A2**). The absorption of hypotonic fluid reduces the osmolality of ECF. This signal inhibits the secretion of ADH, resulting in water diuresis ($\rightarrow$ p. 166) and normalization of plasma osmolality within less than 1 hour.

Water intoxication occurs when excessive volumes of water are absorbed too quickly, leading to symptoms of nausea, vomiting and shock. The condition is caused by an undue drop in the plasma osmolality before adequate inhibition of ADH secretion has occurred.

Volume regulation. Around 8–15 g of NaCl are absorbed each day. The kidneys have to excrete the same amount over time to maintain Na^+ and ECF homeostasis ($\rightarrow$ p. 168). Since Na^+ is the major extracellular ion (Cl^- balance is maintained secondarily), *changes in total body*

Na^+ content lead to changes in ECF volume. It is regulated mainly by the following **factors**:

♦ *Renin–angiotensin system (**RAS**)* ($\rightarrow$ p. 184). Its activation promotes the retention of Na^+ via angiotensin II (AT II; lowers GFR), aldosterone ($\rightarrow$ **A4**) and ADH.

♦ *Atriopeptin (atrial natriuretic peptide; **ANP**)* is a peptide hormone secreted by specific cells of the cardiac atrium in response to rises in ECF volume and hence atrial pressure. ANP promotes the renal excretion of Na^+ by raising the filtration fraction ($\rightarrow$ p. 152) and inhibits Na^+ reabsorption from the collecting duct.

♦ **ADH.** ADH secretion is stimulated by (a) *increased plasma* and *CSF osmolality*; (b) the *Gauer-Henry reflex*, which occurs when stretch receptors in the atrium warn the hypothalamus of a decrease ($> 10\%$) in ECF volume ($\sim$ atrial pressure); (c) *angiotensin II* ($\rightarrow$ p. 184).

♦ **Pressure diuresis** ($\rightarrow$ p. 172), caused by an elevated arterial blood pressure, e.g. due to an elevated ECF volume, results in increased excretion of Na^+ and water, thereby lowering ECF volume and hence blood pressure. This control circuit is thought to be the major mechanism for *long term blood pressure regulation*.

Salt deficit ($\rightarrow$ **A3**). When hyponatremia occurs in the presence of a primarily normal H_2O content of the body, blood osmolality and therefore ADH secretion decrease, thereby increasing transiently the excretion of H_2O. The ECF volume, plasma volume, and blood pressure consequently decrease ($\rightarrow$ **A4**). This, in turn, activates the RAS, which triggers *hypovolemic thirst* by secreting AT II and induces Na^+ retention by secreting aldosterone. The retention of Na^+ increases plasma osmolality leading to secretion of ADH and, ultimately, to the retention of water. The additional intake of fluids in response to *thirst* also helps to normalize the ECF volume.

Salt excess ($\rightarrow$ **A4**). An abnormally high NaCl content of the body in the presence of a normal H_2O volume leads to increased plasma osmolality (thirst) and ADH secretion. Thus, the ECF volume rises and RAS activity is curbed. The additional secretion of ANP, perhaps together with a natriuretic hormone with a longer half-life than ANP (ouabain?), leads to increased excretion of NaCl and H_2O and, consequently, to normalization of the ECF volume.

A. Regulation of salt and water balance

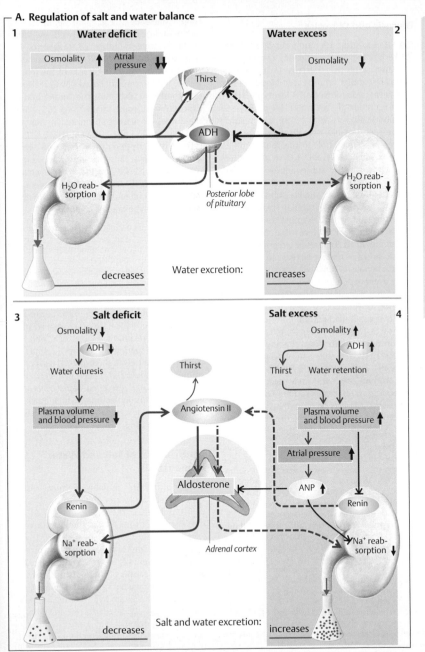

1 Water deficit

Osmolality ↑   Atrial pressure ⬇⬇

Thirst

ADH

H₂O reabsorption ↑

Posterior lobe of pituitary

Water excretion: decreases

2 Water excess

Osmolality ↓

H₂O reabsorption ↓

Water excretion: increases

3 Salt deficit

Osmolality ↓
ADH ↓
Water diuresis
Plasma volume and blood pressure ↓

Thirst

Angiotensin II

Renin

Aldosterone

Adrenal cortex

Na⁺ reabsorption ↑

Salt and water excretion: decreases

4 Salt excess

Osmolality ↑   ADH ↑
Thirst   Water retention
Plasma volume and blood pressure ↑
Atrial pressure ↑
ANP ↑
Renin
Na⁺ reabsorption ↓

Salt and water excretion: increases

Plate 7.12 Salt and Water Regulation

171

Diuresis and Diuretics

Increases in urine excretion above 1 mL/min (**diuresis**) can have the following **causes**:

◆ **Water diuresis**: Decreases in plasma osmolality and/or an increased blood volume lead to the reduction of ADH levels and, thus, to the excretion of "free water" ($\rightarrow$ p. 164).

◆ **Osmotic diuresis** results from the presence of non-reabsorbable, osmotically active substances (e.g., mannitol) in the renal tubules. These substances retain H_2O in the tubule lumen, which is subsequently excreted. Osmotic diuresis can also occur when the concentration of a reabsorbable substance (e.g., glucose) exceeds its tubular reabsorption capacity resulting, for example, in hyperglycemia ($\rightarrow$ p. 158). The glucosuria occurring in diabetes mellitus is therefore accompanied by diuresis and a secondary increase in thirst. Hyperbicarbonaturia can lead to osmotic diuresis to the same reason ($\rightarrow$ p. 176).

◆ **Pressure diuresis** occurs when osmolality in the renal medulla decreases in the presence of increased renal medullary blood flow due, in most cases, to hypertension ($\rightarrow$ p. 170).

◆ **Diuretics** ($\rightarrow$ **A**) are drugs that induce diuresis. Most of them (except osmotic diuretics like mannitol) work primarily by inhibiting NaCl reabsorption (*saluretics*) and, secondarily, by decreasing water reabsorption. The goal of therapeutic diuresis, e.g., in treating edema and hypertension, is to reduce the ECF volume.

Although diuretics basically inhibit NaCl transport throughout the entire body, they have a large degree of **renal "specificity"** because they act from the tubular lumen, where they become highly concentrated due to **tubular secretion** ($\rightarrow$ p. 160) and tubular water reabsorption. Therefore, dosages that do not induce unwanted systemic effects are therapeutically effective in the tubule lumen.

Diuretics of the **carbonic anhydrase inhibitor type** (e.g., acetazolamide, benzolamide) decrease Na^+/H^+ exchange and HCO_3^- reabsorption in the *proximal tubule* ($\rightarrow$ p. 174ff.). The overall extent of diuresis achieved is small because more distal segments of the tubule reabsorb the NaCl not reabsorbed upstream and because the GFR decreases due to tubuloglomerular feedback, TGF ($\rightarrow$ p. 184). In addition, increased HCO_3^- excretion also leads to non-respiratory (metabolic) acidosis. Therefore, this type of diuretic is used only in patients with concomitant alkalosis.

Loop diuretics (e.g., furosemide and bumetanide) are highly effective. They inhibit the *bumetanide-sensitive co-transporter BSC* ($\rightarrow$ p. 162 B6), a Na^+-$2Cl^-$-K^+ symport carrier, in the thick ascending limb (TAL) of the loop of Henle. This not only decreases NaCl reabsorption there, but also stalls the "motor" on the concentration mechanism ($\rightarrow$ p. 166). Since the lumen-positive transepithelial potential (**L**PTP) in the TAL also falls ($\rightarrow$ p. 162 B7), paracellular reabsorption of Na^+, Ca^{2+} and Mg^{2+} is also inhibited. Because increasing amounts of non-reabsorbed Na^+ now arrive at the collecting duct ($\rightarrow$ p. 181 B3), K^+ secretion increases and the simultaneous loss of H^+ leads to *hypokalemia* and hypokalemic *alkalosis*.

Loop diuretics inhibit BSC at the **macula densa**, thereby "tricking" the juxtaglomerular apparatus (JGA) into believing that no more NaCl is present in the tubular lumen. The GFR then rises as a result of the corresponding tubuloglomerular feedback ($\rightarrow$ p. 184), which further promotes diuresis.

Thiazide diuretics inhibit NaCl resorption in the distal tubule (TSC, $\rightarrow$ p. 162 B8). Like loop diuretics, they increase Na^+ reabsorption downstream, resulting in losses of K^+ and H^+.

Potassium-sparing diuretics. *Amiloride* block Na^+ channels in the principal cells of the connecting tubule and collecting duct, leading to a *reduction of K^+ excretion*. *Aldosterone antagonists* (e.g., spironolactone), which block the cytoplasmic aldosterone receptor, also have a potassium-sparing effect.

Disturbances of Salt and Water Homeostasis

When *osmolality* remains *normal*, disturbances of salt and water homeostasis ($\rightarrow$ **B** and p. 170) only affect the ECF volume ($\rightarrow$ **B1** and **4**). When the osmolality of the ECF increases (hyperosmolality) or decreases (hypoosmolality), water in the extra- and intracellular compartments is redistributed ($\rightarrow$ **B2, 3, 5, 6**). The main **causes** of these disturbances are listed in **B** (orange background). The **effects** of these disturbances are hypovolemia in cases 1, 2 and 3, *intracellular edema* (e.g., swelling of the brain) in disturbances 3 and 5, and extracellular edema (pulmonary edema) in disturbances 4, 5 and 6.

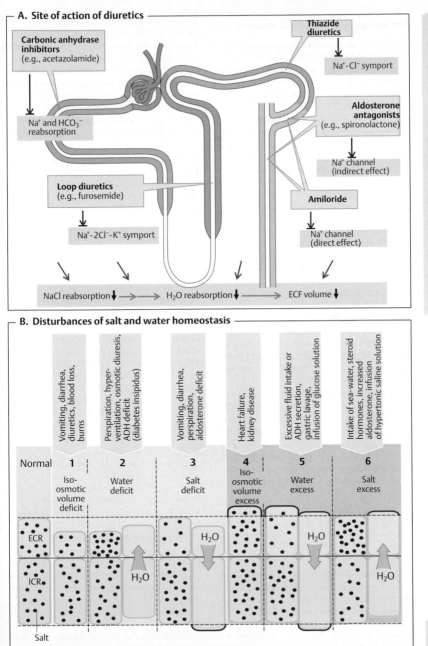

A. Site of action of diuretics

Carbonic anhydrase inhibitors (e.g., acetazolamide)

Thiazide diuretics

Na^+-Cl^- symport

Na^+ and HCO_3^- reabsorption

Aldosterone antagonists (e.g., spironolactone)

Na^+ channel (indirect effect)

Loop diuretics (e.g., furosemide)

Amiloride

Na^+-$2Cl^-$-K^+ symport

Na^+ channel (direct effect)

NaCl reabsorption ↓ → → H₂O reabsorption ↓ → ECF volume ↓

B. Disturbances of salt and water homeostasis

Vomiting, diarrhea, diuretics, blood loss, burns

Perspiration, hyperventilation, osmotic diuresis, ADH deficit (diabetes insipidus)

Vomiting, diarrhea, perspiration, aldosterone deficit

Heart failure, kidney disease

Excessive fluid intake or ADH secretion, gastric lavage, infusion of glucose solution

Intake of sea-water, steroid hormones, increased aldosterone, infusion of hypertonic saline solution

Normal	1	2	3	4	5	6
	Iso-osmotic volume deficit	Water deficit	Salt deficit	Iso-osmotic volume excess	Water excess	Salt excess

ECR

H_2O

H_2O

H_2O

H_2O

ICR

Salt

Plate 7.13 Diuresis and Diuretics

The Kidney and Acid–Base Balance

Main functions of renal **H⁺ secretion** ($\rightarrow$ **A**):
— reabsorption of filtered bicarbonate ($\rightarrow$ **B**),
— excretion of H⁺ ions measurable as titratable acidity ($\rightarrow$ **C**), and
— nonionic transport of NH_4^+, i.e. in the form of NH_3 ($\rightarrow$ **D1, 2**).

1. Very large quantities of H⁺ ions are secreted into the lumen of the **proximal tubule** ($\rightarrow$ **A1**) by (a) primary active transport via *H⁺-ATPase* and (b) by secondary active transport via an electroneutral *Na⁺/H⁺-antiporter* (NHE3 carrier, $\rightarrow$ p. 162). The *luminal pH* then decreases from 7.4 (filtrate) to about 6.6. One OH⁻ ion remains in the cell for each H⁺ ion secreted; OH⁻ reacts with CO_2 to form HCO_3^- (accelerated by carbonic anhydrase-II, see below). HCO_3^- leaves the cell for the blood, where it binds one H⁺ ion. Thus, each H⁺ ion secreted into the lumen (and excreted) results in the elimination of one H⁺ ion from the body, except the secreted H⁺ is accompanied by a secreted NH_3 (see below).

2. In the **connecting tubule** and **collecting duct** ($\rightarrow$ **A2**) *type A intercalated cells* secrete H⁺ ions via H⁺/K⁺-ATPase and H⁺-ATPase, allowing the luminal pH to drop as far as 4.5. In metabolic alkalosis, *type B intercalated cells* can secrete HCO_3^- ($\rightarrow$ **A3**).

Carbonic anhydrase (**CA**) is important in all cases where H⁺ ions exit from one side of a cell and/or HCO_3^- exits from the other, e.g., in renal tubule cells, which contain CA^{II} in the cytosol and CA^{IV} on the outside of the luminal membrane; $\rightarrow$ **A, B, D**), as well as in the stomach, small intestine, pancreatic duct and erythrocytes, etc. CA catalyzes the gross reaction

$$H_2O + CO_2 \rightleftharpoons H^+ + HCO_3^-.$$

Carbonic acid (H_2CO_3) is often considered to be the intermediate product of this reaction, but OH⁻ (not H_2O) probably combines with CA. Therefore, the reactions $H_2O \rightleftharpoons OH^- + H^+$ and $OH^- + CO_2 \rightleftharpoons HCO_3^-$ underlie the aforementioned gross reaction.

Reabsorption of HCO_3^- ($\rightarrow$ **B**). The amount of HCO_3^- filtered each day is 40 times the quantity present in the blood. HCO_3^- must therefore be reabsorbed to maintain acid–base balance ($\rightarrow$ p. 183ff.). The H⁺ ions secreted into the lumen of the proximal convoluted tubule react with about 90% of the filtered HCO_3^- to form

CO_2 and H_2O ($\rightarrow$ **B**). CA^{IV} anchored in the membrane catalyzes this reaction. CO_2 readily diffuses into the cell, perhaps via aquaporin 1 ($\rightarrow$ p. 166). CA^{II} then catalyzes the transformation of $CO_2 + H_2O$ to H⁺ + HCO_3^- within the cell ($\rightarrow$ **B**). The H⁺ ions are again secreted, while HCO_3^- exits through the basolateral membrane of the cell via an electrogenic carrier (hNBC = human Na⁺-bicarbonate co-transporter; $\rightarrow$ **B**). The hNBC co-transports 1 Na⁺ with 3 HCO_3^- (and/or with 1 $HCO_3^- + 1$ CO_3^{2-}?) Thus, HCO_3^- is transported through the luminal membrane in the form of CO_2 (driving force: ΔP_{CO_2}), and exits the cell across the basolateral membrane as HCO_3^- (main driving force: membrane potential).

Hypokalemia leads to a rise in membrane potential (Nernst equation, $\rightarrow$ p. 32) and thus to a rise in basolateral HCO_3^- transport. This results in increased H⁺ secretion and, ultimately, in **hypokalemic alkalosis.**

Urinary acid excretion. If the dietary protein intake is 70 g per day ($\rightarrow$ p. 226), a daily load of about 190 mmol of H⁺ occurs after the amino acids of the protein have been metabolized. HCl (from arginine, lysine and histidine), H_2SO_4 (from methionine and cystine), H_3PO_4, and lactic acid are the main sources of H⁺ ions. They are "fixed" acids which, unlike CO_2, are not eliminated by respiration. Since about 130 mmol H⁺/day are used to break down organic anions (glutamate⁻, aspartate⁻, lactate⁻, etc.), the **net H⁺ production** is about **60** (40–80) **mmol/day**. Although the H⁺ ions are buffered at their production site, they must be excreted to regenerate the buffers.

In extreme cases, the **urinary pH** can rise to about pH 8 (high HCO_3^- excretion) or fall to about pH 4.5 (maximum H⁺ conc. is 0.03 mmol/L). At a daily urine output of 1.5 L, the kidneys will excrete < 1% of the produced H⁺ ions in their free form.

Titratable acids (80% phosphate, 20% uric acid, citric acid, etc.) comprise a significant fraction (10–30 mmol/day) of H⁺ excretion ($\rightarrow$ **C1**). This amount of H⁺ ions can be determined by titrating the urine with NaOH back to the plasma pH value, which is normally pH 7.4 ($\rightarrow$ **C2**). Around 80% of **phosphate** ($pK_a = 6.8$) in the blood occurs in the form of HPO_4^{2-}, whereas about all phosphate in acidic urine occurs as $H_2PO_4^-$ ($\rightarrow$ p. 380), i.e., the secreted

Plate 7.14 The Kidney and Acid–Base Balance I

A. H⁺ secretion

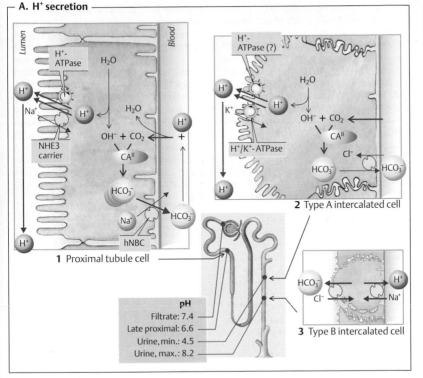

1 Proximal tubule cell

2 Type A intercalated cell

3 Type B intercalated cell

pH
Filtrate: 7.4
Late proximal: 6.6
Urine, min.: 4.5
Urine, max.: 8.2

B. HCO₃⁻ reabsorption

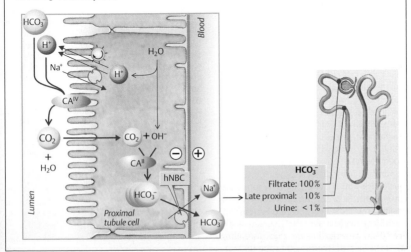

Proximal tubule cell

HCO₃⁻
Filtrate: 100%
Late proximal: 10%
Urine: <1%

H^+ ions are buffered by filtered HPO_4^{2-}. Non-reabsorbed phosphate (5–20% of the filtered quantity, $\rightarrow$ p. 178) is therefore loaded with H^+ ions, about half of it in the proximal tubule (pH 7.4 $\Rightarrow$ ca. 6.6), and the rest in the collecting duct (pH 6.6 $\Rightarrow$ 4.5) ($\rightarrow$ **C1**). When *acidosis* occurs, increased quantities of phosphate are mobilized from the bone and excreted. The resulting increase in H^+ excretion precedes the increased NH_4^+ production associated with acidosis (see below).

Excretion of ammonium ions ($NH_4^+ \rightleftharpoons NH_3 + H^+$), about 25–50 mmol/day on average diet, is *equivalent* to H^+ disposal and is therefore an *indirect* form of H^+ excretion ($\rightarrow$ **D**). NH_4^+ is *not* a titratable form of acidity. Unlike $HPO_4^{2-} + H^+ \rightleftharpoons H_2PO_4^-$, the reaction $NH_3 + H^+ \rightleftharpoons NH_4^+$ does not function in the body as a buffer because of its high pK_a value of ca. 9.2. Nevertheless, for every NH_4^+ excreted by the kidney, one HCO_3^- is spared by the liver. This is equivalent to one H^+ disposed since the spared HCO_3^- ion can buffer a H^+ ion. With an average dietary intake of protein, the amino acid metabolism produces roughly equimolar amounts of HCO_3^- and NH_4^+ (ca. 700–1000 mmol/day). The liver utilizes about 95% of these two products to produce **urea** ($\rightarrow$ **D1**):

$$2\,HCO_3^- + 2\,NH_4^+ \rightleftharpoons \underset{\underset{O}{\|}}{H_2N\text{-}C\text{-}NH_2} + CO_2 + 3\,H_2O \qquad [7.13]$$

Thus, one HCO_3^- less is consumed for each NH_4^+ that passes from the liver to the kidney and is eliminated in the urine. Before exporting NH_4^+ to the kidney, the **liver** incorporates it into glutamate yielding **glutamine**; only a small portion reaches the kidney as free NH_4^+. High levels of $NH_4^+ \rightleftharpoons NH_3$ are toxic.

In the **kidney,** *glutamine* enters proximal tubule cells by Na^+ symport and is cleaved by mitochondrial *glutaminase*, yielding NH_4^+ and glutamate$^-$ (Glu$^-$). Glu$^-$ is further metabolized by *glutamate dehydrogenase* to yield α-ketoglutarate^{2-}, producing a second NH_4^+ ion ($\rightarrow$ **D2**). The NH_4^+ can reach the tubule lumen on two ways: (1) it dissociates within the cell to yield NH_3 and H^+, allowing NH_3 to diffuse (non-ionically, $\rightarrow$ p. 22) into the lumen, where it re-joins the separately secreted H^+ ions; (2) the NHE3 carrier secretes NH_4^+ (instead of H^+). Once NH_4^+ has arrived at *the thick ascending limb* of the loop of Henle ($\rightarrow$ **D4**), the BSC car-

rier ($\rightarrow$ p. 162) reabsorbs NH_4^+ (instead of K^+) so that it remains in the *renal medulla*. Recirculation of NH_4^+ through the loop of Henle yields a very high conc. of $NH_4^+ \rightleftharpoons NH_3 + H^+$ towards the papilla ($\rightarrow$ **D3**). While the H^+ ions are then actively pumped into the lumen of the collecting duct ($\rightarrow$ **A2, D4**), the NH_3 molecules arrive there by non-ionic diffusion ($\rightarrow$ **D4**). The NH_3 gradient required to drive this diffusion can develop because the especially low luminal pH value (about 4.5) leads to a much smaller NH_3 conc. in the lumen than in the medullary interstitium where the pH is about two pH units higher and the NH_3 conc. is consequently about 100-times higher than in the lumen.

Disturbances of acid–base metabolism (see also p. 142ff.). When **chronic non-respiratory acidosis** of non-renal origin occurs, *NH_4^+ excretion rises* to about 3 times the normal level within 1 to 2 days due to a parallel increase in hepatic glutamine production (at the expense of urea formation) and renal glutaminase activity. **Non-respiratory alkalosis** only decreases the renal NH_4^+ production and H^+ secretion. This occurs in conjunction with an increase in filtered HCO_3^- (increased plasma concentration, $\rightarrow$ p. 144), resulting in a sharp rise in HCO_3^- excretion and, consequently, in osmotic diuresis ($\rightarrow$ p. 172). To **compensate for respiratory disturbances** ($\rightarrow$ p. 144), it is important that increased (or decreased) plasma P_{CO_2} levels result in increased (or decreased) H^+ secretion and, thus, in increased (or decreased) HCO_3^- resorption.

The kidney can also be the primary site of an acid–base disturbance (**renal acidosis**), with the defect being either generalized or isolated. In a generalized defect, as observed in *renal failure*, acidosis occurs because of reduced H^+ excretion. In an isolated defect with disturbance of proximal H^+ secretion, large portions of filtered HCO_3^- are not reabsorbed, leading to *proximal renal tubular acidosis*. When impaired renal H^+ secretion occurs in the collecting duct, the urine can no longer be acidified (pH > 6 despite acidosis) and the excretion of titratable acids and NH_4^+ is consequently impaired (*distal renal tubular acidosis*).

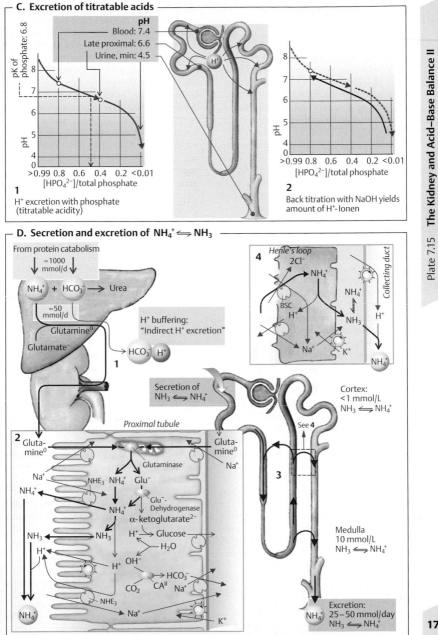

C. Excretion of titratable acids

pH
- Blood: 7.4
- Late proximal: 6.6
- Urine, min: 4.5

1 H⁺ excretion with phosphate (titratable acidity)

2 Back titration with NaOH yields amount of H^+-Ionen

D. Secretion and excretion of $NH_4^+ \rightleftharpoons NH_3$

From protein catabolism
≈1000 mmol/d

NH_4^+ + HCO_3^- → Urea

≈50 mmol/d

Glutamine⁰

Glutamate⁻

H^+ buffering: "Indirect H⁺ excretion"
HCO_3^- H^+

4 Henle's loop 2Cl⁻
NH_4^+
BSC
H^+
Na^+
K^+
Collecting duct
NH_4^+
NH_3
H^+
NH_4^+

Secretion of $NH_3 \rightleftharpoons NH_4^+$

Cortex: <1 mmol/L $NH_3 \rightleftharpoons NH_4^+$

Proximal tubule

2 Glutamine⁰
Glutamine⁰
Na^+
Na^+
NHE₃ NH_4^+ Glu⁻
Glutaminase
NH_4^+
NH_3
NH_3
H^+
H^+
NH_4^+
NHE₃
Na^+
K^+

Glu⁻Dehydrogenase
α-ketoglutarate²⁻
H^+ → Glucose
→ H_2O
OH^-
CO_2 CA^{II} HCO_3^-
Na^+

See **4**

3

Medulla 10 mmol/L $NH_3 \rightleftharpoons NH_4^+$

Excretion: 25–50 mmol/day
NH_4^+ $NH_3 \rightleftharpoons NH_4^+$

Plate 7.15 The Kidney and Acid–Base Balance II

177

Reabsorption and Excretion of Phosphate, Ca²⁺ and Mg²⁺

Phosphate metabolism. The plasma phosphate conc. normally ranges from 0.8–1.4 mmol/L, and a corresponding amount of ca. 150–250 mmol/day of inorganic phosphate P_i ($HPO_4^{2-} \rightleftharpoons H_2PO_4^-$) is filtered each day, a large part of which is reabsorbed. The fractional excretion ($\rightarrow$ **A1**), which ranges between 5 and 20%, functions to balance P_i, H^+, and Ca^{2+}. P_i excretion rises in the presence of a P_i *excess* (elevated P_i levels in plasma) and falls during a P_i *deficit*. *Acidosis* also results in phosphaturia and increased H^+ excretion (titratable acidity, $\rightarrow$ p. 174ff.). This also occurs in phosphaturia of other causes. *Hypocalcemia* and *parathyrin* also induce a rise in P_i excretion ($\rightarrow$ **A3** and p. 290f.).

P_i *is reabsorbed at the proximal tubule* ($\rightarrow$ **A2,3**). Its luminal membrane contains the type 3 Na^+-P_i *symport carrier* (NaPi-3). The carrier accepts $H_2PO_4^-$ and HPO_4^{2-} and cotransports it with Na^+ by secondary active transport ($\rightarrow$ p. 26ff.).

Regulation of P_i reabsorption. P_i deficits, alkalosis, hypercalcemia, and low PTH levels result in the increased incorporation of NaPi-3 transporters into the luminal membrane, whereas P_i excesses, acidosis, hypocalcemia and increased PTH secretion results in internalization (down-regulation) and subsequent lysosomal degradation of NaPi-3 ($\rightarrow$ **A3**).

Calcium metabolism (see also p. 36). Unlike the Na^+ metabolism, the calcium metabolism is regulated mainly by absorption of Ca^{2+} in the gut ($\rightarrow$ p. 290ff.) and, secondarily, by renal excretory function. *Total plasma calcium* (bound calcium + ionized Ca^{2+}) is a mean 2.5 mmol/L. About 1.3 mmol/L of this is present as *free, ionized Ca^{2+}*, 0.2 mmol/L forms *complexes* with phosphate, citrate, etc., and the rest of 1 mmol/L is *bound to plasma proteins* and, thus, not subject to glomerular filtration ($\rightarrow$ p. 154). *Fractional excretion of Ca^{2+}* (FE_{Ca}) in the urine is 0.5%–3% ($\rightarrow$ **A1**).

Ca^{2+} **reabsorption** occurs practically throughout the entire nephron ($\rightarrow$ **A1,2**). The reabsorption of filtered Ca^{2+} occurs to about 60% in the proximal tubule and about 30% in the thick ascending limb (TAL) of the loop of Henle and is paracellular, i.e., passive ($\rightarrow$ **A4a** and p. 163 B5, B7). The lumen-positive transepithelial potential (L**P**TP) provides most of the driving force for this activity. Since Ca^{2+} reabsorption in TAL depends on NaCl reabsorption, *loop diuretics* ($\rightarrow$ p. 172) inhibit Ca^{2+} reabsorption there. **PTH** *promotes Ca^{2+} reabsorption* in TAL as well as in the distal convoluted tubule, where Ca^{2+} is reabsorbed by transcellular active transport ($\rightarrow$ **A4b**). Thereby, Ca^{2+} influx into the cell is passive and occurs via *luminal Ca^{2+} channels*, and Ca^{2+} efflux is active and occurs via *Ca^{2+}-ATPase* (primary active Ca^{2+} transport) and via the *3 Na^+/1 Ca^{2+} antiporter* (secondary active Ca^{2+} transport). Acidosis inhibits Ca^{2+} reabsorption via unclear mechanisms.

Urinary calculi usually consist of *calcium phosphate* or *calcium oxalate*. When Ca^{2+}, Pi or oxalate levels are increased, the solubility product will be exceeded but calcium complex formers (e.g., citrate) and inhibitors of crystallization (e.g., nephrocalcin) normally permit a certain degree of supersaturation. Stone formation can occur if there is a deficit of these substances or if extremely high urinary concentrations of Ca^{2+}, Pi and oxalate are present (applies to all three in pronounced antidiuresis).

Magnesium metabolism and **reabsorption.** Since part of the magnesium in plasma (0.7–1.2 mmol/L) is protein-bound, the Mg^{2+} conc. in the filtrate is only 80% of the plasma magnesium conc. Fractional excretion of Mg^{2+}, FE_{Mg}, is 3–8% ($\rightarrow$ **A1,2**). Unlike Ca^{2+}, however, only about 15% of the filtered Mg^{2+} ions leave the proximal tubule. About 70% of the Mg^{2+} is subject to paracellular reabsorption in the TAL ($\rightarrow$ **A4** and p. 163 B5, B7). Another 10% of the Mg^{2+} is subject to transcellular reabsorption in the distal tubule ($\rightarrow$ **A4b**), probably like Ca^{2+} (see above).

Mg^{2+} **excretion** is stimulated by hypermagnesemia, hypercalcemia, hypervolemia and loop diuretics, and is inhibited by Mg^{2+} deficit, Ca^{2+} deficit, volume deficit, PTH and other hormones that mainly act in the TAL.

The kidney has **sensors** for divalent cations like Ca^{2+} and Mg^{2+} ($\rightarrow$ p. 36). When activated, the sensors inhibit NaCl reabsorption in the TAL which, like loop diuretics, reduces the driving force for paracellular cation resorption, thereby diminishing the normally pronounced Mg^{2+} reabsorption there.

A. Reabsorption of phosphate, Ca²⁺ and Mg²⁺

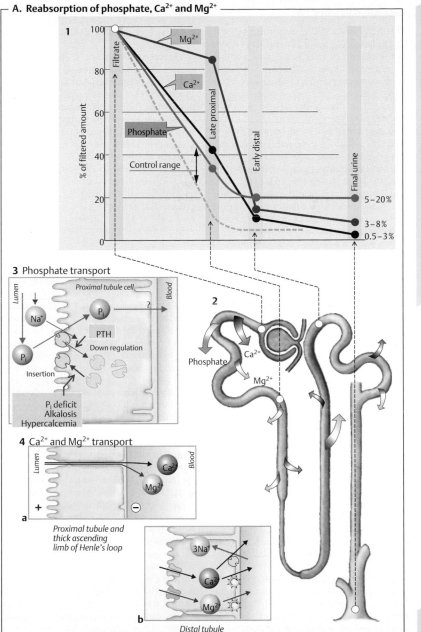

1

Mg²⁺

Ca²⁺

Phosphate

Control range

% of filtered amount

Filtrate

Late proximal

Early distal

Final urine

5–20%

3–8%

0.5–3%

3 Phosphate transport

Lumen — Proximal tubule cell — Blood

Na⁺

Pᵢ

Pᵢ — ?

PTH

Down regulation

Insertion

Pᵢ deficit
Alkalosis
Hypercalcemia

2

Phosphate

Ca²⁺

Mg²⁺

4 Ca²⁺ and Mg²⁺ transport

Lumen — Blood

Ca²⁺

Mg²⁺

+

−

a

Proximal tubule and
thick ascending
limb of Henle's loop

3Na⁺

Ca²⁺

Mg²⁺

b

Distal tubule

Plate 7.16 **Reabsorption and Excretion of Phosphate**

179

Potassium Balance

The dietary intake of K^+ is about 100 mmol/day (minimum requirement: 25 mmol/day). About 90% of intake is excreted in the urine, and 10% is excreted in the feces. The plasma K^+ conc. normally ranges from 3.5 to 4.8 mmol/L, while intracellular K^+ conc. can be more than 30 times as high (due to the activity of Na^+-K^+-ATPase; $\rightarrow$ **A**). Therefore, about 98% of the ca. 3000 mmol of K^+ ions in the body are present in the cells. Although the extracellular K^+ conc. comprises only about 2% of total body K^+, it is still very important because (a) it is needed for regulation of K^+ homeostasis and (b) relatively small changes in cellular K^+ (influx or efflux) can lead to tremendous changes in the plasma K^+ conc. (with an associated risk of cardiac arrhythmias). Regulation of K^+ homeostasis therefore implies *distribution of K^+* through intracellular and extracellular compartments and *adjustment of K^+ excretion according to K^+ intake.*

Acute regulation of the extracellular K^+ conc. is achieved by internal **shifting of K^+** between the extracellular fluid and intracellular fluid ($\rightarrow$ **A**). This relatively rapid process prevents or mitigates dangerous rises in extracellular K^+ (hyperkalemia) in cases where large quantities of K^+ are present due to high dietary intake or internal K^+ liberation (e.g., in sudden hemolysis). The associated K^+ shifting is mainly subject to *hormonal control*. The **insulin** secreted after a meal stimulates Na^+-K^+-ATPase and distributes the K^+ supplied by the food to the animal and vegetable cells of the food to the cells of the body. This is also the case in diet-independent hyperkalemia, which stimulates insulin secretion per se. **Epinephrine** likewise increases cellular K^+ uptake, which is particularly important in muscle work and trauma—two situations that lead to a rise in plasma K^+. In both cases, the increased epinephrine levels allow the re-uptake of K^+ in this and other cells. *Aldosterone* also increases the intracellular K^+ conc. (see below).

Changes in pH affect the intra- and extracellular distribution of K^+ ($\rightarrow$ **A**). This is mainly because the ubiquitous Na^+/H^+ antiporter works faster in alkalosis and more slowly in acidosis ($\rightarrow$ **A**). In acidosis, Na^+ influx therefore decreases, Na^+-K^+-ATPase slows down, and the extracellular K^+ concentration rises (especially in non-respiratory acidosis, i.e., by 0.6 mmol/L per 0.1 unit change in pH). Alkalosis results in hypokalemia.

Chronic regulation of K^+ homeostasis is mainly achieved by the **kidney** ($\rightarrow$ **B**). K^+ is subject to free glomerular filtration, and most of the filtered K^+ is normally reabsorbed (net *reabsorption*). The excreted amount can, in some cases, exceed the filtered amount (net *secretion*, see below). About 65% of the filtered K^+ is reabsorbed before reaching the end of the *proximal tubule*, regardless of the K^+ supply. This is comparable to the percentage of Na^+ and H_2O reabsorbed ($\rightarrow$ **B1** and p. 157, column 2). This type of K^+ transport is mainly paracellular and therefore passive. Solvent drag ($\rightarrow$ p. 24) and the lumen-positive transepithelial potential, LPTP ($\rightarrow$ **B1** and p. 162), in the mid and late proximal segments of the tubule provide the driving forces for it. In the *loop of Henle*, another 15% of the filtered K^+ is reabsorbed by trans- and paracellular routes ($\rightarrow$ **B2**). The amount of K^+ excreted is determined in the *connecting tubule and collecting duct*. Larger or smaller quantities of K^+ are then either reabsorbed or secreted according to need. In extreme cases, the fractional excretion of K^+ (FE_K) can rise to more than 100% in response to a high K^+ intake, or drop to about 3–5% when there is a K^+ deficit ($\rightarrow$ **B**).

Cellular mechanisms of renal K^+ transport. The connecting tubule and collecting duct contain **principal cells** ($\rightarrow$ **B3**) that reabsorb Na^+ and secrete K^+. Accumulated intracellular K^+ can exit the cell through *K^+ channels* on either side of the cell. The electrochemical K^+ gradient across the membrane in question is decisive for the efflux of K^+. The luminal membrane of principal cells also contains *Na^+ channels* through which Na^+ enters the cell ($\rightarrow$ p. 162). This depolarizes the luminal membrane, which reaches a potential of about -20 mV, while the basolateral membrane maintains its normal potential of ca. -70 mV ($\rightarrow$ **B3**). The driving force for K^+ efflux ($E_m - E_K$, $\rightarrow$ p. 32) is therefore higher on the luminal side than on the opposite side. Hence, K^+ preferentially exits the cell toward the lumen (*secretion*). This is mainly why K^+ secretion is coupled with Na^+ reabsorption, i.e., the more Na^+ reabsorbed by the principle cell, the more K^+ secreted.

▶

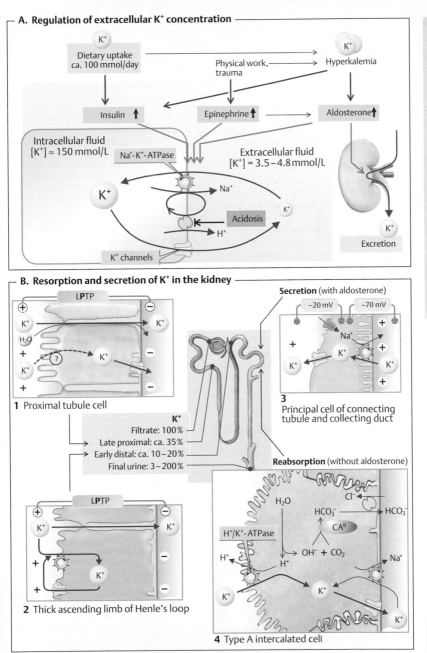

Plate 7.17 Potassium Balance I

A. Regulation of extracellular K⁺ concentration

K⁺

Dietary uptake
ca. 100 mmol/day

Physical work,
trauma

K⁺

Hyperkalemia

Insulin ↑

Epinephrine ↑

Aldosterone ↑

Intracellular fluid
[K⁺] ≈ 150 mmol/L

Na⁺-K⁺-ATPase

Extracellular fluid
[K⁺] = 3.5–4.8 mmol/L

K⁺

Na⁺

K⁺

Acidosis

H⁺

K⁺ channels

K⁺

Excretion

B. Resorption and secretion of K⁺ in the kidney

LPTP

K⁺

H₂O

?

K⁺

K⁺

K⁺

1 Proximal tubule cell

K⁺
Filtrate: 100%
Late proximal: ca. 35%
Early distal: ca. 10–20%
Final urine: 3–200%

Secretion (with aldosterone)

–20 mV   –70 mV

Na⁺

K⁺

K⁺

K⁺

3
Principal cell of connecting
tubule and collecting duct

Reabsorption (without aldosterone)

LPTP

K⁺

K⁺

K⁺

2 Thick ascending limb of Henle's loop

H₂O

Cl⁻

HCO₃⁻

HCO₃⁻

CA^II

H⁺/K⁺-ATPase

OH⁻ + CO₂

H⁺

H⁺

Na⁺

K⁺

K⁺

K⁺

4 Type A intercalated cell

Another apparent reason is that the reabsorption-related increase in intracellular Na^+ concentration decreases the driving force for the $3 Na^+/Ca^{2+}$ exchange at the basolateral cell membrane, resulting in a rise in the **cytosolic Ca^{2+}** concentration. This rise acts as a signal for more frequent opening of luminal K^+ channels.

Type A **intercalated cells** ($\rightarrow$ **B4**) can active reabsorb K^+ in addition to secreting H^+ ions. Like the parietal cells of the stomach, their luminal membrane contains a H^+/K^+-ATPase for this purpose.

Factors that affect K^+ excretion ($\rightarrow$ C):

1. An **increased K^+ intake** raises the intracellular and plasma K^+ concentrations, which thereby increases the chemical driving force for K^+ secretion.

2. Blood pH: The intracellular K^+ conc. in renal cells *rises in alkalosis* and *falls in acute acidosis*. This leads to a simultaneous fall in K^+ excretion, which again rises in chronic acidosis. The reasons for this are that (a) acidosis-related inhibition of Na^+-K^+-ATPase reduces proximal Na^+ reabsorption, resulting in increased distal urinary flow (see no. 3), and (b) the resulting hyperkalemia stimulates aldosterone secretion (see no. 4).

3. If there is **increased urinary flow** in the connecting tubule and collecting duct (e.g., due to a high Na^+ intake, osmotic diuresis, or other factors that inhibit Na^+ reabsorption upstream), larger quantities of K^+ will be excreted. This explains the *potassium-losing effect of certain diuretics* ($\rightarrow$ p. 173). The reason for this is, presumably, that K^+ secretion is limited at a certain luminal K^+ concentration. Hence, the larger the volume/time, the more K^+ taken away over time.

4. Aldosterone leads to retention of Na^+, an increase in extracellular volume ($\rightarrow$ p. 184), a moderate increase in H^+ secretion (cellular pH rises), and *increased K^+ excretion*. It also increases the number of Na^+-K^+-ATPase molecules in the target cells and leads to a chronic increase in mitochondrial density in K^+ adaptation, for example (see below).

Cellular mechanisms of aldosterone effects. The increase in Na^+ reabsorption is achieved by increased production of transport proteins, called aldosterone-induced proteins (AIPs). This is a genome-mediated effect that begins approx. 30 min to 1 hour after al-

dosterone administration or secretion. The maximum effects are observed after several hours. Aldosterone increases Na^+ reabsorption, thereby depolarizing the luminal cell membrane ($\rightarrow$ **B3**). Consequently, it increases the driving force for K^+ secretion and increases K^+ conductance by increasing the pH of the cell. Both effects lead to increased K^+ secretion. Aldosterone also has a very rapid (few seconds to minutes) non-genomic effect on the cell membrane, the physiological significance of which has yet to be explained.

The capacity of the K^+ excretory mechanism increases in response to long-term increases in the K^+ supply (**K^+ adaptation**). Even when renal function is impaired, this largely maintains the K^+ balance in the remaining, intact parts of the tubular apparatus. The colon can then take over more than $1/3$ of the K^+ excretion.

Mineralocortico(stero)ids. Aldosterone is the main mineralocorticoid hormone synthesized and secreted by the *zona glomerulosa of the adrenal cortex* ($\rightarrow$ **D** and p. 294ff.). As with other steroid hormones, aldosterone is not stored, but is synthesized as needed. The principal function of aldosterone is to regulate Na^+ and K^+ transport in the kidney, gut, and other organs ($\rightarrow$ **D**). **Aldosterone secretion** increases in response to (a) drops in blood volume and blood pressure (mediated by angiotensin II; $\rightarrow$ p. 184) and (b) hyperkalemia ($\rightarrow$ **D**). Aldosterone synthesis is inhibited by atriopeptin ($\rightarrow$ p. 171 A4).

Normal **cortisol** concentrations are *not* effective at the aldosterone receptor only because cortisol is converted to cortisone by an 11β-hydroxysteroid oxidoreductase in aldosterone's target cells.

Hyperaldosteronism can be either *primary* (aldosterone-secreting tumors of adrenal cortex, as observed in *Conn's syndrome*) or *secondary* (in volume depletion, $\rightarrow$ p. 184). *Na^+ retention* resulting in high ECF volumes and *hypertension* as well as a simultaneous *K^+ losses* and hypokalemic *alkalosis* are the consequences. When more than about 90% of the adrenal cortex is destroyed, e.g. by autoimmune adrenalitis, metastatic cancer or tuberculosis, primary chronic **adrenocortical insufficiency** develops (*Addison's disease*). The *aldosterone deficit* leads to a sharp increase in Na^+ excretion, resulting in hypovolemia, hypotension and K^+ retention (hyperkalemia). As glucocorticoid deficiency also develops, complications can be life-threatening, especially under severe stress (infections, trauma). If only one gland is destroyed, *ACTH* causes hypertrophy of the other ($\rightarrow$ p. 297 A).

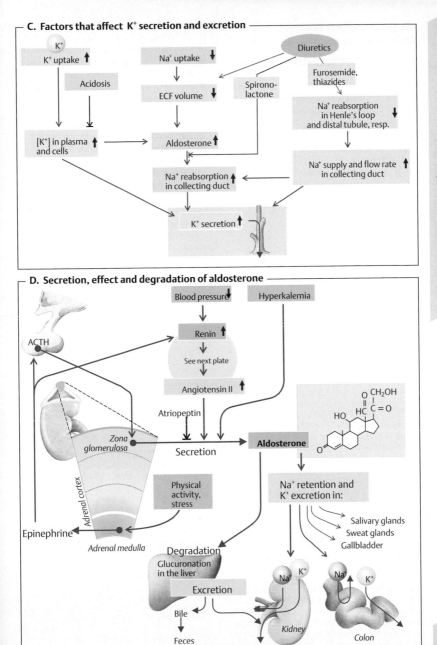

C. Factors that affect K⁺ secretion and excretion

K⁺
K⁺ uptake ↑

Acidosis

[K⁺] in plasma and cells ↑

Na⁺ uptake ↓

ECF volume ↓

Aldosterone ↑

Spirono-lactone

Na⁺ reabsorption in collecting duct ↑

K⁺ secretion ↑

Diuretics

Furosemide, thiazides

Na⁺ reabsorption in Henle's loop and distal tubule, resp. ↓

Na⁺ supply and flow rate in collecting duct ↑

D. Secretion, effect and degradation of aldosterone

ACTH

Blood pressure ↓

Hyperkalemia

Renin ↑

See next plate

Angiotensin II ↑

Atriopeptin

Zona glomerulosa

Secretion

Adrenal cortex

Physical activity, stress

Adrenal medulla

Epinephrine

Aldosterone

Na⁺ retention and K⁺ excretion in:

Salivary glands
Sweat glands
Gallbladder

Degradation
Glucuronation in the liver

Excretion

Bile

Feces

Na⁺ K⁺

Kidney

Na⁺ K⁺

Colon

Plate 7.18 Potassium Balance II

Tubuloglomerular Feedback, Renin–Angiotensin System

The **juxtaglomerular apparatus** (JGA) consists of (a) juxtaglomerular cells of the *afferent arteriole* (including renin-containing and sympathetically innervated granulated cells) and *efferent arteriole*, (b) *macula densa cells* of the thick ascending limb of the loop of Henle and (c) juxtaglomerular mesangial cells (*polkissen*, → **A**) of a given nephron (→ **A**).

JGA functions: (1) local transmission of tubuloglomerular feedback (**TGF**) at its own nephron via angiotensin II (ATII) and (2) systemic production of angiotensin II as part of the renin–angiotensin system (**RAS**).

Tubuloglomerular feedback (TGF). Since the daily GFR is 10 times larger than the total ECF volume (→ p. 168), the excretion of salt and water must be precisely adjusted according to uptake. **Acute changes in the GFR** of the individual nephron (iGFR) and the amount of NaCl filtered per unit time can occur for several reasons. An excessive iGFR is associated with the risk that the distal mechanisms for NaCl reabsorption will be overloaded and that too much NaCl and H_2O will be lost in the urine. A too low iGFR means that too much NaCl and H_2O will be retained. The extent of NaCl and H_2O reabsorption in the proximal tubule determines how quickly the tubular urine will flow through the loop of Henle. When less is absorbed upstream, the urine flows more quickly through the thick ascending limb of the loop, resulting in a lower extent of urine dilution (→ p. 162) and a higher NaCl concentration at the macula densa, $[NaCl]_{MD}$. If the $[NaCl]_{MD}$ becomes too high, the afferent arteriole will constrict to curb the GFR of the affected nephron within 10 s or vice versa (*negative feedback*). It is unclear how the $[NaCl]_{MD}$ results in the signal to constrict, but type 1 A angiotensin II (AT$_{1A}$) receptors play a key role.

If, however, the $[NaCl]_{MD}$ changes due to **chronic shifts in total body NaCl** and an associated change in ECF volume, rigid coupling of the iGFR with the $[NaCl]_{MD}$ through TGF would be fatal. Since long-term increases in the ECF volume reduce proximal NaCl reabsorption, the $[NaCl]_{MD}$ would increase, resulting in a decrease in the GFR and a further increase in the ECF volume. The reverse occurs in ECF volume deficit. To prevent these effects, the *[NaCl]$_{MD}$/iGFR response curve* is shifted in the appropriate direction by certain substances. Nitric oxide (NO) shifts the curve when there is an ECF excess (increased iGFR at same $[NaCl]_{MD}$), and (only locally effective) angiotensin II shifts it in the other direction when there is an ECF deficit.

Renin–angiotensin system (RAS). If the mean renal blood pressure acutely drops below 90 mmHg or so, renal baroreceptors will trigger the release of **renin**, thereby increasing the systemic plasma renin conc. Renin is a peptidase that catalyzes the cleavage of *angiotensin I* from the renin substrate *angiotensinogen* (released from the liver). Angiotensin-converting enzyme (*ACE*) produced in the lung, etc. cleaves two amino acids from angiotensin I to produce **angiotensin II** approx. 30–60 minutes after the drop in blood pressure (→ **B**).

Control of the RAS (→ **B**). The blood pressure *threshold* for renin release is raised by α$_1$-adrenoceptors, and *basal renin secretion* is increased by β$_1$-adrenoceptors. *Angiotensin II* and *aldosterone* are the most important effectors of the RAS. Angiotensin II stimulates the release of aldosterone by adrenal cortex (see below). Both hormones directly (fast action) or indirectly (delayed action) lead to a renewed increase in arterial blood pressure (→ **B**), and renin release therefore decreases to normal levels. Moreover, both hormones *inhibit renin release* (negative feedback).

If the mean blood pressure is decreased in only one kidney (e.g., due to stenosis of the affected renal artery), the affected kidney will also start to release more renin which, in this case, will lead to **renal hypertension** in the rest of the circulation.

Angiotensin II effects: Beside altering the structure of the myocardium and blood vessels (mainly via AT$_2$ receptors), angiotensin II has the following fast or delayed effects mediated by AT$_1$ receptors (→ **A**).

♦ *Vessels*: Angiotensin II has potent vasoconstrictive and hypertensive action, which (via endothelin) takes effect in the arterioles (fast action).
♦ *CNS*: Angiotensin II takes effect in the hypothalamus, resulting in vasoconstriction through the circulatory "center" (rapid action). It also increases ADH secretion in the hypothalamus, which stimulates thirst and a craving for salt (delayed action).
♦ *Kidney*: Angiotensin II plays a major role in regulating renal circulation and GFR by constricting of the afferent and/or efferent arteriole (delayed action; cf. autoregulation, → p. 150). It directly stimulates Na$^+$ reabsorption in the proximal tubule (delayed action).
♦ *Adrenal gland*: Angiotensin II stimulates aldosterone synthesis in the adrenal cortex (delayed action; → p. 182) and leads to the release of epinephrine in the adrenal medulla (fast action).

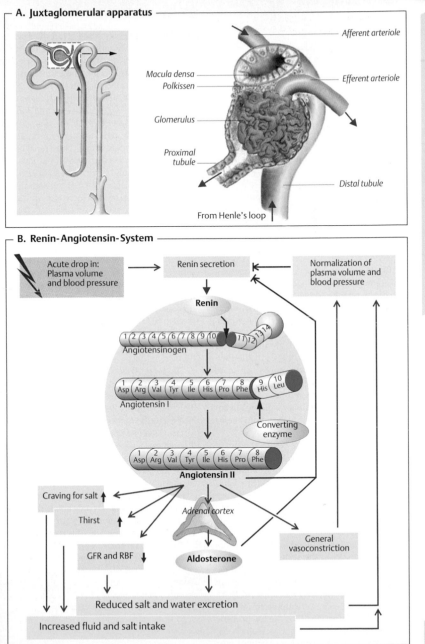

A. Juxtaglomerular apparatus

Afferent arteriole

Macula densa
Polkissen
Efferent arteriole

Glomerulus

Proximal tubule

Distal tubule

From Henle's loop

B. Renin-Angiotensin-System

Acute drop in: Plasma volume and blood pressure → Renin secretion ← Normalization of plasma volume and blood pressure

Renin

1 2 3 4 5 6 7 8 9 10 11 12 13 14
Angiotensinogen

1 Asp 2 Arg 3 Val 4 Tyr 5 Ile 6 His 7 Pro 8 Phe 9 His 10 Leu
Angiotensin I

Converting enzyme

1 Asp 2 Arg 3 Val 4 Tyr 5 Ile 6 His 7 Pro 8 Phe
Angiotensin II

Craving for salt ↑

Thirst ↑

Adrenal cortex

Aldosterone

GFR and RBF ↓

General vasoconstriction

Reduced salt and water excretion

Increased fluid and salt intake

Plate 7.19 Renin–Angiotensin System

Overview

Blood is pumped from the left ventricle of the heart to capillaries in the periphery via the arterial vessels of the *systemic* (or greater) *circulation* and returns via the veins to the right heart. It is then expelled from the right ventricle to the lungs via the *pulmonary* (or lesser) *circulation* and returns to the left heart (→ **A**).

The total **Blood volume** is roughly 4–5 L ($\approx 7\%$ of the fat-free body mass; → table on p. 88). Around 80% of the blood circulates through the veins, right heart and pulmonary vessels, which are jointly referred to as the *low pressure system* (→ **A**, left). These highly distensible *capacitance vessels* function as a *blood reservoir* in which blood is stored and released as needed via venous vasoconstriction (→ e.g., p. 218). When the blood volume increases— due, for example, to a blood transfusion—over 99% of the transfused volume remains in the low-pressure system (*high capacitance*), while only < 1% circulates in the arterial *high-pressure system* (*low capacitance*). Conversely, a decrease will be reflected almost entirely by a decrease in the blood stores in the low-pressure system. **Central venous pressure** (measured in or near to the right atrium; normally 4–12 cm H$_2$O) is therefore a good indicator of blood volume (and ECF volume) in individuals with a normally functioning heart and lungs.

Cardiac output (CO). The cardiac output is calculated as heart rate (HR) times stroke volume (SV). Under normal resting conditions, the CO is approx. 70 [min^{-1}] × 0.08 [L] = 5.6 L/min or, more precisely, a mean 3.4 L/min per m^2 body surface area. An increase in HR (up to about 180 min^{-1}) and/or SV can increase the CO to 15–20 L/min.

The **distribution of blood** to the organs arranged in *parallel* in the **systemic circulation** (→ **A**, $\dot{Q}$ values) is determined by their *functional priority* (vital organs) and by the *current needs* of the body (see also p. 213 A). Maintaining adequate cerebral perfusion (approx. 13% of the resting CO) is the top priority, not only because the *brain* is a major vital organ, but also because it is very susceptible to

hypoxic damage (→ p. 130). *Myocardial* perfusion via coronary arteries (approx. 4% of the CO at rest) must also be maintained, because any disruption of cardiac pumping function will endanger the entire circulation. About 20 to 25% of the CO is distributed to the *kidneys*. This fraction is very large relative to the kidney weight (only 0.5% of body mass). Renal blood flow is primarily used to maintain renal *excretory and control functions*. Thus, renal blood flow may be reduced transiently in favor of cardiac and cerebral perfusion, e.g., to ward off impending shock (→ p. 218). During strenuous physical *exercise*, the CO increases and is alloted mainly to the *skeletal muscle*. During digestion, the *gastrointestinal tract* also receives a relatively high fraction of the CO. Naturally, both of these organ groups cannot receive the maximum blood supply at the same time (→ p. 75 A). Blood flow to the *skin* (approx. 10% of the resting CO) mainly serves the purpose of *heat disposal* (→ p. 222ff.). The cutaneous blood flow rises in response to increased heat production (physical work) and/ or high external temperatures and decreases (pallor) in favor of vital organs in certain situations (e.g., shock; → p. 218).

The total CO flows through the **pulmonary circulation** as it and the systemic circulation are arranged in series (→ **A**). Oxygen-depleted (venous) blood is carried via the pulmonary arteries to the lungs, where it is *oxygenated* or "*arterialized.*" A relatively small quantity of additional oxygenated blood from the systemic circulation reaches the lung tissue via the bronchial arteries. All blood in the pulmonary circulation drains via the pulmonary veins.

Peripheral resistance. Flow resistance in the pulmonary circulation is only about 10% of the *total peripheral resistance (TPR)* in the systemic circulation. Consequently, the mean pressure in the right ventricle (approx. 15 mmHg = 2 kPa) is considerably lower than in the left ventricle (100 mmHg = 13.3 kPa). Since the resistance in the lesser arteries and arterioles amounts to nearly 50% of TPR (→ **A**, top right), they are called *resistance vessels*.

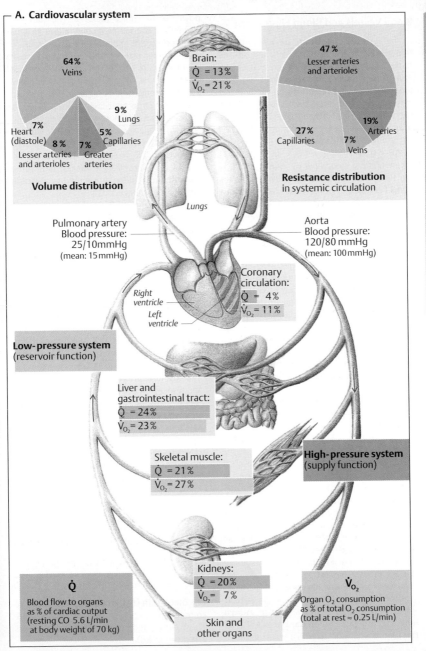

A. Cardiovascular system

64%
Veins

9%
Lungs

5%
Capillaries

7%
Heart
(diastole)

8%
Lesser arteries
and arterioles

7%
Greater
arteries

Volume distribution

Brain:
$\dot{Q}$ = 13%
$\dot{V}_{O_2}$ = 21%

47%
Lesser arteries
and arterioles

19%
Arteries

27%
Capillaries

7%
Veins

Resistance distribution
in systemic circulation

Lungs

Pulmonary artery
Blood pressure:
25/10 mmHg
(mean: 15 mmHg)

Aorta
Blood pressure:
120/80 mmHg
(mean: 100 mmHg)

*Right
ventricle*

*Left
ventricle*

Coronary
circulation:
$\dot{Q}$ = 4%
$\dot{V}_{O_2}$ = 11%

Low-pressure system
(reservoir function)

Liver and
gastrointestinal tract:
$\dot{Q}$ = 24%
$\dot{V}_{O_2}$ = 23%

Skeletal muscle:
$\dot{Q}$ = 21%
$\dot{V}_{O_2}$ = 27%

High-pressure system
(supply function)

$\dot{Q}$

Blood flow to organs
as % of cardiac output
(resting CO 5.6 L/min
at body weight of 70 kg)

Kidneys:
$\dot{Q}$ = 20%
$\dot{V}_{O_2}$ = 7%

Skin and
other organs

$\dot{V}_{O_2}$

Organ O_2 consumption
as % of total O_2 consumption
(total at rest ≈ 0.25 L/min)

Plate 8.1 Overview

Blood Vessels and Blood Flow

In the systemic circulation, blood is ejected from the left ventricle to the aorta and returns to the right atrium via the venae cavae ($\rightarrow$ **A**). As a result, the **mean blood pressure** ($\rightarrow$ p. 206) drops from around 100 mmHg in the aorta to 2–4 mmHg in the venae cavae ($\rightarrow$ **A2**), resulting in a pressure difference (ΔP) of about 97 mmHg (pulmonary circulation; $\rightarrow$ p. 122). According to **Ohm's Law**,

$$\Delta P = \dot{Q} \cdot R \ (mmHg) \qquad [8.1]$$

where $\dot{Q}$ is the blood flow (L $\cdot$ min^{-1}) and R is the flow resistance (mmHg $\cdot$ min $\cdot$ L^{-1}). Equation 8.1 can be used to calculate blood flow in a given organ (R = organ resistance) as well as in the entire cardiovascular system, where $\dot{Q}$ is the cardiac output (CO; $\rightarrow$ p. 186) and R is the total peripheral flow resistance (TPR). The TPR at rest is about 18 mmHg $\cdot$ min $\cdot$ L^{-1}.

The **aorta** and **greater arteries** distribute the blood to the periphery. In doing so, they act as a *hydraulic filter* because (due to their high *compliance*, $\Delta V/\Delta P_{tm}$) they convert the intermittent flow generated by the heart to a nearly steady flow through the capillaries. The high systolic pressures generated during the ejection phase cause the walls of these blood vessels to stretch, and part of the ejected blood is "stored" in the dilated lumen (*windkessel*). Elastic recoil of the vessel walls after aortic valve closure maintains blood flow during diastole. Arterial vessel compliance decreases with age.

Flow velocity ($\dot{V}$) and **flow rate** ($\dot{Q}$) of the blood. Assuming an aortic cross-sectional area (CSA) of 5.3 cm^2 and a total CSA of 20 cm^2 of all downstream arteries ($\rightarrow$ **A5**), the **mean** $\dot{V}$ (during systole and diastole) at rest can be calculated from a resting CO of 5.6 L/min: It equals 18 cm/s in the aorta and 5 cm/s in the arteries ($\rightarrow$ **A3**). As the aorta receives blood only during the ejection phase ($\rightarrow$ p. 90), the maximum resting values for $\dot{V}$ and $\dot{Q}$ in the aortic root are much higher during this phase ($\dot{V}$ = 95 cm/s, $\dot{Q}$ = 500 mL/s).

In the **Hagen–Poiseuille equation**,

$$R = 8 \cdot l \cdot \eta/(\pi \cdot r^4) \qquad [8.2]$$

the flow resistance (R) in a tube of known length (l) is dependent on the viscosity (η) of the fluid in the tube ($\rightarrow$ p. 92) and the fourth power of the inner radius of the tube (r^4). Decreasing the radius by only about 16% will therefore suffice to double the resistance.

The **lesser arteries** and **arterioles** account for nearly 50% of the TPR (*resistance vessels*; $\rightarrow$ **A1** and p. 187 A) since their small radii have a much stronger effect on total TPR (R $\sim$ 1/r^4) than their large total CSA (R $\sim$ r^2). Thus, the blood pressure in these vessels drops significantly. Any change in the radius of the small arteries or arterioles therefore has a radical effect on the TPR ($\rightarrow$ p. 212ff.). Their width and that of the **precapillary sphincter** determines the amount of blood distributed to the capillary beds (exchange area).

Although the **capillaries** have even smaller radii (and thus much higher individual resistances than the arterioles), their total contribution to the TPR is only about 27% because their total CSA is so large ($\rightarrow$ **A1** and p. 187 A). The *exchange of fluid and solutes* takes place across the walls of capillaries and **postcapillary venules**. Both vessel types are particularly suitable for the task because (a) their $\dot{V}$ is very small (0.02–0.1 cm/s; $\rightarrow$ **A3**) (due to the large total CSA), (b) their total surface area is very large (approx. 1000 m^2), (3) and their walls can be very thin as their inner radius (4.5 μm) is extremely small (*Laplace's law*, see below).

Transmural pressure P$_{tm}$ [N/m^2], that is, the pressure difference across the wall of a hollow organ (= internal pressure minus external pressure), causes the wall to stretch. Its materials must therefore be able to withstand this stretch. The resulting tangential *mural tension* T [N/m] is a function of the inner radius r [m] of the organ. According to **Laplace's law** for cylindrical (or spherical) hollow bodies,

$$P_{tm} = T/r \ (or \ P_{tm} = 2\,T/r, \ resp.) \qquad [8.3a/b]$$

Here, T is the total mural tension, regardless how thick the wall is. A thick wall can naturally withstand a given P$_{tm}$ more easily than a thin one. In order to determine the **tension** exerted **per unit CSA of the wall** (i.e., the stress requirements of the wall material in N/m^2), the thickness of the wall (w) must be considered. Equation 8.3 a/b is therefore transformed to

$$P_{tm} = T \cdot w/r \ (or \ P_{tm} = 2\,T \cdot w/r, \ resp.) \qquad [8.4a/b]$$

The blood collects in the **veins**, which can accommodate large volumes of fluid ($\rightarrow$ **A6**). These *capacitance vessels* serve as a *blood reservoir* ($\rightarrow$ p. 186).

Plate 8.2 Blood Vessels and Blood Flow

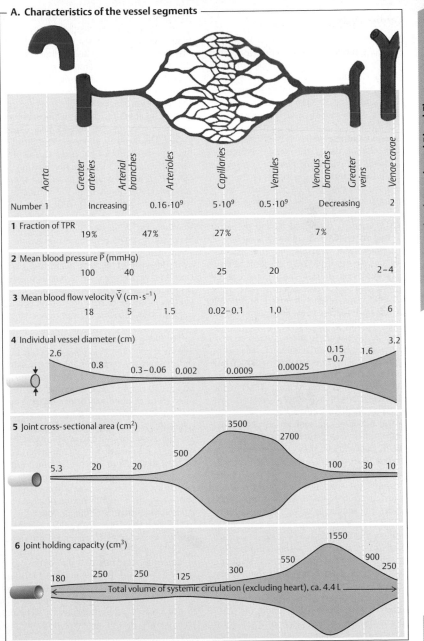

A. Characteristics of the vessel segments

Cardiac Cycle

The resting heart rate is 60–80 beats per minute. A **cardiac cycle** ($\rightarrow$ **A**) therefore takes roughly 1 s. It can be divided into four distinct phases: (I) contraction phase and (II) ejection phase, both occurring in *systole*; (III) relaxation phase and filling phase (IV), both occurring in *diastole*. At the end of phase IV, the atria contract (phase IVc). Electrical excitation of the atria and ventricles precedes their contraction.

The **cardiac valves** determine the direction of blood flow within the heart, e.g., from the atria to the ventricles (phase IV) or from the ventricles to the aorta or pulmonary artery (phase II). All cardiac valves are closed during phases I and III ($\rightarrow$ **A**, top). Opening and closing of the valves is controlled by the pressures exerted on the two sides of the valves.

Cardiac cycle. Near the end of ventricular diastole, the sinoatrial (SA) node emits an **electrical impulse**, marking to the beginning of the P wave of the ECG (phase IVc, $\rightarrow$ **A1** and p. 196ff.). This results in *atrial contraction* ($\rightarrow$ **A4**) and is followed by *ventricular excitation* (QRS complex of the ECG). The ventricular pressure then starts to rise ($\rightarrow$ **A2**, blue line) until it exceeds the atrial pressure, causing the atrioventricular valves (mitral and tricuspid valves) to close. This marks the *end of diastole*. The mean *end-diastolic volume* (**EDV**) in the ventricle is now about 120 mL ($\rightarrow$ **A4**) or, more precisely, $70 \, mL/m^2$ body surface area.

The **isovolumetric contraction** phase now begins (phase I, ca. 50 ms). With all valves are closed, the ventricles now contract, producing the **first heart sound** ($\rightarrow$ **A6**), and the ventricular pressure increases rapidly. The slope of this ascending pressure curve indicates the maximum rate of pressure developed (**maximum dP/dt**). The semilunar valves (aortic and pulmonary valves) now open because the pressure in the left ventricle ($\rightarrow$ **A2**, blue line) exceeds that in the aorta (black broken curve) at about 80 mmHg, and the pressure in the right ventricle exceeds that in the pulmonary artery at about 10 mmHg.

The **ejection** phase (now begins phase II; ca. 210 ms at rest). During this period, the pressure in the left ventricle and aorta reaches a maximum of ca. 120 mmHg (systolic pressure). In the early phase of ejection (IIa or *rapid ejec-

tion phase), a large portion of the stroke volume (SV) is rapidly expelled ($\rightarrow$ **A4**) and the blood flow rate reaches a maximum ($\rightarrow$ **A5**). Myocardial excitation subsequently decreases (T wave of the ECG, $\rightarrow$ **A1**) and ventricular pressure decreases (the remaining SV fraction is *slowly ejected*, phase IIb) until it falls below that of the aorta or pulmonary artery, respectively. This leads to closing of the semilunar valves, producing the **second heart sound** ($\rightarrow$ **A6**). The mean SV at rest is about 80 mL or, more precisely, $47 \, mL/m^2$ body surface area. The corresponding mean **ejection fraction** (**SV/EDV**) at rest is about 0.67. The *end-systolic volume* (**ESV**) remaining in the ventricles at this point is about 40 mL ($\rightarrow$ **A4**).

The first phase of ventricular diastole or **isovolumetric relaxation** now begins (phase III; ca. 60 ms). The atria have meanwhile refilled, mainly due to the suction effect created by the lowering of the valve plane during ejection. As a result, the central venous pressure (CVP) decreases ($\rightarrow$ **A3**, falls from c to x). The ventricular pressure now drops rapidly, causing the atrioventricular valves to open again when it falls short of atrial pressure.

The **filling phase** now begins (phase IV; ca. 500 ms at rest). The blood passes rapidly from the atria into the ventricles, resulting in a drop in CVP ($\rightarrow$ **A3**, point y). Since the ventricles are 80% full by the first quarter of diastole, this is referred to as *rapid ventricular filling* (phase IVa; $\rightarrow$ **A4**). *Ventricular filling slows down* (phase IVb), and the *atrial systole* (phase IVc) and the awave of CVP follows ($\rightarrow$ **A2,3**). At a normal heart rate, the atrial contraction contributes about 15% to ventricular filling. When the heart rate increases, the duration of the cardiac cycle decreases mainly at the expense of diastole, and the contribution of atrial contraction to ventricular filling increases.

The heart beats produce a **pulse wave** (pressure wave) that travels through the arteries at a specific *pulse wave velocity* (PWV): the PWV of the aorta is 3–5 m/s, and that of the radial artery is 5–12 m/s. PWV is much higher than the blood flow velocity ($\dot{V}$), which peaks at 1 m/s in the aorta and increases proportionally to (a) decreases in the compliance of aortic and arterial walls and (b) increases in blood pressure.

A. Action phases of the heart (cardiac cycle)

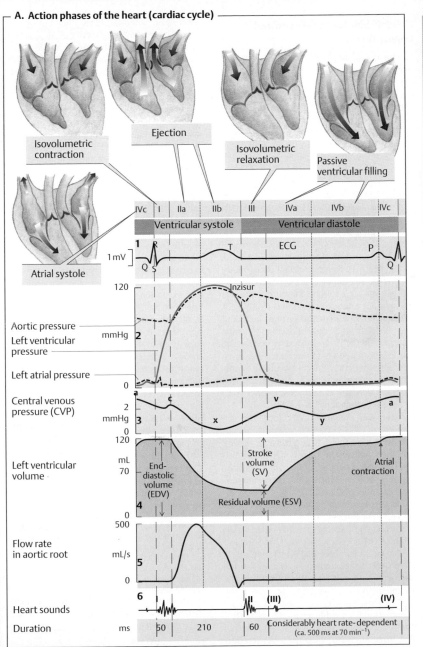

Isovolumetric contraction

Ejection

Isovolumetric relaxation

Passive ventricular filling

Atrial systole

Plate 8.3 Cardiac Cycle

	IVc	I	IIa	IIb	III	IVa	IVb	IVc
		Ventricular systole			**Ventricular diastole**			

1 ECG — 1 mV — R Q S T ECG P Q

2 Aortic pressure — Left ventricular pressure — Left atrial pressure — 120 mmHg 0 — Inzisur

3 Central venous pressure (CVP) — 2 mmHg 0 — a c x v y a

4 Left ventricular volume — 120 mL 70 0 — End-diastolic volume (EDV) — Stroke volume (SV) — Residual volume (ESV) — Atrial contraction

5 Flow rate in aortic root — 500 mL/s 0

6 Heart sounds — I II (III) (IV)

Duration — ms — 50 | 210 | 60 | Considerably heart rate-dependent (ca. 500 ms at 70 min⁻¹)

191

Cardiac Impulse Generation and Conduction

The heart contains muscle cells that generate (*pacemaker system*), conduct (*conduction system*) and respond to electrical impulses (*working myocardium*). Cardiac impulses are generated within the heart (**automaticity**). The frequency and regularity of pacemaker activity are also intrinsic to the heart (**rhythmicity**). Myocardial tissue comprises a *functional* (not truly anatomical) *syncytium* because the cells are connected by gap junctions ($\rightarrow$ p. 16ff.). This also includes the atrioventricular junction ($\rightarrow$ p. 195 A). Thus, an impulse arising in any part of the heart leads to complete contraction of both ventricles and atria or to none at all (**all-or-none response**).

Cardiac contraction is normally stimulated by impulses from the **sinoatrial node** (SA node), which is therefore called the *primary pacemaker*. The impulses are conducted ($\rightarrow$ **A**) through the atria to the *atrioventricular node* (**AV node**). The *bundle of His* is the beginning of the **specialized conduction system**, including also the left and right (Tawara's) *bundle branches* and the *Purkinje fibers*, which further transmit the impulses to the ventricular myocardium, where they travel from inside to outside and from apex to base of the heart. This electrical activity can be tracked in vivo ($\rightarrow$ **C**) by electrocardiography ($\rightarrow$ p. 196ff.).

Pacemaker potential ($\rightarrow$ **B1**, top). The cell potential in the SA node is a *pacemaker potential*. These cells do not have a constant resting potential. Instead, they slowly depolarize immediately after each repolarization, the most negative value of which is the *maximum diastolic potential* (MDP, ca. –70 mV). The slow *diastolic depolarization* or *prepotential* (PP) prevails until the *threshold potential* (TP) has again been reached. Thus triggering another action potential (AP).

The pacemaker potential ($\rightarrow$ **B1**, bottom) is subject to various underlying **changes in ion conductance** (g) and **ionic flow** (I) through the plasma membrane ($\rightarrow$ p. 32 ff.). Starting at the MDP, a hyperpolarization-triggered increase in non-selective conductance and influx (I_f, = "funny") of cations into the cells lead to slow depolarization (PP). When the TP is reached, g_{Ca} increases quickly, and the slope of the

pacemaker potential rises. This upslope is caused by increased influx of Ca^{2+} (I_{Ca}). When the potential rises to the positive range, g_K increases sharply, resulting in the efflux of K^+ (I_K), and the pacemaker cells repolarize to the MDP.

Each action potential in the SA node normally generates one heart beat. The **heart rate** is therefore determined by the rate of impulse generation by the pacemaker. The impulse generation frequency *decreases* ($\rightarrow$ **B3,** broken curves) when (a) the PP slope decreases ($\rightarrow$ **B3a**), (b) the TP becomes less negative ($\rightarrow$ **B3b**), (c) the MDP becomes more negative, resulting in the start of spontaneous depolarization at lower levels ($\rightarrow$ **B3c**), or (d) repolarization after an action potential occurs more slowly (slope flatter).

The first three conditions extend the time required to reach the threshold potential.

All components of the conduction system can depolarize spontaneously, but the SA node is the natural or *nomotopic pacemaker* in cardiac excitation (sinus rhythm normally has a rate of 60 to 100 min^{-1}). The intrinsic rhythms of the other pacemakers are slower than the sinus rhythm ($\rightarrow$ **C**, table) because the slope of their PPs and repolarizations are "flatter" (see above). APs arising in the SA node therefore arrive at subordinate ("lower") levels of the conduction system before spontaneous depolarization has reached the intrinsic TP there. The intrinsic rhythm of the lower components come into play (*ectopic pacemakers*) when (a) their own frequency is enhanced, (b) faster pacemakers are depressed, or (c) the conduction from the SA node is interrupted ($\rightarrow$ p. 200). The heart beats at the AV rhythm (40 to 55 min^{-1})/or even slower (25 to 40 min^{-1}) when controlled by tertiary (ventricular) pacemakers.

Overdrive suppression. The automaticity of lower pacemaker cells (e.g., AV node or Purkinje cells) is suppressed temporarily after they have been driven by a high frequency. This leads to increased Na^+ influx and therefore to increased activity of the Na^+-K^+-ATPase. Because it is electrogenic ($\rightarrow$ p. 28), the cells become hyperpolarized and it takes longer to reach threshold than without prior high-frequency overdrive ($\rightarrow$ **B3c**).

The cells of the **working myocardium** contain voltage-gated *fast Na^+ channels* that permit the brief but rapid influx of Na^+ at the beginning of

$\blacktriangleright$

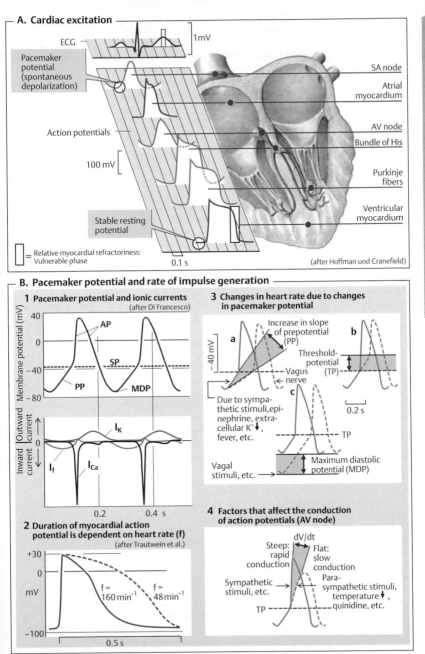

Plate 8.4   Cardiac Impulse Generation and Conduction I

A. Cardiac excitation

ECG

1mV

Pacemaker potential (spontaneous depolarization)

SA node

Atrial myocardium

Action potentials

AV node

Bundle of His

100 mV

Purkinje fibers

Ventricular myocardium

Stable resting potential

= Relative myocardial refractoriness: Vulnerable phase

0.1 s

(after Hoffman und Cranefield)

B. Pacemaker potential and rate of impulse generation

1 Pacemaker potential and ionic currents
(after Di Francesco)

AP

SP

PP

MDP

I_K

I_f

I_{Ca}

0.2   0.4 s

2 Duration of myocardial action potential is dependent on heart rate (f)
(after Trautwein et al.)

+30

0

mV

$f = 160\ min^{-1}$

$f = 48\ min^{-1}$

-100

0.5 s

3 Changes in heart rate due to changes in pacemaker potential

Increase in slope of prepotential (PP)

-40 mV

a

b

Threshold-potential (TP)

Vagus nerve

Due to sympathetic stimuli, epinephrine, extracellular K^+↓, fever, etc.

c

0.2 s

TP

Vagal stimuli, etc.

Maximum diastolic potential (MDP)

4 Factors that affect the conduction of action potentials (AV node)

dV/dt

Steep: rapid conduction

Flat: slow conduction

Sympathetic stimuli, etc.

Parasympathetic stimuli, temperature ↓, quinidine, etc.

TP

193

an AP. The slope of their APs therefore rises more sharply than that of a pacemaker potential ($\rightarrow$ **A**). A **resting potential** prevails between APs, i.e. spontaneous depolarization normally does not occur in the working myocardium. The long-lasting myocardial AP has a characteristic **plateau** ($\rightarrow$ p. 59 A). Thus, the first-stimulated parts of the myocardium are still in a refractory state when the AP reaches the last-stimulated parts of the myocardium. This prevents the cyclic re-entry of APs in the myocardium. This holds true, regardless of whether the heart rate is very fast or very slow since the duration of an AP varies according to heart rate ($\rightarrow$ **B2**).

Role of Ca^{2+}. The incoming AP opens *voltage-gated Ca^{2+} channels* (associated with *dihydropyridine receptors*) on the sarcolemma of myocardial cells, starting an influx of Ca^{2+} from the ECF ($\rightarrow$ p. 63/B3). This produces a local increase in cytosolic Ca^{2+} (*Ca^{2+} "spark"*) which, in turn, triggers the opening of *ligand-gated, ryanodine-sensitive Ca^{2+} channels* in the sarcoplasmic reticulum (Ca^{2+} store). The influx of Ca^{2+} into the cytosol results in electromechanical coupling ($\rightarrow$ p. 62) and myocardial contraction. The cytosolic Ca^{2+} is also determined by active transport of Ca^{2+} ions back (a) into the Ca^{2+} stores via a Ca^{2+}-ATPase, called *SERCA*, which is stimulated by *phospholamban*, and (b) to the ECF. This is achieved with the aid of a Ca^{2+}-ATPase and a *3 Na^+/Ca^{2+} exchange* carrier that is driven by the electrochemical Na^+ gradient established by Na^+-K^+-ATPase.

Although the heart beats autonomously, efferent **cardiac nerves** are mainly responsible for **modulating heart action** according to changing needs. The *autonomic nervous system* (and epinephrine in plasma) can alter the following aspects of heart action: (a) *rate of impulse generation* by the pacemaker and, thus, the heart rate (chronotropism); (b) *velocity of impulse conduction*, especially in the AV node (dromotropism); and (c) *contractility* of the heart, i.e., the force of cardiac muscle contraction at a given initial fiber length (inotropism).

These changes in heart action are induced by **acetylcholine (ACh;** $\rightarrow$ p. 82) released by parasympathetic fibers of the vagus nerve (binds with M_2 cholinoceptors on pacemaker cells), by **norepinephrine (NE)** released by sympathetic nerve fibers, and by plasma **epi-**nephrine (E). NE and E bind with β_1-adrenoceptors ($\rightarrow$ p. 84ff.). The firing frequency of the SA node is increased by NE and E (positive chronotropism) and decreased by ACh (negative chronotropism) because these substances alter the slopes of the PP and the MDP in the SA cells ($\rightarrow$ **B3a** and **c**). Under the influence of ACh, the slope of the PP becomes flatter and the MDP becomes more negative g_K rises. In versely, slope and amplitude of PP rises under the influence of E or sympathetic stimuli (higher I_f) due to a rise in cation (Na^+) conductance and, under certain conditions, a decrease in the g_K. Only NE and E have chronotropic effects in the lesser components of the impulse conduction system. This is decisive when the AV node or tertiary pacemakers take over.

ACh (left branch of vagus nerve) decreases the velocity of **impulse conduction in the AV node**, whereas NE and E increase it due to their negative and positive dromotropic effects, respectively. This is mainly achieved through changes in the amplitude and slope of the upstroke of the AP ($\rightarrow$ **B3c** and **B4**), g_K and g_{Ca}.

In positive inotropism, NE and E have a direct effect on the working myocardium. The resulting increase in **contractility** is based on an *increased influx of Ca^{2+}* ions from the ECF triggered by β_1-adrenoceptors, resulting in an increased cytosolic Ca^{2+}. This Ca^{2+} influx can be inhibited by administering Ca^{2+} channel blockers (*Ca^{2+} antagonists*). Other factors that increase cardiac contractility are an *increase in AP duration*, resulting in a longer duration of Ca^{2+} influx, and *inhibition of Na^+-K^+-ATPase* (e.g., by the cardiac glycosides digitalis and strophanthin). The consequences are: flatter Na^+ gradient across the cell membrane $\Rightarrow$ decreased driving force for 3 Na^+/Ca^{2+} exchange carriers $\Rightarrow$ decreased Ca^{2+} efflux $\Rightarrow$ increased cytosolic Ca^{2+} conc.

When the heart rate is low, Ca^{2+} influx over time is also low (fewer APs per unit time), allowing plenty of time for the efflux of Ca^{2+} between APs. The mean cytosolic Ca^{2+} conc. is therefore reduced and contractility is low. Only by this indirect mechanism are parasympathetic neurons able to elicit a negative inotropic effect (*frequency inotropism*). NE and E can exert their positive inotropic effects either indirectly by increasing the high heart or directly via β_1-adrenoceptors of the working myocardium.

Plate 8.5 Cardiac Impulse Generation and Conduction II

C. Cardiac impulse spreading

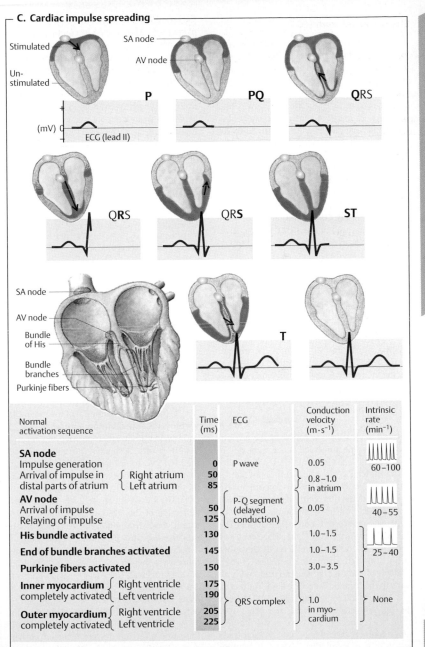

Normal activation sequence		Time (ms)	ECG	Conduction velocity (m·s⁻¹)	Intrinsic rate (min⁻¹)
SA node Impulse generation		0	P wave	0.05	60–100
Arrival of impulse in distal parts of atrium	Right atrium	50		} 0.8–1.0 in atrium	
	Left atrium	85			
AV node Arrival of impulse		50	} P-Q segment (delayed conduction)	0.05	40–55
Relaying of impulse		125			
His bundle activated		130		1.0–1.5	25–40
End of bundle branches activated		145		1.0–1.5	
Purkinje fibers activated		150		3.0–3.5	
Inner myocardium completely activated	Right ventricle	175	} QRS complex	1.0 in myo-cardium	None
	Left ventricle	190			
Outer myocardium completely activated	Right ventricle	205			
	Left ventricle	225			

Electrocardiogram (ECG)

The ECG records **potential differences** (few **m/V**) caused by cardiac excitation. This provides information on heart position, relative chamber size, heart rhythm, impulse origin/propagation and rhythm/conduction disturbances, extent and location of myocardial ischemia, changes in electrolyte concentrations, and drug effects on the heart. However, it does *not* provide data on cardiac contraction or pumping function.

ECG potential differences arise at the interface between stimulated and non-stimulated myocardium. Totally stimulated or unstimulated myocardial tissue does not generate any visible potentials The **migration of the excitatory front** through the heart muscle gives rise to numerous potentials that vary in magnitude and direction.

These *vectors* can be depicted as arrows, where the length of the arrow represents the magnitude of the potential and the direction of the arrow indicates the direction of the potential (arrowhead is +). As in a force parallelogram, the **integral vector** (summation vector) is the sum of the numerous individual vectors at that moment (→ **A**, red arrow).

The magnitude and direction of the integral vector change during the cardiac cycle, producing the typical *vector loop* seen on a **vectorcardiogram**. (In **A**, the maximum or chief vector is depicted by the arrow, called the "electrical axis" of the heart, see below).

Limb and chest leads of the ECG make it possible to visualize the course of the integral vector over time, projected at the plane determined by the leads (scalar ECG). Leads parallel to the integral vector show full deflection (R wave ≈ 1–2 mV), while those perpendicular to it show no deflection. *Einthoven leads I, II, and III* are *bipolar* limb leads positioned in the frontal plane. Lead I records potentials between the left and right arm, lead II those between the right arm and left leg, and lead III those between the left arm and left leg (→ **C1**). *Goldberger leads* are *unipolar* **a**ugmented limb leads in the frontal plane. One lead (right arm, **a**VR, left arm **a**VL, or left leg, **a**VF; → **D2**) acts as the *different electrode*, while the other two limbs are connected and serve as the *indifferent* (reference) *electrode* (→ **D1**). *Wilson leads* (V_1–V_6) are *unipolar chest leads* positioned on the left side of the thorax in a nearly *horizontal plane* (→ **F**). When used in combination with the aforementioned leads in the frontal plane, they provide a *three-dimensional view* of the integral vector. To make recordings with the chest leads (*different electrode*), the three limb leads are connected to form an *indifferent electrode* with high resistances (5 kΩ). The chest leads mainly detect potential vectors directed towards the back. These vectors are hardly detectable in the frontal plane. Since the mean QRS vector (see below) is usually directed downwards and towards the left back region, the QRS vectors recorded by leads V_1–V_3 are usually negative, while those detected by V_5 and V_6 are positive.

Intraesophageal leads and additional leads positioned in the region of the *right chest* (V_{r3}–V_{r6}) and *left back* (V_7–V_9) are useful in certain cases (→ **F2**).

An ECG depicts electrical activity as waves, segments, and intervals (→ **B** and p. 195 C). By convention, upward deflection of the waves is defined as positive (+), and downward deflection as negative (−). The electrical activity associated with *atrial depolarization* is defined as the **P wave** (< 0.3 mV, < 0.1 s). Repolarization of the atria normally cannot be visualized on the ECG since it tends to be masked by the QRS complex. The **QRS complex** (< 0.1 s) consists of one, two or three components: **Q wave** (mV < ¹/₄ of R, < 0.04 s), **R wave** and/or **S wave** (R+S > 0.6 mV). The potential of the **mean QRS vector** is the sum of the amplitudes of the Q, R and S waves (taking their positive and negative polarities into account). The voltage of the mean QRS vector is higher (in most leads) than that of the P wave because the muscle mass of the ventricles is much larger than that of the atria. The R wave is defined as the first positive deflection of the QRS complex, which means that R waves from different leads may not be synchronous. The QRS complex represents the *depolarization of the ventricles*, and the **T wave** represents their *repolarization*. Although opposing processes, the T wave usually points in the same direction as the R wave (+ in most leads). This means that depolarization and repolarization do not travel in the same direction (→ p. 195 C, Q**RS** and **T**: vector arrows point in the same direction despite reversed polarity during repolarization). The **PQ** (or PR) **segment**

▶

Plate 8.6 Electrocardiogram (ECG) I

A. Vector loops of cardiac excitation

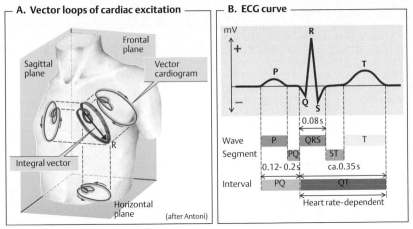

Frontal plane

Sagittal plane

Vector cardiogram

Integral vector

R

Horizontal plane

(after Antoni)

B. ECG curve

mV

+

−

P

R

Q S

T

0.08 s

Wave	P		QRS		T
Segment		PQ		ST	
Interval	PQ			QT	

0.12–0.2 s

ca. 0.35 s

Heart rate-dependent

C. Einthoven leads I, II and III (bipolar)

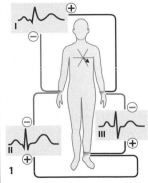

I

+ −

II − +

III − +

1

2

3

D. Goldberger limb leads (unipolar)

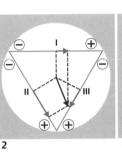

aVR

+ −

1

2

Right arm

Left arm

I

aVR   aVL

II   III

aVF

Foot

E. Cabrera circle

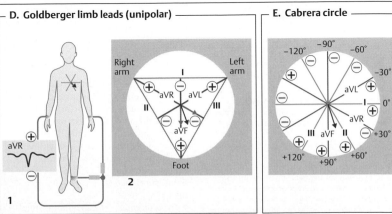

−120°   −90°   −60°

−30°

aVL

I   0°

aVR

+30°

+120°   +90°   +60°

III   aVF   II

197

(complete atrial excitation) and the **ST segment** (complete ventricular excitation) lie approx. on the **isoelectric line** (0 mV). The **PQ** (or PR) **interval** (< 0.2 s) is measured from the beginning of the P wave to the beginning of the Q wave (or to the R wave if Q wave is absent) and corresponds to the time required for *atrioventricular conduction* ($\rightarrow$ **B**). The **QT interval** is measured from the start of the Q wave to the end of the T wave. It represents the overall time required for depolarization and repolarization of the ventricles and is dependent on the heart rate (0.35 to 0.40 s at a heart rate of 75 min^{-1}).

Figure **E** illustrates the six frontal leads (Einthoven and Goldberger leads) on the *Cabrera circle*. Synchronous measurement of the amplitude of Q, R and S from two or more leads can be used to determine any integral vector in the frontal plane ($\rightarrow$ **G**). The direction of the largest mean QRS vector is called the **QRS axis** ($\rightarrow$ **C3** and **G**, red arrows). If the excitation spreads normally, the QRS axis roughly corresponds to the anatomic longitudinal axis of the heart.

The mean QRS axis (**"electrical axis"**) of the heart, which normally lies between +90 degrees to −30 degrees in adults ($\rightarrow$ **G, H**). *Right type* (α = +120° to +90°) is not unusual in children, but is often a sign of abnormality in adults. Mean QRS axes ranging from +90 degrees to +60 degrees are described as the *vertical type* ($\rightarrow$ **G1**), and those ranging from +60 degrees to +30 degrees are classified as the *intermediate type* ($\rightarrow$ **G2**). *Left type* occurs when α = +30 degrees to −30 degrees ($\rightarrow$ **G3**). **Abnormal deviation**: *Right axis deviation* ($> +120°$) can develop due right ventricular hypertrophy, while *left axis deviation* (more negative than −30°) can occur due to left ventricular hypertrophy.

An extensive **myocardial infarction** (**MI**) can shift the electrical axis of the heart. Marked *Q wave abnormality* ($\rightarrow$ **I1**) is typical in *transmural myocardial infarction* (involving entire thickness of ventricular wall): Q wave duration > 0.04 s and Q wave amplitude $> 25\%$ of total amplitude of the QRS complex. These changes appear within 24 hours of MI and are caused by failure of the dead myocardium to conduct electrical impulses. Preponderance of the excitatory vector in the healthy contralateral side of the heart therefore occurs while the affected part of the

myocardium should be depolarizing (first 0.04 s of QRS). The so-called "0.04-sec vector" is therefore said to point away from the infarction. *Anterior MI* is detected as highly negative Q waves (with smaller R waves) mainly in leads V5, V6, I and aVL. Q wave abnormalities can persist for years after MI ($\rightarrow$ **I2/3**), so they may not necessarily be indicative of an acute infarction. *ST elevation* points to ischemic but not (yet) necrotic parts of the myocardium. This can be observed: (1) in myocardial ischemia (angina pectoris), (2) in the initial phase of transmural MI, (3) in nontransmural MI, and (4) along the margins of a transmural MI that occurred a few hours to a few days prior ($\rightarrow$ **I4**). The ST segment normalizes within 1 to 2 days of MI, but the *T wave* remains inverted for a couple of weeks ($\rightarrow$ **I5** and **2**).

Excitation in Electrolyte Disturbances

Hyperkalemia. *Mild hyperkalemia* causes various changes, like elevation of the MDP ($\rightarrow$ p. 192) in the SA node. It can sometimes have positive chronotropic effects ($\rightarrow$ p. 193 B3c). In *severe hyperkalemia*, the more positive MDP leads to the inactivation of Na$^+$ channels ($\rightarrow$ p. 46) and to a reduction in the slope and amplitude of APs in the AV node (negative dromotropic effect; $\rightarrow$ p. 193 B4). Moreover, the K$^+$ conductance (g_K) rises, and the PP slope becomes flatter due to a negative chronotropic effect ($\rightarrow$ p. 193 B3a). Faster myocardial repolarization decreases the cytosolic Ca^{2+} conc. In extreme cases, the pacemaker is also brought to a standstill (*cardiac paralysis*). **Hypokalemia** (moderate) has positive chronotropic and inotropic effects ($\rightarrow$ p. 193 B3a), whereas **hypercalcemia** is thought to raise the g_K and thereby shortens the duration of the myocardial AP.

ECG. Changes in serum K$^+$ and Ca^{2+} induce characteristic changes in myocardial excitation.

◆ *Hyperkalemia* (> 6.5 mmol/L): tall, peaked T waves and conduction disturbances associated with an increased PQ interval and a widened QRS. Cardiac arrest can occur in extreme cases.

◆ *Hypokalemia* (< 2.5 mmol/L): ST depression, biphasic T wave (first positive, then negative) followed by a positive U wave.

◆ *Hypercalcemia* (> 2.75 mmol/L total calcium): shortened QT interval due to a shortened ST segment.

◆ *Hypocalcemia* (< 2.25 mmol/L total calcium): prolonged QT interval.

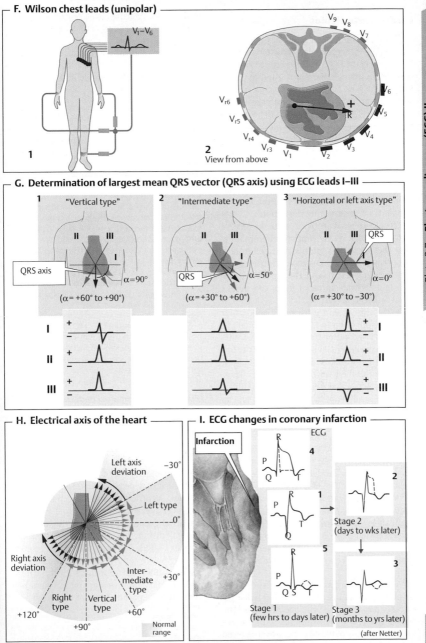

F. Wilson chest leads (unipolar)

V_1–V_6

1

V_9 V_8 V_7

V_{r6}
V_{r5}
V_{r4} V_{r3} V_1

V_6
V_5
V_4
V_3
V_2

+R

2
View from above

G. Determination of largest mean QRS vector (QRS axis) using ECG leads I–III

1 "Vertical type"

II III
I
QRS axis
$\alpha = 90°$
($\alpha = +60°$ to $+90°$)

2 "Intermediate type"

II III
I
QRS
$\alpha = 50°$
($\alpha = +30°$ to $+60°$)

3 "Horizontal or left axis type"

II III QRS
I
$\alpha = 0°$
($\alpha = +30°$ to $-30°$)

I + −
II + −
III + −

H. Electrical axis of the heart

Left axis deviation
−30°
Left type
0°
Right axis deviation
Intermediate type
+30°
Right type | Vertical type
+60°
+120°
+90°
Normal range

I. ECG changes in coronary infarction

Infarction

ECG

4
P Q T

1
P R T Q

2
Stage 2 (days to wks later)

5
P R Q S T

3
Stage 3 (months to yrs later)

Stage 1 (few hrs to days later)

(after Netter)

Plate 8.7 Electrocardiogram (ECG) II

Cardiac Arrhythmias

Arrhythmias are pathological changes in cardiac *impulse generation or conduction* that can be visualized by ECG. **Disturbances of impulse generation** change the sinus rhythm. *Sinus tachycardia* ($\rightarrow$ **A2**): The sinus rhythm rises to 100 min^{-1} or higher e.g., due to physical exertion, anxiety, fever (rise of about 10 beats/min for each 1 °C) or hyperthyroidism. *Sinus bradycardia:* The heart rate falls below 60 min^{-1} (e.g., due to hypothyroidism). In both cases the rhythm is regular whereas in *sinus arrhythmias* the rate varies. In adolescents, sinus arrhythmias can be physiological and respiration-dependent (heart rate increases during inspiration and decreases during expiration).

Ectopic pacemakers. Foci in the atrium, AV node or ventricles can initiate abnormal *ectopic* (heterotopic) impulses, even when normal (*nomotopic*) stimulus generation by the SA node is taking place ($\rightarrow$ **A**). The rapid discharge of impulses from an atrial focus can induce **atrial tachycardia** (serrated baseline instead of normal P waves), which triggers a ventricular response rate of up to 200 min^{-1}. Fortunately, only every second or third stimulus is transmitted to the ventricles because part of the impulses arrive at the Purkinje fibers (longest APs) during their refractory period. Thus, Purkinje fibers act as impulse *frequency filters*. Elevated atrial contraction rates of up to 350 min^{-1} are defined as **atrial flutter**, and all higher rates are defined as **atrial fibrillation** (up to 500 min^{-1}). Ventricular stimulation is then totally irregular (**absolute arrhythmia**). **Ventricular tachycardia** is a rapid train of impulses originating from a ventricular (ectopic) focus, starting with an extrasystole (ES) ($\rightarrow$ **B3**; second ES). The heart therefore fails to fill adequately, and the stroke volume decreases. This can lead to **ventricular fibrillation** (extremely frequent and uncoordinated contractions; $\rightarrow$ **B4**). Because of failure of the ventricle to transport blood, ventricular fibrillation can be fatal.

Ventricular fibrillation mainly occurs when an ectopic focus fires during the **relative refractory period** of the previous AP (called the "*vulnerable phase*" synchronous with T wave on the ECG; $\rightarrow$ p. 193 A). The APs triggered during this period have *smaller slopes, lower propagation velocities, and shorter durations.* This leads to re-excitation of myocardial areas that have already been stimulated (*re-entry cycles*). Ventricular fibrillation can be caused by electrical accidents and can usually be corrected by timely electrical *defibrillation*.

Extrasystoles (ES). The spread of impulses arising from an supraventricular (atrial or nodal) ectopic focus to the ventricles can disturb their sinus rhythm, leading to a *supraventricular arrhythmia*. When **atrial extrasystoles** occur, the P wave on the ECG is distorted while the QRS complex remains normal. **Nodal extrasystoles** lead to retrograde stimulation of the atria, which is why the P wave is negative and is either masked by the QRS complex or appears shortly thereafter ($\rightarrow$ **B1** right). Since the SA node often is discharged by a supraventricular extrasystole, the interval between the R wave of the extrasystole (R_{ES}) and the next normal R wave increases by the amount of time needed for the stimulus to travel from the focus to the SA node. This is called the *post-extrasystole pause.* The *RR intervals* are as follows: $R_{ES}R > RR$ and $(RR_{ES} + R_{ES}R) < 2\ RR$ ($\rightarrow$ **B1**).

Ventricular (or infranodal) **ES** ($\rightarrow$ **B2, B3**) distorts the QRS complex of the ES. If the sinus rate is slow enough, the ES will cause a ventricular contraction between two normal heart beats; this is called an *interpolated* (or interposed) ES ($\rightarrow$ **B2**). If the sinus rate is high, the next sinus stimulus reaches the ventricles while they are still refractory from the ectopic excitation. Ventricular contraction is therefore blocked until the next sinus stimulus arrives, resulting in a *compensatory pause*, where $RR_{ES} + R_{ES}R = 2\ RR$.

Disturbances of impulse conduction: AV block. *First-degree* AV block: prolonged but otherwise normal impulse conduction in the AV node (PQ interval > 0.2 sec); *second-degree* AV block: only every second (2:1 block) or third (3:1 block) impulse is conducted. *Third-degree* AV block: no impulses are conducted; sudden cardiac arrest may occur (*Adam–Stokes attack* or *syncope*). Ventricular atopic pacemakers then take over (ventricular bradycardia with normal atrial excitation rate), resulting in partial or total disjunction of QRS complexes and P waves ($\rightarrow$ **B5**). The heart rate drops to 40 to 55 min^{-1} when the AV node acts as the pacemaker ($\rightarrow$ **B5**), and to a mere 25 to 40 min^{-1} when tertiary (ventricular) pacemakers take over. Artificial pacemakers are then used.

Bundle branch block: disturbance of conduction in a branch of the bundle of His. Severe QRS changes occur because the affected side of the myocardium is activated by the healthy side via abnormal pathways.

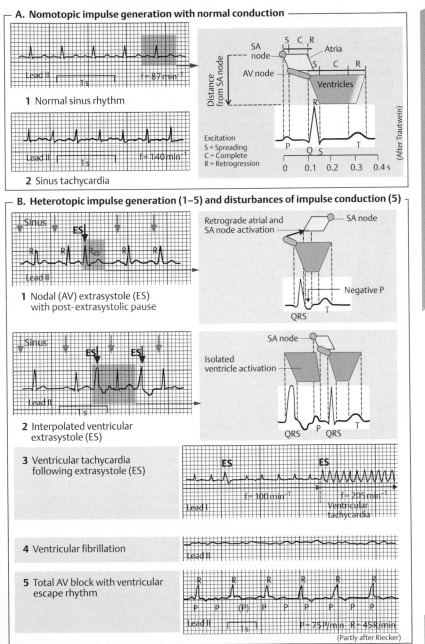

A. Nomotopic impulse generation with normal conduction

1 Normal sinus rhythm

Lead II · 1s · f = 87/min

2 Sinus tachycardia

Lead II · 1s · f = 140/min

SA node — Atria
AV node
Ventricles
S C R
Distance from SA node
Excitation
S = Spreading
C = Complete
R = Retrogression

P Q S T
0 0.1 0.2 0.3 0.4 s

(After Trautwein)

B. Heterotopic impulse generation (1–5) and disturbances of impulse conduction (5)

1 Nodal (AV) extrasystole (ES) with post-extrasystolic pause

Sinus · ES · Lead II
R R R(ES) R R

Retrograde atrial and SA node activation — SA node
Negative P
QRS · T

2 Interpolated ventricular extrasystole (ES)

Sinus · ES · ES · Lead II · 1s

Isolated ventricle activation — SA node
QRS · P · QRS · T

3 Ventricular tachycardia following extrasystole (ES)

ES · ES
f = 100 min⁻¹ · f = 205 min⁻¹
Lead I · Ventricular tachycardia

4 Ventricular fibrillation

Lead II

5 Total AV block with ventricular escape rhythm

R R R R R R R
P P (P) P P P P P P
Lead II · 1s · P = 75 P/min R = 45 R/min

(Partly after Riecker)

Plate 8.8 Cardiac Arrhythmias

201

Ventricular Pressure–Volume Relationships

The relationship between the volume (length) and pressure (tension) of a ventricle illustrates the interdependence between muscle length and force in the specific case of the heart (→ p. 66ff.). The **work diagram** of the heart can be constructed by plotting the changes in *ventricular pressure over volume* during one complete cardiac cycle (→ **A1**, points A-D-S-V-A, pressure values are those for the left ventricle).

The following *pressure–volume curves* can be used to construct a **work diagram of the ventricles**:

◆ **Passive (or resting) pressure–volume curve:** Indicates the pressures that result passively (without muscle contraction) at various ventricular volume loads (→ **A1, 2**; blue curve).

◆ **Isovolumic peak curve** (→ **A1, 2**, green curves): Based on experimental measurements made using an isolated heart. Data are generated for various volume loads by measuring the peak ventricular pressure at a constant ventricular volume during contraction. The contraction is therefore *isovolumetric (isovolumic)*, i.e., ejection does not take place (→ **A2**, vertical arrows).

◆ **Isotonic** (or **isobaric**) **peak curve** (→ **A1, 2**, violet curves). Also based on experimental measurements taken at various volume loads under isotonic (isobaric) conditions, i.e., the ejection is controlled in such a way that the ventricular pressure remains constant while the volume decreases (→ **A2**, horizontal arrows).

◆ **Afterloaded peak curve**: (**A1, 2**, orange curves). Systole (→ p. 190) consists of an *isovolumic* contraction phase (→ **A1**, A–D and p. 191 A, phase I) followed by an *auxotonic* ejection phase (volume decreases while pressure continues to rise) (→ **A1**, D–S and p. 191 A, phase II). This type of mixed contraction is called an *afterloaded contraction* (see also p. 67 B). At a given volume load (preload) (→ **A1**, point A), the afterloaded peak value changes (→ **A1**, point S) depending on the aortic end-diastolic pressure (→ **A1**, point D). All the afterloaded peak values are represented on the curve, which appears as a (nearly) straight line connecting the isovolumic and isotonic peaks for each respective volume load (point A) (→ **A1**, points T and M).

Ventricular work diagram. The pressure–volume relationships observed during the cardiac cycle (→ p. 190) can be plotted as a work diagram, e.g., for the left ventricle (→ **A1**): The *end-diastolic volume* (**EDV**) is 125 mL (→ **A1**, point A). During the *isovolumetric contraction*

phase, the pressure in the left ventricle rises (all valves closed) until the diastolic aortic pressure (80 mmHg in this case) is reached (→ **A1**, point D). The aortic valve then opens. During the *ejection phase*, the ventricular volume is reduced by the *stroke volume* (**SV**) while the pressure initially continues to rise (→ p. 188, *Laplace's law*, Eq. 8.4b: P_{tm} ↑ because r ↓ and w ↑). Once maximum (systolic) pressure is reached (→ **A1**, point S), the volume will remain virtually constant, until the pressure will drop slightly until it falls below the aortic pressure, causing the aortic valve to close (→ **A1**, point K). During the *isovolumetric relaxation* phase, the pressure rapidly decreases to (almost) 0 (→ **A1**, point V). The ventricles now contain only the *end-systolic volume* (**ESV**), which equals 60 mL in the illustrated example. The ventricular pressure rises slightly during the filling phase (passive pressure–volume curve).

Cardiac Work and Cardiac Power

Since work (J = N · m) equals pressure (N · m^{-2} = Pa) times volume (m^3), the area within the working diagram (→ **A1**, pink area) represents the **pressure/volume (P/V) work** achieved by the left ventricle during systole (13,333 Pa· 0.00008 m^3 = 1.07 J; right ventricle: 0.16 J). In systole, the bulk of cardiac work is achieved by active contraction of the myocardium, while a much smaller portion is attributable to passive elastic recoil of the ventricle, which stretches while filling. This represents *diastolic filling work* (→ **A1**, blue area under the blue curve), which is shared by the ventricular myocardium (indirectly), the atrial myocardium, and the respiratory and skeletal muscles (→ p. 204, venous return).

Total cardiac work. In addition to the cardiac work performed by the left and right ventricles in systole (ca. 1.2 J at rest), the heart has to generate 20% more energy (0.24 J) for the pulse wave (→ p. 188, windkessel). Only a small amount of energy is required to accelerate the blood at rest (1% of total cardiac work), but the energy requirement rises with the heart rate. The total **cardiac power** (= work/time, → p. 374) at rest (70 min^{-1} = 1.17 s^{-1}) is approximately 1.45 J · 1.17 s^{-1} = 1.7 W.

Plate 8.9 Ventricular Pressure–Volume Relationships

A. Work diagram of the heart (left ventricle)

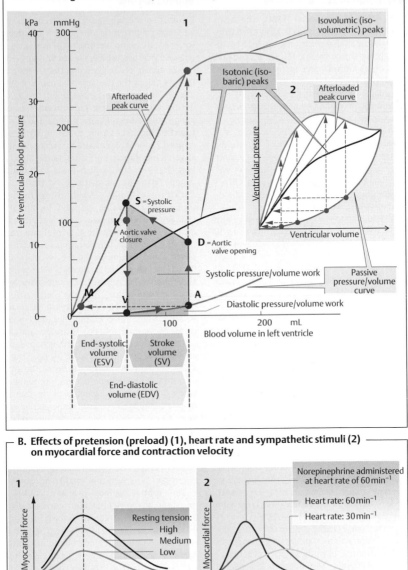

Isovolumic (iso-volumetric) peaks

Isotonic (iso-baric) peaks

2

Afterloaded peak curve

Afterloaded peak curve

Ventricular pressure

Ventricular volume

1

T

Afterloaded peak curve

kPa / mmHg

Left ventricular blood pressure

S = Systolic pressure

K = Aortic valve closure

D = Aortic valve opening

Systolic pressure/volume work

Passive pressure/volume curve

M

V

A

Diastolic pressure/volume work

Blood volume in left ventricle

End-systolic volume (ESV)

Stroke volume (SV)

End-diastolic volume (EDV)

B. Effects of pretension (preload) (1), heart rate and sympathetic stimuli (2) on myocardial force and contraction velocity

1

Myocardial force

Resting tension:
High
Medium
Low

Time (s)

2

Myocardial force

Norepinephrine administered at heart rate of 60 min⁻¹

Heart rate: 60 min⁻¹

Heart rate: 30 min⁻¹

Time (s)

(See text on next page)

(After Sonnenblick)

Regulation of Stroke Volume

Frank–Starling mechanism (**FSM**): The heart *autonomously* responds to changes in ventricular volume load or aortic pressure load by adjusting the stroke volume (SV) in accordance with the myocardial *preload* (resting tension; → p. 66 ff.). The FSM also functions to maintain an equal SV in both ventricles to prevent congestion in the pulmonary or systemic circulation.

Preload change. When the volume load (preload) *increases*, the start of isovolumic contraction shifts to the right along the passive P–V curve (→ **A1**, from point A to point A₁). This increases end-diastolic volume (EDV), stroke volume (SV), cardiac work and end-systolic volume (ESV) (→ **A**).

Afterload change. When the aortic pressure load (afterload) *increases*, the aortic valve will not open until the pressure in the left ventricle has risen accordingly (→ **A2**, point D$_t$). Thus, the SV in the short transitional phase (SV$_t$) will decrease, and ESV will rise (ESV$_t$). Consequently, the start of the isovolumic contraction shifts to the right along the passive P–V curve (→ **A2**, point A₂). SV will then normalize (SV₂) despite the increased aortic pressure (D₂), resulting in a relatively large increase in ESV (ESV₂).

Preload or **afterload-independent changes** in myocardial contraction force are referred to as **contractility** or **inotropism**. It increases in response to norepinephrine (NE) and epinephrine (E) as well as to increases in heart rate (β₁-adrenoceptor-mediated, positive inotropic effect and frequency inotropism, respectively; → p. 194). This causes a number of effects, particularly, an increase in isovolumic pressure peaks (→ **A3**, green curves). The heart can therefore pump against increased pressure levels (→ **A3**, point D₃) and/or eject larger SVs (at the expense of the ESV) (→ **A3**, SV₄).

While changes in the preload only affect the **force** of contraction (→ p. 203 B1), changes in contractility also affect the **velocity** of contraction (→ p. 203/B2). The steepest increase in isovolumic pressure per unit time (*maximum dP/dt*) is therefore used as a measure of contractility in clinical practice. dP/dt is increased E and NE and decreased by bradycardia (→ p. 203 B2) or heart failure.

Venous Return

Blood from the capillaries is collected in the veins and returned to the heart. The **driving forces** for this venous return (→ **B**) are: (a) *vis a tergo*, i.e., the postcapillary blood pressure (BP) (ca. 15 mmHg); (b) the suction that arises due to lowering of the cardiac valve plane in systole; (c) the pressure exerted on the veins during skeletal muscle contraction (*muscle pump*); the valves of veins prevent the blood from flowing in the wrong direction, (d) the increased abdominal pressure together with the lowered intrathoracic pressure during inspiration (P$_{pl}$; → p. 108), which leads to thoracic venous dilatation and suction (→ p. 206).

Orthostatic reflex. When rising from a supine to a standing position (orthostatic change), the blood vessels in the legs are subjected to additional hydrostatic pressure from the blood column. The resulting vasodilation raises blood volume in the leg veins (by ca. 0.4 L). Since this blood is taken from the *central blood volume*, i.e., mainly from pulmonary vessels, venous return to the left atrium decreases, resulting in a decrease in stroke volume and cardiac output. A reflexive increase (*orthostatic reflex*) in heart rate and peripheral resistance therefore occurs to prevent an excessive drop in arterial BP (→ pp. 7 E and 212 ff.); *orthostatic collapse* can occur. The drop in central blood volume is more pronounced when standing than when walking due to muscle pump activity. Conversely, pressure in veins above the heart level, e.g., in the cerebral veins, decreases when a person stands still for prolonged periods of time. Since the venous pressure just below the diaphragm remains constant despite changes in body position, it is referred to as a *hydrostatic indifference point*.

The **central venous pressure** (**CVP**) is measured at the right atrium (normal range: 0–12 cm H$_2$O or 0–9 mmHg). Since it is mainly dependent on the blood volume, the CVP is used to monitor the blood volume in clinical medicine (e.g., during a transfusion). Elevated CVP (> 20 cm H$_2$O or 15 mmHg) may be pathological (e.g., due to heart failure or other diseases associated with cardiac pump dysfunction), or physiological (e.g., in pregnancy).

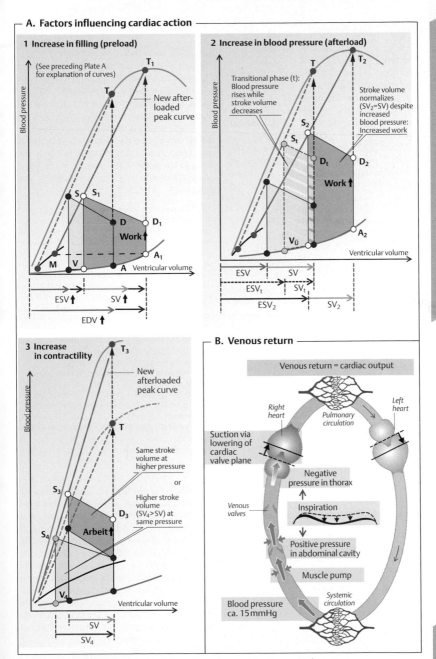

A. Factors influencing cardiac action

1 Increase in filling (preload)

(See preceding Plate A for explanation of curves)

Blood pressure

New afterloaded peak curve

T_1
T

S_1
S

D_1
D

Work ↑

M
V
A_1
A

Ventricular volume

ESV ↑   SV ↑
EDV ↑

2 Increase in blood pressure (afterload)

Blood pressure

Transitional phase (t): Blood pressure rises while stroke volume decreases

Stroke volume normalizes (SV_2=SV) despite increased blood pressure: Increased work

T_2
T

S_2
S_t

D_t D_2

Work ↑

$V_ü$

A_2

Ventricular volume

ESV   SV
ESV_t SV_t
ESV_2 SV_2

3 Increase in contractility

Blood pressure

T_3
T

New afterloaded peak curve

Same stroke volume at higher pressure

or

Higher stroke volume (SV_4>SV) at same pressure

S_3
D_3

Arbeit ↑

S_4

V_4

Ventricular volume

SV
SV_4

B. Venous return

Venous return = cardiac output

Right heart

Pulmonary circulation

Left heart

Suction via lowering of cardiac valve plane

Negative pressure in thorax

Venous valves

Inspiration
↓
Positive pressure in abdominal cavity

Muscle pump

Systemic circulation

Blood pressure ca. 15 mmHg

Plate 8.10   Regulation of Stroke Volume

Arterial Blood Pressure

The term blood pressure (BP) *per se* refers to the arterial BP in the systemic circulation. The maximum BP occurs in the aorta during the systolic ejection phase; this is the **systolic pressure** (P_s); the minimum aortic pressure is reached during the isovolumic contraction phase (while the aortic valves are closed) and is referred to as the **diastolic pressure** (P_d) ($\rightarrow$ **A1** and p. 191, phase I in A2). The systolic–diastolic pressure difference (P_s–P_d) represents the **blood pressure amplitude**, also called **pulse pressure** (**PP**), and is a function of the stroke volume (SV) and arterial compliance (C = dV/dP, $\rightarrow$ p. 188). When C decreases at a constant SV, the systolic pressure P_s will rise more sharply than the diastolic pressure P_d, i.e., the PP will increase (common in the elderly; described below). The same holds true when the SV increases at a constant C.

If the **total peripheral resistance** (TPR, $\rightarrow$ p. 188) **increases** while the SV ejection time remains constant, then P_s and the P_d will increase by the same amount (no change in PP). However, increases in the TPR normally lead to retardation of SV ejection and a decrease in the ratio of arterial volume rise to peripheral drainage during the ejection phase. Consequently, P_s rises less sharply than P_d and PP decreases.

Normal range. In individuals up to 45 years of age, P_d normally range from 60 to 90 mmHg and P_s from 100 to 140 mmHg at rest (while sitting or reclining). A P_s of up to 150 mmHg is considered to be normal in 45 to 60-year-old adults, and a P_s of up to 160 mmHg is normal in individuals over 60 ($\rightarrow$ **C**). Optimal BP regulation ($\rightarrow$ p. 212) is essential for proper tissue perfusion.

Abnormally low BP (**hypotension**) can lead to *shock* ($\rightarrow$ p. 218), *anoxia* ($\rightarrow$ p. 130) and tissue destruction. Chronically elevated BP (**hypertension;** $\rightarrow$ p. 216) also causes damage because important vessels (especially those of the heart, brain, kidneys and retina) are injured.

The **mean BP** (= the average measured over time) is the decisive factor of peripheral perfusion ($\rightarrow$ p. 188).

The mean BP can be determined by continuous BP measurement using an arterial catheter, etc. ($\rightarrow$ **A**).

By *attenuating* the pressure signal, only the mean BP is recorded.

Although the mean BP falls slightly as the blood travels from the aorta to the arteries, the P_s in the greater arteries (e.g., femoral artery) is usually higher than in the aorta (**A1** *v.* **A2**) because their compliance is lower than that of the aorta (see pulse wave velocity, p. 190).

Direct invasive BP measurements show that the BP curve in arteries distal to the heart is not synchronous with that of the aorta due to the time delay required for passage of the pulse wave (3–10 m/s; $\rightarrow$ p. 190); its shape is also different ($\rightarrow$ **A1/A2**).

The **BP is routinely measured** externally (at the level of the heart) according to the *Riva-Rocci* method by **sphygmomanometer** ($\rightarrow$ **B**). An inflatable cuff is snugly wrapped around the arm and a stethoscope is placed over the brachial artery at the crook of the elbow. While reading the manometer, the cuff is inflated to a pressure higher than the expected P_s (the radial pulse disappears). The air in the cuff is then slowly released (2–4 mmHg/s). The first sounds synchronous with the pulse (*Korotkoff sounds*) indicate that the cuff pressure has fallen below the P_s. This value is read from the manometer. These sounds first become increasingly louder, then more quiet and muffled and eventually disappear when the cuff pressure falls below the P_d (second reading).

Reasons for false BP readings. When re-measuring the blood pressure, the cuff pressure must be completely released for 1 to 2 min. Otherwise venous pooling can mimic elevated P_d. The cuff of the sphygmomanometer should be 20% broader than the diameter of the patient's upper arm. Falsely high P_d readings can also occur if the cuff is too loose or too small relative to the arm diameter (e.g., in obese or very muscular patients) or if measurement has to be made at the thigh.

The **blood pressure in the pulmonary artery** is much lower than the aortic pressure ($\rightarrow$ p. 186). The pulmonary vessels have relatively thin walls and their environment (air-filled lung tissue) is highly compliant. Increased cardiac output from the right ventricle therefore leads to expansion and thus to decreased resistance of the pulmonary vessels ($\rightarrow$ **D**). This prevents excessive rises in pulmonary artery pressure during physical exertion when cardiac output rises. The pulmonary vessels also function to buffer short-term fluctuations in blood volume ($\rightarrow$ p. 204).

Plate 8.11 Arterial Blood Pressure

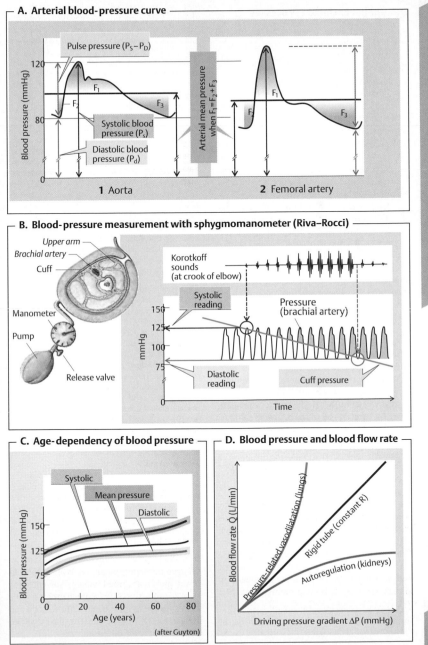

A. Arterial blood-pressure curve

Pulse pressure ($P_S - P_D$)

Blood pressure (mmHg)

120

80

F_1

F_2

F_3

Systolic blood pressure (P_s)

Diastolic blood pressure (P_d)

Arterial mean pressure when $F_1 = F_2 + F_3$

F_1

F_2

F_3

0

1 Aorta

2 Femoral artery

B. Blood-pressure measurement with sphygmomanometer (Riva–Rocci)

Upper arm

Brachial artery

Cuff

Manometer

Pump

Release valve

Korotkoff sounds (at crook of elbow)

Systolic reading

Pressure (brachial artery)

150

125

100

75

mmHg

Diastolic reading

Cuff pressure

Time

C. Age-dependency of blood pressure

Systolic

Mean pressure

Diastolic

Blood pressure (mmHg)

150

125

75

0   20   40   60   80

Age (years)

(after Guyton)

D. Blood pressure and blood flow rate

Blood flow rate Q̇ (L/min)

Pressure-related vasodilatation (lungs)

Rigid tube (constant R)

Autoregulation (kidneys)

Driving pressure gradient ΔP (mmHg)

Endothelial Exchange Processes

Nutrients and waste products are exchanged across the walls of the capillaries and post-papillary venules (*exchange vessels*; → p. 188). Their endothelia contain small (ca. 2–5 nm) or large (20–80 nm, especially in the kidneys and liver) *functional pores:* permeable, intercellular fissures or endothelial fenestrae, respectively. The degree of endothelial permeability varies greatly from one organ to another. Virtually all endothelia allow water and inorganic ions to pass, but most are largely impermeable to blood cells and large protein molecules. Transcytosis and carriers (→ p. 26f.) allow for passage of certain larger molecules.

Filtration and reabsorption. About 20 L/day of fluid is filtered (excluding the kidneys) into the interstitium from the body's exchange vessels. About 18 L/day of this fluid is thought to be reabsorbed by the venous limb of these vessels. The remaining 2 L/day or so make up the **lymph flow** and thereby return to the bloodstream (→ **A**). The *filtration or reabsorption rate* Q_f is a factor of the endothelial **filtration coefficient K_f** (= water permeability k · exchange area A) and the **effective filtration pressure P_{eff}** ($Q_f = K_f · P_{eff}$). P_{eff} is calculated as the hydrostatic pressure difference ΔP minus the oncotic pressure difference $\Delta \pi$ across the capillary wall (*Starling's relationship;* → **A**), where ΔP = capillary pressure (P_{cap}) minus interstitial pressure (P_{int}, normally ≈ 0 mmHg). At the level of the heart, ΔP at the arterial end of the systemic capillaries is about 30 mmHg and decreases to about 22 mmHg at the venous end. Since $\Delta \pi$ (ca. 24 mmHg; → **A**) counteracts ΔP, the initially high filtration rate (P_{eff} = + 6 mmHg) is thought to change into reabsorption whenever P_{eff} becomes negative. (Since ΔP is only 10 mmHg in the lungs, the pulmonary P_{eff} is very low). $\Delta \pi$ occurs because the concentration of proteins (especially albumin) in the plasma is much higher than their interstitial concentration. The closer the *reflection coefficient* of the plasma proteins (σ_{prot}) to 1.0, the higher $\Delta \pi$ and, consequently, the lower the permeability of the membrane to these proteins (→ p. 377).

According to Starling's relationship, **water reabsorption** should occur as long as P_{eff} is negative. However, recent data suggest that a negative P_{eff} re-

sults in only transient reabsorption. After several minutes it stops because the interstitial oncotic pressure rises due to "self-regulation". Thus, a major part of the 18 L/d expected to be reabsorbed from the exchange vessels (see above) might actually be reabsorbed in the lymph nodes. Rhythmic contraction of the arterioles (**vasomotion**) may also play a role by decreasing P_{eff} and thus by allowing intermittent capillary reabsorption.

In parts of the body below the heart, the effects of **hydrostatic pressure** from the blood column increase the pressure in the capillary lumen (in the feet ≈ 90 mmHg). The filtration rate in these regions therefore rise, especially when standing still. This is counteracted by two "self-regulatory" mechanisms: (1) the outflow of water results in an increase in the luminal protein concentration (and thus $\Delta \pi$) along the capillaries (normally the case in glomerular capillaries, → p. 152); (2) increased filtration results in an increase in P_{int} and a consequent decrease in ΔP.

Edema. Fluid will accumulate in the interstitial space (*extracellular edema*), portal venous system (*ascites*), and pulmonary interstice (*pulmonary edema*) if the volume of filtered fluid is higher than the amount returned to the blood.

Causes of edema (→ B):

◆ *Increased capillary pressure* (→ **B1**) due to precapillary vasodilatation (P_{cap} ↑), especially when the capillary permeability to proteins also increases (σ_{prot} ↓ and $\Delta \pi$ ↓) due, for example, to infection or anaphylaxis (histamine etc.). Hypertension in the portal vein leads to ascites.

◆ *Increased venous pressure* (P_{cap} ↑, → **B2**) due, for example, to venous thrombosis or cardiac insufficiency (*cardiac edema*).

◆ *Decreased concentration of plasma proteins*, especially albumin, leading to a drop in $\Delta \pi$ (→ **B3** and p. 379 A) due, for example, to loss of proteins (proteinuria), decreased hepatic protein synthesis (e.g., in liver cirrhosis), or to increased breakdown of plasma proteins to meet energy requirements (*hunger edema*).

◆ *Decreased lymph drainage* due, e.g., to lymph tract compression (tumors), severance (surgery), obliteration (radiation therapy) or obstruction (bilharziosis) can lead to *localized edema* (→ **B4**).

◆ *Increased hydrostatic pressure* promotes edema formation in lower regions of the body (e.g., in the ankles; → **B**).

Diffusion. Although dissolved particles are dragged through capillary walls along with filtered and reabsorbed water (solvent drag; → p. 24), diffusion plays a much greater role in the exchange of solutes. *Net diffusion* of a substance (e.g., O_2, CO_2) occurs if its plasma and interstitial conc. are different.

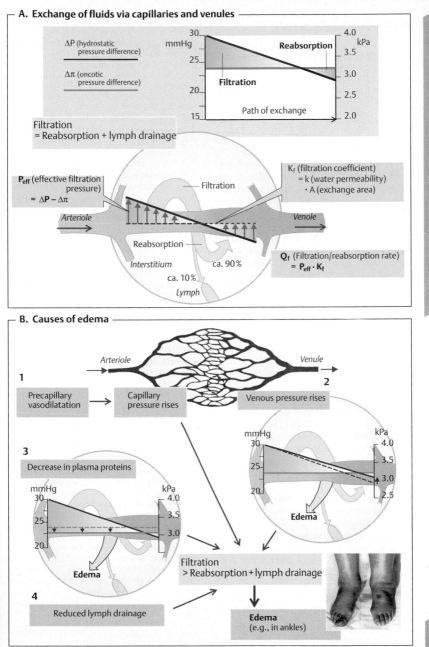

A. Exchange of fluids via capillaries and venules

ΔP (hydrostatic pressure difference)

Δπ (oncotic pressure difference)

Reabsorption

Filtration

Path of exchange

Filtration = Reabsorption + lymph drainage

P_{eff} (effective filtration pressure) = $\Delta P - \Delta\pi$

Filtration

K_f (filtration coefficient) = k (water permeability) · A (exchange area)

Arteriole

Venole

Reabsorption

Interstitium   ca. 90%

Q_f (Filtration/reabsorption rate) = $P_{eff} \cdot K_f$

ca. 10%

Lymph

B. Causes of edema

Arteriole

Venule

1

Precapillary vasodilatation → Capillary pressure rises

2

Venous pressure rises

3

Decrease in plasma proteins

Edema

Edema

Filtration > Reabsorption + lymph drainage

↓

Edema (e.g., in ankles)

4

Reduced lymph drainage

Plate 8.12 Endothelial Exchange Processes

Myocardial Oxygen Supply

Coronary arteries. The blood flow to the myocardium is supplied by the two coronary arteries that arise from the aortic root. The right coronary artery (approx. 1/7th of the blood) usually supplies the greater portion of the right ventricle, while the left coronary artery (6/7th of the blood) supplies the left ventricle ($\rightarrow$ **A**). The contribution of both arteries to blood flow in the septum and posterior wall of the left ventricle varies.

Coronary blood flow ($\dot{Q}_{cor}$) is phasic, i.e., the amount of blood in the coronary arteries fluctuates during the cardiac cycle due to extremely high rises in extravascular tissue pressure during systole ($\rightarrow$ **B, C**). The blood flow in the epicardial coronary artery branches and subepicardial vessels remains largely unaffected by these pressure fluctuations. However, the *subendocardial vessels* of the *left ventricle* are compressed during systole when the extravascular pressure in that region ($\approx$ pressure in left ventricle, P_{LV}) exceeds the pressure in the lumen of the vessels ($\rightarrow$ **C**). Consequently, the left ventricle is mainly supplied during diastole ($\rightarrow$ **B** middle). The fluctuations in right ventricular blood flow are much less distinct because right ventricular pressure (P_{RV}) is lower ($\rightarrow$ **B, C**).

Myocardial O_2 consumption ($\dot{V}O_2$) is defined as $\dot{Q}_{cor}$ times the arteriovenous O_2 concentration difference, $(C_a–C_v)O_2$. The myocardial $(C_a–C_v)O_2$ is relatively high (0.12 L/L blood), and *oxygen extraction* at rest ($[C_a–C_v]O_2/CaO_2 = 0.12/0.21$) is almost 60% and, thus, not able to rise much further. Therefore, an increase in $\dot{Q}_{cor}$ is practically the only way to increase myocardial $\dot{V}O_2$ when the O_2 demand rises ($\rightarrow$ **D**, right side).

Adaptation of the myocardial O_2 supply according to need is therefore primarily achieved by *adjusting vascular resistance* ($\rightarrow$ **D**, left side). The (distal) coronary vessel resistance can normally be reduced to about $^1/_4$ the resting value (*coronary reserve*). The coronary blood flow $\dot{Q}_{cor}$ (approx. 250 mL/min at rest) can therefore be increased as much as 4–5 fold. In other words, approx. 4 to 5 times more O_2 can be supplied during maximum physical exertion.

Arteriosclerosis (atherosclerosis) of the coronary arteries leads to luminal narrowing and a resultant decrease in poststenotic pressure. Dilatation of the distal vessels then occurs as an autoregulatory response (see below). Depending on the extent of the stenosis, it may be necessary to use a fraction of the coronary reserve, even during rest. As a result, lower or insufficient quantities of O_2 will be available to satisfy increased O_2 demand, and *coronary insufficiency* may occur ($\rightarrow$ **D**)

Myocardial O_2 demand increases with cardiac output (increased pressure–volume–work/time), i.e., in response to increases in heart rate and/or contractility, e.g., during physical exercise ($\rightarrow$ **D**, right). It also increases as a function of mural tension (T_{ventr}) times the duration of systole (*tension–time index*). Since $T_{ventr} = P_{ventr} \cdot r_{ventr}/2w$ (*Laplace's law* $\rightarrow$ Eq. 8.4b, p. 188), O_2 demand is greater when the ventricular pressure (P_{ventr}) is high and the stroke volume small than when P_{ventr} is low and the stroke volume high, even when the same amount of work (P $\times$ V) is performed. In the first case, the **efficiency of the heart** is reduced. When the ventricular pressure P_{ventr} is elevated, e.g., in *hypertension*, the myocardium therefore requires more O_2 to perform the same amount of work ($\rightarrow$ **D**, right).

Since the myocardial metabolism is aerobic, an increased O_2 demand quickly has to lead to vasodilatation. The following factors are involved in the **coronary vasodilatation**:

◆ **Metabolic factors**: (a) oxygen deficiency since O_2 acts as a vasoconstrictor; (b) Adenosine; oxygen deficiencies result in insufficient quantities of AMP being re-converted to ATP, leading to accumulation of *adenosine*, a degradation product of AMP. This leads to A_2 receptor-mediated vasodilatation; (c) Accumulation of lactate and H^+ ions (from the anaerobic myocardial metabolism); (d) prostaglandin I_2.

◆ **Endothelial factors**: ATP (e.g., from platelets), bradykinin, histamine and acetylcholine are vasodilators. They liberate nitric oxide (NO) from the endothelium, which diffuses into vascular muscle cells to stimulate vasodilatation ($\rightarrow$ p. 279 E).

◆ **Neurohumoral factors**: Norepinephrine released from sympathetic nerve endings and adrenal epinephrine have a vasodilatory effect on the distal coronary vessels via β_2 adrenoceptors.

Myocardial energy sources. The myocardium can use the available glucose, free fatty acids, lactate and other molecules for ATP production. The oxidation of each of these three energy substrates consumes a certain fraction of myocardial O_2 (O_2 extraction coefficient); accordingly, each contributes approx. one-third of the produced ATP at rest. The myocardium consumes increasing quantities of *lactate* from the skeletal muscles during physical exercise ($\rightarrow$ **A**, $\rightarrow$ p. 72 and 282).

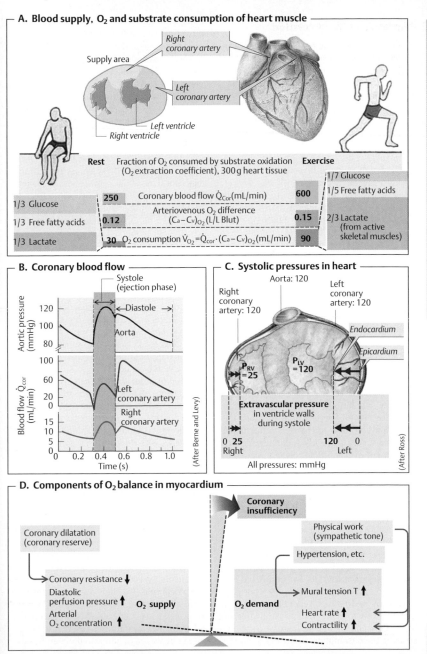

A. Blood supply, O₂ and substrate consumption of heart muscle

Right coronary artery

Left coronary artery

Supply area

Left ventricle
Right ventricle

	Rest	Fraction of O₂ consumed by substrate oxidation (O₂ extraction coefficient), 300 g heart tissue	Exercise	
1/3 Glucose			1/7 Glucose	
			1/5 Free fatty acids	
1/3 Free fatty acids	250	Coronary blood flow $\dot{Q}_{Cor}$ (mL/min)	600	
	0.12	Arteriovenous O₂ difference $(C_a - C_v)_{O_2}$ (L/L Blut)	0.15	2/3 Lactate (from active skeletal muscles)
1/3 Lactate	30	O₂ consumption $\dot{V}_{O_2} = \dot{Q}_{cor} \cdot (C_a - C_v)_{O_2}$ (mL/min)	90	

B. Coronary blood flow

Systole (ejection phase)

Diastole

Aortic pressure (mmHg): 120, 100, 80

Aorta

Blood flow $\dot{Q}_{cor}$ (mL/min)

Left coronary artery

Right coronary artery

100, 60, 20, 0
15, 10, 5, 0

Time (s): 0 0.2 0.4 0.6 0.8 1.0

(After Berne and Levy)

C. Systolic pressures in heart

Aorta: 120

Right coronary artery: 120

Left coronary artery: 120

Endocardium

Epicardium

$P_{RV} = 25$ $P_{LV} = 120$

Extravascular pressure in ventricle walls during systole

0 25   120   0
Right   Left

All pressures: mmHg

(After Ross)

D. Components of O₂ balance in myocardium

Coronary insufficiency

Coronary dilatation (coronary reserve)

Physical work (sympathetic tone)

Hypertension, etc.

Coronary resistance ↓

Diastolic perfusion pressure ↑   **O₂ supply**

Arterial O₂ concentration ↑

Mural tension T ↑

O₂ demand

Heart rate ↑

Contractility ↑

Plate 8.13 Myocardial Oxygen Supply

Regulation of the Circulation

The blood flow must be regulated to ensure an adequate blood supply, even under changing environmental conditions and stress (cf. p.74). This implies (a) optimal regulation of cardiac activity and blood pressure (*homeostasis*), (b) adequate perfusion of all organ systems, and (c) shunting of blood to active organ systems (e.g., muscles) at the expense of the resting organs (e.g., gastrointestinal tract) to keep from overtaxing the heart ($\rightarrow$ **A**).

Regulation of blood flow to the organs is mainly achieved by changing the *diameter of blood vessels*. The muscle tone (tonus) of the vascular smooth musculature changes in response to (1) *local stimuli* ($\rightarrow$ **B2a/b**), (2) *hormonal signals* ($\rightarrow$ **B3 a/b**) and (3) *neuronal signals* ($\rightarrow$ **B1 a/b**). Most blood vessels have an intermediary muscle tone at rest (*resting tone*). Many vessels dilate in response to denervation, resulting in a *basal tone*. This occurs due to spontaneous depolarization of smooth muscle in the vessels (see also p. 70).

Local Regulation of Blood Flow (Autoregulation)

Autoregulation has two **functions**:

$\blacklozenge$ Autoregulatory mechanisms help to maintain a *constant blood flow* to certain organs when the blood pressure changes (e.g., renal vessels constrict in response to rises in blood pressure; $\rightarrow$ p. 150).

$\blacklozenge$ Autoregulation also functions to *adjust the blood flow* according to changes in metabolic activity of an organ (*metabolic autoregulation*); the amount of blood flow to the organ (e.g., cardiac and skeletal muscle; $\rightarrow$ **A** and p.201) can thereby increase many times higher than the resting level.

Types of autoregulatory **mechanism:**

$\blacklozenge$ **Myogenic effects** arising from the vascular musculature of the lesser arteries and arterioles (*Bayliss effect*) ensure that these vessels *contract* in response to blood pressure-related dilatation ($\rightarrow$ **B2a**) in certain organs (e.g., kidneys, gastrointestinal tract and brain), but not in others (e.g., skin and lungs).

$\blacklozenge$ **Oxygen deficiencies** generally cause the blood vessels to *dilate*. Hence, the degree of blood flow and O_2 uptake increase with increasing O_2 consumption. In the lungs, on the other hand, a low P_{O_2} in the surrounding alveoli causes the vessels to *contract* (*hypoxic vasoconstriction*; $\rightarrow$ p. 122).

$\blacklozenge$ **Local metabolic (chemical) effects:** An increase in local concentrations of *metabolic products* such as CO_2, H^+, ADP, AMP, adenosine, and K^+ in the interstitium has a *vasodilatory effect*, especially in precapillary arterioles. The resulting rise in blood flow not only improves the supply of substrates and O_2, but also accelerates the efflux of these metabolic products from the tissue. The blood flow to the *brain* and *myocardium* ($\rightarrow$ p. 210) is almost entirely subject to local metabolic control. Both local metabolic effects and O_2 deficiencies lead to an up to 5-fold increase in blood flow to an affected region in response to the decreased blood flow (*reactive hyperemia*).

$\blacklozenge$ **Vasoactive substances:** A number of vasoactive substances such as prostaglandins play a role in autoregulation (see below).

Hormonal Control of Circulation

Vasoactive substances. Vasoactive hormones either have a direct effect on the vascular musculature (e.g., epinephrine) or lead to the local release of vasoactive substances (e.g., nitric oxide, endothelin) that exert local paracrine effects ($\rightarrow$ **B**).

$\blacklozenge$ **Nitric (mon)oxide (NO)** acts as a *vasodilatory* agent. NO is released from the *endothelium* when acetylcholine (M receptors), ATP, endothelin (ET_B receptors), or histamine (H_1 receptors) binds with an endothelial cell ($\rightarrow$ p.278). NO then diffuses to and relaxes vascular myocytes in the vicinity.

$\blacklozenge$ **Endothelin-1** can lead to *vasodilatation* by inducing the release of NO from the *endothelium* by way of ET_B receptors (see above), or can cause *vasoconstriction* via ET_A receptors in the vascular musculature. When substances such as angiotensin II or ADH (= vasopressin; V_1 receptor) bind to an endothelial cell, they release endothelin-1, which diffuses to and constricts the adjacent vascular muscles with the aid of ET_A receptors.

$\blacklozenge$ **Epinephrine (E):** High concentrations of E from the adrenal medulla ($\rightarrow$ p.86) have a *vasoconstrictive* effect (α_1-adrenoceptors), whereas low concentrations exert *vasodilatory* effects by way of β_2 adrenoceptors in the *myo-*

$\blacktriangleright$

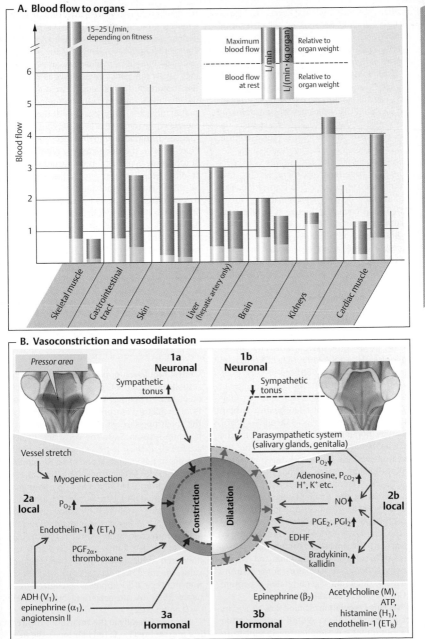

A. Blood flow to organs

15–25 L/min, depending on fitness

Blood flow

Maximum blood flow — L/min — Relative to organ weight
Blood flow at rest — L/(min · kg organ) — Relative to organ weight

Skeletal muscle
Gastrointestinal tract
Skin
Liver (hepatic artery only)
Brain
Kidneys
Cardiac muscle

B. Vasoconstriction and vasodilatation

Pressor area

1a Neuronal
Sympathetic tonus ↑

1b Neuronal
Sympathetic tonus ↓

Parasympathetic system (salivary glands, genitalia)

Vessel stretch
Myogenic reaction

2a local
P_{O_2} ↑

Endothelin-1 ↑ (ET_A)

$PGF_{2\alpha}$, thromboxane

ADH (V_1), epinephrine (α_1), angiotensin II
3a Hormonal

Constriction | Dilatation

P_{O_2}↓
Adenosine, P_{CO_2} ↑ H^+, K^+ etc.

NO ↑

PGE_2, PGI_2 ↑

EDHF

Bradykinin, kallidin

2b local

Epinephrine (β_2)
3b Hormonal

Acetylcholine (M), ATP, histamine (H_1), endothelin-1 (ET_B)

Plate 8.14 Regulation of the Circulation I

cardium, skeletal muscle and *liver* (→ **C**). The effect of E mainly depends on which type of adrenoceptor is predominant in the organ. α_1-adrenoceptors are predominant in the blood vessels of the *kidney* and *skin.*

◆ **Eicosanoids** (→ p. 269): Prostaglandin (PG) $F_{2\alpha}$ and thromboxane A_2 (released from platelets, → p. 102) and B_2 have *vasoconstrictive* effects, while PGI_2 (= prostacyclin, e.g. released from endothelium) and PGE_2 have vasodilatory effects. Another vasodilator released from the endothelium (e.g., by bradykinin; see below) opens K^+ channels in vascular myocytes and hyperpolarizes them, leading to a drop in the cytosolic Ca^{2+} concentration. This endothelium-derived hyperpolarizing factor (EDHF), has been identified as a 11,12-epoxy-eicosatrienoic acid (11,12-EET).

◆ **Bradykinin and kallidin** are *vasodilatory* agents cleaved from kininogens in blood plasma by the enzyme kallikrein. **Histamine** also acts as a vasodilator. All three substances influence also vessel permeability (e.g., during infection) and blood clotting.

Neuronal Regulation of Circulation

Neuronal regulation of *blood flow* (→ **B1a/b**) mainly involves the *lesser arteries* and *greater arterioles* (→ p. 188), while that of *venous return* to the heart (→ p. 188) can be controlled by dilating or constricting the *veins* (changes in their blood storage capacity). Both mechanisms are usually controlled by the **sympathetic nervous system** (→ **B1a** and p. 78ff.), whereby **norepinephrine** (NE) serves as the postganglionic transmitter (except in the sweat glands). NE binds with the α_1 adrenoceptors on blood vessels, causing them to constrict (→ **B**). Vasodilatation is usually achieved by decreasing the tonus of the sympathetic system (→ **B1b**). This does not apply to blood vessels in salivary glands (increased secretion) or the genitals (erection), which dilate in response to *parasympathetic* stimuli. In this case, vasoactive substances (bradykinin and NO, respectively) act as the mediators. Some neurons release calcitonin gene-related peptide (CGRP), a potent vasodilator.

Neuronal regulation of blood flow to organs occurs mainly: (a) via *central co-innervation* (e.g., an impulse is simultaneously sent from the cerebral cortex to circulatory centers when a muscle group is activated, or (b) via *neuronal feedback* from the organs whose activity level and metabolism have changed. If the neuronal and local metabolic mechanisms are conflicting (e.g., when sympathetic nervous stimulation occurs during skeletal muscle activity), the metabolic factors will predominate. Vasodilatation therefore occurs in the active **muscle** while the sympathetic nervous system reduces the blood flow to the inactive muscles. Blood flow to the **skin** is mainly regulated by neuronal mechanisms for the purpose of controlling heat disposal (*temperature control*; → p. 224). Hypovolemia and hypotension lead to *centralization of blood flow*, i.e., vasoconstriction in the kidney (oliguria) and skin (pallor) occurs to increase the supply of blood to vital organs such as the heart and central nervous system (→ p. 218).

During exposure to extremely low temperatures, the cold-induced vasoconstriction of cutaneous vessels is periodically interrupted to supply the skin with blood to prevent tissue damage (**Lewis response**). **Axoaxonal reflexes** in the periphery play a role in this response, as afferent cutaneous nerve fibers transmit signals to efferent vasomotor axons. Skin reddening in response to scratching (*dermatographism*) is also the result of axoaxonal reflexes.

Central regulation of blood flow (→ **C**) is the responsibility of the CNS areas in the medulla oblongata and pons. They receive information from **circulatory sensors (S)** or **receptors** (a) in the high-pressure system (barosensors or pressure sensors, S_P, in the aorta and carotid artery); (b) in the low-pressure system (stretch sensors in the vena cava and atria, S_A and S_B); and (c) in the left ventricle (S_V). The sensors measure *arterial blood pressure* (S_P), *pulse rate* (S_P and S_V) and filling pressure in the low pressure system (indirect measure of *blood volume*). The A sensors (S_A) mainly react to atrial contraction, whereas the B sensors (S_B) react to passive filling stretch (→ **C2**). If the measured values differ from the set-point value, the *circulatory control centers* of the CNS transmit regulatory impulses through efferent nerve fibers to the heart and blood vessels (→ **D** and p. 5 C2).

Situated laterally in the circulatory "center" is a **pressor area** (→ **C**, reddish zone), the neu-

Plate 8.15 Regulation of the Circulation II

C. Central regulation of blood flow

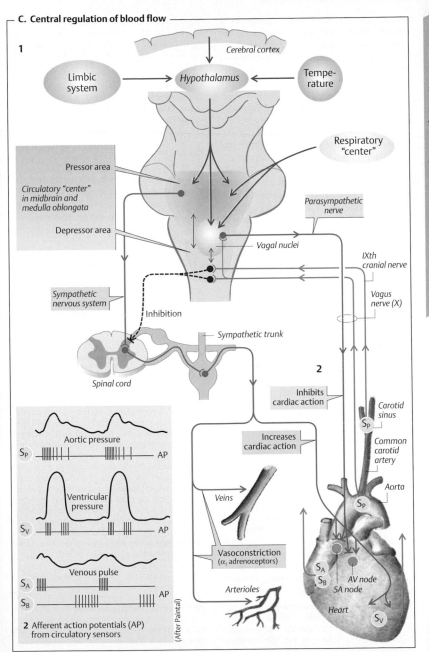

1

Cerebral cortex

Limbic system → Hypothalamus ← Temperature

Respiratory "center"

Pressor area

Circulatory "center" in midbrain and medulla oblongata

Depressor area

Vagal nuclei

Parasympathetic nerve

IXth cranial nerve

Vagus nerve (X)

Sympathetic nervous system

Inhibition

Sympathetic trunk

Spinal cord

2

Inhibits cardiac action

Increases cardiac action

Carotid sinus

S_P

Common carotid artery

Aorta

S_P

Veins

Vasoconstriction (α_1 adrenoceptors)

S_A
S_B
SA node

AV node

Arterioles

Heart

S_V

Aortic pressure

S_P ———|||||||———|||||||——— AP

Ventricular pressure

S_V ——||||||———||||||——— AP

Venous pulse

S_A —|||||————|||||——— AP

S_B ————|||||||||————||||— AP

(After Paintal)

2 Afferent action potentials (AP) from circulatory sensors

215

rons of which (blue arrows) continuously transmit sympathetic nerve impulses to the heart to increase its activity (heart rate, conduction and contractility). Their effects on vessels are predominantly vasoconstrictive (resting tone). The pressor area is in close contact with more medial neurons (**depressor area**, light blue area in **C**). The pressor and depressor areas are connected to the dorsal nuclei of the vagus nerve ($\rightarrow$ **C**, green), the stimulation of which reduces the heart rate and cardiac impulse conduction rate ($\rightarrow$ **C**, orange arrows).

Homeostatic circulatory reflexes include signals along afferent nerve tracts ($\rightarrow$ **D3a/b**) that extend centrally from the pressosensors in the aorta and carotid sinus ($\rightarrow$ **C**, green tracts). The main purpose of homeostatic control is to maintain the *arterial blood pressure* at a stable level. Acute increases in blood pressure heighten the rate of afferent impulses and activate the depressor area. By way of the vagus nerve, parasympathetic neurons ($\rightarrow$ **C**, orange tract) elicit the depressor reflex response, i.e., they decrease the *cardiac output* (*CO*). In addition, inhibition of sympathetic vessel innervation causes the vessels to dilate, thereby reducing the *peripheral resistance* (TPR; $\rightarrow$ **D4a/b**). Both of these mechanisms help to lower acute increases in blood pressure. Conversely, an acute drop in blood pressure leads to activation of pressor areas, which stimulates a rise in CO and TPR as well as venous vasoconstriction ($\rightarrow$ **C**, blue tracts), thereby raising the blood pressure back to normal.

Due to the *fast adaptation* of pressosensors (*differential characteristics,* $\rightarrow$ p. 312ff.), these regulatory measures apply to *acute changes in blood pressure*. Rising, for example, from a supine to a standing position results in rapid redistribution of the blood volume. Without homeostatic control (*orthostatic reflex*; $\rightarrow$ p. 204), the resulting change in venous return would lead to a sharp drop in arterial blood pressure. The circulatory centers also respond to falling P_{O_2} or rising P_{CO_2} in the blood (cross-links from respiratory center) to raise the blood pressure as needed.

In individuals with chronic **hypertension**, the input from the pressosensors is normal because they are fully adapted. Therefore, circulatory control centers cannot respond to and decrease the high pressures. On the contrary, they may even help to "fix" the blood pressure at the high levels. Chronic hypertension leads to stiffening of the carotid sinus. This may also contribute to decreasing the sensitivity of carotid pressosensors in hypertension.

A temporary *increase in venous return* (e.g., after an intravenous infusion) also leads to an increase in heart action ($\rightarrow$ **D**, right). This mechanism is known as the **Bainbridge reflex**. The physiological significance of this reflex is, however, not entirely clear, but it may complement the Frank–Starling mechanism ($\rightarrow$ p. 202ff.).

Hypertension

Hypertension is defined as a chronic increase in the systemic arterial blood pressure. The general criterion for diagnosis of hypertension is consistent elevation of resting blood pressure to more than 90 mmHg diastolic ($\rightarrow$ p. 206). Untreated or inadequately managed hypertension results in stress and compensatory hypertrophy of the left ventricle which can ultimately progress to *left heart failure*. Individuals with hypertension are also at risk for arteriosclerosis and its sequelae (myocardial infarction, stroke, renal damage, etc.). Therefore, hypertension considerably shortens the life expectancy of a large fraction of the population.

The **main causes of hypertension** are (a) increased extracellular fluid (ECF) volume with increased venous return and therefore increased cardiac output (*volume hypertension*) and (b) increased total peripheral resistance (*resistance hypertension*). As hypertension always leads to vascular changes resulting in increased peripheral resistance, type *a* hypertension eventually proceeds to type *b* which, regardless of how it started, ends in a vicious circle.

The ECF volume increases when more NaCl (and water) is absorbed than excreted. The usually high intake of dietary salt may therefore play a role in the development of **essential hypertension** (primary hypertension), the most common type of hypertension, at least in patients sensitive to salt. Volume hypertension can even occur when a relatively low salt intake can no longer be balanced. This can occur in renal insufficiency or when an adrenocortical tumor produces uncontrolled amounts of aldosterone, resulting in Na$^+$ retention.

Other important cause of hypertension is **pheochromocytoma**, a tumor that secretes epinephrine and norepinephrine and therefore raises the CO and TPR. **Renal hypertension** can occur due to renal artery stenosis and renal disease. This results in the increased secretion of *renin*, which in turn raises the blood pressure via the renin–angiotensin–aldosterone (RAA) system ($\rightarrow$ p. 184).

Plate 8.16 Regulation of the Circulation III

D. Circulatory reflexes

Carotid and aortic
sinus reflex (depressant)

Atrial (Bainbridge) reflex
(excitatory)

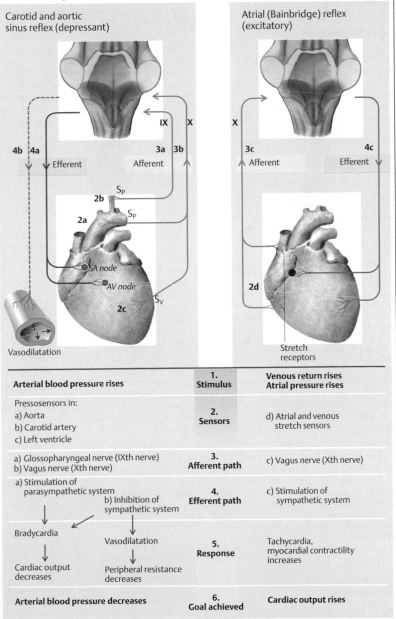

4b 4a
Efferent

3a 3b
Afferent

IX   X

X

3c
Afferent

4c
Efferent

S_P
2b

S_P

2a

SA node

AV node

S_V

2c

2d

Vasodilatation

Stretch
receptors

Arterial blood pressure rises	1. Stimulus	Venous return rises Atrial pressure rises
Pressosensors in: a) Aorta b) Carotid artery c) Left ventricle	2. Sensors	d) Atrial and venous stretch sensors
a) Glossopharyngeal nerve (IXth nerve) b) Vagus nerve (Xth nerve)	3. Afferent path	c) Vagus nerve (Xth nerve)
a) Stimulation of parasympathetic system b) Inhibition of sympathetic system	4. Efferent path	c) Stimulation of sympathetic system
Bradycardia Vasodilatation	5. Response	Tachycardia, myocardial contractility increases
Cardiac output decreases Peripheral resistance decreases		
Arterial blood pressure decreases	6. Goal achieved	Cardiac output rises

217

Circulatory Shock

Shock is characterized by acute or subacute progressive *generalized failure of the circulatory system* with disruption of the microcirculation and *failure to maintain adequate blood flow to vital organs*. In most cases, the **cardiac output (CO) is insufficient** due to a variety of reasons, which are explained below.

◆ **Hypovolemic shock** is characterized by reduced central venous pressure and *reduced venous return*, resulting in an inadequate stroke volume (Frank–Starling mechanism). The blood volume can be reduced due to bleeding (*hemorrhagic shock*) or any other conditions associated with the external **loss of fluids** from the gastrointestinal tract (e.g., severe vomiting, chronic diarrhea), the kidneys (e.g., in diabetes mellitus, diabetes insipidus, high-dose diuretic treatment) or the skin (burns, profuse sweating without fluid intake). An *internal loss* of blood can also occur, e.g., due to bleeding into soft tissues, into the mediastinum or into the pleural and abdominal space.

◆ **Cardiogenic shock**: Acute *heart failure* can be caused by acute myocardial infarction, acute decompensation of heart failure or impairment of cardiac filling, e.g. in pericardial tamponade. The central venous pressure is higher than in hypovolemic shock.

◆ Shock can occur due to **hormonal causes**, such as adrenocortical insufficiency, diabetic coma or insulin overdose (hypoglycemic shock).

◆ **Vasogenic shock**: Reduced cardiac output can also be due to peripheral vasodilatation (absence of pallor) and a resultant drop of venous return. This occurs in Gram-positive septicemia (*septic shock*), *anaphylactic shock*, an immediate hypersensitivity reaction (food or drug allergy, insect bite/sting) in which vasoactive substances (e.g., histamines) are released.

Symptoms. Hypovolemic and cardiovascular shock are characterized by *decreased blood pressure* (weak pulse) *increased heart rate, pallor* with cold sweats (not observed with shock caused by vasodilatation), reduced urinary output (*oliguria*) and extreme *thirst.*

Shock index. The ratio of pulse rate (beats/min) to systolic blood pressure (mmHg), or *shock index*, provides a rough estimate of the extent of volume loss. An index of up to 0.5 indicates normal or < 10% blood loss; up to 1.0 = < 20–30% blood loss and impending shock; up to 1.5 = > 30–50% blood loss and manifest shock..

Most of the symptoms described reflect the **counterregulatory measures** taken by the body during the *non-progressive phase* of shock in order to ward off progressive shock (→ **A**). Rapid-acting mechanisms for *raising the blood pressure* and slower-acting mechanisms to *compensate for volume losses* both play a role.

Blood pressure compensation (→ **A** left): A drop in blood pressure *increases sympathetic tonus* (→ **A1** and p. 214). *Arterial vasoconstriction* (absent in shock due to vasodilatation) shunts the reduced cardiac output from the skin (pallor), abdominal organs and kidneys (oliguria) to vital organs such as the coronary arteries and brain. This is known as **centralization of blood flow** (→ **A2**). Sympathetic constriction of venous capacitance vessels (which raises ventricular filling), tachycardia and positive inotropism increase the diminished cardiac output to a limited extent.

Compensation for volume deficits (→ **A**, right): When shock is imminent, the resultant drop in blood pressure and peripheral vasoconstriction lead to a *reduction of capillary filtration pressure*, allowing interstitial fluid to enter the bloodstream. Atrial stretch sensors detect the decrease in ECF volume (reduced atrial filling) and transmit signals to stop the atria from secreting atriopeptin (= ANP) and to start the secretion of antidiuretic hormone (ADH) from the posterior lobe of the pituitary (*Gauer–Henry reflex*; → p. 170). ADH induces vasoconstriction (V_1 receptors) and fluid retention (V_2 receptors). The drop in renal blood pressure triggers an increase in *renin secretion* and activation of the renin–angiotensin–aldosterone (RAA) system (→ p. 184). If these measures are successful in warding off the impending shock, the lost red blood cells are later replaced (via increased renal erythropoietin secretion, → p. 88) and the plasma protein concentration is normalized by increased hepatic synthesis.

Manifest (or progressive) shock will develop if these homeostatic compensation mechanisms are unable to prevent impending shock and the patient does not receive medical treatment (infusion, etc.). Severe hypotension (< 90 mmHg systolic or < 60 mmHg mean blood pressure) can persist for extended periods, even in spite of volume replacement. The resulting development of hypoxia leads to organ damage and **multiple organ failure**, ultimately culminating in **irreversible shock** and death.

Plate 8.17 Circulatory Shock

A. Compensation mechanisms for impending hypovolemic shock

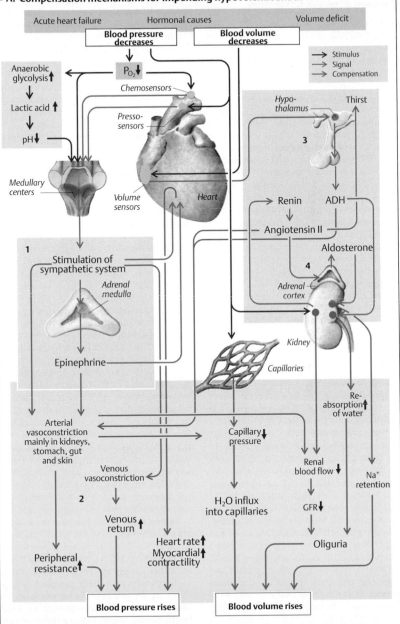

Acute heart failure   Hormonal causes   Volume deficit

Blood pressure decreases

Blood volume decreases

→ Stimulus
→ Signal
→ Compensation

Anaerobic glycolysis ↑

P_{O_2} ↓

Lactic acid ↑

Chemosensors

Hypo-thalamus   Thirst

pH ↓

Presso-sensors

3

Medullary centers

Volume sensors   *Heart*

Renin   ADH

Angiotensin II

Aldosterone

1

Stimulation of sympathetic system

4

Adrenal medulla

Adrenal cortex

Kidney

Epinephrine

Capillaries

Re-absorption of water

Arterial vasoconstriction mainly in kidneys, stomach, gut and skin

Capillary ↓ pressure

Renal blood flow ↓

Venous vasoconstriction

Na⁺ retention

2

Venous return ↑

H_2O influx into capillaries

GFR ↓

Peripheral resistance ↑

Heart rate ↑
Myocardial contractility ↑

Oliguria

Blood pressure rises

Blood volume rises

Plate 8.17 Circulatory Shock

Fetal and Neonatal Circulation

Placenta. The maternal placenta acts as the "gut" (absorption of nutrients), "kidneys" (removal of waste products) and "lungs" of the fetus (uptake of O_2 and elimination of CO_2). Although the *fetal O_2-hemoglobin dissociation curve* is shifted to the left compared to that of adults ($\rightarrow$ p. 129 C), only 60% (0.6) of placental hemoglobin is saturated with O_2 ($\rightarrow$ **A**).

Fetal blood is distributed according to need. Inactive or hardly active organs receive little blood. The **fetal cardiac output** (from both ventricles) is about 0.2 L/min per kg body weight. The **fetal heart rate** rises from an initial 65 min^{-1} (week 5) to 130–160 min^{-1} in later weeks. Approx. 50% of the blood ejected from the heart flows through the placenta, the other half supplies the body (35%) and lungs (15%) of the fetus. This is supplied by the left and right heart, which function essentially *in parallel*. The serial connection of the systemic circulation to the pulmonary circuit (as in adults) is not fully necessary in the fetus.

Fetal circulation. The blood flows through the fetal body as follows ($\rightarrow$ **A**): After being *arterialized in the* **placenta**, the blood passes into the fetus via the *umbilical vein* and part of it travels through the **ductus venosus** (Arantii), thereby bypassing the liver. When entering the inferior vena cava, the blood *mixes with venous blood* from the lower half of the body. Guided by special folds in the vena cava, the mixed blood passes directly from right atrium to the left atrium through an opening in the atrial septum (**foramen ovale**). From the left atrium, it then proceeds to the left ventricle. While in the right atrium, the blood mingles with venous blood from the superior vena cava (only slight mixing), which is received by the right ventricle. Only about one-third of this blood reaches the lungs (due to high flow resistance since the lungs are not yet expanded, and due to hypoxic vasoconstriction, $\rightarrow$ **C** and p. 122). The other two-thirds of the blood travels through the **ductus arteriosus** (Botalli) to the aorta (*right-to-left shunt*). Due to the low peripheral resistance (placenta), the **blood pressure** in the aorta is relatively low—only about 65 mmHg towards the end of pregnancy.

The arteries of the head and upper body are supplied with partly arterialized blood from the left ventricle ($\rightarrow$ **A**). This is important since brain tissue is susceptible to hypoxia. The remaining blood leaves the aorta and mixes with venous blood from the ductus arteriosus. As a result, the blood supplied to the lower half of the body has a relatively low O_2 concentration (O_2 saturation = 0.3; $\rightarrow$ **A**). The majority of this blood returns via the *umbilical arteries* to the placenta, where it is oxygenated again.

Circulation during birth. The exchange of O_2, nutrients, and waste materials through the placenta stops abruptly during birth. This leads to a rise in blood P_{CO_2}, triggering chemosensors ($\rightarrow$ p. 132) that induce a strong breathing reflex. The resultant *inspiratory movement* causes negative pressure (suction) in the thoracic cavity, which removes the blood from the placenta and umbilical vein (*placental transfusion*) and expands the lungs. The unfolding of the lungs and the rise in alveolar P_{O_2} reduces the resistance in the pulmonary circulation, and the blood flow increases while the pressure decreases ($\rightarrow$ **B1, 2**). Meanwhile, the resistance in the systemic circulation increases due to occlusion or clamping of the umbilical cord. This changes the direction of blood flow in the **ductus arteriosus**, resulting in a *left-to-right shunt*. The pulmonary circulation therefore receives aortic blood for a few days after birth. The right atrial filling volume decreases due to the lack of placental blood, while that of the left atrium increases due to the increased pulmonary blood flow. Due to the resultant pressure gradient from the left to right atrium and to a decrease in vasodilatory prostaglandins, the **foramen ovale** usually closes within about 2 weeks after birth. The ductus arteriosus and ductus venosus also close, and the systemic and pulmonary circulation now form serial circuits.

Shunts occur when the foramen ovale or ductus arteriosus remains open, placing a strain on the heart. In **patent foramen ovale** (atrial septum defect), the blood flows from left atrium $\rightarrow$ right atrium (left-to-right shunt) $\rightarrow$ right ventricle (*volume overload*) $\rightarrow$ lungs $\rightarrow$ left atrium. In **patent ductus arteriosus**, the blood flows from aorta $\rightarrow$ pulmonary artery (= left-to-right shunt) $\rightarrow$ lungs (*pressure overload*) $\rightarrow$ aorta.

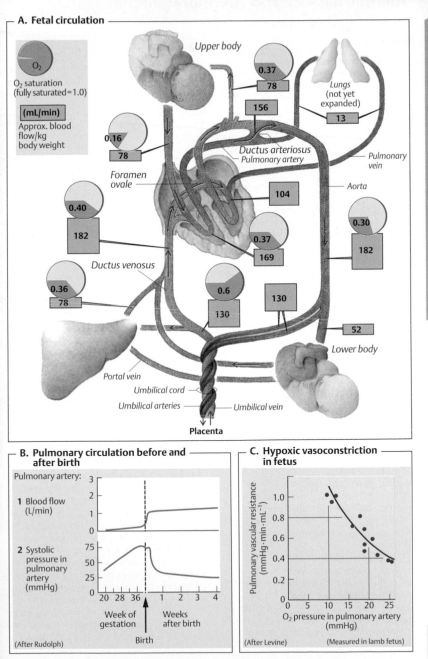

A. Fetal circulation

O₂ saturation (fully saturated = 1.0)

(mL/min) Approx. blood flow/kg body weight

Upper body

0.37 / 78

156

Lungs (not yet expanded)

13

Pulmonary vein

0.16 / 78

Ductus arteriosus Pulmonary artery

Foramen ovale

104

Aorta

0.40 / 182

0.37 / 169

0.30 / 182

Ductus venosus

0.36 / 78

0.6 / 130

130

130

52

Lower body

Portal vein

Umbilical cord

Umbilical arteries — *Umbilical vein*

Placenta

B. Pulmonary circulation before and after birth

Pulmonary artery:

1 Blood flow (L/min)

2 Systolic pressure in pulmonary artery (mmHg)

Week of gestation

Weeks after birth

Birth

(After Rudolph)

C. Hypoxic vasoconstriction in fetus

Pulmonary vascular resistance (mmHg·min·mL⁻¹)

O₂ pressure in pulmonary artery (mmHg)

(After Levine)   (Measured in lamb fetus)

Plate 8.18   Fetal and Neonatal Circulation

Thermal Balance

The body temperature of humans remains relatively constant despite changes in the environmental temperature. This **homeothermy** applies only to the **core temperature** ($\approx 37\,°C$) of the body. The extremities and skin ("*shell*") exhibit **poikilothermy**, i.e., their temperature varies to some extent with environmental temperature. In order to maintain a constant core temperature, the body must balance the *amount of heat it produces and absorbs* with the *amount it loses*; this is **thermoregulation** ($\rightarrow$ p. 224).

Heat production. The amount of heat produced is determined by *energy metabolism* ($\rightarrow$ p. 228). At rest, approximately 56% of total heat production occurs in the internal organs and about 18% in the muscles and skin ($\rightarrow$ **A2**, top). During *physical exercise*, heat production increases several-fold and the percentage of heat produced by muscular work can rise to as much as 90% ($\rightarrow$ **A2**, bottom). To keep warm, the body may have to generate additional voluntary (limb movement) and involuntary (shivering) muscle contractions. Newborns also have tissue known as **brown fat**, which enables them to produce additional heat without shivering ($\rightarrow$ p. 225). Cold stimulates a reflex pathway resulting in norepinephrine release (β_3-adrenergic receptors) in fatty tissues, which in turn stimulates (1) lipolysis and (2) the expression of *lipoprotein lipase (LPL)* and *thermogenin*. LPL increases the supply of free fatty acids ($\rightarrow$ p. 254). Thermogenin localized in the inner mitochondrial membrane is an uncoupling protein that functions as an H^+ uniporter (UCP1, $\rightarrow$ p. 230). It short-circuits the H^+ gradient across the inner mitochondrial membrane ($\rightarrow$ p. 17/B2), thereby uncoupling the (heat-producing) respiratory chain of ATP production.

Heat produced in the body is absorbed by the bloodstream and conveyed to the body surface. In order for this **internal flow of heat** to occur, the temperature of the body surface must be lower than that of the body interior. The *blood supply to the skin* is the chief determinant of heat transport to the skin ($\rightarrow$ p. 224).

Heat loss occurs by the physical processes of radiation, conduction, convection, and evaporation ($\rightarrow$ **B**).

1. Radiation ($\rightarrow$ **B1, C**). The amount of heat lost by radiation from the skin is chiefly determined by the temperature of the radiator (fourth power of its absolute temperature). Heat net–radiates from the body surface to objects or individuals when they are cooler than the skin, and net–radiates to the body from objects (sun) that are warmer than the skin. Heat radiates from the body into the environment when no radiating object is present (night sky). Heat radiation does not require the aid of any vehicle and is hardly affected by the air temperature (air itself is a poor radiator). Therefore, the body loses heat to a cold wall (despite warm air in between) and absorbs radiation from the sun or an infrared radiator without air (space) or cold air, respectively, in between.

2. Conduction and **convection** ($\rightarrow$ **B2, C**). These processes involve the transfer of heat from the skin to cooler air or a cooler object (e.g. sitting on rock) in contact with the body (*conduction*). The amount of heat lost by conduction to air increases greatly when the warmed air moves away from the body by natural *convection* (heated air rises) or forced convection (wind).

3. Evaporation ($\rightarrow$ **B3, C**). The first two mechanisms alone are unable to maintain adequate temperature homeostasis at high environmental temperatures or during strenuous physical activity. Evaporation is the means by which the body copes with the additional heat. The water lost by evaporation reaches the skin surface by diffusion (*insensible perspiration*) and by neuron-activated **sweat** glands ($\rightarrow$ **B3**, pp. 73ff. and 225 D). About 2428 kJ (580 kcal) of heat are lost for each liter of water evaporating and thereby cooling the skin. At temperatures above 36 °C or so, heat loss occurs by evaporation only ($\rightarrow$ **C**, right). At even higher environmental temperatures, heat is absorbed by radiation and conduction/convection. The body must lose larger amounts of heat by evaporation to make up for this. The surrounding air must be relatively dry in order for heat loss by evaporation to occur. Humid air retards evaporation. When the air is extremely humid (e.g., in a tropical rain forest), the average person cannot tolerate temperatures above 33 °C, even under resting conditions.

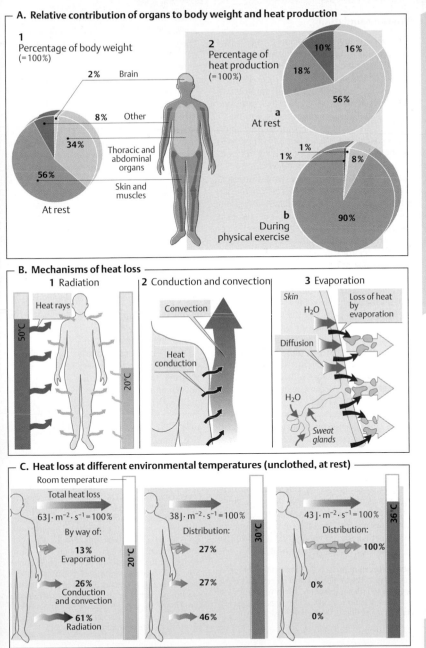

A. Relative contribution of organs to body weight and heat production

1 Percentage of body weight (=100%)

2% Brain
8% Other
34% Thoracic and abdominal organs
56% Skin and muscles

At rest

2 Percentage of heat production (=100%)

a At rest
10%
16%
18%
56%

b During physical exercise
1%
1%
8%
90%

B. Mechanisms of heat loss

1 Radiation

Heat rays

50°C
20°C

2 Conduction and convection

Convection
Heat conduction

3 Evaporation

Skin
H_2O
Loss of heat by evaporation
Diffusion
H_2O
Sweat glands

C. Heat loss at different environmental temperatures (unclothed, at rest)

Room temperature

Total heat loss
$63 J \cdot m^{-2} \cdot s^{-1} = 100\%$

By way of:

13% Evaporation
26% Conduction and convection
61% Radiation

20°C

$38 J \cdot m^{-2} \cdot s^{-1} = 100\%$

Distribution:

27%
27%
46%

30°C

$43 J \cdot m^{-2} \cdot s^{-1} = 100\%$

Distribution:

100%
0%
0%

36°C

Plate 9.1 Thermal Balance

Thermoregulation

Thermoregulation maintains the **core temperature** ($\rightarrow$ **A**) at a constant **set point** ($\approx$ **37°C**) despite fluctuations in heat absorption, production, and loss ($\rightarrow$ p. 222). The core temperature exhibits *circadian variation*. It fluctuates by about 0.6 °C and is lowest around 3 a.m., and highest around 6 p.m. ($\rightarrow$ p. 381 C). The set point changes are controlled by an intrinsic *biological clock* ($\rightarrow$ p. 334). Extended set-point fluctuations happen during the menstrual cycle ($\rightarrow$ p. 299/A3) and fever.

The *control center* for body temperature and *central thermosensors* are located in the **hypothalamus** ($\rightarrow$ p. 330). Additional thermosensors are located in the spinal cord and skin ($\rightarrow$ p. 314). The control center compares the actual core temperature with the set-point value and initiates measures to counteract any deviations ($\rightarrow$ **D** and p. 4f.).

When the **core temperature rises** above the set point (e.g., during exercise), the body increases the internal heat flow ($\rightarrow$ p. 222) by *dilating the blood vessels* of the skin. Moreover, arteriovenous anastomoses open in the periphery, especially in the fingers. The blood volume transported per unit time then not only conveys more heat, but also reduces the countercurrent exchange of heat between the arteries and their accompanying veins ($\rightarrow$ **B**). In addition, venous return in the extremities is *re-routed* from the deep, accompanying veins to the superficial veins. **Sweat secretion** also increases. The evaporation of sweat cools the skin, thereby creating the core/skin temperature gradient needed for the internal heat flow. *Central warm sensors* emit the signals that activate the sweat glands. (In this case, the thermosensors of the skin do not detect warmth because their environment is cooler than the core temperature). The efferent nerve fibers to the sweat glands are cholinergic fibers of the sympathetic nervous system ($\rightarrow$ **D**).

Acclimatization to high environmental temperatures (e.g., in the tropics) is a slow process that often takes years. Characteristically, the sweat secretion rate rises, the salt content of the sweat decreases, and thirst and thus H_2O intake increase.

When the **core temperature falls** below set point, the body checks heat loss by constricting the blood vessels in the shell ($\rightarrow$ **A**, left) and increases heat production by generating voluntary and involuntary (*shivering*) muscle activity ($\rightarrow$ **D**). Although infants can quickly become hypothermic because of their high surface/volume ratio, their *brown fat* allows them to produce additional heat (*non-shivering thermogenesis*; $\rightarrow$ p. 222). Upon exposure to low ambient temperatures, these three mechanisms are activated by the cold receptors of the skin ($\rightarrow$ p. 314) *before* the core temperature falls.

The range of ambient temperatures between the sweating and shivering thresholds is known as the **thermoneutral zone**. It lies between ca. 27 °C and 32 °C in the nearly unclothed test subject. The only thermoregulatory measure necessary within this range is variation of blood flow to the skin. The narrow range of this zone shows the thermoregulatory importance of **behavior**. It involves choosing the appropriate clothing, seeking shade, heating or cooling our dwellings, etc. Behavioral adaptation is the chief factor in survival at extreme ambient temperatures ($\rightarrow$ **C**).

The thermoneutral zone is subjectively perceived as the **comfort zone**. 95% of all subjects wearing normal office attire and performing normal office activities perceive an indoor climate with the following conditions to be comfortable: ambient and radiant (wall) temperature $\approx$ 23 °C, wind velocity < 0.1 m/s, and relative humidity $\approx$ 50%. A resting, unclothed subject feels comfortable at about 28 °C and ca. 31 °C to 36 °C in water depending on the thickness of subcutaneous fat (heat isolator).

Fever. *Exogenous* (e.g., bacteria) and *endogenous pyrogens* (various interleukins and other cytokines from macrophages) can cause the set-point temperature to rise above normal. This is triggered by prostaglandin PGE_2 in the hypothalamus. In the initial phase of fever, the core temperature (although at its normal level) is too low compared to the elevated set-point. This results in shivering to raise the core temperature. As the fever decreases, i.e. the set-point returns toward the normal temperature, the core temperature is now too warm compared to the normalized set-point, resulting in vasodilatation and sweating to lower the core temperature again.

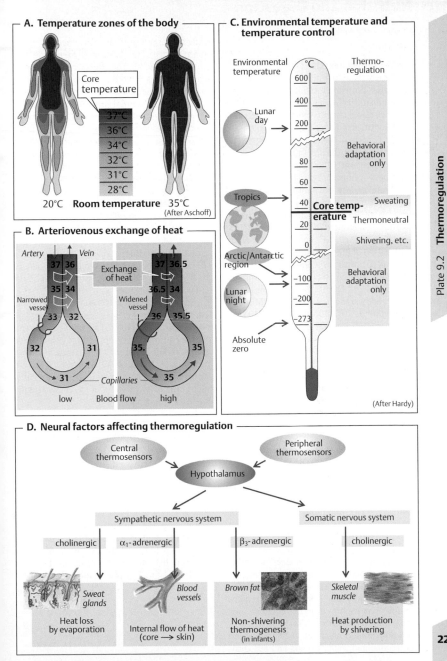

A. Temperature zones of the body

Core temperature

37°C
36°C
34°C
32°C
31°C
28°C

20°C **Room temperature** 35°C

(After Aschoff)

B. Arteriovenous exchange of heat

Artery   *Vein*

37 | 36
Exchange of heat
35 | 34
Narrowed vessel   33 | 32
32 | 31
31 — *Capillaries*

37 | 36.5
36.5 | 34
Widened vessel   36 | 35.5
35. | 35
35 — *Capillaries*

low   **Blood flow**   high

C. Environmental temperature and temperature control

Environmental temperature   °C   Thermo-regulation

600
400
200

Lunar day

80
60
40 **Core temp-erature**
20
0
-100
-200
-273

Tropics

Arctic/Antarctic region

Lunar night

Absolute zero

Behavioral adaptation only

Sweating
Thermoneutral
Shivering, etc.

Behavioral adaptation only

(After Hardy)

D. Neural factors affecting thermoregulation

Central thermosensors   Peripheral thermosensors

Hypothalamus

Sympathetic nervous system   Somatic nervous system

cholinergic | α_1-adrenergic | β_3-adrenergic | cholinergic

Sweat glands | *Blood vessels* | *Brown fat* | *Skeletal muscle*

Heat loss by evaporation | Internal flow of heat (core → skin) | Non-shivering thermogenesis (in infants) | Heat production by shivering

Plate 9.2 **Thermoregulation**

225

Nutrition

An **adequate diet** must meet the body's energy requirements and provide a minimum of carbohydrates, proteins (incl. all essential amino acids) and fats (incl. essential fatty acids). Minerals (incl. trace elements), vitamins, and sufficient quantities of water are also essential. To ensure a normal passage time, especially through the colon, the diet must also provide a sufficient amount of *roughage* (indigestible plant fibers—cellulose, lignin, etc.).

The **total energy expenditure** (**TEE**) or *total metabolic rate* consists of (1) the *basal metabolic rate* (**BMR**), (2) the *activity energy costs,* and the (3) *diet-induced thermogenesis* (DIT; → p. 228, 231 A). TEE equals BMR when measured (a) in the morning (b) 20 h after the last meal, (3) resting, reclining, (4) at normal body temp., and (5) at a comfortable ambient temp. (→ p. 224). The BMR varies according to sex, age, body size and weight. The BMR for a young adult is ca. 7300 kJ/day (≈ 1740 kcal/day; see p. 374 for units) in men, and ca. 20% lower in women. During **physical activity**, TEE increases by the following factors: 1.2-fold for sitting quietly, 3.2-fold for normal walking, and 8-fold for forestry work. Top athletes can perform as much as 1600 W (=J/s) for two hours (e.g., in a marathon) but their daily TEE is much lower. TEE also increases at various degrees of **injury** (1.6-fold for sepsis, 2.1-fold for burns). 1 °C of **fever** increases TEE 1.13-fold.

Protein, fats and carbohydrates are the three **basic energy substances** (→ **B**).

An adequate intake of **protein** is needed to maintain a proper *nitrogen balance*, i.e., balance of dietary intake and excretory output of nitrogen. The minimum requirement for protein is 0.5 g/kg BW per day (*functional minimum*). About half of dietary protein should be animal protein (meat, fish, milk and eggs) to ensure an adequate supply of *essential amino acids* such as histidine, isoleucine, leucine, lysine, methionine, phenylalanine, threonine, tryptophan and valine (children also require arginine). The content of most vegetable proteins is only about 50% of animal protein.

Carbohydrates (starch, sugar, glycogen) and **fats** (animal and vegetable fats and oils) provide the largest portion of the energy requirement. They are basically interchangeable sources of energy. The energy contribution of carbohydrates can fall to about 10% (normally 60%) before metabolic disturbances occur.

Fat is not essential provided the intake of *fat-soluble vitamins* (vitamins E, D, K and A) and *essential fatty acids* (linoleic acid) is sufficient. About 25–30% of dietary energy is supplied by fat (one-third of which is supplied as essential fatty acids; →**A**), although the proportion rises according to energy requirements (e.g., about 40% during heavy physical work). Western diets contain generally too much energy (more fats than carbohydrates!) considering the generally low level of physical activity of the Western lifestyle. Alcohol also contains superfluous energy (ca. 30 kJ/g = 7.2 kcal/g). The excessive intake of dietary energy leads to weight gain and obesity (→ p. 230).

An adequate intake of **minerals** (inorganic compounds), especially calcium (800 mg/day; → p. 290ff.), iron (10–20 mg/day; → p. 90) and iodine (0.15 mg/day; → p. 288), is essential for proper body function. Many **trace elements** (As, F, Cu, Si, V, Sn, Ni, Se, Mn, Mo, Cr, Co) are also essential. The normal diet provides sufficient quantities of them, but excessive intake has toxic effects.

Vitamins (A, B_1, B_2, B_6, B_{12}, C, D_2, D_3, E, H (biotin), K_1, K_2, folic acid, niacinamide, pantothenic acid) are compounds that play a vital role in metabolism (usually function as coenzymes). However, the body cannot produce (or sufficient quantities of) them. A deficiency of vitamins (hypovitaminosis) can lead to specific conditions such as night blindness (vit. A), scurvy (vit. C), rickets (vit. D = calciferol; → p. 292), anemia (vit. B_{12} = cobalamin; folic acid; → p. 90), and coagulation disorders (vit. K; → p. 104). An excessive intake of certain vitamins like vitamin A and D, on the other hand, can be toxic *(hypervitaminosis).*

A. Energy content of foodstuffs and energy requirement

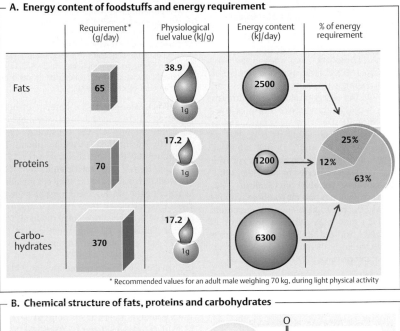

	Requirement* (g/day)	Physiological fuel value (kJ/g)	Energy content (kJ/day)	% of energy requirement
Fats	65	38.9 (1g)	2500	25%
Proteins	70	17.2 (1g)	1200	12%
Carbo-hydrates	370	17.2 (1g)	6300	63%

* Recommended values for an adult male weighing 70 kg, during light physical activity

B. Chemical structure of fats, proteins and carbohydrates

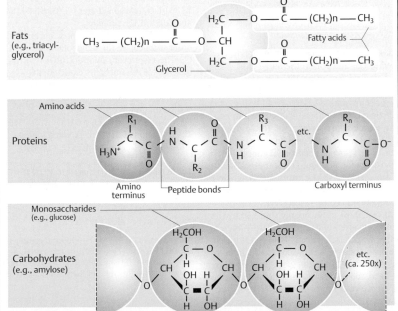

Fats (e.g., triacyl-glycerol)

$$CH_3 — (CH_2)n — \overset{O}{\underset{\|}{C}} — O — CH$$

$$H_2C — O — \overset{O}{\underset{\|}{C}} — (CH_2)n — CH_3$$

$$H_2C — O — \overset{O}{\underset{\|}{C}} — (CH_2)n — CH_3$$

Fatty acids

Glycerol

Proteins

Amino acids

$$H_3N^+ \quad R_1 \quad R_2 \quad R_3 \quad R_n$$

Amino terminus — Peptide bonds — Carboxyl terminus

Carbohydrates (e.g., amylose)

Monosaccharides (e.g., glucose)

etc. (ca. 250x)

Plate 10.1 Nutrition

Energy Metabolism and Calorimetry

The chemical energy of foodstuffs is first converted into energy-rich substances such as **creatine phosphate** and **adenosine triphosphate (ATP)**. Energy, work and amount of heat are expressed in joules (J) or calories (cal) ($\rightarrow$ p. 374). The energy produced by hydrolysis of ATP ($\rightarrow$ p. 41) is then used to fuel muscle activity, to synthesize many substances, and to create concentration gradients (e.g., Na^+ or Ca^{2+} gradients; $\rightarrow$ p. 26ff.). During all these energy conversion processes, part of the energy is always converted to *heat* ($\rightarrow$ p. 38ff.).

In **oxidative (aerobic) metabolism** ($\rightarrow$ p. 39 C), carbohydrates and fat combine with O_2 to yield CO_2, water, high-energy compounds (ATP etc.) and heat. When a foodstuff is completely oxidized, its biologically useable energy content is therefore equivalent to its physical **caloric value (CV)**.

The **bomb calorimeter** ($\rightarrow$ A), a device consisting of an insulated combustion chamber in a tank of water, is used to *measure the CV* of foodstuffs. A known quantity of a foodstuff is placed in the combustion chamber of the device and incinerated in pure O_2. The surrounding water heats up as it absorbs the heat of combustion. The degree of warming is equal to the caloric value of the foodstuff.

Fats and carbohydrates are completely oxidized to CO_2 and H_2O in the body. Thus, their **physiological fuel value (PFV)** is identical to their CV. The mean PFV is 38.9 kJ/g (= 9.3 kcal/g) for fats and 17.2 kJ/g (= 4.1 kcal/g) for digestible carbohydrates ($\rightarrow$ p. 227 A). In contrast, **proteins** are not completely broken down to CO_2 and water in the human body but yield urea, which provides additional energy when it is oxidized in the bomb calorimeter. The CV of proteins (ca. 23 kJ/g) is therefore greater than their PFV, which is a mean of about 17.2 kJ/g or 4.1 kcal/g ($\rightarrow$ p. 227 A).

At rest, most of the energy supplied by the diet is converted to **heat**, since hardly any external mechanical work is being performed. The heat produced is equivalent to the internal energy turnover (e.g., the work performed by the heart and respiratory muscles or expended for active transport or synthesis of substances).

In **direct calorimetry** ($\rightarrow$ **B**), the amount of heat produced is measured directly. The test subject, usually an experimental animal, is placed in a small chamber immersed in a known volume of ice. The amount of heat produced is equivalent to the amount of heat absorbed by the surrounding water or ice. This is respectively calculated as the rise in water temperature or the amount of ice that melts.

In **indirect calorimetry**, the amount of heat produced is determined indirectly by measuring the amount of O_2 consumed ($\dot{V}_{O_2}$; $\rightarrow$ p. 120). This method is used in humans. To determine the total metabolic rate (or TEE; $\rightarrow$ p. 226) from $\dot{V}_{O_2}$, the **caloric equivalent (CE)** of a foodstuff oxidized in the subject's metabolism during the measurement must be known. The CE is calculated from the PFV and the amount of O_2 needed to oxidize the food. The PFV of *glucose* is 15.7 kJ/g and 6 mol of O_2 (6×22.4 L) are required to oxidize 1 mol (= 180 g) of glucose ($\rightarrow$ **C**). The oxidation of 180 g of glucose therefore generates 2827 kJ of heat and consumes 134.4 L of O_2 resulting in a CE of 21 kJ/L. This value represents the CE for glucose under standard conditions ($0\,^\circ$C; $\rightarrow$ **C**). The mean CE of the basic nutrients at $37\,^\circ$C is 18.8 kJ/L O_2 (carbohydrates), 17.6 kJ/L O_2 (fats) and 16.8 kJ/L O_2 (proteins).

The oxidized nutrients must be known in order to calculate the metabolic rate from the CE. The **respiratory quotient (RQ)** is a rough measure of the nutrients oxidized. RQ = $\dot{V}_{CO_2}/\dot{V}_{O_2}$ ($\rightarrow$ p. 120). For pure carbohydrates oxidized, RQ = 1.0. This can be illustrated for *glucose* as follows:
$$C_6H_{12}O_6 + 6\,O_2 \rightleftharpoons 6\,CO_2 + 6\,H_2O \qquad [10.1]$$
The oxidation of the fat *tripalmitin* yields:
$$2\,C_{51}H_{98}O_6 + 145\,O_2 \rightleftharpoons 102\,CO_2 + 98\,H_2O \qquad [10.2]$$
The RQ of tripalmitin is therefore $102/145 = 0.7$. Since the protein fraction of the diet stays relatively constant, each RQ between 1 and 0.7 can be assigned a CE ($\rightarrow$ **D**). Using the known CE, the TEE can be calculated as CE $\cdot \dot{V}_{O_2}$.

Food increases the TEE (**diet–induced thermogenesis, DIT**) because energy must be consumed to absorb and store the nutrients. The DIT of protein is higher than that of other substances, e.g., glucose.

Plate 10.2   Energy Metabolism and Calorimetry

A. Bomb calorimeter

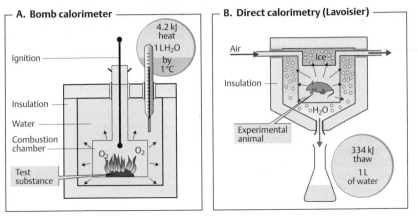

Ignition

Insulation

Water

Combustion chamber

Test substance

O_2 O_2

4.2 kJ heat 1 L H_2O by 1 °C

B. Direct calorimetry (Lavoisier)

Air

Ice

Insulation

Experimental animal

H_2O

334 kJ thaw 1 L of water

C. Oxidation of glucose: fuel value, CE and RQ

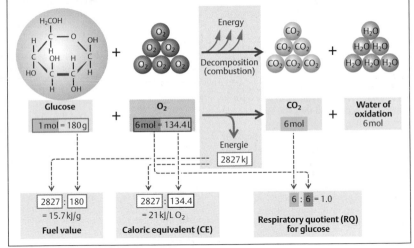

Energy

Decomposition (combustion)

CO_2 CO_2 CO_2 CO_2 CO_2 CO_2 + H_2O H_2O H_2O H_2O H_2O H_2O

Glucose
1 mol = 180 g

+

O_2
6 mol = 134.4 L

Energie
2827 kJ

CO_2
6 mol

+

Water of oxidation
6 mol

2827 : 180
= 15.7 kJ/g
Fuel value

2827 : 134.4
= 21 kJ/L O_2
Caloric equivalent (CE)

6 : 6 = 1.0
Respiratory quotient (RQ) for glucose

D. RQ and caloric equivalent relative to nutrient composition

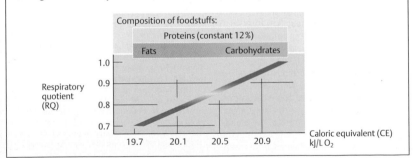

Composition of foodstuffs:

Proteins (constant 12 %)

Fats   Carbohydrates

Respiratory quotient (RQ)

1.0
0.9
0.8
0.7

19.7   20.1   20.5   20.9

Caloric equivalent (CE) kJ/L O_2

Energy Homeostasis and Body Weight

Fat depots are by far the body's largest **energy reserve**. Accurate long-term homeostasis of energy absorption and consumption is necessary to keep the size of the fat depots constant, i.e., to maintain *lipostasis*. Considering that a person's body weight (BW) mainly varying with the weight of the fat depots, it is obvious that energy homeostasis is synonymous with the **regulation of body weight** ($\rightarrow$ **A**).

The **body mass index (BMI)** is commonly used to determine whether an individual is underweight, overweight or in the normal weight range. BMI is calculated from body weight (kg) and height (m) as follows:

$$BMI = (weight/height)^2 \qquad [10.3]$$

The normal BMI range is 19–24 in women and 20–25 in men. The "normal" BMI range is defined as the values at which the mean life expectancy is highest. An abnormally high body mass index (BMI > 24 or 25 = overweight; BMI > 30 = obesity) reduces the life expectancy since this is often associated with diabetes mellitus (type II), hypertension and cardiac disease.

The following **regulatory mechanisms** serve to keep the fat depots and thus the body weight constant ($\rightarrow$ **B**):

◆ The **hypothalamus**, the center responsible for feedback control of BW, maintains communication with the limbic system, cerebral cortex and brain stem. It sends and receives:

◆ *Afferent messages* concerning the size of fat depots. **Leptin**, a 16-kDa proteohormone produced by fat cells, is the main indicator for this. The plasma leptin concentration rises as fat cell mass increases.

◆ *Efferent commands* (a) to reduce nutrient absorption and increase energy consumption when plasma leptin are high ("fat reserve high!"), and (b) to increase nutrient absorption and decrease energy consumption when plasma leptin levels are low ("fat reserve low!") ($\rightarrow$ **B**).

Leptin receptors. Leptin binds with *type b leptin receptors* (**Ob-Rb**) of the hypothalamus (mainly in dorsomedial, ventromedial, lateral, paraventricular and arcuate nuclei). Certain neurons with Ob-Rb lie in front of the blood–brain barrier. (T lymphocytes and B cells of the pancreas are also equipped with Ob-Rb receptors.)

Effects of leptin. In contrast to primary starvation, the *weight loss* induced by leptin is restricted to the body's fat depots and completes the feedback loop of the regulatory process. The effects of leptin are chiefly mediated by two neurotransmitters located in the hypothalamus: α-*MSH* and *NPY* ($\rightarrow$ **B**).

◆ **α-MSH.** Leptin stimulates the release of α-MSH (α-melanocyte-stimulating hormone), one of the melanocortins (MC) synthesized from POMC ($\rightarrow$ p. 280). α-MSH inhibits the absorption of nutrients and increases sympathetic nervous activity and energy consumption via MC4 receptors in various areas of the hypothalamus.

The **mechanism** by which α-MSH increases energy consumption is not entirely clear, but an involuntary increase in ordinary skeletal muscle activity and tone appears to occur. In addition, **uncoupling proteins** (type UCP2 and UCP3) that were recently discovered in skeletal muscle and white fat make the membranes of the mitochondria more permeable to H+, thereby uncoupling the respiratory chain. As a result, chemical energy is converted into more heat and less ATP. The action of these UCPs, the expression of which may be stimulated by α-MSH, is therefore similar to that of thermogenin (UCP1; $\rightarrow$ p. 222).

◆ **NPY.** Leptin inhibits the hypothalamic release of NPY (neuropeptide Y), a neuropeptide that stimulates hunger and appetite, increases the parasympathetic activity and reduces energy consumption.

Leptin deficiency. Since NPY increases the secretion of gonadotropin-releasing hormone (GnRH), extreme weight loss results in *amenorrhea* ($\rightarrow$ **B**). Certain *genetic defects* can lead to impaired leptin production (*ob* [obesity] *gene*) or impaired leptin receptor function (*db* [diabetes] *gene*). The symptoms include arrested development in puberty and childhood *obesity*.

Other neurotransmitters and neuropeptides also play a role in the long-term regulation of fat depots. Some like *orexin A/B* and *norepinephrine* (α2-adrenoceptors) are **orexigenic**, i.e. they stimulate the appetite, while others like CCK, CRH, CART (cocaine and amphetamine-regulated transcript), insulin, and serotonin are **anorexigenic**. Peptides like CCK, GLP-1 (glucagon-like peptide amides), somatostatin, glucagon, and GRP (gastrin-releasing peptide) signal *satiety*, i.e., that one has had enough to eat. Together with gustatory stimuli and stretch receptors of the stomach wall, these **satiety peptides** help to limit the amount of food consumed with each meal.

Plate 10.3 Energy Homeostasis and Body Weight

A. Energy homeostasis

Energy reserve
(body weight)

Decrease Increase

Physical activity

Diet-induced
thermogenesis

Basal
metabolic rate

Ingestion of:

Proteins

Fats

Carbo-
hydrates

B. Regulation of body weight by leptin, α-MSH and NPY

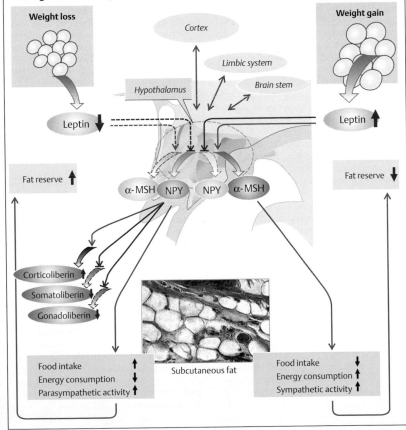

Weight loss

Weight gain

Cortex

Limbic system

Hypothalamus

Brain stem

Leptin ↓

Leptin ↑

Fat reserve ↑

Fat reserve ↓

α-MSH NPY NPY α-MSH

Corticoliberin

Somatoliberin

Gonadoliberin

Subcutaneous fat

Food intake ↑
Energy consumption ↕
Parasympathetic activity ↑

Food intake ↓
Energy consumption ↑
Sympathetic activity ↑

Gastrointestinal (GI) Tract: Overview, Immune Defense, Blood Flow

Food covering the body's energy and nutrient requirements ($\rightarrow$ p. 228ff.) must be swallowed, processed and broken down (**digestion**) before it can be **absorbed** from the intestines. The three-layered GI **musculature** ensures that the GI contents are properly mixed and transported. The **passage time** through the different GI segments varies and is largely dependent on the composition of the food (see **A** for mean passage times).

Solid food is chewed and mixed with **saliva**, which lubricates it and contains immunocompetent substances (see below) and enzymes. The **esophagus** rapidly transports the food bolus to the stomach. The lower esophageal sphincter opens only briefly to allow the food to pass. The **proximal stomach** mainly serves as a food reservoir. Its tone determines the rate at which food passes to the **distal stomach**, where it is further processed (chyme formation) and its proteins are partly broken down. The distal stomach (including the pylorus) is also responsible for portioning chyme delivery to the small intestine. The stomach also secretes *intrinsic factor* ($\rightarrow$ p. 90).

In the **small intestine**, enzymes from the **pancreas** and small intestinal mucosa break down the nutrients into absorbable components. HCO_3^- in pancreatic juices neutralizes the acidic chyme. Bile salts in **bile** are essential for fat digestion. The products of digestion (monosaccharides, amino acids, dipeptides, monoglycerides and free fatty acids) as well as water and vitamins are absorbed in the small intestine.

Waste products (e.g. bilirubin) to be excreted reach the feces via bile secreted by the **liver**. The liver has various other metabolic functions. It serves, for example, as an obligatory relay station for metabolism and distribution of substances reabsorbed from the intestine (via the portal vein, see below), synthesizes plasma proteins (incl. albumin, globulins, clotting factors, apolipoproteins etc.) and detoxifies foreign substances (*biotransformation*) and metabolic products (e.g., ammonia) before they are excreted.

The **large intestine** is the last stop for water and ion absorption. It is colonized by *bacteria* and contains storage areas for feces (**cecum, rectum**).

Immune defense. The large internal surface area of the GI tract (roughly 100 m^2) requires a very effective immune defense system. Saliva contains *mucins, immunoglobulin A (IgA)* and *lysozyme* that prevent the penetration of pathogens. *Gastric juice* has a bactericidal effect. *Peyer's patches* supply the GI tract with immunocompetent lymph tissue. *M cells* (special membranous cells) in the mucosal epithelium allow antigens to enter Peyer's patches. Together with *macrophages*, the Peyer's patches can elicit immune responses by secreting **IgA** ($\rightarrow$ p. 98). IgA is transported to the intestinal lumen by transcytosis ($\rightarrow$ p. 30). In the epithelium, IgA binds to a secretory component, thereby protecting it from digestive enzymes. Mucosal epithelium also contains *intraepithelial lymphocytes* (**IEL**) that function like T killer cells ($\rightarrow$ p. 98). Transmitter substances permit reciprocal communication between IEL and neighboring enterocytes. Macrophages of the hepatic sinusoids (*Kupffer's cells*) are additional bastions of immune defense. The physiological colonies of **intestinal flora** in the large intestine prevent the spread of pathogens. IgA from breast milk protects the GI mucosa of neonates.

Blood flow to the stomach, gut, liver, pancreas and spleen (roughly 30% of cardiac output) is supplied by the three main branches of the abdominal aorta. The intestinal circulation is regulated by local reflexes, the autonomic nervous system, and hormones. Moreover, it is *autoregulatory*, i.e., largely independent of systemic blood pressure fluctuations. Blood flow to the intestines rises sharply after meals (acetylcholine, vasoactive intestinal peptide VIP, etc. function as vasodilatory transmitters) and falls during physical activity (transmitters: norepinephrine, etc.). The venous blood carries substances reabsorbed from the intestinal tract and enters the liver via the **portal vein**. Some components of reabsorbed fat are absorbed by the *intestinal lymph*, which transports them to the greater circulation while bypassing the liver.

Plate 10.4 **GI Tract: Overviews, Defense, Blood Flow**

A. Function of gastrointestinal organs

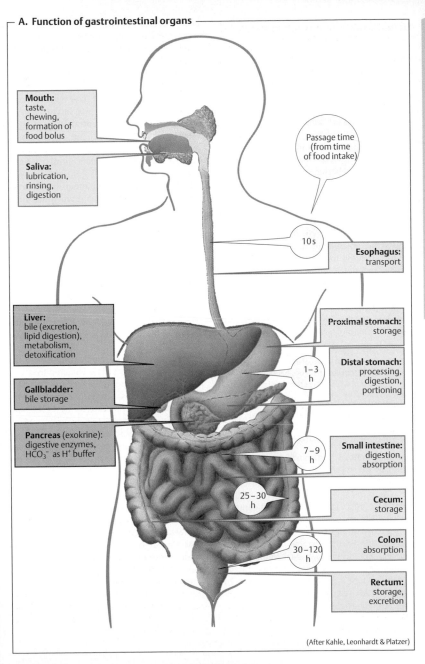

Mouth:
taste,
chewing,
formation of
food bolus

Saliva:
lubrication,
rinsing,
digestion

Passage time
(from time
of food intake)

10 s

Esophagus:
transport

Liver:
bile (excretion,
lipid digestion),
metabolism,
detoxification

Gallbladder:
bile storage

Pancreas (exocrine):
digestive enzymes,
HCO_3^- as H^+ buffer

Proximal stomach:
storage

1–3
h

Distal stomach:
processing,
digestion,
portioning

7–9
h

Small intestine:
digestion,
absorption

25–30
h

Cecum:
storage

Colon:
absorption

30–120
h

Rectum:
storage,
excretion

(After Kahle, Leonhardt & Platzer)

Neural and Hormonal Integration

Endocrine and paracrine hormones and neurotransmitters control GI *motility, secretion, perfusion and growth*. Reflexes proceed within the mesenteric and submucosal plexus (*enteric nervous system, ENS*), and external innervation modulates ENS activity.

Local reflexes are triggered by stretch sensors in the walls of the esophagus, stomach and gut or by chemosensors in the mucosal epithelium and trigger the contraction or relaxation of neighboring smooth muscle fibers. *Peristaltic reflexes* extend further towards the oral (ca. 2 mm) and anal regions (20–30 mm). They are mediated in part by interneurons and help to propel the contents of the lumen through the GI tract (*peristalsis*).

External innervation of the GI tract (cf. p. 78ff.) comes from the *parasympathetic nervous system* (from lower esophagus to ascending colon) and *sympathetic nervous system*. Innervation is also provided by *visceral afferent fibers* (in sympathetic or parasympathetic nerves) through which the afferent impulses for *supraregional reflexes* flow.

ENS function is largely independent of external innervation, but **external innervation** has some advantages (a) rapid transfer of signals between relatively distant parts of the GI tract via the abdominal ganglia (short visceral afferents) or CNS (long visceral afferents); (b) GI tract function can be ranked subordinate to overall body function (c) GI tract activity can be processed by the brain so the body can become aware of them (e.g., stomach ache).

Neurotransmitters. *Norepinephrine* (**NE**) is released by the adrenergic postganglionic neurons, and *acetylcholine* (**ACh**) is released by pre- and postganglionic (enteric) fibers ($\rightarrow$ p. 78ff.). **VIP** (vasoactive intestinal peptide) mediates the relaxation of circular and vascular muscles of the GI tract. **Met- and leu-enkephalin** intensify contraction of the pyloric, ileocecal and lower esophageal sphincters by binding to opioid receptors. **GRP** (gastrin-releasing peptide) mediates the release of gastrin. **CGRP** (calcitonin gene-related peptide) stimulates the release of somatostatin (**SIH**).

All **endocrine hormones** effective in the GI tract are *peptides* produced in endocrine cells of the mucosa. (a) *Gastrin* and *cholecystokinin* (**CCK**) and (b) *secretin* and **GIP** are structurally similar; so are glucagon ($\rightarrow$ p. 282ff.) and **VIP**. High concentrations of hormones from the same family therefore have very similar effects.

Gastrin occurs in short (G17 with 17 amino acids, AA) and long forms (G34 with 34 AA). G17 comprises 90% of all antral gastrin. Gastrin is secreted in the antrum and duodenum. Its release ($\rightarrow$**A1**) via *gastrin-releasing peptide* (**GRP**) is subject to neuronal control; gastrin is also released in response to stomach wall stretching and protein fragments in the stomach. Its secretion is inhibited when the pH of the gastric/duodenal lumen falls below 3.5 ($\rightarrow$**A1**). The main effects of gastrin are acid secretion and gastric mucosal growth ($\rightarrow$**A2**).

Cholecystokinin, CCK (33 AA) is produced throughout small intestinal mucosa. Long-chain fatty acids, AA and oligopeptides in the lumen stimulate the release of CCK ($\rightarrow$**A1**). It causes the gallbladder to contract and inhibits emptying of the stomach. In the pancreas, it stimulates growth, production of enzymes and secretion of HCO_3^- (via secretin, see below) ($\rightarrow$**A2**).

Secretin (27 AA) is mainly produced in the duodenum. Its release is stimulated by acidic chyme ($\rightarrow$**A1**). Secretin inhibits acid secretion and gastric mucosal growth and stimulates HCO_3^- secretion (potentiated by CCK), pancreatic growth and hepatic bile flow ($\rightarrow$**A2**).

GIP (glucose-dependent insulinotropic peptide, 42 AA; formerly called gastric inhibitory polypeptide = enterogastrone) is produced in the duodenum and jejunum and released via protein, fat and carbohydrate fragments (e.g., glucose) ($\rightarrow$**A1**). GIP inhibits acid secretion ($\rightarrow$**A2**) and stimulates insulin release (this is why oral glucose releases more insulin than intravenous glucose).

Motilin (22 AA) is released by neurons in the small intestine and regulates interdigestive motility ($\rightarrow$**A1, 2**).

Paracrine transmitters. Histamine, somatostatin and prostaglandin are the main paracrine transmitters in the GI tract.

Plate 10.5 Neural and Hormonal Integration

A. Gastrointestinal hormones

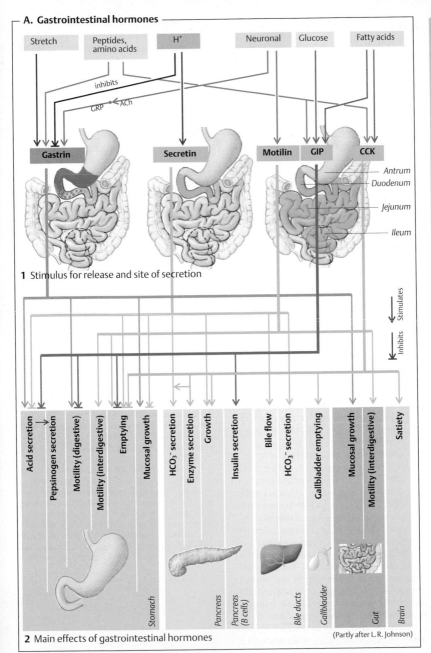

1 Stimulus for release and site of secretion

Stretch | Peptides, amino acids | H⁺ | Neuronal | Glucose | Fatty acids

inhibits

GRP ← ACh

Gastrin | **Secretin** | **Motilin** | **GIP** | **CCK**

Antrum
Duodenum
Jejunum
Ileum

Stimulates
Inhibits

Acid secretion
Pepsinogen secretion
Motility (digestive)
Motility (interdigestive)
Emptying
Mucosal growth
HCO₃⁻ secretion
Enzyme secretion
Growth
Insulin secretion
Bile flow
HCO₃⁻ secretion
Gallbladder emptying
Mucosal growth
Motility (interdigestive)
Satiety

Stomach
Pancreas
Pancreas (B cells)
Bile ducts
Gallbladder
Gut
Brain

2 Main effects of gastrointestinal hormones

(Partly after L. R. Johnson)

Saliva

The **functions** of saliva are reflected by its constituents. *Mucins* serve to lubricate the food, making it easier to swallow, and to keep the mouth moist to facilitate masticatory and speech-related movement. Saliva *dissolves* compounds in food, which is a prerequisite for taste buds stimulation ($\rightarrow$ p. 338) and for dental and oral hygiene. Saliva has a *low NaCl* concentration and is *hypotonic*, making it suitable for rinsing of the taste receptors (NaCl) while eating. Infants need saliva to seal the lips when suckling. Saliva also contains α-amylase, which starts the digestion of starches in the mouth, while *immunoglobulin A* and *lysozyme* are part of the immune defense system ($\rightarrow$ p. 94ff.). The high HCO_3^- concentration in saliva results in a pH of around 7, which is optimal for α-amylase-catalyzed digestion. Swallowed saliva is also important for buffering the acidic gastric juices refluxed into the esophagus ($\rightarrow$ p. 238). The secretion of profuse amounts of saliva before vomiting also prevents gastric acid from damaging the enamel on the teeth. Saliva secretion is very dependent on the body water content. A low content results in decreased saliva secretion—the mouth and throat become dry, thereby evoking the sensation of *thirst*. This is an important mechanism for maintaining the fluid balance ($\rightarrow$ pp. 168 and 184).

Secretion rate. The rate of saliva secretion varies from 0.1 to 4 mL/min (10–250 μL/min per gram gland tissue), depending on the degree of stimulation ($\rightarrow$ **B**). This adds up to about 0.5 to 1.5 L per day. At 0.5 mL/min, 95% of this rate is secreted by the *parotid gland* (serous saliva) and *submandibular gland* (mucin-rich saliva). The rest comes from the sublingual glands and glands in the buccal mucosa.

Saliva secretion occurs in *two steps*: The *acini* (*end pieces*) produce **primary saliva** ($\rightarrow$ **A, C**) which has an electrolyte composition similar to that of plasma ($\rightarrow$ **B**). Primary saliva secretion in the acinar cells is the result of *transcellular Cl$^-$ transport*: Cl$^-$ is actively taken up into the cells (secondary active transport) from the blood by means of a Na+-K$^+$-2Cl$^-$ cotransport carrier and is released into the lumen (together with HCO_3^-) via anion channels, resulting in a lumen-negative transepithelial potential (**LNTP**) that drives Na$^+$ paracellularly into the lumen. Water also follows passively (osmotic effect). Primary saliva is modified in **excretory ducts**, yielding **secondary saliva**. As the saliva passes through the excretory ducts, Na$^+$ and Cl$^-$ are reabsorbed and K$^+$ and (carbonic anhydrase-dependent) HCO_3^- is secreted into the lumen. The saliva becomes *hypotonic* (far below 100 mOsm/kg H_2O; $\rightarrow$ **B**) because Na$^+$ and Cl$^-$ reabsorption is greater than K$^+$ and HCO_3^- secretion and the ducts are relatively impermeable to water ($\rightarrow$ **B**). If the secretion rate rises to values much higher than 100 μL/(min·g), these processes lag behind and the composition of secondary saliva becomes similar to that of primary saliva ($\rightarrow$ **B**).

Salivant stimuli. *Reflex stimulation* of saliva secretion occurs in the larger salivary glands ($\rightarrow$ **D**). Salivant stimuli include the smell and taste of food, tactile stimulation of the buccal mucosa, mastication and nausea. *Conditioned reflexes* also play a role. For instance, the routine clattering of dishes when preparing a meal can later elicit a salivant response. Sleep and dehydration inhibit saliva secretion. Saliva secretion is stimulated via the sympathetic and parasympathetic nervous systems ($\rightarrow$ **C2**):

◆ **Norepinephrine** triggers the secretion of highly viscous saliva with a high concentration of mucin via β_2 adrenoreceptors and cAMP. *VIP* also increases the cAMP concentration of acinar cells.

◆ **Acetylcholine**: (a) With the aid of M_1 cholinoceptors and IP_3 ($\rightarrow$ pp. 82 and 274), acetylcholine mediates an increase in the cytosolic Ca^{2+} concentration of acinar cells. This, in turn, increases the conductivity of luminal anion channels, resulting in the production of watery saliva and increased exocytosis of salivary enzymes. (b) With the aid of M_3 cholinergic receptors, ACh mediates the contraction of *myoepithelial cells* around the acini, leading to emptying of the acini. (c) ACh enhances the production of kallikreins, which cleave *bradykinin* from plasma kininogen. Bradykinin and *VIP* ($\rightarrow$ p. 234) dilate the vessels of the salivary glands. This is necessary because maximum saliva secretion far exceeds resting blood flow.

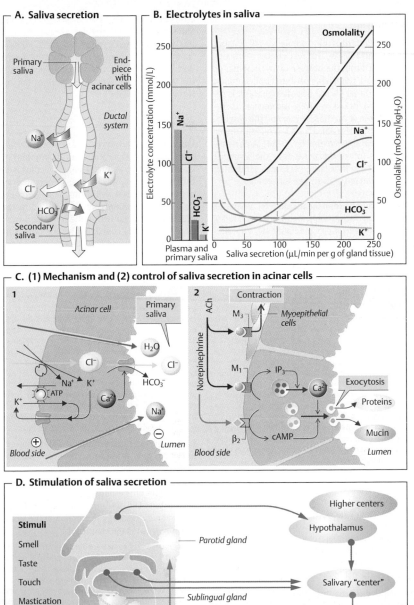

A. Saliva secretion

Primary saliva

End-piece with acinar cells

Ductal system

Na^+

K^+

Cl^-

HCO_3^-

Secondary saliva

B. Electrolytes in saliva

Electrolyte concentration (mmol/L)

Osmolality

Osmolality (mOsm/kgH₂O)

Na^+

Cl^-

HCO_3^-

K^+

Plasma and primary saliva

Saliva secretion (µL/min per g of gland tissue)

C. (1) Mechanism and (2) control of saliva secretion in acinar cells

1

Acinar cell

Primary saliva

H_2O

Cl^-

Cl^-

HCO_3^-

Na^+

K^+

ATP

Ca^{2+}

Na^+

K^+

$\oplus$

Blood side

$\ominus$

Lumen

2

ACh

Contraction

M_3

Myoepithelial cells

Norepinephrine

M_1

IP_3

Ca^{2+}

Exocytosis

Proteins

β_2

cAMP

Mucin

Blood side

Lumen

D. Stimulation of saliva secretion

Stimuli

Smell

Taste

Touch

Mastication

Nausea etc.

Parotid gland

Sublingual gland

Submandibular gland

Higher centers

Hypothalamus

Salivary "center"

Sympathetic and parasympathetic activation

Plate 10.6 **Saliva**

237

Deglutition

The upper third of the esophageal wall consists of striated muscle, the rest contains smooth muscle. During the process of swallowing, or deglutition, the tongue pushes a bolus of food into the throat (→ **A1**). The nasopharynx is reflexively blocked, (→ **A2**), respiration is inhibited, the vocal chords close and the epiglottis seals off the trachea (→ **A3**) while the *upper esophageal sphincter* opens (→ **A4**). A peristaltic wave forces the bolus into the stomach (→ **A5, B1,2**). If the bolus gets stuck, stretching of the affected area triggers a *secondary peristaltic wave*.

The **lower esophageal sphincter** opens at the start of deglutition due to a *vagovagal reflex* (*receptive relaxation*) mediated by VIP- and NO-releasing neurons (→ **B3**). Otherwise, the lower sphincter remains closed to prevent the reflux of aggressive gastric juices containing pepsin and HCl.

Esophageal motility is usually checked by measuring pressure in the lumen, e.g., during a peristaltic wave (→ **B1–2**). The resting pressure within the lower sphincter is normally 20–25 mmHg. During receptive relaxation, esophageal pressure drops to match the low pressure in the proximal stomach (→ **B3**), indicating opening of the sphincter. In *achalasia*, receptive relaxation fails to occur and food collects in the esophagus.

Pressure in the lower esophageal sphincter is *decreased* by VIP, CCK, NO, GIP, secretin and progesterone (→ p. 234) and *increased* by acetylcholine, gastrin and motilin. Increased abdominal pressure (external pressure) also increases sphincter pressure because part of the lower esophageal sphincter is located in the abdominal cavity.

Gastroesophageal reflux. The sporadic reflux of gastric juices into the esophagus occurs fairly often. Reflux can occur while swallowing (lower esophageal sphincter opens for a couple of seconds), due to unanticipated pressure on a full stomach or to *transient opening of the sphincter* (lasts up to 30 seconds and is part of the *eructation reflex*). Gastric reflux greatly reduces the pH in the distal esophagus.

Protective mechanisms to prevent damage to the esophageal mucosa after gastroesophageal reflux include 1. **Volume clearance**, i.e., the rapid return of refluxed fluid to the stomach via the esophageal peristaltic reflex. A refluxed volume of 15 mL, for example, remains in the esophagus for only 5 to 10 s (only a small amount remains). 2. **pH clearance.** The pH of the residual gastric juice left after volume clearance is still low, but is gradually increased during each act of swallowing. In other words, the saliva that is swallowed buffers the residual gastric juice.

Vomiting

Vomiting mainly serves as a *protective reflex* but is also an important clinical symptom of conditions such as intracranial bleeding and tumors. The act of vomiting is heralded by **nausea**, increased salivation and retching (→ **C**). The **vomiting center** is located in the medulla oblongata within the reticular formation. It is mainly controlled by chemosensors of the *area postrema*, which is located on the floor of the fourth ventricle; this is called the *chemosensory trigger zone* (CTZ). The blood-brain barrier is less tight in the area postrema.

The **CTZ** is **activated** by nicotine, other toxins, and dopamine agonists like apomorphine (used as an emetic). Cells of the CTZ have receptors for neurotransmitters responsible for their neuronal control. The vomiting center can also be activated independent of the CTZ, for example, due to abnormal stimulation of the organ of balance (*kinesia, motion sickness*), overextension of the stomach or intestines, delayed gastric emptying and inflammation of the abdominal organs. Nausea and vomiting often occur during the first trimester of pregnancy (*morning sickness*) and can exacerbate to *hyperemesis gravidarum* leading to vomiting–related disorders (see below).

During the **act of vomiting**, the diaphragm remains in the inspiratory position and the abdominal muscles quickly contract exerting a high pressure on the stomach. Simultaneous contraction of the duodenum blocks the way to the gut; the lower esophageal sphincter then relaxes, resulting in ejection of the stomach contents via the esophagus.

The sequelae of **chronic vomiting** are attributable to reduced food intake (*malnutrition*) and the related loss of gastric juices, swallowed saliva, fluids and intestinal secretions. In addition to *hypovolemia*, nonrespiratory *alkalosis* due to the loss of gastric acid (10–100 mmol H^+/L gastric juice) also develops. This is accompanied by *hypokalemia* due to the loss of K^+ in the vomitus (nutrients, saliva, gastric juices) and urine (hypovolemia-related *hyperaldosteronism*; → p. 180 ff.).

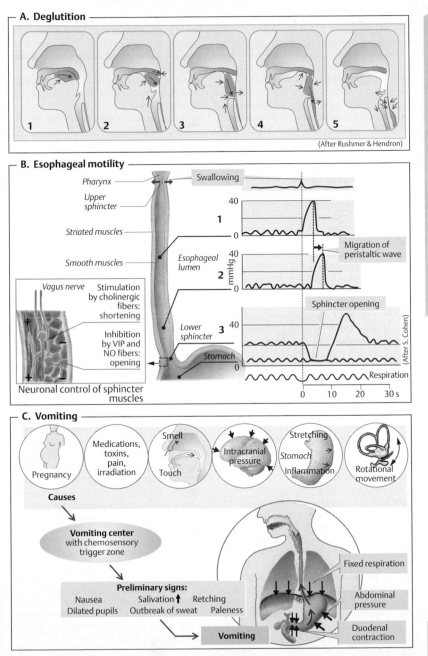

A. Deglutition

1 2 3 4 5

(After Rushmer & Hendron)

B. Esophageal motility

Pharynx

Upper sphincter

Striated muscles

Smooth muscles

Swallowing

Esophageal lumen

Migration of peristaltic wave

Sphincter opening

Lower sphincter

Stomach

mmHg

40 0

40 0

40 0

1

2

3

Vagus nerve

Stimulation by cholinergic fibers: shortening

Inhibition by VIP and NO fibers: opening

Neuronal control of sphincter muscles

Respiration

0 10 20 30 s

(After S. Cohen)

Plate 10.7 Deglutition

C. Vomiting

Pregnancy

Medications, toxins, pain, irradiation

Smell

Touch

Intracranial pressure

Stretching

Stomach Inflammation

Rotational movement

Causes

Vomiting center with chemosensory trigger zone

Preliminary signs:

Nausea   Salivation ↑   Retching
Dilated pupils   Outbreak of sweat   Paleness

Vomiting

Fixed respiration

Abdominal pressure

Duodenal contraction

Stomach Structure and Motility

Structure. The *cardia* connects the esophagus to the upper stomach (*fundus*), which merges with the body (*corpus*) followed by the *antrum* of the stomach. The lower outlet of the stomach (*pylorus*) merges with the duodenum (→ **A**). Stomach **size** is dependent on the degree of gastric filling, but this distension is mainly limited to the proximal stomach (→ **A, B**). The stomach **wall** has an outer layer of longitudinal muscle fibers (only at curvatures; regulates stomach length), a layer of powerful circular muscle fibers, and an inner layer of oblique muscle fibers. The mucosa of the **tubular glands** of the fundus and corpus contain *chief cells* (**CC**) and *parietal cells* (**PC**) (→ **A**) that produce the constituents of gastric juice (→ p. 242). The gastric mucosa also contains *endocrine cells* (that produce gastrin in the antrum, etc.) and *mucous neck cells* (**MNC**).

Functional anatomy. The stomach can be divided into a proximal and a distal segment (→ **A**). A vagovagal reflex triggered by swallowing a bolus of food causes the lower esophageal sphincter to open (→ p. 238) and the **proximal stomach** to dilate for a short period (*receptive relaxation*). This continues when the food has entered the stomach (vagovagal *accommodation reflex*). As a result, the internal pressure hardly rises in spite of the increased filling. Tonic contraction of the proximal stomach, which mainly serves as a *reservoir*, slowly propel the gastric contents to the **distal stomach**. Near its upper border (middle third of the corpus) is a *pacemaker zone* (see below) from which peristaltic waves of contraction arise due mainly to local stimulation of the stomach wall (in response to reflex stimulation and gastrin; → **D1**). The peristaltic waves are strongest in the antrum and spread to the pylorus. The chyme is thereby driven towards the pylorus (→ **C5, 6, 1**), then compressed (→ **C2, 3**) and propelled back again after the pylorus closes (→ **C3, 4**). Thereby, the food is *processed*, i.e., *ground*, mixed with gastric juices and digested, and fat is *emulsified*.

The distal stomach contains **pacemaker cells** (*interstitial Cajal cells*), the membrane potential of which oscillates roughly every 20 s, producing characteristic *slow waves* (→ p. 244). The velocity (0.5–4 cm/s) and amplitude (0.5–4 mV) of the waves increases as they spread to the pylorus. Whether and how often contraction follows such an excitatory wave depends on the sum of all neuronal and hormonal influences. Gastrin increases the response frequency and the pacemaker rate. Other hormones like GIP inhibit this motility directly, whereas somatostatin (SIH) does so indirectly by inhibiting the release of GRP (→ **D1** and p. 234).

Gastric emptying. Solid food remains in the stomach until it has been broken down into small particles (diameter of < 1 mm) and suspended in chyme. The chyme then passes to the duodenum. The time required for 50% of the ingested volume to leave the stomach varies, e.g., 10–20 min for water and 1–4 hours for solids (carbohydrates < proteins < fats). Emptying is mainly dependent on the *tone of the proximal stomach* and *pylorus*. Motilin *stimulates* emptying of the stomach (tone of proximal stomach rises, pylorus dilates), whereas decreases in the pH or osmolality of chyme or increases in the amount of long-chain free fatty acids or (aromatic) amino acids *inhibit* gastric emptying. Chemosensitive enterocytes and brush cells of the small intestinal mucosa, enterogastric reflexes and certain hormones (CCK, GIP, secretin and gastrin; → p. 234) mediate these regulatory activities (→ **D2**). The **pylorus** is usually slightly open during the process (free flow of "finished" chyme). It contracts only 1) at the end of "antral systole" (see above) in order to retain solid food and 2) when the duodenum contracts in order to prevent the reflux of harmful bile salts. If such reflex does occur, refluxed free amino acids not normally present in the stomach elicit reflex closure of the pylorus (→ **D2**).

Indigestible substances (bone, fiber, foreign bodies) do not leave the stomach during the digestive phase. Special contraction waves called *migrating motor complexes* (**MMC**) pass through the stomach and small intestine roughly every 1.5 hours during the ensuing **interdigestive phase**, as determined by an intrinsic "biological clock." These peristaltic waves transport indigestible substances from the stomach and bacteria from the small intestine to the large intestine. This "clearing phase" is controlled by *motilin*.

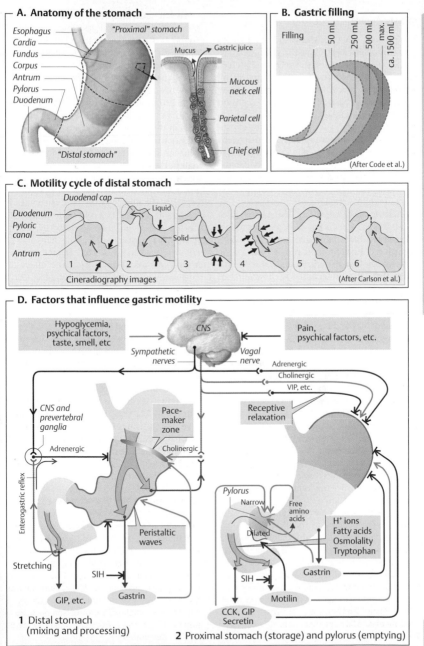

A. Anatomy of the stomach

Esophagus
Cardia
Fundus
Corpus
Antrum
Pylorus
Duodenum

"Proximal" stomach

"Distal stomach"

Mucus → Gastric juice

Mucous neck cell
Parietal cell
Chief cell

B. Gastric filling

Filling

50 mL
250 mL
500 mL
max. ca. 1500 mL

(After Code et al.)

C. Motility cycle of distal stomach

Duodenal cap
Duodenum
Pyloric canal
Antrum

Liquid
Solid

1 2 3 4 5 6

Cineradiography images

(After Carlson et al.)

D. Factors that influence gastric motility

Hypoglycemia, psychical factors, taste, smell, etc

CNS

Pain, psychical factors, etc.

Sympathetic nerves

Vagal nerve

Adrenergic
Cholinergic
VIP, etc.

CNS and prevertebral ganglia

Pace-maker zone

Receptive relaxation

Adrenergic

Cholinergic

Enterogastric reflex

Pylorus
Narrow

Free amino acids

H+ ions
Fatty acids
Osmolality
Tryptophan

Dilated

Peristaltic waves

Gastrin

Stretching

SIH

GIP, etc.

Gastrin

SIH

CCK, GIP
Secretin

Motilin

1 Distal stomach (mixing and processing)

2 Proximal stomach (storage) and pylorus (emptying)

Plate 10.8 Stomach Structure and Motility

Gastric Juice

The *tubular glands* of the gastric fundus and corpus secrete 3–4 L of gastric juice each day. *Pepsinogens* and *lipases* are released by chief cells and *HCl* and *intrinsic factor* (→ p. 260) by parietal cells. *Mucins* and HCO_3^- are released by mucous neck cells and other mucous cells on the surface of the gastric mucosa.

Pepsins function as endopeptidases in protein digestion. They are split from pepsinogens exocytosed from chief cells in the glandular and gastric lumen at a pH of < 6. Acetylcholine (ACh), released locally in response to H^+ (and thus indirectly also to gastrin) is the chief activator of this reaction.

Gastric acid. The pH of the gastric juice drops to ca. 0.8 during peak HCl secretion. Swallowed food buffers it to a pH of 1.8–4, which is optimal for most pepsins and gastric lipases. The low pH contributes to the *denaturation* of dietary proteins and has a *bactericidal effect*.

HCl secretion (→ **A**). The *H^+/K^+-ATPase* in the luminal membrane of parietal cells drives H^+ ions into the glandular lumen in exchange for K^+ (primary active transport, → **A1** and p. 26), thereby raising the H^+ conc. in the lumen by a factor of ca. 10^7. K^+ taken up in the process circulates back to the lumen via *luminal K^+ channels*. For every H^+ ion secreted, one HCO_3^- ion leaves the blood side of the cell and is exchanged for a Cl^- ion via an *anion antiporter* (→ **A2**). (The HCO_3^- ions are obtained from CO_2 + OH^-, a reaction catalyzed by carbonic anhydrase, CA). This results in the intracellular accumulation of Cl^- ions, which diffuse out of the cell to the lumen via *Cl^- channels* (→ **A3**). Thus, one Cl^- ion reaches the lumen for each H^+ ion secreted.

The **activation of parietal cells** (see below) leads to the opening of *canaliculi*, which extend deep into the cell from the lumen of the gland (→ **B**). The canaliculi are equipped with a brush border that greatly increases the luminal surface area which is densely packed with membrane-bound H^+/K^+ ATPase molecules. This permits to increase the secretion of H^+ ions from 2 mmol/hour at rest to over 20 mmol/hour during digestion.

Gastric acid secretion is **stimulated** in phases by neural, local gastric and intestinal factors (→ **B**). Food intake leads to reflex secretion of gastric juices, but deficient levels of glucose in the brain can also trigger the reflex. The optic, gustatory and olfactory nerves are the afferents for this partly *conditioned reflex* (→ p. 236), and efferent impulses flow via the *vagus nerve*. *ACh* directly activates parietal cells in the fundus (M_3 cholinoceptors → **B2**). *GRP* (gastrin-releasing peptide) released by neurons stimulates *gastrin* secretion from G cells in the antrum (→ **B3**). Gastrin released in to the systemic circulation in turn activates the parietal cells via CCK_B receptors (= gastrin receptors). The glands in the fundus contain H (histamine) cells or ECL cells (enterochromaffin–like cells), which are activated by gastrin (CCK_B receptors) as well as by ACh and β_3 adrenergic substances (→ **B2**). The cells release *histamine*, which has a paracrine effect on neighboring parietal cells (H_2 receptor). Local gastric and intestinal factors also influence gastric acid secretion because chyme in the antrum and duodenum stimulates the secretion of gastrin (→ **B1** and p. 235, A).

Factors that inhibit gastric juice secretion: (a) A pH of < 3.0 in the antral lumen inhibits **G cells** (negative feedback, → **B1, 3**) and activates antral **D cells**, which secrete SIH (→ p. 234), which in turn has a paracrine effect. SIH inhibits **H cells** in the fundus as well as G cells in the antrum (→ **B2, 3**). CGRP released by neurons (→ p. 234) activates D cells in the antrum and fundus, (→ **B2, 3**). (c) Secretin and GIP released from the small intestine have a retrograde effect on gastric juice secretion (→ **B1**). This adjusts the composition of chyme from the stomach to the needs of the small intestine.

Protection of the gastric mucosa from destructive gastric juices is chiefly provided by (a) a *layer of mucus* and (b) HCO_3^- *secretion* by the underlying mucous cells of the gastric mucosa. HCO_3^- diffuses through the layer of mucus and buffers the acid that diffuses into it from the lumen. *Prostaglandins* PGE_2 and PGI_2 promote the secretion of HCO_3^-. Anti-inflammatory drugs that inhibit cyclooxygenase 1 and thus prostaglandin production (→ p. 269) impair this mucosal protection and can result in ulcer development.

Plate 10.9   **Gastric Juice**

A. HCl secretion by parietal cells

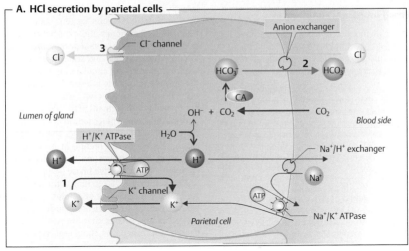

Anion exchanger

Cl⁻ channel

3

Cl⁻ → Cl⁻

HCO_3^- → **2** → HCO_3^-

CA

Lumen of gland

$OH^- + CO_2$ ← CO_2

H_2O

Blood side

H⁺/K⁺ ATPase

Na^+/H^+ exchanger

H⁺ → H⁺

1   ATP

K⁺ channel

Na⁺

K⁺ ← K⁺

ATP

Parietal cell

Na^+/K^+ ATPase

B. Regulation of gastric acid secretion

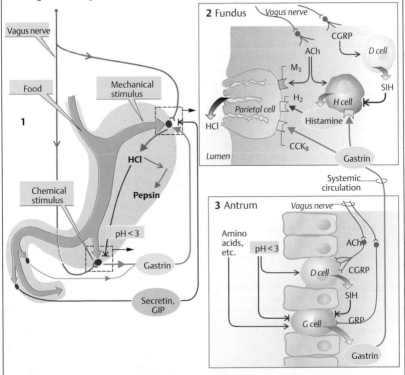

2 Fundus   *Vagus nerve*

Vagus nerve

CGRP → D cell

ACh

Food

Mechanical stimulus

M_3

SIH

1

H_2 *H cell*

Parietal cell

HCl

Histamine

Chemical stimulus

CCK_B

Lumen

Gastrin

HCl

Pepsin

Systemic circulation

$pH < 3$

3 Antrum   *Vagus nerve*

Gastrin

Amino acids, etc.

ACh

$pH < 3$

D cell   CGRP

Secretin, GIP

SIH

G cell   GRP

Gastrin

243

Small Intestinal Function

The main function of the small intestine (SI) is to finish digesting the food and to absorb the accumulated breakdown products as well as water, electrolytes and vitamins.

Structure. The SI of live human subjects is about 2 m in length. It arises from the pylorus as the *duodenum* and continues as the *jejunum*, and ends as the *ileum*, which merges into the large intestine. From outside inward, the SI consists of an outer serous coat (*tunica serosa*, → **A1**), a layer of *longitudinal muscle fibers* (→ **A2**), the *myenteric plexus* (Auerbach's plexus, → **A3**), a layer of circular muscle fibers (→ **A4**), the *submucous plexus* (Meissner's plexus, → **A5**) and a *mucous layer* (tunica mucosa, → **A6**), which is covered by epithelial cells (→ **A13–15**). The SI is supplied with blood vessels (→ **A8**), lymph vessels (→ **A9**), and nerves (→ **A10**) via the *mesentery* (→ **A7**). The surface area of the epithelial-luminal interface is roughly 300–1600 times larger (> 100 m^2) than that of a smooth cylindrical pipe because of the *Kerckring's folds* (→ **A11**), the *intestinal villi* (→ **A12**), and the *enterocytic microvilli*, or the *brush border* (→ **A13**).

Ultrastructure and function. *Goblet cells* (→ **A15**) are interspersed between the resorbing *enterocytes* (→ **A14**). The mucus secreted by goblet cells acts as a protective coat and lubricant. *Intestinal glands* (crypts of Lieberkühn, → **A16**) located at the bases of the villi contain (a) undifferentiated and mitotic cells that differentiate into villous cells (see below), (b) mucous cells, (c) endocrine and paracrine cells that receive information about the composition of chyme from chemosensor cells, and (d) immune cells (→ p. 232). The chyme composition triggers the secretion of endocrine hormones and of paracrine mediators (→ p. 234). The tubuloacinar *duodenal glands* (Brunner's glands), located deep in the intestinal wall (tela submucosa) secrete a HCO$_3^-$-rich fluid containing urogastrone (*human epidermal growth factor*), an important stimulator of epithelial cell proliferation.

Cell replacement. The tips of the villi are continually shed and replaced by new cells from the crypts of Lieberkühn. Thereby, the entire SI epithelium is renewed every 3–6 days. The dead cells disintegrate in the lumen, thereby releasing enzymes, stored iron, etc.

Intestinal motility is autonomously regulated by the enteric nervous system, but is influenced by hormones and external innervation (→ p. 234). Local *pendular movements* (by longitudinal muscles) and *segmentation* (contraction/relaxation of circular muscle fibers) of the SI serve to mix the intestinal contents and bring them into contact with the mucosa. This is enhanced by movement of the intestinal villi (lamina muscularis mucosae). *Reflex peristaltic waves* (30–130 cm/min) propel the intestinal contents towards the rectum at a rate of ca. 1 cm/min. These waves are especially strong during the interdigestive phase (→ p. 240).

Peristaltic reflex. Stretching of the intestinal wall during the passage of a bolus (→ **B**) triggers a reflex that constricts the lumen behind the bolus and dilates that ahead of it. Controlled by interneurons, cholinergic type 2 motoneurons with prolonged excitation simultaneously activate circular muscle fibers behind the bolus and longitudinal musculature in front of it. At the same time the circular muscle fibers in front of the bolus are inhibited (accommodation) while those behind it are disinhibited (→ **B** and p. 234).

Pacemakers. The intestine also contains pacemaker cells (*interstitial Cajal cells*). The membrane potential of these cells oscillates between 10 and 20 mV every 3–15 min, producing *slow waves* (→ **C1**). Their amplitude can rise (less negative potential) or fall in response to neural, endocrine or paracrine stimuli. A series of action potentials (spike bursts) are fired once the membrane potential rises above a certain threshold (ca. –40 mV) (→ **C2**). *Muscle spasms* occur if the trough of the wave also rises above the threshold potential (→ **C3**).

Impulse conduction. The spike bursts are conducted to myocytes via *gap junctions* (→ p. 70). The myocytes then contract rhythmically at the same frequency (or slower). Conduction in the direction of the anus dwindles after a certain distance (→ **D**, pacemaker zone), so more distal cells (with a lower intrinsic rate) must assume the pacemaker function. Hence, peristaltic waves of the small intestine only move in the anal direction.

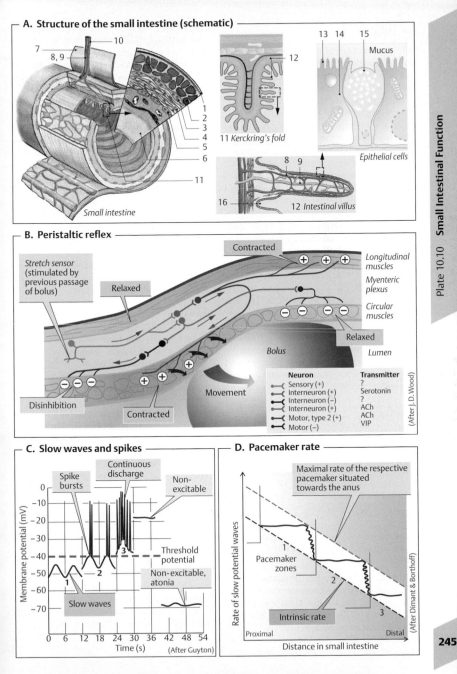

Plate 10.10 Small Intestinal Function

A. Structure of the small intestine (schematic)

7
8, 9
10
1
2
3
4
5
6
11

Small intestine

12

11 *Kerckring's fold*

13 14 15
Mucus

Epithelial cells

8 9
16
12 *Intestinal villus*

B. Peristaltic reflex

Contracted

Stretch sensor (stimulated by previous passage of bolus)

Relaxed

Longitudinal muscles

Myenteric plexus

Circular muscles

⊕ ⊕ ⊕

⊖ ⊖ ⊖ ⊖

Relaxed

Bolus

Lumen

Disinhibition

⊖ ⊖ ⊖

⊕ ⊕

⊕

Contracted

Movement

Neuron	Transmitter
Sensory (+)	?
Interneuron (+)	Serotonin
Interneuron (−)	?
Interneuron (+)	ACh
Motor, type 2 (+)	ACh
Motor (−)	VIP

(After J. D. Wood)

C. Slow waves and spikes

Spike bursts

Continuous discharge

Non-excitable

Membrane potential (mV)

0
−10
−20
−30
−40
−50
−60
−70

Threshold potential

Non-excitable, atonia

3

1 2

Slow waves

0 6 12 18 24 30 36 42 48 54
Time (s)

(After Guyton)

D. Pacemaker rate

Maximal rate of the respective pacemaker situated towards the anus

Rate of slow potential waves

Pacemaker zones

1

2

3

Intrinsic rate

Proximal

Distal

Distance in small intestine

(After Dimant & Borthoff)

Pancreas

The *exocrine part* of the pancreas secretes 1–2 L of pancreatic juice into the duodenum each day. The pancreatic juice contains *bicarbonate* (HCO_3^-), which neutralizes (pH 7–8) HCl-rich chyme from the stomach, and mostly inactive precursors of *digestive enzymes* that break down proteins, fats, carbohydrates and other substances in the small intestine.

Pancreatic secretions are similar to saliva in that they are produced in two stages: (1) Cl^- is secreted in the **acini** by active secondary transport, followed by passive transport of Na^+ and water ($\rightarrow$ p. 237 C1). The electrolyte composition of these *primary secretions* corresponds to that of plasma ($\rightarrow$ **A1** and **A2**). Primary pancreatic secretions also contain *digestive proenzymes* and other proteins (exocytosis; $\rightarrow$ p. 30). (2) HCO_3^- is added to the primary secretions (in exchange for Cl^-) in the **secretory ducts**; Na^+ and water follow by passive transport. As a result, the HCO_3^- concentration of pancreatic juice rises to over 100 mmol/L, while the Cl^- concentration falls ($\rightarrow$ **A3**). Unlike saliva ($\rightarrow$ p. 237 B), the osmolality and Na^+/K^+ concentrations of the pancreatic juice remain constant relative to plasma ($\rightarrow$ **A1** and **A2**). Most of the pancreatic juice is secreted during the digestive phase ($\rightarrow$ **A3**).

HCO_3^- **is secreted** from the luminal membrane of the ductules via an anion exchanger that simultaneously reabsorbs Cl^- from the lumen ($\rightarrow$ **B1**). Cl^- returns to the lumen via a Cl^- channel, which is more frequently opened by **secretin** to ensure that the amount of HCO_3^- secreted is not limited by the availability of Cl^- ($\rightarrow$ **B2**). In **cystic fibrosis** (mucoviscidosis), impairment of this CFTR channel (**c**ystic **f**ibrosis **t**ransmembrane **c**onductance **r**egulator) leads to severe disturbances of pancreatic function. The HCO_3^- involved is the product of the $CO_2 + OH^-$ reaction catalyzed by carbonic anhydrase (CA). For each HCO_3^- molecule secreted, one H^+ ion leaves the cell on the blood side via an Na^+/H^+ exchanger ($\rightarrow$ **B3**).

Pancreatic juice **secretion is controlled** by cholinergic (vagal) and hormonal mechanisms (CCK, secretin). Vagal stimulation seems to be enhanced by CCK_A receptors in cholinergic fibers of the acini ($\rightarrow$ **A2,3, B, C** and p. 234). Fat in the chyme stimulates the release of CCK, which, in turn, increases the *(pro)enzyme con-*

tent of the pancreatic juice ($\rightarrow$ **C**). Trypsin in the small intestinal lumen deactivates CCK release via a feedback loop ($\rightarrow$ **D**). *Secretin* increases HCO_3^- and water secretion by the ductules. CCK and acetylcholine (ACh) potentiate this effect by raising the cytosolic Ca^{2+} concentration. Secretin and CCK also affect the pancreatic enzymes.

Pancreatic enzymes are essential for digestion. They have a pH optimum of 7–8. Insufficient HCO_3^- secretion (e.g., in cystic fibrosis) results in inadequate neutralization of chyme and therefore in impaired digestion.

Proteolysis is catalyzed by proteases, which are secreted in their inactive form, i.e., as *proenzymes*: *trypsinogen 1–3, chymotrypsinogen A and B, proelastase 1 and 2* and *procarboxypeptidase A1, A2, B1 and B2*. They are not activated until they reach the intestine, where an *enteropeptidase* first converts trypsinogen to **trypsin** ($\rightarrow$ **D**), which in turn converts chymotrypsinogen into active *chymotrypsin*. Trypsin also activates many other pancreatic proenzymes including proelastases and procarboxypeptidases. Pathological activation of the proenzymes within the pancreas causes the organ to digest itself (*acute pancreatic necrosis*). Trypsins, chymotrypsins and elastases are *endoproteases*, i.e., they split certain peptide bonds within protein chains. Carboxypeptidases A and B are *exopeptidases*, i.e., they split amino acids off the carboxyl end of the chain.

Carbohydrate catabolism. α-*Amylase* is secreted in active form and splits starch and glycogen into maltose, maltotriose and α-limit dextrin. These products are further digested by enzymes of the intestinal epithelium ($\rightarrow$ p. 259).

Lipolysis. *Pancreatic lipase* (see p. 252ff.) is the most important enzyme for lipolysis. It is secreted in its active form and breaks triacylglycerol to 2-monoacylglycerol and free fatty acids. Pancreatic lipase activity depends on the presence of *colipases*, generated from pro-colipases in pancreatic secretions (with the aid of trypsin). *Bile salts* are also necessary for fat digestion ($\rightarrow$ p. 248).

Other important pancreatic enzymes include (pro-) phospholipase A_2, RNases, DNases, and a carboxylesterase.

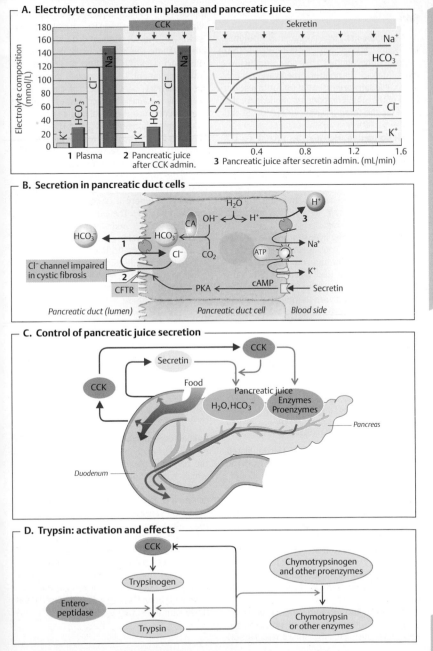

A. Electrolyte concentration in plasma and pancreatic juice

CCK

Sekretin

Electrolyte composition (mmol/L)

1 Plasma

2 Pancreatic juice after CCK admin.

3 Pancreatic juice after secretin admin. (mL/min)

Na⁺

HCO₃⁻

Cl⁻

K⁺

B. Secretion in pancreatic duct cells

H_2O

CA

HCO_3^-

OH^- ← → H^+

H^+

3

HCO_3^-

Cl^-

CO_2

Na⁺

ATP

K⁺

1

Cl⁻ channel impaired in cystic fibrosis

2

CFTR

PKA ← cAMP ← Secretin

Pancreatic duct (lumen)

Pancreatic duct cell

Blood side

C. Control of pancreatic juice secretion

CCK

Secretin

CCK

Food

Pancreatic juice

H_2O, HCO_3^-

Enzymes Proenzymes

Pancreas

Duodenum

D. Trypsin: activation and effects

CCK

Trypsinogen

Chymotrypsinogen and other proenzymes

Entero-peptidase

Trypsin

Chymotrypsin or other enzymes

Bile

Bile components. Bile contains electrolytes, bile salts (bile acids), cholesterol, lecithin (phosphatidylcholine), bilirubin diglucuronide, steroid hormones, medications etc. ($\rightarrow$ **A**). *Bile salts* are essential for fat digestion. Most of the other components of bile leave the body via the feces (excretory function of the liver $\rightarrow$ p. 250).

Bile formation. Hepatocytes secrete ca. 0.7 L/day of bile into *biliary canaliculi* ($\rightarrow$ **A**), the fine canals formed by the cell membranes of adjacent of hepatocytes. The sinusoidal and canalicular membranes of the hepatocytes contain numerous carriers that absorb bile components from the blood and secrete them into the canaliculi, resp.

Bile salts (BS). The liver synthesizes *cholate* and *chenodeoxycholate* (primary bile salts) from *cholesterol.* The intestinal bacteria convert some of them into secondary bile salts such as *deoxycholate* and *lithocholate.* Bile salts are conjugated with taurine or glycine in the liver and are secreted into the bile in this form ($\rightarrow$ **A**). This conjugation is essential for micelle formation in the bile and gut.

Hepatic bile salt carriers. *Conjugated bile salts* in sinusoidal blood are actively taken up by **NTCP** (Na^+ taurocholate cotransporting polypeptide; secondary active transport), and transported against a steep concentration gradient into the canaliculi (primary active transport) by the ATP-dependent carrier **hBSEP** (human bile salt export pump), also referred to as **cBAT** (canalicular bile acid transporter).

Enterohepatic circulation of BS. Unconjugated bile salts are immediately reabsorbed from the bile ducts (cholehepatic circulation). Conjugated bile salts enter the duodenum and are reabsorbed from the terminal ileum by the Na^+ symport carrier ISBT (= **i**leal **s**odium **b**ile acid co**t**ransporter) and circulated back to the liver (*enterohepatic circulation*; $\rightarrow$ **B**) once they have been used for fat digestion ($\rightarrow$ p. 252). The total bile pool (2–4 g) recirculates about 6–10 times a day, depending on the fat content of the diet. Ca. 20–30 g of bile salts are required for daily fat absorption.

Choleresis. Enterohepatic circulation raises the bile salt concentration in the portal vein to a high level during the digestive phase. This (a)

inhibits the hepatic synthesis of bile salts (cholesterol-7α-hydroxylase; negative feedback; $\rightarrow$ **B**) and (b) stimulates the secretion of bile salts into the biliary canaliculi. The latter effect increases the bile flow due to osmotic water movement, i.e., causes *bile salt-dependent choleresis* ($\rightarrow$ **C**). *Bile salt-independent choleresis* is, caused by secretion of other bile components into the canaliculi as well as of HCO_3^- (in exchange for Cl^-) and H_2O into the *bile ducts* ($\rightarrow$ **C**). The latter form is increased by the vagus nerve and secretin.

Gallbladder. When the sphincter of Oddi between the common bile duct and duodenum is closed, *hepatic bile* (*C bile*) is diverted to the gallbladder, where it is concentrated (1 : 10) and stored ($\rightarrow$ **D**). The gallbladder epithelium reabsorbs Na^+, Cl^- and water ($\rightarrow$ **D1**) from the stored bile, thereby greatly raising the concentration of specific bile components (bile salts, bilirubin-*di*-glucuronide, cholesterol, phosphatidylcholine, etc.). If bile is used for fat digestion (or if a peristaltic wave occurs in the interdigestive phase, $\rightarrow$ p. 240), the gallbladder contracts and its contents are mixed in portions with the duodenal chyme ($\rightarrow$ **D2**).

Cholesterol in the bile is transported inside *micelles* formed by aggregation of cholesterol with lecithin and bile salts. A change in the ratio of these three substances in favor of cholesterol ($\rightarrow$ **E**) leads to the precipitation of cholesterol crystals responsible for gallstone development in the highly concentrated *gallbladder bile* (*B bile*). The red and green dots in **E** show the effects of two different ratios.

Gallbladder contraction is *triggered* by CCK ($\rightarrow$ p. 234), which binds to CCK$_A$ receptors, and the neuronal plexus of the gallbladder wall, which is innervated by preganglionic parasympathetic fibers of the vagus nerve ($\rightarrow$ **D2**). CGRP ($\rightarrow$ p. 234) and substance P ($\rightarrow$ p. 86) released by sensory fibers appear to stimulate the gallbladder musculature indirectly by increasing acetylcholine release. The sympathetic nervous system *inhibits* gallbladder contractions via α$_2$ adrenoreceptors located on cholinergic fiber terminals. As *cholagogues*, fatty acids and products of protein digestion ($\rightarrow$ p. 234) as well as egg yolk and $MgSO_4$ effectively stimulate CCK secretion.

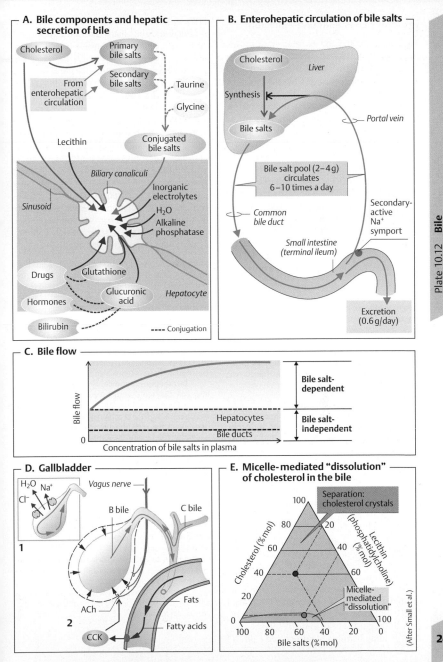

A. Bile components and hepatic secretion of bile

Cholesterol

Primary bile salts

Secondary bile salts

From enterohepatic circulation

Taurine

Glycine

Conjugated bile salts

Lecithin

Biliary canaliculi

Inorganic electrolytes

H_2O

Alkaline phosphatase

Sinusoid

Drugs

Glutathione

Glucuronic acid

Hormones

Hepatocyte

Bilirubin

---- Conjugation

B. Enterohepatic circulation of bile salts

Cholesterol

Liver

Synthesis

Bile salts

Portal vein

Bile salt pool (2–4 g) circulates 6–10 times a day

Secondary-active Na^+ symport

Common bile duct

Small intestine (terminal ileum)

Excretion (0.6 g/day)

C. Bile flow

Bile flow

Bile salt-dependent

Hepatocytes

Bile salt-independent

Bile ducts

Concentration of bile salts in plasma

D. Gallbladder

H_2O Na^+

Cl^-

Vagus nerve

B bile

C bile

1

ACh

2

Fats

Fatty acids

CCK

E. Micelle-mediated "dissolution" of cholesterol in the bile

Separation: cholesterol crystals

Cholesterol (% mol)

Lecithin (phosphatidylcholine) (% mol)

Micelle-mediated "dissolution"

Bile salts (% mol)

(After Small et al.)

Plate 10.12 **Bile**

Excretory Liver Function—Bilirubin

The **liver detoxifies** and **excretes** many mostly lipophilic substances, which are either generated during metabolism (e.g., bilirubin or steroid hormones) or come from the intestinal tract (e.g., the antibiotic chloramphenicol). However, this requires prior **biotransformation** of the substances. In the *first step* of the process, reactive OH, NH_2 or COOH groups are enzymatically added (e.g., by monooxygenases) to the hydrophobic substances. In the *second step*, the substances are *conjugated* with glucuronic acid, acetate, glutathione, glycine, sulfates, etc. The conjugates are now water-soluble and can be either further processed in the kidneys and excreted in the urine, or secreted into bile by liver cells and excreted in the feces. Glutathione conjugates, for example, are further processed in the kidney excreted as mercapturic acids in the urine.

Carriers. The canalicular membrane of hepatocytes contains various carriers, most of which are directly fueled by ATP (see also p. 248). The principal carriers are: **MDR1** (multidrug resistance protein 1) for relatively hydrophobic, mainly cationic metabolites, **MDR3** for phosphatidylcholine ($\rightarrow$ p. 248), and **cMOAT** (canalicular multispecific organic anion transporter = multidrug resistance protein MRP2) for conjugates (formed with glutathione, glucuronic acid or sulfate) and many other organic anions.

Bilirubin sources and conjugation. Ca. 85% of all bilirubin originates from the *hemoglobin* in erythrocytes; the rest is produced by other hemoproteins like cytochrome ($\rightarrow$ **A** and **B**). When degraded, the globulin and iron components ($\rightarrow$ p. 90) are cleaved from hemoglobin. Via intermediate steps, *biliverdin* and finally *bilirubin*, the yellow bile pigment, are then formed from the porphyrin residue. Each gram of hemoglobin yields ca. 35 mg of bilirubin. Free unconjugated bilirubin (*"indirect"* *bilirubin*) is poorly soluble in water, yet lipid-soluble and toxic. It is therefore *complexed with albumin* when present in the blood (2 mol bilirubin : 1 mol albumin), but not when absorbed by hepatocytes ($\rightarrow$ **A**). Bilirubin is conjugated (catalyzed by *glucuronyltransferase*) with 2 molecules of *UDP-glucuronate* (synthesized from glucose, ATP and UTP) in the liver cells yielding **bilirubin diglucuronide** (*"direct"*

bilirubin). It is a water-soluble substance secreted into the biliary canaliculi by primary active transport mechanisms (cMOAT, see above).

Bilirubin excretion. 200–250 mg of bilirubin is excreted in the bile each day. Ca. 90% of it is *excreted in the feces*. In the gut, bacteria break bilirubin down into the colorless compound, *stercobilinogen* ($\rightarrow$ **B**). It is partly oxidized into *stercobilin*, the brown compound that colors the stools. About 10% of all bilirubin diglucuronide is deconjugated by intestinal bacteria and returned to the liver in this lipophilic form (partly as stercobilinogen) via enterohepatic circulation. A small portion (ca. 1%) reaches the systemic circulation and is excreted by the kidneys as *urobilinogen* = stercobilinogen (see below) ($\rightarrow$ **B**). The renal excretion rate increases when the liver is damaged.

Jaundice. The *plasma bilirubin concentration* normally does not exceed 17 µmol/L (= 1 mg/dL). Concentrations higher than 30 µmol/L (1.8 mg/dL) lead to yellowish discoloration of the sclera and skin, resulting in **jaundice** (**icterus**). **Types of jaundice:**

1. **Prehepatic jaundice**. When excessive amounts of bilirubin are formed, for example, due to increased hemolysis, the liver can no longer cope with the higher load unless the plasma bilirubin concentration rises. Thus, *unconjugated (indirect) bilirubin* is mainly elevated in these patients.

2. **Intrahepatic jaundice**. The main causes are (a) liver cell damage due to toxins (*Amanita*) or infections (*viral hepatitis*) resulting in the impairment of bilirubin transport and conjugation; (b) deficiency or absence of the glucuronyltransferase system in the newborn (Crigler–Najjar syndrome); (c) inhibition of glucuronyltransferase, e.g., by steroids; (d) impaired secretion of bilirubin into the biliary canaliculi due to a congenital defect (Dubin–Johnson syndrome) or other reasons (e.g., drugs, steroid hormones).

3. **Posthepatic jaundice**: Impairment of the flow of bile occurs due to an obstruction (e.g., stone or tumor) in the bile ducts, usually accompanied by elevated serum concentrations of *conjugated (direct) bilirubin* and alkaline phosphatase—both of which are normal components of bile.

Types 2a, 2d and 3 jaundice are associated with increased urinary concentrations of conjugated bilirubin, leading to brownish discoloration of the urine. In type 3 jaundice, the stools are gray due to the lack of bilirubin in the intestine and the resulting absence of stercobilin formation.

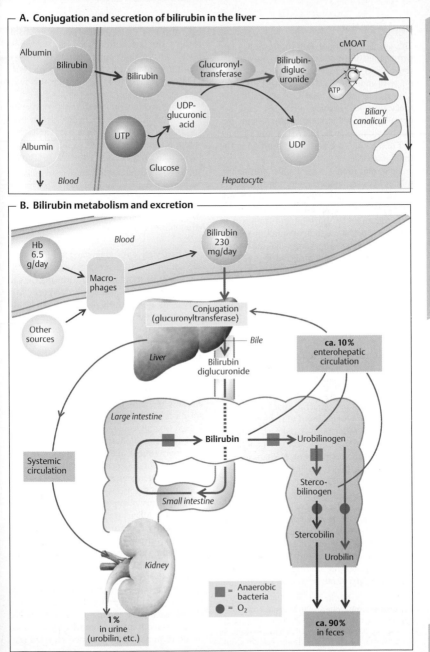

A. Conjugation and secretion of bilirubin in the liver

Albumin

Bilirubin → Bilirubin → Glucuronyl-transferase → Bilirubin-diglucuronide → cMOAT / ATP

UDP-glucuronic acid

UTP

Glucose

UDP

Biliary canaliculi

Albumin

Blood

Hepatocyte

B. Bilirubin metabolism and excretion

Hb 6.5 g/day

Macro-phages

Other sources

Blood

Bilirubin 230 mg/day

Conjugation (glucuronyltransferase)

Liver

Bile

Bilirubin diglucuronide

ca. 10 % enterohepatic circulation

Large intestine

Bilirubin → Urobilinogen

Stercobilinogen

Systemic circulation

Small intestine

Stercobilin

Urobilin

Kidney

= Anaerobic bacteria
= O₂

1 % in urine (urobilin, etc.)

ca. 90 % in feces

Plate 10.13 Excretory Liver Function—Bilirubin

Lipid Digestion

The average **intake of fats** (butter, oil, margarine, milk, meat, sausages, eggs, nuts etc.) is roughly 60–100 g/day, but there is a wide range of individual variation (10–250 g/day). Most fats in the diet (90%) are neutral fats or *triacylglycerols* (triglycerides). The rest are phospholipids, cholesterol esters, and fat-soluble vitamins (vitamins A, D, E and K). Over 95% of the lipids are normally absorbed in the small intestine.

Lipid digestion ($\rightarrow$ **A**). Lipids are poorly soluble in water, so special mechanisms are required for their digestion in the watery environment of the gastrointestinal tract and for their subsequent absorption and transport in plasma ($\rightarrow$ p. 254). Although small quantities of undegraded triacylglycerol can be absorbed, dietary fats must be hydrolyzed by enzymes before they can be efficiently absorbed. Optimal enzymatic activity requires the prior mechanical *emulsification* of fats (mainly in the distal stomach, $\rightarrow$ p. 240) because emulsified *lipid droplets* (1–2 µm; $\rightarrow$ **B1**) provide a much larger surface (relative to the mass of fat) for lipases.

Lipases, the fat digesting enzymes, originate from the lingual glands, gastric fundus (chief and mucous neck cells) and *pancreas* ($\rightarrow$ **A** and p. 246). About 10–30% of dietary fat intake is hydrolyzed in the stomach, while the remaining 70–90% is broken down in the duodenum and upper jejunum. Lingual and gastric lipases have an acid pH optimum, whereas pancreatic lipase has a pH optimum of 7–8. Lipases become active at the fat/oil and water interface ($\rightarrow$ **B**). **Pancreatic lipase** (triacylglycerol hydrolase) develops its lipolytic activity (max. 140 g fat/min) in the presence of **colipase** and Ca^{2+}. *Pro-colipase* in pancreatic juice yields colipase after being activated by trypsin. In most cases, the pancreatic lipases split *triacylglycerol* (TG) at the 1st and 3rd ester bond. This process requires the addition of water and yields free *fatty acids (FFA)* and *2-monoacylglycerol*.

A **viscous-isotropic phase** with aqueous and hydrophobic zones then forms around the enzyme ($\rightarrow$ **B2**). *Ca^{2+} excesses* or *monoacylglycerol deficiencies* result in the conversion of the fatty acids into *calcium soaps*, which are later excreted.

Phospholipase A₂ (from pro-phospholipase A_2 in pancreatic juice—activated by trypsin) cleaves the 2nd ester bond of the phospholipids (mainly phosphatidylcholine = lecithin) contained in micelles. The presence of *bile salts* and Ca^{2+} is required for this reaction.

An **unspecific carboxylesterase** (= unspecific lipase = cholesterol ester hydrolase) in pancreatic secretions also acts on cholesterol esters on micelles as well as all three ester bonds of TG and esters of vitamins, A, D and E.

This lipase is also present in human breast **milk** (but not cow's milk), so breast-fed infants receive the digestive enzyme required to break down milk fat along with the milk. Since the enzyme is heat-sensitive, pasteurization of human milk significantly reduces the infant's ability to digest milk fat to a great extent.

2-Monoacylglycerols, long-chain free fatty acids and other lipids aggregate with *bile salts* ($\rightarrow$ p. 248) to spontaneously form **micelles** in the small intestine ($\rightarrow$ **B3**). (Since short-chain fatty acids are relatively polar, they can be absorbed directly and do not require bile salts or micelles). The micelles are only about 20–50 nm in diameter, and their surface-to-volume ratio is roughly 50 times larger than that of the lipid droplets in emulsions. They facilitate close contact between the products of fat digestion and the wall of the small intestine and are therefore essential for lipid absorption. The polar side of the substances involved (mainly conjugated bile salts, 2-monoacylglycerol and phospholipids) faces the watery environment, and the non-polar side faces the interior of the micelle. Totally apolar lipids (e.g., cholesterol esters, fat-soluble vitamins and lipophilic poisons) are located inside the micelles. Thus, the apolar lipids remain in the lipophilic milieu (*hydrocarbon continuum*) during all these processes until they reach the lipophilic brush border membrane of the epithelium. They are then absorbed by the mucosa cells via dissolution in the membrane or by a passive transport mechanism (e.g., carriers in the case of free fatty acids). Although fat absorption is completed by the time the chyme reaches the end of the jejunum, the *bile salts* released from micelles are only absorbed in the terminal ileum and then recycled (enterohepatic circulation; $\rightarrow$ p. 249 B).

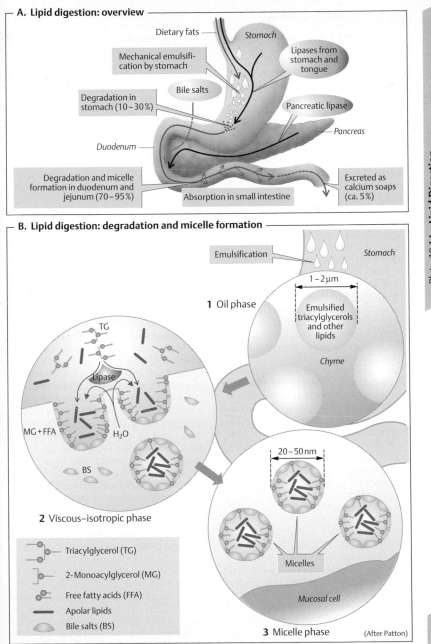

A. Lipid digestion: overview

Dietary fats
Stomach
Mechanical emulsification by stomach
Lipases from stomach and tongue
Bile salts
Degradation in stomach (10–30%)
Pancreatic lipase
Duodenum
Pancreas
Degradation and micelle formation in duodenum and jejunum (70–95%)
Absorption in small intestine
Excreted as calcium soaps (ca. 5%)

B. Lipid digestion: degradation and micelle formation

Emulsification
Stomach
1 Oil phase
1–2 μm
Emulsified triacylglycerols and other lipids
Chyme
TG
Lipase
MG + FFA
H_2O
BS
2 Viscous–isotropic phase

20–50 nm
Micelles
Mucosal cell

Triacylglycerol (TG)
2-Monoacylglycerol (MG)
Free fatty acids (FFA)
Apolar lipids
Bile salts (BS)

3 Micelle phase

(After Patton)

Plate 10.14 Lipid Digestion

Lipid Distribution and Storage

Lipids in the blood are transported in **lipoproteins, LPs** (→ **A**), which are molecular aggregates (microemulsions) with a core of very hydrophobic lipids such as *triacylglycerols* (**TG**) and *cholesterol esters* (**CHO-esters**) surrounded by a layer of amphipathic lipids (phospholipids, cholesterol). LPs also contain several types of proteins, called apolipoproteins. LPs are differentiated according to their size, density, lipid composition, site of synthesis, and their apolipoprotein content. **Apolipoproteins (Apo)** function as *structural elements* of LPs (e.g. ApoAII and ApoB48), *ligands* (ApoB100, ApoE, etc.) for LP receptors on the membranes of LP target cells, and as *enzyme activators* (e.g. ApoAI and ApoCII).

Chylomicrons transport lipids (mainly *triacylglycerol, TG*) from the gut to the periphery (via intestinal lymph and systemic circulation; → **D**), where their ApoCII activates endothelial lipoprotein lipase (**LPL**), which *cleaves FFA* from TG. The FFA are mainly absorbed by myocytes and fat cells (→ **D**). With the aid of ApoE, the *chylomicron remnants* deliver the rest of their TG, cholesterol and cholesterol ester load to the hepatocytes by receptor-mediated endocytosis (→ **B, D**).

Cholesterol (CHO) and the TG imported from the gut and newly synthesized in the liver are exported inside **VLDL** (very low density lipoproteins) from the liver to the periphery, where they by means of their ApoCII also activate **LPL**, resulting in the *release of FFA* (→ **D**). This results in the loss of ApoCII and exposure of ApoE. VLDL remnants or **IDL** (intermediate-density lipoproteins) remain. Ca. 50% of the IDL returns to the liver (mainly bound by its ApoE on LDL receptors; see below) and is reprocessed and exported from the liver as VLDL (→ **B**).

The other 50% of the IDL is converted to **LDL** (low density lipoprotein) after coming in contact with hepatic lipase (resulting in loss of ApoE and exposure of ApoB100). Two-thirds of the LDLs deliver their CHO and CHO-esters to the liver, the other third transfers its CHO to extrahepatic tissue (→ **B**). Binding of ApoB100 to **LDL receptors** is essential for both processes (see below).

High-density lipoproteins (**HDL**) exchange certain apoproteins with chylomicrons and VLDL and absorb superfluous CHO from the extrahepatic cells and blood (→ **B**). With their ApoAI, they activate the plasma enzyme **LCAT** (lecithin–cholesterol acyltransferase), which is responsible for the partial esterification of CHO. HDL also deliver cholesterol and CHO-esters to the liver and steroid hormone-producing glands with *HDL receptors* (ovaries, testes, adrenal cortex).

Triacylglycerol (TG)

Dietary TGs are broken down into *free fatty acids* (**FFA**) and *2-monoacylglycerol* (**MG**) in the gastrointestinal tract (→ **C** and p. 252). Since short-chain FFAs are water-soluble, they can be absorbed and transported to the liver via the portal vein. Long-chain FFAs and 2-monoacylglycerols are not soluble in water. They are re-synthesized to TG in the mucosa cells (→ **C**). (The FFAs needed for TG synthesis are carried by FFA-binding proteins from the cell membrane to their site of synthesis, i.e., the smooth endoplasmic reticulum.) Since TGs are not soluble in water, they are subsequently loaded onto **chylomicrons**, which are exocytosed into the extracellular fluid, then passed on to the intestinal lymph (thereby by-passing the liver), from which they finally reach the greater circulation (→ **C, D**). (Plasma becomes cloudy for about 20–30 minutes after a fatty meal due to its chylomicron content). The *liver* also synthesizes TGs, thereby taking the required FFAs from the plasma or synthesizing them from glucose. Hepatic TGs are loaded onto **VLDL** (see above) and subsequently secreted into the plasma (→ **D**). Since the export capacity of this mechanism is limited, an excess of FFA or glucose (→ **D**) can result in the accumulation of TGs in the liver (*fatty liver*).

Free fatty acids (FFAs) are high-energy substrates used for energy metabolism (→ p. 228). Fatty acids circulating in the blood are mainly transported in the form of TG (in *lipoproteins*) whereas plasma FFA are *complexed with albumin*. Fatty acids are removed from TGs of chylomicrons and VLDL by lipoprotein lipase (**LPL**) localized on the luminal surface of the capillary endothelium of many organs (mainly in fat tissue and muscles) (→ **D**). ApoCII on the

▶

Plate 10.15 Lipid Distribution and Storage I

A. Lipoproteins

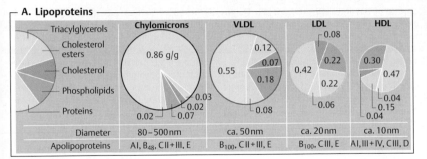

	Chylomicrons	**VLDL**	**LDL**	**HDL**
Triacylglycerols				
Cholesterol esters	0.86 g/g	0.12	0.08	0.30
Cholesterol		0.55 0.07	0.42 0.22	0.47
Phospholipids	0.03	0.18	0.22	0.04
Proteins	0.02 0.07	0.08	0.06	0.15 0.04
Diameter	80–500 nm	ca. 50 nm	ca. 20 nm	ca. 10 nm
Apolipoproteins	AI, B$_{48}$, CII+III, E	B$_{100}$, CII+III, E	B$_{100}$, CIII, E	AI, III+IV, CIII, D

B. Sources and fate of cholesterol

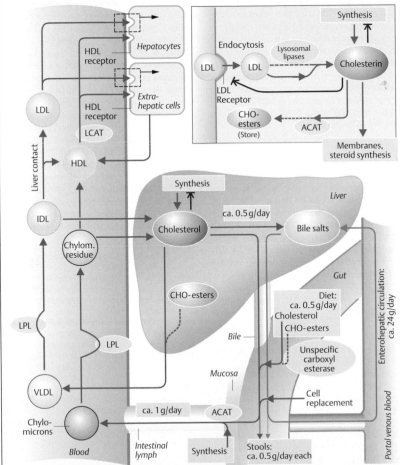

Hepatocytes

HDL receptor

Extra-hepatic cells

HDL receptor

LCAT

Endocytosis

Synthesis

LDL → LDL → Lysosomal lipases → Cholesterin

LDL Receptor

CHO-esters (Store) ← ACAT

Membranes, steroid synthesis

LDL

HDL

Liver contact

IDL

Chylom. residue

Synthesis

Cholesterol   ca. 0.5 g/day   Bile salts

Liver

CHO-esters

Gut

Diet: ca. 0.5 g/day
Cholesterol
CHO-esters

Unspecific carboxyl esterase

Enterohepatic circulation: ca. 24 g/day

Bile

Mucosa

Cell replacement

Portal venous blood

LPL

LPL

VLDL

Chylo-microns

ca. 1 g/day   ACAT

Intestinal lymph   Synthesis

Stools: ca. 0.5 g/day each

Blood

▶

surface of TGs and VLDL activates LPL. The **insulin** secreted after a meal *induces LPL* ($\rightarrow$ **D**), which promotes the rapid degradation of reabsorbed dietary TGs. LPL is also activated by *heparin* (from endothelial tissue, mast cells, etc.), which helps to eliminate the chylomicrons in cloudy plasma; it therefore is called a *clearance factor*. Albumin-complexed FFAs in plasma are mainly transported to the following **target sites** ($\rightarrow$ **D**):

◆ *Cardiac muscle, skeletal muscle, kidneys and other organs*, where they are oxidized to CO_2 and H_2O in the mitochondria (β oxidation) and used as a **source of energy**.

◆ *Fat cells* ($\rightarrow$ **D**), which either **store** the FFAs or use them to synthesize TG. When energy requirements increase or intake decreases, the FFAs are cleaved from triacylglycerol in the fat cells (lipolysis) and transported to the area where they are needed ($\rightarrow$ **D**). *Lipolysis is stimulated* by epinephrine, glucagon and cortisol and *inhibited* by insulin ($\rightarrow$ p.282ff.).

◆ The *liver*, where the FFAs are oxidized or used to synthesize TG.

Cholesterol (CHO)

Cholesterol esters (CHO-esters), like TGs, are apolar lipids. In the watery milieu of the body, they can only be transported when incorporated in lipoproteins (or bound to proteins) and can be used for metabolism only after they have been converted to CHO, which is more polar ($\rightarrow$ **B**). CHO-esters serve as *stores* and in some cases the transported form of CHO. CHO-esters are present in all lipoproteins, but are most abundant (42%) in LDL ($\rightarrow$ **A**).

Cholesterol is an important constituent of cell membranes ($\rightarrow$ p. 14). Moreover, it is a precursor for *bile salts* ($\rightarrow$ **B** and p.248), *vitamin D* ($\rightarrow$ p. 292), and *steroid hormones* ($\rightarrow$ p. 294ff.). Each day ca. 0.6 g of CHO is lost in the feces (reduced to *coprosterol*) and sloughed off skin. The bile salt loss amounts to about 0.5 g/day. These losses (minus the dietary CHO intake) must be compensated for by continuous resynthesis of CHO in the intestinal tract and liver ($\rightarrow$ **B**). CHO supplied by the diet is absorbed in part as such and in part in esterified form ($\rightarrow$ **B**, lower right). Before it is reabsorbed, CHO-esters are split by *unspecific pancreatic carboxylesterase* to CHO, which is absorbed in

the upper part of the small intestine ($\rightarrow$ **B**, bottom). *Mucosal cells* contain an enzyme that re-esterifies part of the absorbed CHO: **ACAT** (acyl-CoA-cholesterol acyltransferase) so that both cholesterol and CHO-esters can be integrated in chylomicrons ($\rightarrow$ **A**). CHO and CHO-esters in the *chylomicron remnants* (see above) are transported to the liver, where lysosomal acid lipases again break the CHO-esters down into CHO. This CHO and that taken up from other sources (LDL, HDL) leave the liver ($\rightarrow$ **B**): 1. by excretion into the bile ($\rightarrow$ p.248), 2. by conversion into bile salts which also enter the bile ($\rightarrow$ p.249 B), and 3. by incorporation into VLDL, the hepatic lipoprotein for export of lipids to other tissues. Under the influence of LPL (see above), the VLDL yield IDL and later **LDL** ($\rightarrow$ **B**, left). The LDL transport CHO and CHO-esters to cells with **LDL receptors** (hepatic and extrahepatic cells; $\rightarrow$ **B**, top). The receptor density on the cell surface is adjusted according to the prevailing CHO requirement. Like hepatic cells (see above) extrahepatic cells take up the LDL by receptor-mediated endocytosis, and lysosomal acid lipases reduce CHO-esters to CHO ($\rightarrow$ **B**, top right). The cells can then insert the CHO in their cell membranes or use it for steroid synthesis. A *cholesterol excess* leads to (a) inhibition of CHO synthesis in the cells (3-HMG-CoA-reductase) and (b) activation of ACAT, an enzyme that esterifies and stores CHO in the form of its ester (see above).

Hyperlipoproteinemia. An excess of lipids in the blood can be reflected by elevation of triacylglycerol levels and/or CHO levels ($>$ 200–220 mg/dL serum; affects about one in five adults in Western countries). In the most severe form, *familial hypercholesterolemia*, a genetic defect causes elevated plasma CHO concentrations from birth on, which can result in myocardial infarction in juvenile age. The disease is caused by genetic defects of the high-affinity LDL receptors. The serum CHO level rises since the cells take up smaller quantities of cholesterol-rich LDLs. Extrahepatic tissues synthesize larger quantities of CHO because 3-HMG-CoA-reductase fails to inhibit CHO synthesis due to the decreased absorption of LDLs. As a result, more LDLs bind to the low-affinity *scavenger receptors* that mediate the storage of CHO in macrophages, cutaneous tissues, and blood vessels. Hypercholesterolemia therefore increases the risk of arteriosclerosis and coronary disease.

C. Fat absorption

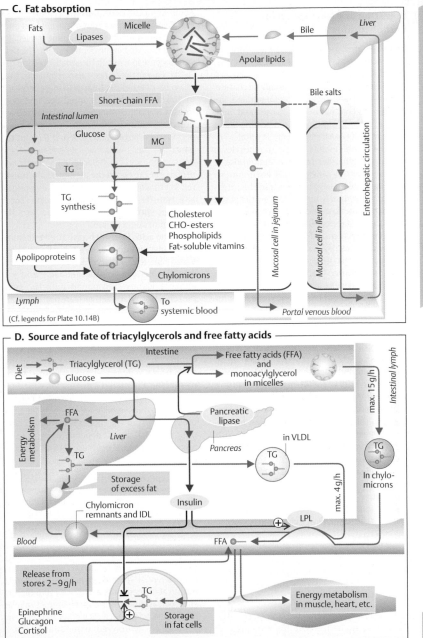

Fats
Lipases
Micelle
Bile
Liver
Apolar lipids
Short-chain FFA
Intestinal lumen
Bile salts
Glucose
MG
TG
TG synthesis
Cholesterol
CHO-esters
Phospholipids
Fat-soluble vitamins
Apolipoproteins
Chylomicrons
Lymph
To systemic blood
Mucosal cell in jejunum
Mucosal cell in ileum
Enterohepatic circulation
Portal venous blood
(Cf. legends for Plate 10.14B)

D. Source and fate of triacylglycerols and free fatty acids

Diet
Intestine
Triacylglycerol (TG)
Glucose
Free fatty acids (FFA) and monoacylglycerol in micelles
Intestinal lymph
max. 15 g/h
Energy metabolism
FFA
Liver
Pancreatic lipase
Pancreas
in VLDL
TG
TG
In chylomicrons
TG
Storage of excess fat
Insulin
max. 4 g/h
Chylomicron remnants and IDL
LPL
Blood
FFA
Release from stores 2–9 g/h
TG
Epinephrine
Glucagon
Cortisol
Storage in fat cells
Energy metabolism in muscle, heart, etc.

Plate 10.16 Lipid Distribution and Storage II

257

Digestion and Absorption of Carbohydrates and Protein

Carbohydrates provide half to two-thirds of the energy requirement ($\rightarrow$ p. 226). At least 50% of dietary carbohydrates consist of *starch* (amylose and amylopectin), a polysaccharide; other important dietary carbohydrates are cane sugar (saccharose = sucrose) and milk sugar (lactose). Carbohydrate digestion starts in the mouth ($\rightarrow$ **A1** and p. 236). *Ptyalin*, an **α-amylase** found in saliva, breaks starches down into oligosaccharides (*maltose, maltotriose, α limit dextrins*) in a neutral pH environment. This digestive process continues in the proximal stomach, but is interrupted in the distal stomach as the food is mixed with acidic gastric juices. A pancreatic α-amylase, with a pH optimum of 8 is mixed into the chyme in the duodenum. Thus, polysaccharide digestion can continue to the final oligosaccharide stage mentioned above. The carbohydrates can be only **absorbed** in the form of monosaccharides. Thus, the enzymes *maltase* and *isomaltase* integrated in the luminal brush border membrane of enterocytes break down *maltose, maltotriose* and α *limit dextrins* into **glucose** as the final product. As in the renal tubules ($\rightarrow$ p. 158), glucose is first actively taken up by the Na^+ symport carrier SGLT1 into mucosal cells ($\rightarrow$ A2, p. 29 B1) before passively diffusing into the portal circulation via *GLUT2*, the glucose uniport carrier (facilitated diffusion; $\rightarrow$ p. 22). The hydrolysis of *saccharose, lactose, and trehalose* is catalyzed by other brush border enzymes: lactase, saccharase (sucrase) and trehalase. In addition to glucose, these reactions release *galactose* (from lactose), which is absorbed by the same carriers as glucose, and *fructose*, which crosses the enterocytes by passive uniporters, GLUT5 in the luminal and GLUT2 in the basolateral membrane ($\rightarrow$ **A2**).

Lactase deficiency. Lactose cannot be broken down and absorbed unless sufficient lactase is available. Lactase deficiencies lead to diarrhea 1) because water is retained in the intestinal lumen due to osmotic mechanisms, and 2) because intestinal bacteria convert the lactose into toxic substances.

Protein digestion starts in the stomach ($\rightarrow$ **B1**). *HCl* in the stomach denatures proteins and converts the three secreted *pepsinogens* into about eight different **pepsins**. At a pH of 2–5, these *endopeptidases* split off proteins at sites where tyrosine or phenylalanine molecules are incorporated in the peptide chain. The pepsins become inactive in the small intestine (pH 7–8). Pancreatic juice also contains proenzymes of other peptidases that are activated in the duodenum ($\rightarrow$ p. 246). The endopeptidases *trypsin, chymotrypsin* and *elastase* hydrolyze the protein molecules into short-chain peptides. *Carboxypeptidase A and B* (from the pancreas) as well as *dipeptidases* and *aminopeptidase* (brush border enzymes) act on proteins at the end of the peptide chain, breaking them down into tripeptides, dipeptides, and (mostly) individual amino acids. These cleavage products are absorbed in the duodenum and jejunum.

Amino acids (AA) are transported by a number of specific carriers ($\rightarrow$ **B2**) similar to those found in the kidneys ($\rightarrow$ p. 158). Neutral (without net charge) and anionic ("acid") L-amino acids are transported with Na^+ symporters (secondary active transport; $\rightarrow$ p. 28) from the intestinal lumen into mucosal cells, from which they passively diffuse with carriers into the blood. Cationic ("basic") L-amino acids such as L-arginine⁺, L-lysine⁺ and L-ornithine⁺ are partly taken up into the enterocytes by Na^+ independent mechanisms, as the membrane potential is a driving force for their uptake. *Anionic amino acids* like L-glutamate⁻ and L-aspartate⁻ which, for the most part, are broken down in the mucosal cells, also have their own (Na^+ and K^+ dependent) carrier systems. *Neutral amino acids* use several different transporters.

AA absorption disorders can be congenital and affect various amino acid groups. These disorders are often associated with defects of renal tubular reabsorption (renal aminoaciduria, e.g. cystinuria).

Dipeptides and tripeptides can be absorbed as intact molecules by a symport carrier (PepT1). The carrier is driven by an H^+ gradient ($\rightarrow$ **B2**), which in turn is generated by H^+ secretion (tertiary active H^+-peptide symport, $\rightarrow$ p. 29 B5). Amino acids generally are much more rapidly absorbed as dipeptides and tripeptides than as free amino acids. Once they enter the cells, the peptides are hydrolyzed to free amino acids.

Plate 10.17 Digestion of Carbohydrates and Protein

A. Carbohydrate digestion and absorption of monosaccharides

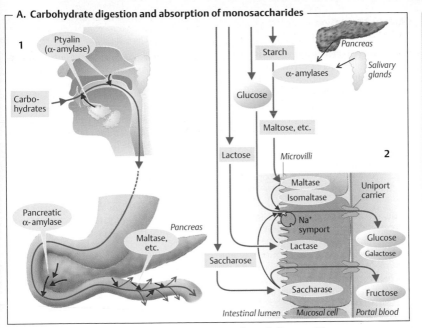

B. Protein digestion and absorption of amino acids and oligopeptides

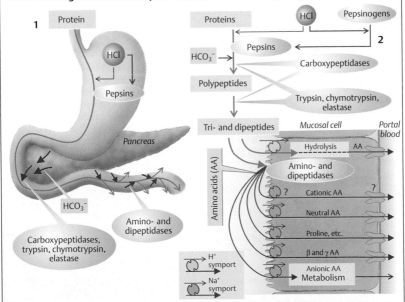

Vitamin Absorption

Since higher animals cannot synthesize **cobalamins** (**vitamin B₁₂**), they must obtain this *cobalt-containing coenzyme* from the diet. Animal products (liver, kidneys, fish, eggs, milk) are the main source.

Cobalamin biochemistry. Aqua- and OH-cobalamin are precursors of the two active forms, methyl- and adenosylcobalamin. *Methylcobalamin* is needed to form methionine from homocysteine; cobalamin transfers the methyl group required for this from N^5-methyltetrahydrofolate (see below) to homocysteine. Some enzymes, e.g. methyl-malonyl-CoA mutase, need *adenosylcobalamin* to break and form carbon–carbon bonds.

Cobalamins are relatively large and lipophobic molecules that require **transport proteins** ($\rightarrow$ **A**). During passage through the GI tract, plasma and other compartments, cobalamins bind to (1) *intrinsic factor* (**IF**), which is secreted by gastric parietal cells; (2) *trans-cobalamin II* (**TC II**) in plasma; and (3) *R proteins* in plasma (**TC I**), and granulocytes (**TC III**), saliva, bile, milk, etc. Gastric acid releases cobalamin from **dietary proteins**. In most cases, the cobalamin then binds to R protein in saliva or (if the pH is high) to IF ($\rightarrow$ **A1**). The R protein is digested by trypsin in the duodenum, resulting in the release of cobalamin, which is then bound by (trypsin-resistant) **intrinsic factor**. The mucosa of the *terminal ileum* has highly specific receptors for the cobalamin-IF complex. IT binds to these receptors and is absorbed by *receptor-mediated endocytosis*, provided a pH of > 5.6 and Ca^{2+} ions are available ($\rightarrow$ **A2**). The receptor density and, thus, the absorption rate increases during pregnancy. Cobalamin binds to TC I, II and III in **plasma** ($\rightarrow$ **A3**). TC II mainly distributes cobalamin to all cells undergoing division (TC II receptors, endocytosis). TC III (from granulocytes) transports excess cobalamin and unwanted cobalamin derivatives to the **liver** (TC III receptors), where it is either stored or excreted in the bile. TC I has a half-life of roughly 10 days and serves as a short-term depot for cobalamin in the plasma.

A vegan diet or disturbed cobalamin absorption can lead to severe **deficiency symptoms** like pernicious anemia and spinal cord damage (funicular myelosis).

It takes years for these symptoms to manifest as the body initially has a reserve of 1000 times the daily requirement of 1 µg ($\rightarrow$ p. 90).

Folic acid/folate (= pteroylglutamic acid). N^5, N^{10}-*methylenetetrahydrofolate*, the metabolically active form of folic acid (daily requirement: 0.1–0.2 mg) is needed for *DNA synthesis* (formation of deoxythymidylate from deoxyuridylate). Folic acid in the **diet** usually occurs in forms that contain up to seven glutamyl residues (γ-linked peptide chain; *Pte-Glu₇*) instead of pteroylglutamic acid (*Pte-Glu₁*). Since only Pte-Glu₁ can be absorbed from the lumen of the proximal jejunum ($\rightarrow$ **B**), its polyglutamyl chain must be shortened before absorption. This is done by *pteroylpolyglutamate hydrolases* located in the luminal membrane of enterocytes. The **absorption** of Pte-Glu₁ is mediated by a specific active transporter. In mucosal cells, Pte-Glu₁ is than broken down to yield N^5-methyltetrahydrofolate (5-Me-H₄-folate) and other metabolites. If already present in the ingested food, these metabolites are absorbed from the intestinal lumen by the aforementioned mechanism. (The same applies to the cytostatic drug, *methotrexate*.) *Methylcobalamin* is needed to convert 5-Me-H₄-folate to tetrahydrofolate (see above). The body stores about 7 mg of folic acid, enough for several months (cf. *folic acid deficiency*, $\rightarrow$ p. 90).

The **other water-soluble vitamins**—B₁ (thiamin), B₂ (riboflavin), C (ascorbic acid), and H (biotin, niacin)—are absorbed via Na^+ symport carriers ($\rightarrow$ **C**). Vitamin C is absorbed from the ileum, whereas vitamins B₁, B₂, and H are absorbed from the jejunum. Members of the vitamin B₆ group (pyridoxal, pyridoxine, pyridoxamine) are probably absorbed by passive mechanisms.

Fat-soluble vitamins—A (retinol), D₃ (cholecalciferol), E (tocopherol), K₁ (phylloquinone), and K₂ (menaquinone)—must be incorporated into *micelles* for absorption (cf. lipid digestion, p. 252). The exact absorption mechanism has not yet been explained, though it is known to be partly saturation- and energy-dependent. Fat-soluble vitamins are incorporated into chylomicrons and VLDL for transport in plasma ($\rightarrow$ p. 254ff.).

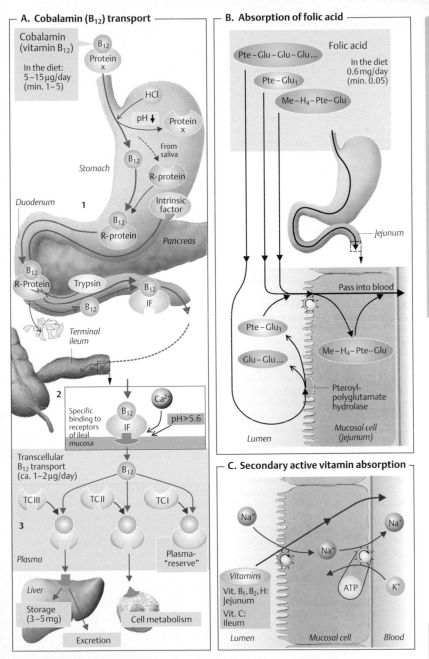

A. Cobalamin (B₁₂) transport

Cobalamin (vitamin B₁₂)

In the diet:
5–15 µg/day
(min. 1–5)

B₁₂
Protein X

HCl

pH↓

Protein X

From saliva

Stomach

R-protein

Intrinsic factor

B₁₂
R-protein

Duodenum

1

Pancreas

B₁₂
R-Protein

Trypsin

B₁₂

B₁₂
IF

Terminal ileum

2

Specific binding to receptors of ileal mucosa

B₁₂
IF

Ca²⁺

pH > 5.6

Transcellular B₁₂ transport (ca. 1–2 µg/day)

TC III   TC II   TC I

3

Plasma

Plasma-"reserve"

Liver

Storage (3–5 mg)

Cell metabolism

Excretion

B. Absorption of folic acid

Folic acid

Pte – Glu – Glu – Glu ...

In the diet 0.6 mg/day (min. 0.05)

Pte – Glu₁

Me – H₄ – Pte – Glu

Jejunum

Pass into blood

Pte – Glu₁

Me – H₄ – Pte – Glu

Glu – Glu ...

Pteroyl-polyglutamate hydrolase

Mucosal cell (jejunum)

Lumen

C. Secondary active vitamin absorption

Na⁺

Na⁺

Na⁺

ATP

K⁺

Vitamins

Vit. B₁, B₂, H: Jejunum

Vit. C: Ileum

Lumen   *Mucosal cell*   *Blood*

Plate 10.18   Vitamin Absorption

Water and Mineral Absorption

The average intake of water (in beverages and foodstuffs) is roughly 1.5 L per day. An additional 7 L of fluid are secreted into the gastrointestinal (GI) tract (saliva, gastric juices, bile, pancreatic juice and intestinal secretions), whereas only about 0.1 L/day is eliminated in the feces. The digestive tract must therefore absorb a net volume of at least 8.4 L of water per day. GI **absorption of water** occurs mainly in the jejunum and ileum, with smaller quantities being absorbed by the colon (→ **A**). Water is driven through the intestinal epithelium by osmosis. When solutes (Na^+, Cl^-, etc.) are absorbed in the intestine, water follows (→ **B**). (The stool contains only small quantities of Na^+, Cl^- and water.) Conversely, the secretion of substances into the lumen or the ingestion of non-absorbable substances leads to water fluxes into the lumen. Poorly absorbable substances therefore act as *laxatives* (e.g. sulfate, sorbitol, polyethylene glycol).

Water absorption is mainly driven by the **absorption of Na^+, Cl^-** and **organic compounds** (→ **B**). The luminal concentration of Na^+ and Cl^- steadily decreases from the duodenum to the colon. That of Na^+, for example, is approximately 145 mmol/L in the duodenum, 125 mmol/L in the ileum and only 40 mmol/L in the colon (→ **C**). Na^+ is absorbed by various **mechanisms**, and the *Na^+-K^+-ATPase* on the basolateral cell membrane is the primary driving mechanism (→ p. 26) for all of them (→ **B, D**).

◆ **Symport of Na^+ and organic substances** (→ see pp. 26ff. and 258): Na^+ passively influxes into cells of the duodenum and jejunum via symporter carriers, which actively cotransport glucose, amino acids, phosphates and other compounds (secondary active transport; → **D1**). Since this is an electrogenic transport mechanism (→ p. 28), a *lumen-negative transepithelial potential* (LNTP; → p. 162) that drives Cl^- out of the lumen forms (→ **D2**).

◆ **Parallel transport of Na^+ and Cl^-**: Na^+ ions in the lumen of the ileum are exchanged for H^+ ions (→ **D3**) while Cl^- is exchanged for HCO_3^- at the same time (→ **D4**). The H^+ ions combine with HCO_3^- to yield $H_2O + CO_2$, which diffuse out of the lumen. Most Na^+, Cl^- and, by subsequent osmosis, H_2O is absorbed by this electroneutral transport mechanism.

◆ **Na^+ diffusion**: Na^+ in the colon is mainly absorbed through luminal *Na^+ channels* (→ **D5**). This type of Na^+ transport is electrogenic and aldosterone-dependent (→ p. 182). The related lumen-negative transepithelial potential (L**N**TP, see above) either leads to K^+ secretion or drives Cl^- out of the lumen (→ **D2**).

The **Cl^- secretion** mechanism of epithelial cells (mainly *Lieberkühn's crypts*, → p. 245, A16) is similar to that of the acini of salivary glands (→ p. 236). The efflux of Cl^- into the lumen and the associated efflux of Na^+ and water are stimulated by cAMP and regulated by neurons and hormones such as VIP (vasoactive intestinal peptide) and prostaglandins. The physiological function of this form of H_2O secretion could be to dilute viscous chyme or to ensure the recirculation of water (from the crypts → lumen → villi → crypts) to promote the absorption of poorly soluble substances. **Cholera toxin** inhibits the GTPase of the G_s proteins (→ p. 274), thereby maintaining a maximal cAMP concentration and therefore a marked increase in Cl^- secretion. In response to it, large quantities of water and Na^+ are secreted into the lumen, which can lead to severe diarrhea (up to 1 L/hour!).

In addition to **HCO_3^-** from pancreatic juice, HCO_3^- is also secreted into the intestinal lumen by the small and large intestine (→ **A**). **K^+** is secreted (aldosterone-dependent) by crypt cells of the colon (luminal K^+ concentration ≈ 90 mmol/l!) and reabsorbed via the H^+-K^+ pump of the surface epithelium (similar to the mechanism in the stomach). The aldosterone-dependent K^+ secretion/absorption ratio determines the net amount of K^+ excreted (→ **A** and p. 180). *Diarrhea* results in losses of K^+ and HCO_3^- (hypokalemia and metabolic acidosis; → p. 142).

Ca^{2+}. The stool contains one-third of the dietary Ca^{2+} intake. Ca^{2+} is absorbed in the upper part of the small intestine (→ **A**) with the aid of intracellular *calcium-binding protein* (CaBP). Calcitriol increases CaBP synthesis, thereby increasing Ca^{2+} absorption (→ p. 292). Deficiencies of vitamin D or substances that form water-insoluble compounds with Ca^{2+} (phytin, oxalate, fatty acids) decrease Ca^{2+} absorption. **Mg^{2+}** is absorbed by similar mechanisms, but **iron (Fe)** is absorbed by a different mechanism (→ p. 90).

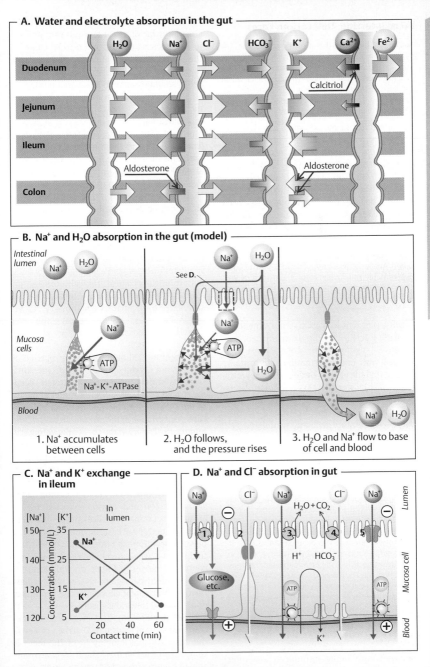

A. Water and electrolyte absorption in the gut

	H_2O	Na^+	Cl^-	HCO_3^-	K^+	Ca^{2+}	Fe^{2+}
Duodenum						Calcitriol	
Jejunum							
Ileum							
Colon		Aldosterone			Aldosterone		

B. Na⁺ and H₂O absorption in the gut (model)

Intestinal lumen

Na^+ H_2O Na^+ H_2O

See **D.**

Na^+ Na^+

Mucosa cells

ATP   ATP

Na^+-K^+-ATPase H_2O

Blood

Na^+ H_2O

1. Na⁺ accumulates between cells
2. H₂O follows, and the pressure rises
3. H₂O and Na⁺ flow to base of cell and blood

C. Na⁺ and K⁺ exchange in ileum

In lumen

[Na⁺] [K⁺]

Na⁺

K⁺

Concentration (mmol/L)

150 — 35
140 — 25
130 — 15
120 — 5

20   40   60
Contact time (min)

D. Na⁺ and Cl⁻ absorption in gut

Lumen

Na^+ Cl^- Na^+ Cl^- Na^+

⊖ 1 2 ⊕ 3 $H_2O + CO_2$ 4 5 ⊖

Glucose, etc.

H^+ HCO_3^-

Mucosa cell

ATP   ATP

⊕ K^+ ⊕

Blood

Plate 10.19 Water and Mineral Absorption

Large Intestine, Defecation, Feces

Anatomy. The terminal end of the gastrointestinal tract includes the *large intestine* (*cecum* and *colon*, ca. 1.3 m in length) and *rectum*. The large intestinal mucosa has characteristic pits (crypts), most of which are lined with mucus-forming cells (*goblet cells*). Some of the surface cells are equipped with a brush border membrane and reabsorb ions and water.

The large intestine has two **main functions**: (1) It serves as a *reservoir* for the intestinal contents (cecum, ascending colon, rectum). (2) It absorbs water and electrolytes ($\rightarrow$ p. 262), so the ca. 500–1500 mL of chyme that reaches the large intestine can be reduced to about 100–200 mL. The large intestine is not an essential organ; therefore, large segments of the intestine can be removed—e.g., for treatment of cancer.

Water instilled into the **rectum** via an *enema* is reabsorbed. Anally delivered drugs (*suppositories*) also diffuse through the intestinal wall into the bloodstream. Substances administered by this route bypass the liver and also escape the effects of gastric acid and digestive enzymes.

Motility. Different local mixing movements of the large intestine can be distinguished, e.g., powerful segmentation contractions associated with pouch formation (*haustration*) and anterograde or retrograde *peristaltic waves* (pacemaker located in transverse colon). Thus, stool from the colon can also be transported to the cecum. *Mass movements* occur 2–3 times daily ($\rightarrow$ **A**). They are generally stimulated by a meal and are caused by the gastrocolic reflex and gastrointestinal hormones.

The typical sequence of **mass movement** can be observed on X-ray films after administration of a barium meal, as shown in the diagrams ($\rightarrow$ **A1 – 8**). **A1**, barium meal administered at 7:00 a.m. **A2**, 12 noon: the barium mass is already visible in the last loop of the ileum and in the cecum. Lunch accelerates the emptying of the ileum. **A3**, about 5 minutes later, the tip of the barium mass is choked off. **A4**, shortly afterwards, the barium mass fills the transverse colon. **A5**, haustration divides the barium mass in the transverse colon, thereby mixing its contents. **A6–8**, a few minutes later (still during the meal), the transverse colon suddenly contracts around the leading end of the intestinal contents and rapidly propels them to the sigmoid colon.

Intestinal bacteria. The intestinal tract is initially sterile at birth, but later becomes colonized with orally introduced anaerobic bacteria during the first few weeks of life. The large intestine of a healthy adult contains 10^{11} to 10^{12} bacteria per mL of intestinal contents; the corresponding figure for the ileum is roughly 10^6/mL. The low pH inside the stomach is an important barrier against pathogens. Consequently, there are virtually no bacteria in the upper part of the small intestine (0–10^4/mL). Intestinal bacteria increase the activity of intestinal immune defenses (*"physiological inflammation"*), and their metabolic activity is useful for the host. The bacteria synthesize vitamin K and convert indigestible substances (e.g. cellulose) or partially digested saccharides (e.g. lactose) into absorbable short-chain fatty acids and *gases* (methane, H_2, CO_2).

The **anus** is normally closed. Anal closure is regulated by Kohlrausch's valve (transverse rectal fold), the puborectal muscles, the (involuntary) internal and (voluntary) external anal sphincter muscles, and a venous spongy body. Both sphincters contract tonically, the internal sphincter (smooth muscle) intrinsically or stimulated by sympathetic neurons (L_1, L_2) via α-adrenoceptors, the external sphincter muscle (striated muscle) by the pudendal nerve.

Defecation. Filling of the upper portion of the rectum (*rectal ampulla*) with intestinal contents stimulates the rectal stretch receptors ($\rightarrow$ **B2**), causing *reflex relaxation of the internal sphincter* (accommodation via VIP neurons), *constriction of the external sphincter*, and an *urge to defecate*. If the (generally voluntary) decision to defecate is made, the rectum shortens, the puborectal and external anal sphincter muscles relax, and (by a spinal parasympathetic reflex via S_2–S_4) annular contractions of the circular muscles of the descending colon, sigmoid colon and rectum—assisted by increased abdominal pressure—propel the feces out of the body ($\rightarrow$ **B**). The normal **frequency of bowel evacuation** can range from 3 times a day to 3 times a week, depending on the dietary content of indigestible fiber (e.g. cellulose, lignin). Frequent passage of watery stools (*diarrhea*) or infrequent stool passage (*constipation*) can lead to various disorders.

Stool (feces; $\rightarrow$ C). The average adult excretes 60–80 g of feces/day. Diarrhea can raise this over 200 g/d. Roughly $1/4$ of the feces is composed of *dry matter*, about $1/3$ is attributable to bacteria from the large intestine.

Plate 10.20 Large Intestine, Defecation, Feces

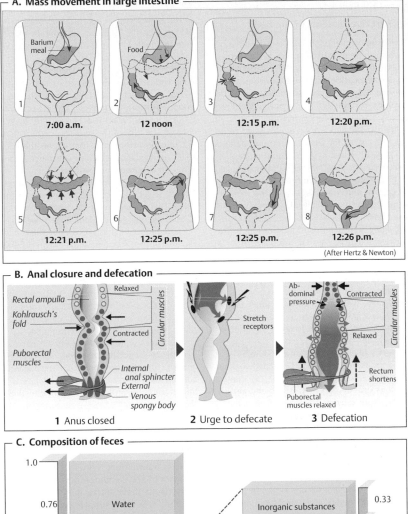

A. Mass movement in large intestine

Barium meal

7:00 a.m.

Food

12 noon

12:15 p.m.

12:20 p.m.

12:21 p.m.

12:25 p.m.

12:25 p.m.

12:26 p.m.

(After Hertz & Newton)

B. Anal closure and defecation

Rectal ampulla

Relaxed

Kohlrausch's fold

Contracted

Circular muscles

Puborectal muscles

Internal anal sphincter

External

Venous spongy body

1 Anus closed

Stretch receptors

2 Urge to defecate

Abdominal pressure

Contracted

Relaxed

Circular muscles

Puborectal muscles relaxed

Rectum shortens

3 Defecation

C. Composition of feces

1.0

0.76

Water

Inorganic substances — 0.33

N-containing substances — 0.33

0.24

0.08 Intestinal epithelium cells

0.08 Bacteria

Cellulose, etc. — 0.17

0.08 Food residue

Lipids — 0.17

0

Feces (60–180 g/day = 1.0)

Dry matter (25–40 g/day = 1.0)

11 Hormones and Reproduction

Integrative Systems of the Body

Unlike unicellular organisms, multicellular organisms have numerous specialized groups of cells and organs, the many different functions of which must be expediently *integrated* and *coordinated* (see also p. 2). In mammals, the **nervous system** and **endocrine system** are chiefly responsible for control and integration, while the **immune system** serves as an information system for corporal immune defense (→ p. 94ff.). These systems communicate by way of *electrical and/or chemical* **signals** (→ **A**).

Nerve impulses and hormonal signals serve to **control and regulate** (→ p. 4) the metabolism and internal milieu (blood pressure, pH, water and electrolyte balance, temperature, etc.), physical growth and maturation, reproductive functions, sexual response, and responses to the social environment. The signals received by sensors (= sensory receptors) in the inner organs, musculoskeletal system, skin and the sensory organs, as well as psychological factors, skeletal muscles and other factors also play a part in regulation and control. The signals are used by many *feedback mechanisms* in the body (→ p. 4).

Nerve fibers are specifically adapted for rapid transmission of finely graded signals. The nervous system consists of the **central nervous system** (CNS; → p. 310ff.) and **peripheral nervous system** . The latter consists of:

◆ The **somatic nervous system**, which conducts impulses from non-visceral sensors to a center (afferent neurons) and controls the skeletal musculature (efferent neurons).

◆ The **peripheral autonomic nervous system** (→ p. 78ff.), which consists of efferent neurons and mainly functions to control the circulatory system, inner organs and sexual functions. It is supplemented by:

◆ **Visceral afferent neurons**, i.e., nerve fibers that conduct signals from inner organs to a center. They are usually located in the same nerves as autonomous fibers (e.g., in vagus nerve); and the

◆ **Enteric nervous system**, which integrates the local functions of the esophagus, stomach and gut (→ p. 234).

Hormones. Like neurotransmitters (see below) and the immune system's cytokines and chemokines (→ p. 94ff.), hormones serve as *messenger substances* that are mainly utilized for *slower, long-term* transmission of signals. **Endocrine hormones** are carried by the blood to target structures great distances away. **Paracrine hormones** (and other paracrine transmitters) only act on cells in the immediate vicinity of the cells from which they are released. Hormones that act on the cells that produced the messenger substance are referred to as **autocrine hormones**.

Hormones are synthesized in specialized glands, tissues and cells (e.g., *neuroendocrine cells*). Their **target organ** is either a subordinate endocrine gland (*glandotropic hormone*) or non-endocrine tissue (*aglandotropic hormone*). The **target cells** have high-affinity binding sites (**receptors**) for their specific hormone, so very low concentrations of the hormone suffice for signal transduction (10^{-6} to 10^{-12} mol/L). The receptors on the target cells pick out the substances specifically intended for them from a wide variety of different messenger substances in their environment.

Hormones work closely with the nervous system to regulate *digestion, metabolism, growth, maturation, physical* and *mental development, maturation, reproduction, adaptation,* and the internal milieu of the body (*homeostasis*) (→ **A**). Most of these actions are predominately autonomous functions subject to central control by the **hypothalamus**, which is controlled by higher centers of the brain (→ p. 330).

Neurotransmitters released at chemical **synapses** of nerve endings transmit signals to postsynaptic nerve fibers, muscles or glands (→ p. 50ff.). Some neuropeptides released by presynaptic neurons also exert their effects in neighboring synapses, resulting in a kind of "paracrine" action.

Neurons can also secrete hormones, e.g., epinephrine, oxytocin and antidiuretic hormone. Some transmitter substances of the immune system, e.g. thymosin and various cytokines, also have endocrine effects.

Plate 11.1 Integrative Systems of the Body

A. Regulation of autonomic nervous system functions (overview)

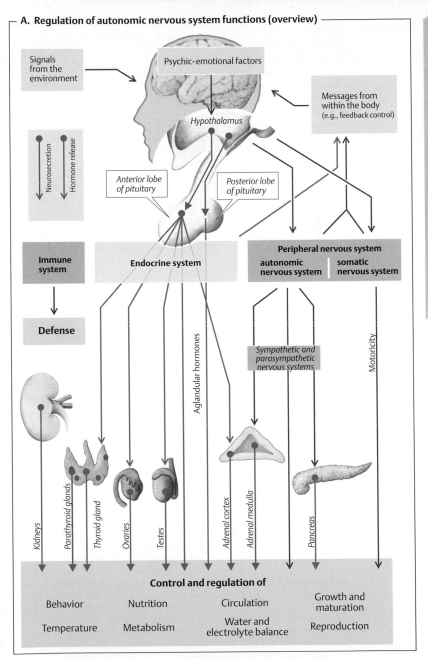

Signals from the environment

Psychic-emotional factors

Messages from within the body (e.g., feedback control)

Hypothalamus

Neurosecretion

Hormone release

Anterior lobe of pituitary

Posterior lobe of pituitary

Immune system

Endocrine system

Peripheral nervous system

autonomic nervous system | somatic nervous system

Defense

Aglandular hormones

Sympathetic and parasympathetic nervous systems

Motoricity

Kidneys

Parathyroid glands

Thyroid gland

Ovaries

Testes

Adrenal cortex

Adrenal medulla

Pancreas

Control and regulation of

Behavior

Nutrition

Circulation

Growth and maturation

Temperature

Metabolism

Water and electrolyte balance

Reproduction

Hormones

Hormones are messenger substances that convey information signals relevant to cell function ($\rightarrow$ p. 266). **Endocrine hormones**, i.e., those transported in the bloodstream, are produced in *endocrine glands* such as the hypothalamus, thyroid, parathyroid glands, adrenal medulla, pancreatic islets, ovaries and testes. They are also synthesized in diffusely scattered *endocrine cells* of the CNS, in C cells of the thyroid, and in the thymus, atria, kidneys, liver, gastrointestinal tract, etc. **Paracrine hormones**, i.e., those that affect nearby cells only (tissue hormones or *mediators*; see below) are secreted by cells widely distributed throughout the body.

Types of hormone.

1. **Peptide hormones** ($\rightarrow$ **A**, dark blue areas) and **glycoprotein hormones** ($\rightarrow$ **A**, light blue areas) are hydrophilic hormones stored in *secretory granules* and released by exocytosis as required. Multiple hormones can be produced from a single gene ($\rightarrow$ e.g., POMC gene, p. 280) by variable splicing and posttranslational modification ($\rightarrow$ p. 8 ff.) .

2. **Steroid hormones** ($\rightarrow$ **A**, yellow areas) and **calcitriol** are chemically related lipophilic hormones metabolized from *cholesterol* ($\rightarrow$ pp. 292ff). They are not stored, but are synthesized as needed.

3. **Tyrosine derivatives** ($\rightarrow$ **A**, orange areas) include (a) the *hydrophilic catecholamines* dopamine, epinephrine and norepinephrine ($\rightarrow$ p. 84) and (b) *lipophilic thyroid hormones* (T_3, T_4; $\rightarrow$ p. 286).

The **lipophilic hormones** in (2) and (3b) are **transported** in the blood while bound to plasma proteins. Corticosteroids are carried bound to globulin and albumin, testosterone and estrogen to sex hormone-binding globulin and T_3 and T_4 to albumin and two other plasma proteins ($\rightarrow$ p. 286).

Hormone receptors. The receptors (docking sites) for glycoprotein hormones, peptide hormones and catecholamines are transmembrane proteins ($\rightarrow$ p. 14) that bind to their specific hormone on the outer cell surface. Many of these hormones induce the release of intracellular **second messengers** that transmit the hormone signal inside the cell. cAMP, cGMP, IP_3, DAG, Ca^{2+} and NO function as second messengers (and sometimes as third messengers; $\rightarrow$ p. 274ff.). Some peptide hormones like insulin, prolactin, atriopeptin and numerous growth factors bind to cell surface receptors with cytosolic domains with *enzymatic activity* ($\rightarrow$ p. 278). Steroid hormones, on the other hand, enter the cells themselves ($\rightarrow$ p. 278). Once they bind to *cytosolic receptor proteins*, steroid hormones (as well as calcitriol, T_3 and T_4) are transported to the cell nucleus, where they influence transcription (genomic action). A target cell can have different receptors for different hormones (e.g., insulin and glucagon) or different receptors for a single hormone (e.g., α_1 and β_2 adrenoceptors for epinephrine).

Hierarchy of hormones ($\rightarrow$ **A**). The secretion of hormones is often triggered by neural impulses from the CNS. The *hypothalamus* is the main neurohormonal control center ($\rightarrow$ p. 280 and 330). Hypothalamic neurons extend to the posterior pituitary (neurohypophysis). The hormones are secreted either by the hypothalamus itself or by the posterior pituitary. Hypothalamic hormones also control hormone release from the *anterior pituitary* (adenohypophysis). Anterior pituitary *glandotropic hormones* control *peripheral endocrine glands* ($\rightarrow$ **A top**, green areas), which release the end-hormone ($\rightarrow$ **A**). The original signal can be *amplified* or *modulated* at these relay sites ($\rightarrow$ p. 272).

Pituitary hormones. Hypothalamic hormones control *anterior pituitary* hormone secretion by either stimulating or inhibiting hormone production. They are therefore called *releasing hormones (RH)* or *release-inhibiting hormones (IH)*, resp. ($\rightarrow$ **A** and table). Most anterior pituitary hormones are glandotropic ($\rightarrow$ p. 280). The *posterior pituitary* hormones are released by neuronal signals and are mainly aglandotropic ($\rightarrow$ p. 280).

Other **endocrine hormones** are secreted largely independent of the hypothalamic–pituitary axis, e.g., *pancreatic hormones, parathyroid hormone (PTH), calcitonin* and *calcitriol, angiotensin II, aldosterone* ($\rightarrow$ p. 182ff.), *erythropoietin* ($\rightarrow$ p. 88) and *gastrointestinal hormones* ($\rightarrow$ p. 234). *Atriopeptin* is secreted from the heart atrium in response to stretch stimuli ($\rightarrow$ p. 170), whereas the release of *melatonin* is subject to afferent neuron control ($\rightarrow$ p. 334).

Some of these hormones (e.g., angiotensin II) and *tissue hormones* or **mediators** exert *paracrine effects* within endocrine and exocrine glands, the stomach wall, other organs, and on inflammatory processes. Bradykinin ($\rightarrow$ pp. 214 and 236), histamine ($\rightarrow$ pp. 100 and 242), serotonin (5-hydroxytryptamine, $\rightarrow$ p. 102) and eicosanoids are members of this group.

Eicosanoids. Prostaglandins (PG), thromboxane (TX), leukotrienes and epoxyeicosatrienoates are eicosanoids (Greek εικοσι = twenty [C atoms]) derived in humans from the fatty acid **arachidonic acid (AA)**. (Prostaglandins derived from AA have the index number 2). AA occurs as an ester in the phospholipid layer of the cell membranes and is obtained from dietary sources (meat), synthesized from linoleic acid, an essential fatty acid, and released by *phospholipase A_2* ($\rightarrow$ p. 252).

Pathways of **eicosanoid synthesis** from arachidonic acid (AA):
1. *Cyclooxygenase pathway:* Cyclooxygenase (COX)-1 and COX-2 convert AA into PGG_2, which gives rise to PGH_2, the primary substance of the biologically active compounds PGE_2, PGD_2, $PGF_{2\alpha}$, PGI_2 (prostacyclin) and TXA_2. COX-1 and 2 are inhibited by nonsteroidal anti-inflammatory drugs (e.g., Aspirin®).
2. *Lipoxygenase pathway:* Leukotriene A_4 is synthesized from AA (via the intermediate 5-HPETE = 5-hydroperoxyeicosatetraenoate) by way of 5-lipoxygenase (especially in neutrophilic granulocytes). Leukotriene A_4 is the parent substance of the leukotrienes C_4, D_4 and E_4. The significance of 12-lipoxygenase (especially in thrombocytes) is not yet clear, but *15-lipoxygenase* is known to produce vasoactive *lipoxins* (LXA_4, LXB_4).
3. *Cytochrome P450-epoxygenase* produces epoxyeicosatrienoates (EpETrE = EE).

Typical **effects of eicosanoids:**

PGE_2 dilates the bronchial and vascular musculature (and keeps the lumen of the fetal ductus arteriosus and foramen ovale open; $\rightarrow$ p. 220), stimulates intestinal and uterine contractions, protects the gastric mucosa ($\rightarrow$ p. 242), inhibits lipolysis, increases the glomerular filtration rate (GFR), plays a role in fever development ($\rightarrow$ p. 224), sensitizes nociceptive nerve endings (pain) and increases the permeability of blood vessels (inflammation). **PGD_2** induces bronchoconstriction. **PGI_2** (**prostacyclin**) synthesized in the endothelium, is vasodilatory and inhibits platelet aggregation. **TXA_2**, on the other hand, occurs in platelets, promotes platelet aggregation and acts as a vasoconstrictor ($\rightarrow$ p. 102). **11,12-EpETrE** has a vasodilatory effect (= EDHF, $\rightarrow$ p. 214).

Hormones (h.) of the hypothalamus and pituitary

Name*	Abbreviation/synonyme
Hypothalamus	

The suffix "-liberin" denotes releasing h. (RH) or factor (RF); "-statin" is used for release-inhibiting h. (IH) or factors (IF)

Name	Abbreviation/synonyme
Corticoliberin	Corticotropin RH, CRH, CRF
Gonadoliberin	Gonadotropin RH, Gn-RH; ICSH
Prolactostatin	Prolactin IH, PIH, PIF, dopamine
Somatoliberin	Somatotropin RH, SRH, SRF, GHRH, GRH
Somatostatin**	Somatotropin (growth h.) IH, SIH
Thyroliberin	Thyrotropin RH, TRH, TRF

Anterior lobe of the pituitary

Name	Abbreviation/synonyme
Corticotropin	Adrenocorticotropic h. (ACTH)
Follitropin	Follicle-stimulating h. (FSH)
Lutropin	Luteinizing h. (LH), interstitial cell-stimulating h. (ICSH)
Melanotropin	α-Melanocyte-stimulating h. (α-MSH), α-melanocortin
Somatotropin	Somatotropic h. (STH), growth h. (GH)
Thyrotropin	Thyroid stimulating h. (TSH)
Prolactin	PRL, lactogenic (mammotropic) h.

Posterior lobe of the pituitary

Name	Abbreviation/synonyme
Oxytocin	–
Adiuretin	Anti-diuretic h. ADH, (arginine-) vasopressin (AVP)

* Names generally recommended by IUPAC-IUB Committee on Biochemical Nomenclature.

** Also synthesized in gastrointestinal organs, etc.

A. The hormones (simplified overview excluding tissue hormones)

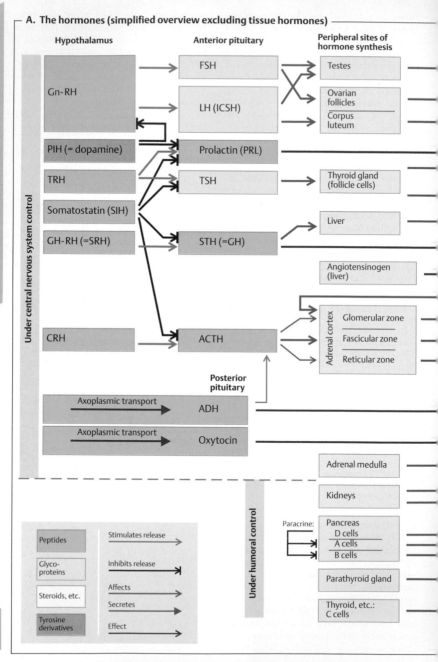

Hypothalamus	Anterior pituitary	Peripheral sites of hormone synthesis
Gn-RH	FSH	Testes
	LH (ICSH)	Ovarian follicles / Corpus luteum
PIH (= dopamine)	Prolactin (PRL)	
TRH	TSH	Thyroid gland (follicle cells)
Somatostatin (SIH)		
GH-RH (=SRH)	STH (=GH)	Liver
		Angiotensinogen (liver)
CRH	ACTH	Adrenal cortex — Glomerular zone / Fascicular zone / Reticular zone

Under central nervous system control

Posterior pituitary

| Axoplasmic transport | ADH |
| Axoplasmic transport | Oxytocin |

Under humoral control

Adrenal medulla

Kidneys

Paracrine: Pancreas
D cells
A cells
B cells

Parathyroid gland

Thyroid, etc.:
C cells

Peptides

Glyco-proteins

Steroids, etc.

Tyrosine derivatives

Stimulates release

Inhibits release

Affects

Secretes

Effect

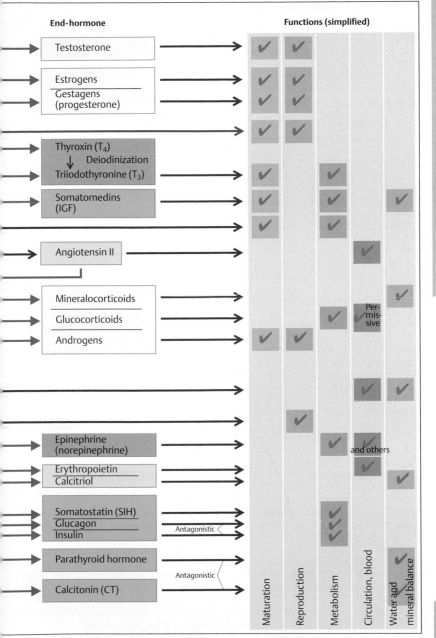

Plate 11.2 u. 11.3   Hormones

End-hormone

Testosterone

Estrogens
Gestagens
(progesterone)

Thyroxin (T$_4$)
↓ Deiodinization
Triiodothyronine (T$_3$)

Somatomedins
(IGF)

Angiotensin II

Mineralocorticoids

Glucocorticoids

Androgens

Epinephrine
(norepinephrine)

Erythropoietin
Calcitriol

Somatostatin (SIH)
Glucagon
Insulin

Antagonistic

Parathyroid hormone

Antagonistic

Calcitonin (CT)

Functions (simplified)

Per-mis-sive

and others

Maturation

Reproduction

Metabolism

Circulation, blood

Water and mineral balance

Humoral Signals: Control and Effects

Hormones and other humoral signals function to provide **feedback control**, a mechanism in which the response to a signal feeds back on the signal generator (e.g., endocrine gland). The speed at which control measures are implemented depends on the rate at which the signal substance is broken down—the quicker the degradation process, the faster and more flexible the control.

In **negative feedback control**, the response to a feedback signal opposes the original signal. In the example shown in **A1**, a rise in plasma cortisol in response to the release of corticoliberin (corticotropin-releasing hormone, CRH) from the hypothalamus leads to down-regulation of the signal cascade "CRH ⇒ ACTH ⇒ adrenal cortex," resulting in a decrease in cortisol secretion. In shorter feedback loops, ACTH can also negatively feed back on the hypothalamus (→ **A2**), and cortisol, the *end-hormone*, can negatively feed back on the anterior pituitary (→ **A3**). In some cases, the *metabolic parameter* regulated by a hormone (e.g., plasma glucose concentration) rather then the hormone itself represents the feedback signal. In the example (→ **B**), glucagon increases blood glucose levels (while insulin decreases them), which in turn inhibits the secretion of glucagon (and stimulates that of insulin). *Neuronal signals* can also serve as feedback (*neuroendocrine feedback*) used, for example, to regulate plasma osmolality (→ p. 170).

In **positive feedback control**, the response to the feedback amplifies the original signal and heightens the overall response (e.g., in *autocrine regulation*; see below).

The higher hormone not only controls the synthesis and excretion of the end-hormone, but also controls the **growth of peripheral endocrine gland**. If, for example, the end-hormone concentration in the blood is too low despite maximum synthesis and secretion of the existing endocrine cells, the gland will enlarge to increase end-hormone production. This type of **compensatory hypertrophy** is observed for instance in goiter development (→ p. 288) and can also occur after surgical excision of part of the gland.

Therapeutic administration of a hormone (e.g., cortisone, a cortisol substitute) have the same effect on higher hormone secretion (ACTH and CRH in the example) as that of the end-hormone (cortisol in the example) normally secreted by the peripheral gland (adrenal cortex in this case). *Long-term* administration of an end-hormone would therefore lead to inhibition and atrophy of the endocrine gland or cells that normally produce that hormone. This is known as **compensatory atrophy**.

A **rebound effect** can occur if secretion of the higher hormone (e.g., ACTH) is temporarily elevated after discontinuation of end-hormone administration.

The **principal functions** of *endocrine hormones, paracrine hormones* and other humoral transmitter substances are to control and regulate:

◆ *enzyme activity* by altering the conformation (*allosterism*) or inhibiting/stimulating the synthesis of the enzyme (induction);

◆ *transport processes*, e.g., by changing the rate of insertion and synthesis of ion channels/carriers or by changing their opening probability or affinity;

◆ *growth* (see above), i.e., increasing the rate of mitosis (*proliferation*), "programmed cell death" (*apoptosis*) or through cell differentiation or dedifferentiation;

◆ *secretion of other hormones.* Regulation can occur via endocrine pathways (e.g., ACTH-mediated cortisol secretion; → **A5**), a short portal vein-like circuit within the organ (e.g., effect of CRH on ACTH secretion, → **A4**), or the effect of cortisol from the adrenal cortex on the synthesis of epinephrine in the adrenal medulla, (→ **A6**), or via paracrine pathways (e.g., the effect of somatostatin, SIH, on the secretion of insulin and glucagon; → **B**).

Cells that have receptors for their own humoral signals transmit **autocrine signals** that function to

◆ exert *negative feedback control* on a target cell, e.g., to discontinue secretion of a transmitter (e.g., norepinephrine; → p. 84);

◆ *coordinate* cells of the same type (e.g., in growth);

◆ exert *positive feedback control* on the secreting cell or to cells of the same type. These mechanisms serve to amplify weak signals as is observed in the eicosanoid secretion or in T cell clonal expansion (→ p. 96ff.).

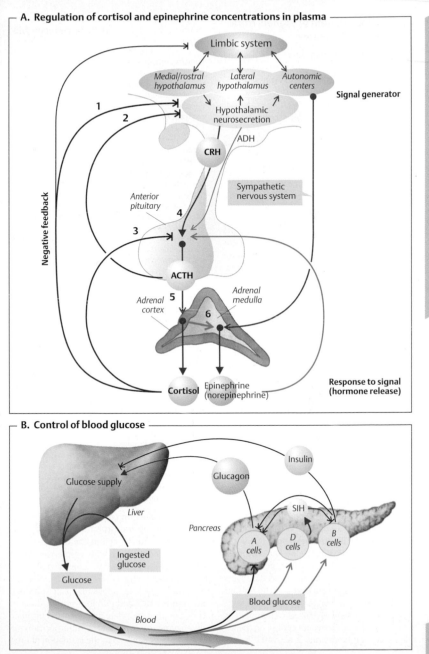

A. Regulation of cortisol and epinephrine concentrations in plasma

Limbic system

Medial/rostral hypothalamus   Lateral hypothalamus   Autonomic centers   **Signal generator**

Hypothalamic neurosecretion

1

2

ADH

CRH

Anterior pituitary

Sympathetic nervous system

4

3

ACTH

Adrenal cortex   5   Adrenal medulla

6

Negative feedback

Cortisol   Epinephrine (norepinephrine)   **Response to signal (hormone release)**

B. Control of blood glucose

Glucose supply

Liver

Glucagon   Insulin

SIH

Pancreas   A cells   D cells   B cells

Ingested glucose

Glucose

Blood glucose

Blood

Plate 11.4   Humoral Signals: Control and Effects

273

Cellular Transmission of Signals from Extracellular Messengers

Hormones, neurotransmitters ($\rightarrow$ p. 55 and p. 82), cytokines and chemokines ($\rightarrow$ p. 94ff.) act as *messenger substances* (**first messengers**) that are transported to their respective target cells by extracellular pathways. The target cell has a high-affinity binding site (**receptor**) for its *specific* messenger substance.

Glycoprotein and peptide messengers as well as **catecholamines** bind to cell surface receptors on the target cell. *Binding of the messenger to its receptor* usually triggers certain protein-protein interactions (and sometimes protein-phospholipid interactions). This leads to the release of secondary messenger substances (**second messengers**) that forward the signal within the cell. Cyclic adenosine monophosphate (cAMP), cyclic guanosine monophosphate (cGMP), inositol 1,4,5-trisphosphate (IP_3), 1,2-diacylglycerol (DAG) and Ca^{2+} are such second messengers. Since the molecular structure of the receptor ensures that the effect of the first messenger will be specific, multiple first messengers can use the same second messenger. Moreover, the intracellular concentration of the second messenger can be raised by one messenger and lowered by another. In many cases, different types of receptors exist for a single first messenger.

cAMP as a Second Messenger

For a cAMP-mediated response to occur, the cell membrane must contain stimulatory (G_s) or inhibitory (G_i) **G proteins** (guanyl nucleotide-binding proteins) ($\rightarrow$ **A1**). These G proteins consist of three subunits—alpha (α_S or α_i), beta (β) and gamma (γ)—and are therefore *heterotrimers*. Guanosine diphosphate (GDP) is bound to the α-subunit of an inactive G protein. Once the first **messenger (M)** binds to the **receptor (Rec.)**, the M–Rec. complex conjugates with the G_s-GDP (or G_i-GDP) molecule ($\rightarrow$ **A2**). GDP is then replaced by cytosolic GTP, and the $\beta\gamma$-subunit and the M–Rec. complex dissociate from the α-subunit if Mg^{2+} is present ($\rightarrow$ **A3**). α_S-GTP or α_i-GTP remain as the final products. **Adenylate cyclase** on the inside of the cell membrane is activated by α_S-GTP (cytosolic *cAMP concentration rises*) and inhibited by α_i-GTP (*cAMP concentration falls*; $\rightarrow$ **A3**).

G_s-activating messengers. ACTH, adenosine (A_{2A} and A_{2B} rec.), antidiuretic hormone = vasopressin (V_2 rec.), epinephrine and norepinephrine (β_1, β_2, β_3 adrenoceptors), calcitonin, CGRP, CRH, dopamine (D_1 and D_5 rec.), FSH, glucagon, histamine (H_2 rec.), oxytocin (V_2 rec., see above), many prostaglandins (DP, IP, EP_2 and EP_4 rec.), serotonin = 5-hydroxytryptamine (5-HT_4 and 5-HT_7 rec), secretin and VIP activate G_s proteins, thereby **raising cAMP levels**. TRH and TSH induce partial activation.

G_i-activating messengers. Some of the above messenger substances also activate G_i proteins (thereby **lowering cAMP levels**) using a different binding receptor. Acetylcholine (M_2 and M_4 rec.), adenosine (A_1 and A_3 rec.), epinephrine and norepinephrine (α_2 adrenoceptors), angiotensin II, chemokines, dopamine (D_2, D_3 and D_4 rec.), GABA ($GABA_B$ rec.), glutamate ($mGLU_{2-4}$ and $mGLU_{6-8}$ rec.), melatonin, neuropeptide Y, opioids, serotonin = 5-hydroxytryptamine (5-HT_i rec.), somatostatin and various other substances activate G_i proteins.

Effects of cAMP. cAMP activates type A protein kinases (**PKA = protein kinase A**) which then activate other proteins (usually enzymes and membrane proteins, but sometimes the receptor itself) by **phosphorylation** ($\rightarrow$ **A4**). The specific response of the cell depends on the type of protein phosphorylated, which is determined by the type of protein kinases present in the target cell. Phosphorylation converts the proteins from an inactive to an active form or vice versa.

Hepatic glycogenolysis, for instance, is dually increased by cAMP and PKA. Glycogen synthase catalyzing glycogen synthesis is inactivated by phosphorylation whereas glycogen phosphorylase stimulating glycogenolysis is activated by cAMP-mediated phosphorylation.

Signal transduction comprises the entire signaling pathway from the time the first messenger binds to the cell to the occurrence of cellular effect, during which time the signal can be (a) *modified* by other signals and (b) *amplified* by many powers of ten. A single adenylate cyclase molecule can produce numerous cAMP and PKA molecules, which in turn can phosphorylate an enormous number of enzyme molecules. The interposition of more kinases can lead to the formation of long

▶

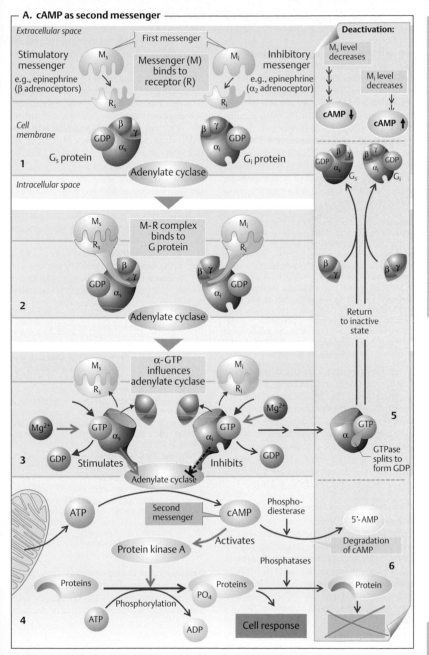

A. cAMP as second messenger

Extracellular space

First messenger

Stimulatory messenger
e.g., epinephrine
(β adrenoceptors)

Messenger (M) binds to receptor (R)

M_s R_s

Inhibitory messenger
e.g., epinephrine
($α_2$ adrenoceptor)

M_i R_i

Deactivation:

M_s level decreases
cAMP ↓

M_i level decreases
cAMP ↑

Cell membrane

GDP β γ $α_s$
G_s protein

β γ GDP $α_i$
G_i protein

Adenylate cyclase

1

Intracellular space

2

M–R complex binds to G protein

M_s R_s
β γ
GDP $α_s$

M_i R_i
β γ
$α_i$ GDP

Adenylate cyclase

β γ GDP $α_s$ G_s
β γ GDP $α_i$ G_i

β γ

Return to inactive state

3

α-GTP influences adenylate cyclase

M_s R_s

M_i R_i

Mg^{2+}

GTP $α_s$

GTP $α_i$

Mg^{2+}

GDP Stimulates

GDP Inhibits

Adenylate cyclase

5

α GTP
GTPase splits to form GDP

4

ATP

Second messenger cAMP

Phosphodiesterase

Protein kinase A

Activates

Proteins

PO_4 Proteins

Phosphatases

Phosphorylation

ATP ADP

Cell response

5′-AMP

Degradation of cAMP

6

Protein

Plate 11.5 Cellular Signal Transmission I

kinase cascades that additionally amplify the original signal while receiving further regulatory signals.

Deactivation of the signaling cascade (→ **A**, **right** panel) is induced by the α-subunit in that its GTP molecule splits off GDP and P_i after reacting with its *GTPase* (→ **A5**), and the subunit subsequently binds to a βγ subunit to again form the trimeric G protein. *Phosphodiesterase* also converts cAMP into inactive 5′-AMP (→ **A4, A6**), and *phosphatases* dephosphorylate the protein previously phosphorylated by protein kinase A (→ **A4**). Another way to inactivate a receptor in the presence of high messenger concentrations is to make the receptor insensitive by phosphorylating it (*desensitization*).

Cholera toxin inhibits the GTPase, thereby blocking its deactivating effect on adenylate cyclase (→ **A5**). This results in extremely high levels of intracellular cAMP. When occurring in intestinal cells, this can lead to severe diarrhea (→ p. 262). **Pertussis** (whooping cough) **toxin** and **forskolin** also lead to an increase in the cytosolic cAMP concentration. Pertussis toxin does this by inhibiting G_i protein and thereby blocking its inhibitory effect on adenylate cyclase, while forskolin directly activates adenylate cyclase. **Theophylline** and **caffeine** inhibit the conversion of cAMP to 5′-AMP, which extends the life span of cAMP and prolongs the effect of the messenger.

Certain **ion channels** are regulated by G_s, G_i and other G proteins (G_o) with or without the aid of adenylate cyclase. Some Ca^{2+} channels are activated by G_s proteins and inactivated by G_o proteins, whereas some K^+ channels are activated by G_o proteins and (the βγ subunits of) G_i proteins (→ p. 83 B). G_{olf} in olfactory receptors, **transducin** in retinal rods (→ p. 348ff.), and **α-gustducin** in gustatory sensors are also members of the G protein family (→ p. 338).

IP_3 and DAG as Second Messengers

As in the case of G_s proteins, once the first messenger using this transduction pathway binds to its receptor outside the cell, the $α_q$ subunit dissociates from the heterotrimeric G_q **protein** and activates phospholipase C-β (**PLC-β**) on the inside of the cell membrane (→ **B1**). PLC-β converts phosphatidylinositol 4,5-bisphosphate (PIP_2), to **inositol 1,4,5-trisphosphate** (IP_3) and **diacylglycerol** (**DAG**). IP_3 and DAG function as parallel second messengers with different actions that are exerted either independently or jointly (→ **B1**).

IP_3 is a hydrophilic molecule carried via the cytosol to Ca^{2+} stores within the cell (mainly in the endoplasmic reticulum; → p. 36). IP_3 binds there to Ca^{2+} channels to open them (→ **B2**), leading to an efflux of Ca^{2+} from the intracellular stores into the cytosol. In the cytosol, Ca^{2+} acts as a *third messenger* that regulates various cell functions, e.g., by interacting with the cAMP signaling chain. Many Ca^{2+}-related activities are mediated by *calmodulin*, a calcium-binding protein (→ pp. 36 and 70).

DAG is a lipophilic molecule that remains in the cell membrane and has two main functions:

◆ DAG is broken down by phospholipase A2 (PLA-2) to yield *arachidonic acid*, a precursor of eicosanoids (→ **B3** and p. 269).

◆ DAG activates *protein kinase C* (**PKC**). PKC is Ca^{2+}-dependent (hence the "C") because the Ca^{2+} released by IP_3 (see above) is needed to transfer PKC from the cytosol to the intracellular side of the cell membrane (→ **B4**). Thus activated PKC phosphorylates the serine or threonine residues of many proteins.

PKC triggers a series of other phosphorylation reactions (high signal amplification) that ultimately lead to the phosphorylation of **MAP kinase** (mitogen-activated protein kinase). It enters the cell nucleus and activates *Elk-1*, a gene-regulating protein. *NF-κB*, another gene-regulating protein, is also released in response to PKC phosphorylation. In addition, PKC activates Na^+/H^+ *antiporters*, thereby raising the **cellular pH**—a stimulus that triggers many other cellular reactions.

IP_3 **and DAG activating messengers** include acetylcholine (M_1 and M_3 cholinoceptors), antidiuretic hormone = vasopressin (V_1 rec.), epinephrine and norepinephrine ($α_1$ adrenoceptor), bradykinin, CCK, endothelin, gastrin, glutamate ($mGLU_1$ and $mGLU_5$ rec.), GRP, histamine (H_1 rec.), leukotrienes, neurotensin, oxytocin and various prostaglandins (FP, TP, and Ep_1 rec.), serotonin = 5-hydroxytryptamine (5-HT_2 rec.), tachykinin, thromboxane A_2, TRH and TSH induce partial activation.

Deactivation of the signaling cascade can also be achieved through *self-inactivation* of the G proteins involved (GTP cleavage) and phosphatase (see above) as well as by degradation of IP_3.

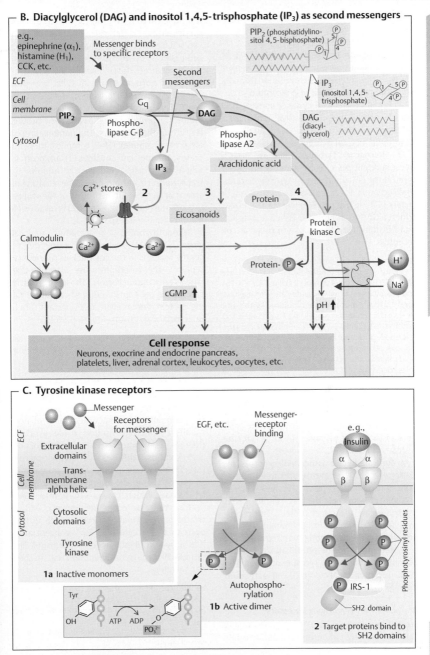

B. Diacylglycerol (DAG) and inositol 1,4,5-trisphosphate (IP₃) as second messengers

e.g., epinephrine (α_1), histamine (H_1), CCK, etc.

Messenger binds to specific receptors

PIP₂ (phosphatidylinositol 4,5-bisphosphate)

Second messengers

ECF

Cell membrane

Cytosol

PIP₂ — G_q → DAG

Phospholipase C-β

1

IP₃

IP₃ (inositol 1,4,5-trisphosphate)

DAG (diacylglycerol)

Phospholipase A2

Arachidonic acid

Ca²⁺ stores

2

3

4

Protein

Calmodulin

Protein kinase C

Ca²⁺

Ca²⁺

Eicosanoids

cGMP ↑

Protein- P

H⁺

Na⁺

pH ↑

Cell response
Neurons, exocrine and endocrine pancreas, platelets, liver, adrenal cortex, leukocytes, oocytes, etc.

C. Tyrosine kinase receptors

Messenger

Receptors for messenger

EGF, etc.

Messenger-receptor binding

e.g., Insulin

ECF

Extracellular domains

Cell membrane

Transmembrane alpha helix

Cytosol

Cytosolic domains

Tyrosine kinase

α α

β β

P P

P P

P P

1a Inactive monomers

Autophosphorylation

1b Active dimer

IRS-1

SH2 domain

Phosphotyrosinyl residues

Tyr OH

ATP ADP

PO_3^{2-}

2 Target proteins bind to SH2 domains

Plate 11.6 Cellular Signal Transmission II

Enzyme-Linked Cell Surface Receptors For Messenger Substances

These (G protein-independent) receptors, together with their cytosolic domains, act as enzymes that are activated when a messenger binds to the receptor's extracellular domain. There are five classes of these receptors:

1. Receptor guanylyl cyclases convert GTP into the second messenger cGMP, which activates protein kinase G (PKG; see below). The *atriopeptin receptor* belongs to this class.

2. Receptor tyrosine kinases ($\rightarrow$ **C**), phosphorylate proteins (of same or different type) at the OH group of their tyrosyl residues. The receptors for *insulin* and various *growth factors* (GF) such as e.g., E[epidermal]GF, PDGF, N[nerve]GF, F[fibroblast]GF, H[hepatocyte]GF, and I[insulin-like]GF-1 belong to this class of receptors.

Signals regarding first messenger binding (e.g., EGF and PDGF) are often transferred inside the cell via binding of two receptors (**dimerization**; **C1a** $\Rightarrow$ **C1b**) and subsequent mutual phosphorylation of their cytosolic domain (**autophosphorylation**, $\rightarrow$ **C1b**). The receptor for certain hormones, like *insulin* and *IGF-1*, is from the beginning a heterotetramer ($\alpha_2\beta_2$) that undergoes autophosphorylation before phosphorylating another protein (*insulin receptor substrate-1, IRS-1*) that in turn activates intracellular target proteins containing SH2 domains ($\rightarrow$ **C2**).

3. Receptor serine/threonine kinases, which like the *TGF-β receptor*, function similar to kinases in Group 2, the only difference being that they phosphorylate serine or threonine residues of the target protein instead of tyrosine residues (as with PKC; see above).

4. Tyrosine kinase-associated receptors are those where the receptor works in combination with non-receptor tyrosine kinases (chiefly proteins of the Src family) that phosphorylate the target protein. The receptors for *STH*, *prolactin*, *erythropoietin* and numerous *cytokines* belong to this group.

5. Receptor tyrosine phosphatases remove phosphate groups from tyrosine residues. The CD45 receptor involved in T cell activation belongs to this group.

Hormones with Intracellular Receptors

Steroid hormones ($\rightarrow$ p. 270ff., yellow areas), calcitriol and thyroid hormones are like other hormones in that they induce a *specific cell response* with the difference being that they activate a different type of signaling cascade in the cell. They are *lipid-soluble* substances that freely penetrate the cell membrane.

Steroid hormones bind to their respective *cytoplasmic receptor protein* in the target cell ($\rightarrow$ **D**). This binding leads to the dissociation of inhibitory proteins (e.g., heat shock protein, HSP) from the receptors. The *hormone–receptor protein complex* (H–R complex) then migrates to the cell nucleus (*translocation*), where it activates (**induces**) or inhibits the transcription of certain genes. The resulting increase or decrease in synthesis of the respective protein (e.g., AIPs; $\rightarrow$ p. 182) is responsible for the actual cell response ($\rightarrow$ **D**).

Triiodothyronine (T$_3$; $\rightarrow$ p. 286ff.) and **calcitriol** ($\rightarrow$ p. 292) bind to their respective receptor proteins in the cell nucleus (*nuclear receptors*). These receptors are *hormone-activated transcription factors*. Those of calcitriol can induce the transcription of calcium-binding protein, which plays an important role in interstitial Ca^{2+} absorption ($\rightarrow$ p. 262).

Recent research indicates that steroid hormones and calcitriol also regulate cell function by **non-genomic control mechanisms**.

Nitric Oxide as a Transmitter Substance

In nitrogenergic neurons and endothelial tissues, **nitric (mon)oxide (NO)** is released by Ca^{2+}/calmodulin-mediated activation of neuronal or endothelial **nitric oxide synthase (NOS)** ($\rightarrow$ **E**). Although NO has a half-life of only a few seconds, it diffuses into neighboring cells (e.g., from endothelium to vascular myocytes) so quickly that it activates **cytoplasmic guanylyl cyclase**, which converts GTP into **cGMP** ($\rightarrow$ **E**). Acting as a second messenger, cGMP activates *protein kinase G* (**PKG**), which in turn decreases the cytosolic Ca^{2+} concentration [Ca^{2+}]i by still unexplained mechanisms. This ultimately leads to *vasodilatation* (e.g., in coronary arteries).

Penile erections are produced by cGMP-mediated vasodilatation of the deep arteries of the penis ($\rightarrow$ p. 308). The erection can be prolonged by drugs that inhibit cGMP-specific phosphodiesterase type 5, thereby delaying the degradation of cGMP (e.g., sildenafil citrate = Viagra®).

Plate 11.7 Cellular Signal Transmission III

D. Mode of action of steroid hormones

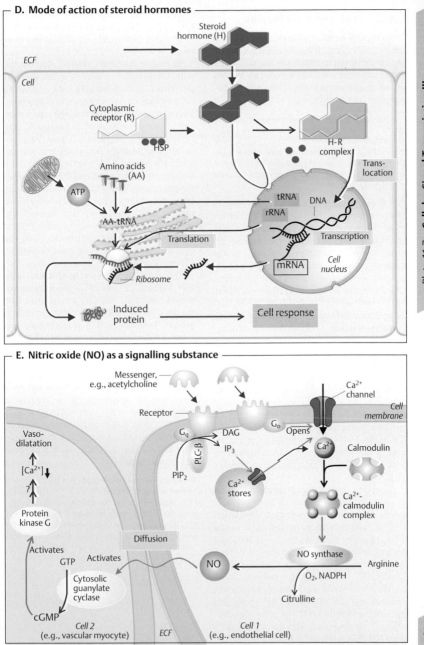

E. Nitric oxide (NO) as a signalling substance

Hypothalamic–Pituitary System

In the **hypothalamus**, (1) humoral signals from the periphery (e.g., from circulating cortisol) can be converted to efferent neuronal signals, and (2) afferent neuronal signals can be converted to endocrine messengers (*neurosecretion*).

The first case is possible because the hypothalamus is situated near **circumventricular organs** like the organum vasculosum laminae terminalis (OVLT), subfornical organ, the median eminence of the hypothalamus, and the neurohypophysis. Since there is no blood-brain barrier there, hydrophilic peptide hormones can also enter.

The hypothalamus is closely connected to other parts of the CNS ($\rightarrow$ p. 330). It controls many autonomous regulatory functions and its neuropeptides influence higher brain functions. The hypothalamus is related to the sleeping-waking rhythm ($\rightarrow$ p. 334) and to psychogenic factors. Stress, for example, stimulates the release of cortisol (via CRH, ACTH) and can lead to the cessation of hormone-controlled menstruation (amenorrhea).

Neurosecretion. Hypothalamic neurons synthesize hormones, incorporate them in granules that are transported to the ends of the axons (*axoplasmic transport* $\rightarrow$ p. 42), and secrete them into the *bloodstream*. In this way, oxytocin and ADH are carried from magnocellular hypothalamic nuclei to the neurohypophysis, and RHs and IHs (and ADH) reach the *median eminence* of the hypothalamus ($\rightarrow$ **A**). The *action potential-triggered* exocytotic release of the hormones into the bloodstream results in Ca^{2+} influx into the nerve endings ($\rightarrow$ p. 50ff.).

Oxytocin (Ocytocin) and **antidiuretic hormone** (ADH) are two posterior pituitary hormones that enter the systemic circulation directly. **ADH** induces water retention in the renal collecting ducts (V_2-rec.; $\rightarrow$ p. 166) and induces vasoconstriction (endothelial V_1 rec.) by stimulating the secretion of endothelin-1 ($\rightarrow$ p. 212ff.). ADH-bearing neurons also secrete ADH into the portal venous circulation (see below). The ADH and CRH molecules regulate the secretion of ACTH by the adenohypophysis. **Oxytocin** promotes uterine contractions and milk ejection ($\rightarrow$ p. 304). In nursing mothers, suckling stimulates nerve endings in the nipples, triggering the secretion of oxytocin (and prolactin, $\rightarrow$ p. 303) via neurohumoral reflexes.

Releasing hormones (RH) or **liberins** that stimulate hormone release from the adenohypophysis (Gn-RH, TRH, SRH, CRH; $\rightarrow$ p. 270ff.) are secreted by hypothalamic neurons into a kind of **portal venous system** and travel only a short distance to the anterior lobe ($\rightarrow$ **A**). Once in its vascular network, they trigger the release of anterior pituitary hormones into the systemic circulation ($\rightarrow$ **A**). Some anterior pituitary hormones are regulated by release-**inhibiting hormones (IH)** or **statins**, such as SIH and PIH = dopamine ($\rightarrow$ p. 270ff.). Peripheral hormones, ADH (see above) and various neurotransmitters such as neuropeptide Y (NPY), norepinephrine (NE), dopamine, VIP and opioids also help to regulate anterior pituitary functions ($\rightarrow$ p. 272).

The four **glandotropic hormones** (ACTH, TSH, FSH and LH) and the **aglandotropic hormones** (prolactin and GH) are secreted from the anterior pituitary ($\rightarrow$ **A**). The secretion of **growth hormone (GH** = somatotropic hormone, STH) is subject to control by GH-RH, SIH and IGF-1. GH stimulates protein synthesis (*anabolic action*) and skeletal growth with the aid of *somatomedins* (growth factors formed in the liver), which play a role in sulfate uptake by cartilage. Somatomedin C = insulin-like growth factor-1 (IGF-1) inhibits the release of GH by the anterior pituitary via negative feedback control. GH has lipolytic and glycogenolytic actions that are independent of somatomedin activity.

Proopiomelanocortin (POMC) is a peptide precursor not only of ACTH, but (inside or outside the anterior pituitary) also of β-endorphin and α-melanocyte-stimulating hormone (α-MSH = α-melanocortin). β-endorphin has analgesic effects in the CNS and immunomodulatory effects, while α-MSH in the hypothalamus helps to regulate the body weight ($\rightarrow$ p. 230) and stimulates peripheral melanocytes.

Plate 11.8 Hypothalamic–Pituitary System

A. Hypothalamic-pituitary hormone secretion (schematic)

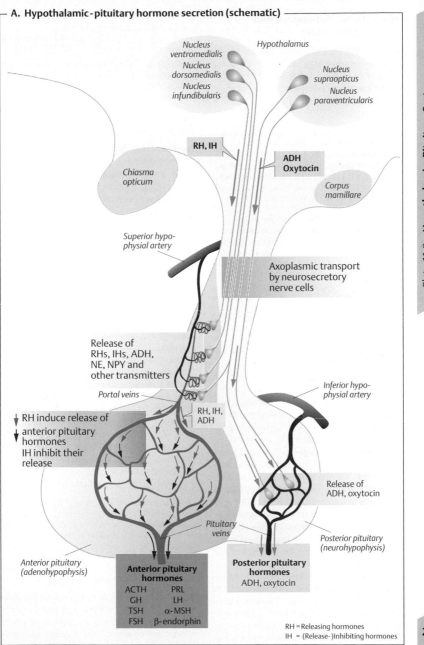

Hypothalamus

Nucleus ventromedialis
Nucleus dorsomedialis
Nucleus infundibularis

Nucleus supraopticus
Nucleus paraventricularis

RH, IH

ADH Oxytocin

Chiasma opticum

Corpus mamillare

Superior hypo-physial artery

Axoplasmic transport by neurosecretory nerve cells

Release of RHs, IHs, ADH, NE, NPY and other transmitters

Inferior hypo-physial artery

Portal veins

RH, IH, ADH

↓RH induce release of
↓anterior pituitary hormones
IH inhibit their release

Release of ADH, oxytocin

Pituitary veins

Anterior pituitary (adenohypophysis)

Posterior pituitary (neurohypophysis)

Anterior pituitary hormones

ACTH	PRL
GH	LH
TSH	α-MSH
FSH	β-endorphin

Posterior pituitary hormones
ADH, oxytocin

RH = Releasing hormones
IH = (Release-)Inhibiting hormones

Carbohydrate Metabolism and Pancreatic Hormones

Glucose is the *central energy carrier* of the human metabolism. The brain and red blood cells are fully glucose-dependent. The **plasma glucose concentration** (blood sugar level) is determined by the level of glucose *production* and *consumption*.

The following terms are important for proper understanding of carbohydrate metabolism (→ **A, C**):

1. Glycolysis generally refers to the anaerobic conversion of glucose to lactate (→ p. 72). This occurs in the red blood cells, renal medulla, and skeletal muscles (→ p. 72). Aerobic oxidation of glucose occurs in the CNS, heart, skeletal muscle and in most other organs.

2. Glycogenesis, i.e., the synthesis of glycogen from glucose (in liver and muscle), facilitates the storage of glucose and helps to maintain a constant plasma glucose concentration. Glycogen stored in a muscle can only be used by that muscle.

3. Glycogenolysis is the breakdown of glycogen to glucose, i.e., the opposite of glycogenesis.

4. Gluconeogenesis is the production of glucose (in liver and renal cortex) from non-sugar molecules such as amino acids (e.g., glutamine), lactate (produced by anaerobic glycolysis in muscles and red cells), and glycerol (from lipolysis).

5. Lipolysis is the breakdown of triacylglycerols into glycerol and free fatty acids.

6. Lipogenesis is the synthesis of triacylglycerols (for storage in fat depots).

Islets of Langerhans in the *pancreas* play a primary role in carbohydrate metabolism. Three cell types (A, B, D) have been identified so far (→ p. 273 B). 25% of all islet cells are type A (α) cells that produce **glucagon**, 60% are B (β) cells that synthesize **insulin**, and 10% are D (δ) cells that secrete somatostatin (**SIH**). These hormones mutually influence the synthesis and secretion of each other (→ p. 273 B). Islet cells in the pancreas head synthesize *pancreatic polypeptide*, the physiological function of which is not yet clear. High concentrations of these hormones reach the liver by way of the portal venous circulation.

Function. Pancreatic hormones (1) ensure that ingested food is stored as glycogen and fat (insulin); (2) mobilize energy reserves in response to food deprivation, physical activity or stress (glucagon and the non-pancreatic hormone epinephrine); (3) maintain the plasma glucose concentration as constant as possible (→ **A**); and (4) promote growth.

Insulin

Synthesis. Insulin is a 6 kDa peptide (51 amino acids, AA) formed by the C chain cleaved from *proinsulin* (84 AA), the precursor of which is *preproinsulin*, a preprohormone. Insulin contains two peptide chains (A and B) held together by disulfide bridges **Degradation**: Insulin has a half-life of about 5–8 min and is degraded mainly in liver and kidneys.

Secretion. Insulin is secreted in pulsatile bursts, mainly in response to *increases in the blood levels of glucose* (→ **B right**), as follows: plasma glucose ↑ → glucose in B cells ↑ → glucose oxidation ↑ → cytosolic ATP ↑ → closure of ATP-gated K⁺ channels → depolarization → opening of voltage-gated Ca^{2+} channels → cytosolic Ca^{2+} ↑. The rising Ca^{2+} in B cells leads to (a) *exocytosis* of insulin and (b) re-opening of K⁺ channels (deactivated by feedback control). **Stimulation.** Insulin secretion is stimulated mainly during food digestion via acetylcholine (vagus nerve), gastrin, secretin, GIP (→ p. 234) and GLP-1 (glucagon-like peptide = enteroglucagon), a peptide that dissociates from intestinal proglucagon. Certain amino acids (especially arginine and leucine), free fatty acids, many pituitary hormones and some steroid hormones also increase insulin secretion. **Inhibition.** Epinephrine and norepinephrine (α_2-adrenoceptors; → **A, B**), SIH (→ p. 273 B) and the neuropeptide *galanin* inhibit insulin secretion. When *hypoglycemia* occurs due, e.g., to fasting or prolonged physical exercise, the low blood glucose concentration is sensed by central chemosensors for glucose, leading to reflex activation of the sympathetic nervous system.

The **insulin receptor** is a heterotetramer ($\alpha_2\beta_2$) consisting of two extracellular α subunits and two transmembranous β subunits. The α subunits bind the hormone. Once the β subunits are autophosphorylated, they act as *receptor tyrosine kinases* that phosphorylate *insulin receptor substrate*-1 (IRS-1). Intracellular proteins with SH2 domains are phosphorylated by IRS-1 and pass on the signal (→ p. 277 C3).

Action of insulin (→ **A, B, C**). Insulin has *anabolic* and *lipogenic* effects, and promotes the *storage of glucose*, especially in the liver, where it activates enzymes that *promote glycolysis*

▶

A. Glucose metabolism (simplified overview)

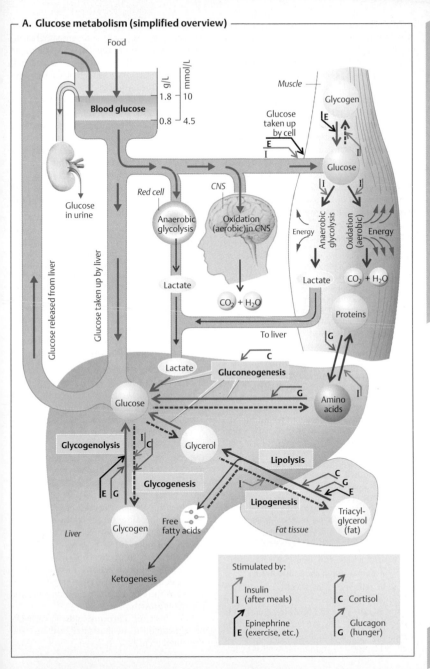

Plate 11.9 Carbohydrate Metabolism I

and *glycogenesis* and *suppresses* those involved in *gluconeogenesis*. Insulin also *increases the number of GLUT-4* uniporters in skeletal myocytes. All these actions serve to **lower the plasma glucose concentration** (which increases after food ingestion). About two-thirds of the glucose absorbed by the intestines after a meal (postprandial) is temporarily stored and kept ready for mobilization (via glucagon) during the interdigestive phase. This provides a relatively constant supply of glucose for the glucose-dependent CNS and vital organs in absence of food ingestion. Insulin increases the *storage of amino acids* (AA) in the form of proteins, especially in the skeletal muscles (*anabolism*). In addition, it promotes growth, *inhibits extrahepatic lipolysis* ($\rightarrow$ p. 257, D) and affects K^+ distribution ($\rightarrow$ p. 180).

Hypoglycemia develops when the insulin concentration is too high. Glucose levels of < 2 mmol/L (35 mg/dL) produce glucose deficiencies in the brain, which can lead to coma and *hypoglycemic shock*.

The **excessive intake of carbohydrates** can overload glycogen stores. The liver therefore starts to convert glucose into fatty acids, which are transported to and stored in fatty tissues in the form of *triacylglycerols* ($\rightarrow$ p. 257 D).

Diabetes mellitus (DM). One type of DM is *insulin-dependent diabetes mellitus* **(IDDM)**, or type 1 DM, which is caused by an insulin deficiency. Another type is *non-insulin-dependent DM* **(NIDDM)**, or type 2 DM, which is caused by the decreased efficacy of insulin and sometimes occurs even in conjunction with increased insulin concentrations. DM is characterized by an abnormally high plasma glucose concentration (*hyperglycemia*), which leads to *glucosuria* ($\rightarrow$ p. 158). Large quantities of *fatty acids* are liberated since lipolysis is no longer inhibited ($\rightarrow$ p. 257 D). The fatty acids can be used to produce energy via acetylcoenzyme A (acetyl-CoA); however, this leads to the formation of acetoacetic acid, acetone (*ketosis*), and β-oxybutyric acid (*metabolic acidosis*, $\rightarrow$ p. 142). Because hepatic fat synthesis is insulin-independent and since so many fatty acids are available, the liver begins to store triacylglycerols, resulting in the development of *fatty liver*.

Glucagon, Somatostatin and Somatotropin

Glucagon released from A cells is a peptide hormone (29 AA) derived from *proglucagon* (glicentin). The granules in which glucagon is stored are secreted by exocytosis. Secretion is **stimulated** by AA from digested proteins (especially alanine and arginine) as well as by hypoglycemia (e.g., due to fasting, prolonged physical exercise; $\rightarrow$ **B**) and sympathetic impulses (via β_2 adrenoceptors; $\rightarrow$ **A**). Glucagon secretion is **inhibited** by glucose and SIH ($\rightarrow$ p. 273, B) as well as by high plasma concentrations of free fatty acids.

The **actions** of glucagon ($\rightarrow$ **A, B, C**) (via cAMP; $\rightarrow$ p. 274) mainly antagonize those of insulin. Glucagon maintains a normal *blood glucose level between meals* and during phases of increased glucose consumption to ensure a constant energy supply. It does this (a) by increasing glycogenolysis (in liver not muscle) and (b) by stimulating gluconeogenesis from lactate, AA (protein degradation = catabolism) and glycerol (from *lipolysis*).

Increased plasma concentrations of **amino acids (AA)** stimulate insulin secretion which would lead to hypoglycemia without the simultaneous ingestion of glucose. Hypoglycemia normally does not occur, however, since AA also stimulate the release of glucagon, which increases the blood glucose concentration. Glucagon also stimulates gluconeogenesis from AA, so some of the AA are used for energy production. In order to increase protein levels in patients, glucose must therefore be administered simultaneously with therapeutic doses of AA to prevent their metabolic degradation.

Somatostatin (SIH). Like insulin, SIH stored in D cells (**SIH 14** has 14 AA) is released in response to increased plasma concentrations of glucose and arginine (i.e., after a meal). Through *paracrine* pathways (via G_i-linked receptors), SIH inhibits the release of insulin ($\rightarrow$ p. 273, B). Therefore, SIH inhibits not only the release of gastrin, which promotes digestion ($\rightarrow$ p. 243, B3), but also interrupts the insulin-related storage of nutrients. SIH also inhibits glucagon secretion ($\rightarrow$ p. 273 B). This effect does not occur in the presence of a glucose deficiency because of the release of catecholamines that *decrease* SIH secretion.

Somatotropin (STH) = growth hormone (GH). The short-term effects of GH are similar to those of insulin; its action is mediated by somatomedins ($\rightarrow$ p. 280). In the long-term, GH increases the blood glucose concentration and promotes growth.

The effects of **glucocorticoids** on carbohydrate metabolism are illustrated on plate **C** and explained on p. 296.

Plate 11.10 Carbohydrate Metabolism II

B. Hormonal control of blood glucose concentration

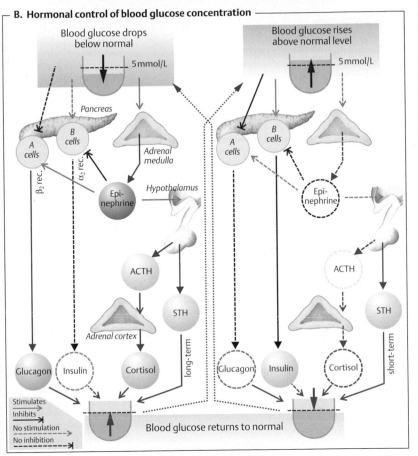

Blood glucose drops below normal

5 mmol/L

Pancreas

A cells

B cells

β₂ rec.

α₂ rec.

Adrenal medulla

Epi-nephrine

Hypothalamus

ACTH

STH

Adrenal cortex

Glucagon

Insulin

Cortisol

long-term

Blood glucose rises above normal level

5 mmol/L

A cells

B cells

Epi-nephrine

ACTH

STH

Glucagon

Insulin

Cortisol

short-term

Stimulates
Inhibits
No stimulation
No inhibition

Blood glucose returns to normal

C. Hormonal effects on carbohydrate and fat metabolism

Hormone **Function**	Insulin Satiated ← Buffer	Glucagon Buffer → Hungry	Epinephrine Stress, exercise	Cortisol Supply
Glucose Uptake by cell	**+** Muscle, fat		**+** Muscle	**−** Muscle, fat
Glycolysis	**+**	**−**	**+**	**−**
Gluconeogenesis (liver)	**−**	**+**	**+**	**+**
Glycogen Synthesis ⇄ Lysis	← Liver, muscle	→ Liver	→ Liver, muscle	← Liver
Fat Synthesis ⇄ Lysis	← Liver, fat	→ Fat	→ Fat	→ Fat

Thyroid Hormones

The thyroid gland contains spherical follicles (50–500 μm in diameter). Follicle cells synthesize the two iodine-containing thyroid hormones **thyroxine** (T_4, tetraiodothyronine) and **triiodothyronine** (T_3). T_3 and T_4 are bound to the glycoprotein *thyroglobulin* ($\rightarrow$ **B2**) and stored in the colloid of the follicles ($\rightarrow$ **A1, B1**). The synthesis and release of T_3/T_4 is controlled by the *thyroliberin* (= *thyrotropin-releasing hormone*, **TRH**)–*thyrotropin* (**TSH**) *axis* ($\rightarrow$ **A**, and p. 270ff.). T_3 and T_4 influence physical growth, maturation and metabolism. The *parafollicular cells* (*C cells*) of the thyroid gland synthesize *calcitonin* ($\rightarrow$ p. 292).

Thyroglobulin, a dimeric glycoprotein (660 kDa) is synthesized in the thyroid cells. TSH stimulates the transcription of the thyroglobulin gene. Thyroglobulin is stored in vesicles and released into the colloid by exocytosis ($\rightarrow$ **B1** and p. 30).

Iodine uptake. The iodine needed for hormone synthesis is taken up from the bloodstream as iodide (I^-). It enters thyroid cells through secondary active transport by the Na^+-I^- symport carrier (NIS) and is concentrated in the cell ca. 25 times as compared to the plasma ($\rightarrow$ **B2**). Via cAMP, **TSH** increases the transport capacity of basolateral I^- uptake up to 250 times. Other anions competitively inhibit I^- uptake; e.g., ClO_4^-, SCN^- and NO_2^-.

Hormone synthesis. I^- ions are continuously transported from the intracellular *I^- pool* to the apical (colloidal) side of the cell by a I^-/Cl^- antiporter, called *pendrin*, which is stimulated by **TSH**. With the aid of *thyroid peroxidase* (TPO) and an H_2O_2 generator, they are oxidized to elementary I_2^0 along the microvilli on the colloid side of the cell membrane. With the help of TPO, the I^0 reacts with about 20 of the 144 tyrosyl residues of *thyroglobulin* ($\rightarrow$ **C**). The phenol ring of the tyrosyl residue is thereby iodinated at the 3 and/or 5 position, yielding a protein chain containing either *diiodotyrosine (DIT)* residues and/or *monoiodotyrosine (MIT)* residues. These steps of synthesis are stimulated by TSH (via IP_3) and inhibited by thiouracil, thiocyanate, glutathione, and other reducing substances. The structure of the thyroglobulin molecule allows the iodinated tyrosyl residues to react with each other in the thyrocolloid. The phenol ring of one DIT (or MIT) molecule links with another DIT molecule (ether bridges). The resulting thyroglobulin chain now contains *tetraiodothyronine residues* and (less) *triiodothyronine* residues ($\rightarrow$ **C**). These are the *storage* form of T_4 and T_3.

TSH also stimulates T_3 and T_4 secretion. The iodinated thyroglobulin in thyrocolloid are reabsorbed by the cells via endocytosis ($\rightarrow$ **B3, C**). The endosomes fuse with primary lysosomes to form phagolysosomes in which thyroglobulin is hydrolyzed by proteases. This leads to the release of T_3 and T_4 (ca. 0.2 and 1–3 mol per mol of thyroglobulin, respectively). T_3 and T_4 are then secreted into the bloodstream ($\rightarrow$ **B3**). With the aid of *deiodase*, I^- meanwhile is split from concomitantly released MIT and DIT and becomes reavailable for synthesis.

Control of T_3/T_4 secretion. TSH secretion by the anterior pituitary is stimulated by **TRH**, a hypothalamic tripeptide ($\rightarrow$ p. 280) and inhibited by **somatostatin (SIH)** ($\rightarrow$ **A** and p. 270). The effect of TRH is modified by T_4 in the plasma. As observed with other target cells, the T_4 taken up by the thyrotropic cells of the anterior pituitary is converted to T_3 by *5′-deiodase*. T_3 reduces the density of TRH receptors in the pituitary gland and inhibits TRH secretion by the hypothalamus. The secretion of TSH and consequently of T_3 and T_4 therefore decreases (*negative feedback circuit*). In neonates, cold seems to stimulate the release of TRH via neuronal pathways (thermoregulation, $\rightarrow$ p. 224). **TSH** is a heterodimer (26 kDa) consisting of an α subunit (identical to that of LH and FSH) and a β subunit. *TSH controls all thyroid gland functions*, including the uptake of I^-, the synthesis and secretion of T_3 and T_4 ($\rightarrow$ **A–C**), the blood flow and growth of the thyroid gland.

Goiter (struma) is characterized by diffuse or nodular enlargement of the thyroid gland. Diffuse goiter can occur due to an iodine deficiency, resulting in T_3/T_4 deficits that ultimately lead to increased secretion of TSH. Chronic elevation of TSH leads to the proliferation of follicle cells, resulting in goiter development (*hyperplastic goiter*). This prompts an increase in T_3/T_4 synthesis, which sometimes normalizes the plasma concentrations of T_3/T_4 (*euthyroid goiter*). This type of goiter often persists even after the iodide deficiency is rectified.

▶

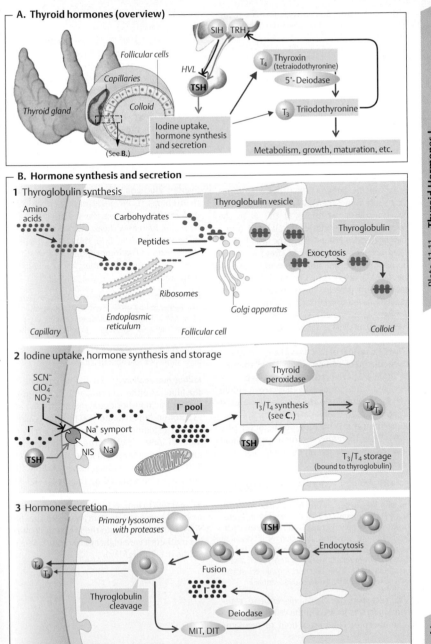

A. Thyroid hormones (overview)

Thyroid gland

Follicular cells

Capillaries

Colloid

(See **B.**)

SIH TRH

HVL

TSH

T_4 Thyroxin (tetraiodothyronine)

5'-Deiodase

T_3 Triiodothyronine

Iodine uptake, hormone synthesis and secretion

Metabolism, growth, maturation, etc.

B. Hormone synthesis and secretion

1 Thyroglobulin synthesis

Amino acids

Carbohydrates

Peptides

Thyroglobulin vesicle

Thyroglobulin

Exocytosis

Ribosomes

Endoplasmic reticulum

Golgi apparatus

Capillary

Follicular cell

Colloid

2 Iodine uptake, hormone synthesis and storage

SCN^-
ClO_4^-
NO_2^-

I^-

Na^+ symport

NIS Na^+

TSH

I^- pool

Thyroid peroxidase

T_3/T_4 synthesis (see **C.**)

T_4 T_3

TSH

T_3/T_4 storage (bound to thyroglobulin)

3 Hormone secretion

Primary lysosomes with proteases

TSH

Endocytosis

T_4
T_3

Fusion

Thyroglobulin cleavage

I^-

Deiodase

MIT, DIT

Plate 11.11   Thyroid Hormones I

287

Hypothyroidism occurs when TSH-driven thyroid enlargement is no longer able to compensate for the T_3/T_4 deficiency (*hypothyroid goiter*). This type of goiter can also occur due to a congenital disturbance of T_3/T_4 synthesis (see below) or thyroid inflammation. **Hyperthyroidism** occurs when a thyroid tumor (*hot node*) or diffuse struma (e.g., in *Grave's disease*) results in the overproduction of T_3/T_4, independent of TSH. In the latter case, an autoantibody against the TSH receptor binds to the TSH receptor. Its effects mimic those of TSH, i.e., it stimulates T_3/T_4 synthesis and secretion.

T_3/T_4 transport. T_3 and T_4 occur at a ratio of $1:40$ in the plasma, where $>99\%$ of them (mainly T_4) are bound to plasma proteins: *thyroxine-binding globulin* (**TBG**), thyroxine-binding prealbumin (**TBPA**), and *serum albumin*. TBG transports two-thirds of the T_4 in the blood, while TBPA and serum albumin transport the rest. Less than 0.3% of the total T_3/T_4 in blood occurs in an unbound (free) form, although only the unbound molecules have an effect on the target cells. Certain drugs split T_3 and T_4 from protein bonds, resulting in increased plasma concentrations of the free hormones.

Potency of T_3/T_4. T_3 is 3–8 times more potent than T_4 and acts more rapidly (half-life of T_3 is 1 day, that of T_4 7 days). Only ca. 20% of all circulating T_3 originate from the thyroid; the other 80% are produced by the liver, kidneys, and other target cells that cleave iodide from T_4. The conversion of T_4 to T_3 is catalyzed by microsomal **5′-deiodase**, which removes iodine from the 5′ position on the outer ring ($\rightarrow$ **D**). T_3 is therefore the more potent hormone, while T_4 is mainly ascribed a *storage function* in plasma.

The inactive form of T_3 called **reverse T_3 (rT_3)** is produced from T_4 when the iodine is split from the inner ring with the aid of a 5- (not 5′-)deiodase. Approximately equal amounts of T_3 and rT_3 are normally produced in the periphery (ca. 25 µg/day). When a person fasts, the resulting *inhibition of 5′-deiodase* decreases T_3 synthesis (to *save energy*, see below) while rT_3 synthesis increases. *Pituitary 5′-deiodase* is not inhibited, so TSH secretion (unwanted in this case) is suppressed by the negative feedback.

T_3/T_4 receptors are *hormone-sensitive transcription factors* located in the cell nuclei. Hormone–receptor complexes bind to regulator proteins of certain genes in the nuclei and influence their transcription.

The **actions of T_3/T_4** are numerous and mainly involve the *intermediate metabolism*. The thyroid hormones increase the number of *mitochondria* and its cristae, increase Na^+-K^+-ATPase activity and modulate the cholesterol metabolism. This results in an increase in *energy turnover* and a corresponding rise in O_2 *consumption* and *heat production*. T_3 also specifically stimulates heat production by increasing the expression of the uncoupling protein *thermogenin* in brown fat ($\rightarrow$ p.222). T_3 also influences the efficacy of other hormones. Insulin, glucagon, GH and epinephrine lose their energy turnover-increasing effect in *hypothyroidism*, whereas the sensitivity to epinephrine increases (heart rate increases, etc.) in *hyperthyroidism*. T_3 is thought to increase the density of β-adrenoceptors. T_3 also stimulates *growth* and *maturation*, especially of the brain and bones.

Cretinism occurs due to neonatal T_3/T_4 deficiencies and is marked by growth and maturation disorders (dwarfism, delayed sexual development, etc.) and central nervous disorders (intelligence deficits, seizures, etc.). The administration of thyroid hormones in the first six months of life can prevent or reduce some of these abnormalities.

Iodine metabolism ($\rightarrow$ **D**). Iodine circulates in the blood as either (1) inorganic I^- (2–10 µg/L), (2) organic non-hormonal iodine (traces) and (3) protein-bound iodine (PBI) within T_3 and T_4 (35–80 µg iodine/L). The average daily requirement of iodine is ca. 150 µg; larger quantities are required in fever and hyperthyroidism (ca. 250–500 µg/day). Iodine excreted from the body must be replaced by the diet ($\rightarrow$ **D**). Sea salt, seafood, and cereals grown in iodine-rich soil are rich in iodine. Iodized salt is often used to supplement iodine deficiencies in the diet. Since iodine passes into the breast milk, nursing mothers have a higher daily requirement of iodine (ca. 200 µg/day).

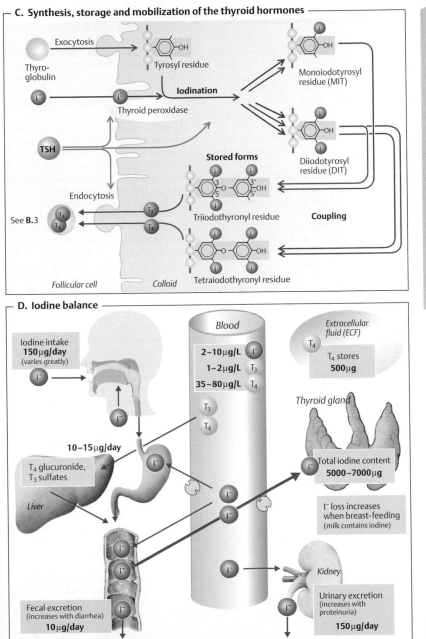

C. Synthesis, storage and mobilization of the thyroid hormones

Thyro-globulin

Exocytosis

Tyrosyl residue

I^-

Iodination

Thyroid peroxidase

Monoiodotyrosyl residue (MIT)

Diiodotyrosyl residue (DIT)

TSH

Endocytosis

Stored forms

See **B.3**

T_3
T_4

T_3

Triiodothyronyl residue

T_4

Coupling

Tetraiodothyronyl residue

Follicular cell   *Colloid*

Plate 11.12 Thyroid Hormones II

D. Iodine balance

Blood

Extracellular fluid (ECF)

Iodine intake
150 µg/day
(varies greatly)

I^-

$2-10$ µg/L I^-
$1-2$ µg/L T_3
$35-80$ µg/L T_4

T_3
T_4

T_4

T_4 stores
500 µg

Thyroid gland

$10-15$ µg/day

T_4 glucuronide,
T_3 sulfates

Liver

I^-

I^-

I^-

Total iodine content
5000–7000 µg

I^- loss increases
when breast-feeding
(milk contains iodine)

Fecal excretion
(increases with diarrhea)
10 µg/day

I^-

I^-

I^-

I^-

Kidney

Urinary excretion
(increases with
proteinuria)
150 µg/day

Calcium and Phosphate Metabolism

Calcium, particularly ionized calcium (Ca^{2+}), plays a central role in the regulation of numerous cell functions ($\rightarrow$ pp. 36, 62ff., 192, 276). Calcium accounts for 2% of the body weight. Ca. 99% of the calcium occurs in bone while 1% is dissolved in body fluids. The total calcium conc. in serum is normally 2.1–2.6 mmol/L. Ca. 50% of it is **free Ca^{2+} (1.1–1.3 mmol/L)** while ca. 10% is bound in complexes and 40% is bound to proteins (mainly albumin; $\rightarrow$ p. 178). **Calcium protein binding** increases as the pH of the blood rises since the number of Ca^{2+} binding sites on protein molecules also rises with the pH. The Ca^{2+} conc. accordingly *decreases in alkalosis* and rises in acidosis (by about 0.21 mmol/L Ca^{2+} per pH unit). Alkalosis (e.g., due to hyperventilation) and hypocalcemia (see below) can therefore lead to *tetany*.

The **calcium metabolism** is tightly regulated to ensure a balanced intake and excretion of Ca^{2+} ($\rightarrow$ **A**). The *dietary intake of Ca^{2+}* provides around 12–35 mmol of Ca^{2+} each day (1 mmol = 2 mEq = 40 mg). Milk, cheese, eggs and "hard" water are particularly rich in Ca^{2+}. When calcium homeostasis is maintained, most of the ingested Ca^{2+} is excreted in the feces, while the remainder is excreted in the urine ($\rightarrow$ p. 178). When a calcium deficiency exists, up to 90% of the ingested Ca^{2+} is absorbed by the intestinal tract ($\rightarrow$ **A** and p. 262).

Pregnant and nursing mothers have higher Ca^{2+} requirements because they must also supply the fetus or newborn infant with calcium. The fetus receives ca. 625 mmol/day of Ca^{2+} via the placenta, and nursed infants receive up to 2000 mmol/day via the breast milk. In both cases, the Ca^{2+} is used for bone formation. Thus, many women develop a Ca^{2+} deficiency during or after pregnancy.

Phosphate metabolism is closely related to calcium metabolism but is less tightly controlled. The daily intake of phosphate is about 1.4 g; 0.9 g of intake is absorbed and usually excreted by the kidneys ($\rightarrow$ p. 178). The **phosphate concentration** in serum normally ranges from 0.8–1.4 mmol/L.

Calcium phosphate salts are sparingly soluble. When the product of Ca^{2+} conc. times phosphate conc. (*solubility product*) exceeds a certain threshold, calcium phosphate starts to precipitate in solutions, and the deposition of calcium phosphate salts occurs. The salts are chiefly deposited in the bone, but can also precipitate in other organs. The infusion of phosphate leads to a decrease in the serum calcium concentration since calcium phosphate accumulates in bone. Conversely, hypophosphatemia leads to hypercalcemia (Ca^{2+} is released from bone).

Hormonal control. Calcium and phosphate homeostasis is predominantly regulated by *parathyroid hormone* and *calcitriol*, but also by calcitonin to a lesser degree. These hormones mainly affect three organs: the *intestines*, the *kidneys* and the *bone* ($\rightarrow$ **B** and **D**).

Parathyrin or parathyroid hormone (PTH) is a peptide hormone (84 AA) secreted by the *parathyroid glands*. **Ca^{2+} sensors** in cells of the *parathyroid glands* regulate PTH synthesis and secretion in response to changes in the plasma concentration of *ionized Ca^{2+}* ($\rightarrow$ p. 36). More PTH is secreted into the bloodstream whenever the Ca^{2+} conc. falls below normal (*hypocalcemia*). Inversely, PTH secretion decreases when the Ca^{2+} level rises ($\rightarrow$ **D**, left panel). The primary function of PTH is to **normalize decreased Ca^{2+}** conc. in the blood ($\rightarrow$ **D**). This is accomplished as follows: (1) PTH activates osteoclasts, resulting in bone breakdown and the *release of Ca^{2+}* (and phosphate) from the **bone**; (2) PTH accelerates the final step of calcitriol synthesis in the kidney, resulting in *increased reabsorption of Ca^{2+}* from the **gut**; (3) in the **kidney,** PTH increases calcitriol synthesis and Ca^{2+} reabsorption, which is particularly important due to the increased Ca^{2+} supply resulting from actions (1) and (2). PTH also *inhibits renal phosphate reabsorption* ($\rightarrow$ p. 178), resulting in hypophosphatemia. This, in turn, stimulates the release of Ca^{2+} from the bone or prevents the precipitation of calcium phosphate in tissue (solubility product; see above).

Hypocalcemia occurs due to a deficiency (*hypoparathyroidism*) or lack of efficiency (*pseudohypoparathyroidism*) of PTH, which can destabilize the resting potential enough to produce *muscle spasms* and *tetany*. These deficiencies can also lead to a secondary calcitriol deficiency. An excess of PTH (*hyperparathyroidism*) and malignant osteolysis overpower the Ca^{2+} control mechanisms, leading to **hypercalcemia**. The long-term elevation of Ca^{2+} results in cal-

▶

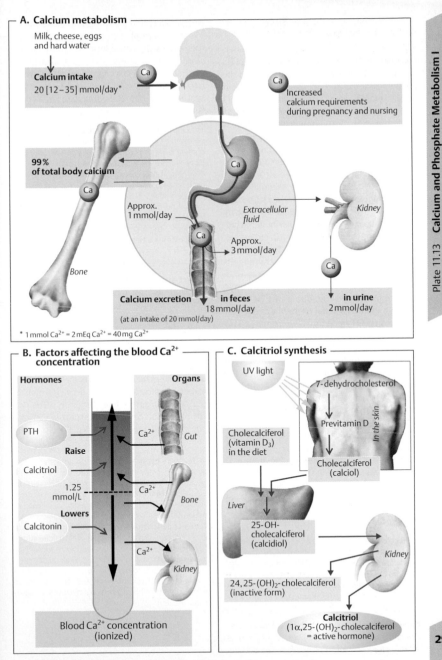

Plate 11.13 Calcium and Phosphate Metabolism I

A. Calcium metabolism

Milk, cheese, eggs and hard water

Calcium intake
20 [12–35] mmol/day*

Increased calcium requirements during pregnancy and nursing

99% of total body calcium

Bone

Approx. 1 mmol/day

Extracellular fluid

Approx. 3 mmol/day

Kidney

Calcium excretion   in feces
18 mmol/day
(at an intake of 20 mmol/day)

in urine
2 mmol/day

* 1 mmol Ca²⁺ = 2 mEq Ca²⁺ = 40 mg Ca²⁺

B. Factors affecting the blood Ca²⁺ concentration

Hormones

Organs

PTH

Ca²⁺ *Gut*

Raise

Calcitriol

Ca²⁺ *Bone*

1.25 mmol/L

Lowers

Calcitonin

Ca²⁺ *Kidney*

Blood Ca²⁺ concentration (ionized)

C. Calcitriol synthesis

UV light

7-dehydrocholesterol

Previtamin D

In the skin

Cholecalciferol (vitamin D₃) in the diet

Cholecalciferol (calciol)

Liver

25-OH-cholecalciferol (calcidiol)

Kidney

24,25-(OH)₂-cholecalciferol (inactive form)

Calcitriol
(1α,25-(OH)₂-cholecalciferol = active hormone)

▶ cium deposition (e.g., in the kidneys). Ca^{2+} conc. exceeding 3.5 mmol/L lead to coma, renal insufficiency and cardiac arrhythmias.

Calcitonin (CT), or *thyrocalcitonin*, is a peptide hormone (32). It is mainly synthesized in the parafollicular cells (C cells) of the thyroid gland, which also contain Ca^{2+} sensors ($\rightarrow$ p. 36). *Hypercalcemia* increases the plasma calcitonin conc. ($\rightarrow$ **D, right** panel), whereas calcitonin can no longer be detected when the calcium conc. [Ca^{2+}] falls below 2 mmol/L. *Calcitonin normalizes elevated serum Ca^{2+} conc.* mainly by acting on **bone**. Osteoclast activity is inhibited by calcitonin (and stimulated by PTH). Calcitonin therefore increases the uptake of Ca^{2+} by the bone—at least temporarily ($\rightarrow$ **D5**). Some *gastrointestinal hormones* accelerate calcitonin secretion, thereby enhancing the postprandial absorption of Ca^{2+} by bone. These effects (and perhaps the restraining effect of calcitonin on digestive activities) function to prevent postprandial hypercalcemia and the (unwanted) inhibition of PTH secretion and increased renal excretion of the just absorbed Ca^{2+}. Calcitonin also acts on the kidneys ($\rightarrow$ **D6**).

Calcitriol (1,25-$(OH)_2$-cholecalciferol) is a lipophilic, steroid-like hormone synthesized as follows ($\rightarrow$ **C**): **Cholecalciferol** (vitamin D_3) is produced from hepatic *7-dehydrocholesterol* in the **skin** via an intermediate product (*previtamin D*) in response to UV light (sun, tanning lamps). Both substances bind to *vitamin D-binding protein* (DBP) in the blood, but cholecalciferol is preferentially transported because of its higher affinity. Previtamin D therefore remains in the skin for a while after UV light exposure (short-term storage). Calcidiol (25-OH-cholecalciferol) and calcitriol bind to DBP. An estrogen-dependent rise in DBP synthesis occurs during pregnancy.

Cholecalciferol (vitamin D_3) is administered to compensate for inadequate UV exposure. The recommended daily dosage in children is approximately 400 units = 10 µg; adults receive half this amount. Plant-derived **ergocalciferol** (vitamin D_2) is equally effective as animal-derived vitamin D_3. The following actions apply for both forms.

Cholecalciferol is converted to **calcidiol** (25-OH-cholecalciferol) in the liver. Vitamin D is mainly *stored as calcidiol* because the plasma conc. of calcidiol is 25 µg/L, and its half-life is 15 days. **Calcitriol** (1,25-$(OH)_2$-cholecalciferol), the hormonally active form, is mainly synthesized in the **kidneys** ($\rightarrow$ C), but also in the placenta. The plasma conc. of **calcitriol is regulated** by renal 1-α-hydroxylase (final step of synthesis) and by 24-hydroxylase, an enzyme that deactivates calcitriol.

The **calcitriol concentration rises** in response to *hypocalcemia*-related **PTA** secretion ($\rightarrow$ **D2**), to *phosphate deficiency* and to *prolactin* (lactation). All three inhibit 24-hydroxylase and activate 1-α-hydroxylase. It **decreases** due to several negative feedback loops, i.e. due to the fact that calcitriol (a) directly inhibits 1-α-hydroxylase, (b) inhibits parathyroid hormone secretion, and (c) normalizes the (decreased) plasma conc. of Ca^{2+} and phosphate by increasing the intestinal absorption of Ca^{2+} and phosphate (see below). Calcium and phosphate inhibit 1-α-hydroxylase, while phosphate activates 24-hydroxylase.

Target organs. Calcitriol's primary target is the **gut**, but it also acts on the *bone, kidneys, placenta, mammary glands, hair follicles, skin* etc. It binds with its nuclear receptor and induces the expression of calcium-binding protein and Ca^{2+}-ATPase ($\rightarrow$ pp. 278 and 36). Calcitriol has also genomic effects. Calcitriol *increases the intestinal absorption of Ca^{2+}* ($\rightarrow$ **D4**) and *promotes mineralization of the bone*, but an **excess of calcitriol** leads to *decalcification* of the bone, an effect heightened by PTH. Calcitriol also increases the transport of Ca^{2+} and phosphate at the kidney ($\rightarrow$ p. 178), placenta and mammary glands.

In transitory **hypocalcemia**, the bones act as a temporary Ca^{2+} buffer ($\rightarrow$ **D**) until the Ca^{2+} deficit has been balanced by a calcitriol-mediated increase in Ca^{2+} absorption from the gut. If too little calcitriol is available, skeletal demineralization will lead to **osteomalacia** in adults and **rickets** in children. **Vitamin D deficiencies** are caused by inadequate dietary intake, reduced absorption (fat maldigestion), insufficient UV light exposure, and/or reduced 1-α-hydroxylation (renal insufficiency). Skeletal demineralization mostly occurs due to the prolonged increase in parathyroid hormone secretion associated with chronic hypocalcemia (*compensatory hyperparathyroidism*).

D. Hormonal regulation of the blood Ca²⁺ concentration

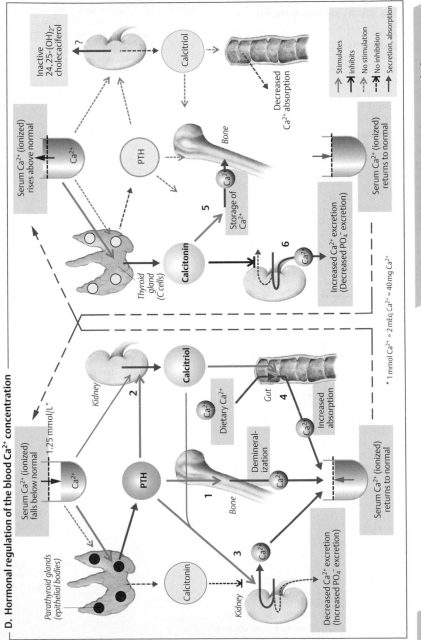

Inactive 24,25-(OH)₂-cholecalciferol

Calcitriol

Decreased Ca²⁺ absorption

Stimulates
Inhibits
No stimulation
No inhibition
Secretion, absorption

Serum Ca²⁺ (ionized) rises above normal

Ca²⁺

PTH

Bone

Calcitonin

Storage of Ca²⁺

5

6

Ca²⁺

Thyroid gland (C cells)

Increased Ca²⁺ excretion (Decreased PO₄⁻ excretion)

Serum Ca²⁺ (ionized) returns to normal

* 1 mmol Ca²⁺ = 2 mEq Ca²⁺ = 40 mg Ca²⁺

Kidney

Calcitriol

1.25 mmol/L*

Ca²⁺

PTH

Calcitonin

Decreased Ca²⁺ excretion (Increased PO₄⁻ excretion)

Kidney

3

Ca²⁺

Serum Ca²⁺ (ionized) falls below normal

Parathyroid glands (epithelial bodies)

2

Dietary Ca²⁺

Ca²⁺

Gut

4

Ca²⁺

Increased absorption

Demineral-ization

Ca²⁺

Bone

1

Serum Ca²⁺ (ionized) returns to normal

Plate 11.14   Calcium and Phosphate Metabolism II

293

Biosynthesis of Steroid Hormones

Cholesterol is the precursor of steroid hormones ($\rightarrow$ **A**). Cholesterol is mainly synthesized in the **liver**. It arises from acetyl-coenzyme A (acetyl-CoA) via a number of intermediates (e.g., squalene, lanosterol) and is transported to the endocrine glands by lipoproteins ($\rightarrow$ p. 256). Cholesterol can be synthesized de novo also in the adrenal cortex, but not in the placenta ($\rightarrow$ p. 304). Since only small quantities of steroid hormones are stored in the organs of origin, i.e., the adrenal cortex, ovaries, testes and placenta ($\rightarrow$ p. 304), they must be synthesized from the cellular cholesterol pool as needed.

Cholesterol contains 27 carbon atoms. **Pregnenolone** (21 C atoms; $\rightarrow$ **A, a**), the precursor of steroid hormones, arises from cholesterol via numerous intermediates. Pregnenolone also yields **progesterone** ($\rightarrow$ **A, b**), which is not only a potent hormone itself (female sex hormone; $\rightarrow$ p. 298ff.), but can act as the precursor of all other steroid hormones, i.e., (1) the adrenocortical hormones with 21 carbon atoms ($\rightarrow$ **A**, yellow and orange fields); (2) male sex hormones (androgens, 19 carbon atoms) synthesized in the testes ($\rightarrow$ p. 306), ovaries and adrenal cortex ($\rightarrow$ **A**, green and blue fields); and (3) female sex hormones (*estrogens*, 18 carbon atoms; $\rightarrow$ p. 29 B ff.) synthesized in the ovaries ($\rightarrow$ **A**, red zones).

The precursors for steroid hormone synthesis are present in all steroid hormone glands. The type of hormone produced and the site of hormone synthesis depend on (1) the type of receptors available for the superordinate control hormones (ACTH, FSH, LH, etc.) and (2) the dominant enzyme responsible for changing the structure of the steroid molecule in the hormone-producing cells of the gland in question. The **adrenal cortex** contains 11-, 17- and 21-hydroxylases—enzymes that introduce an OH group at position C21, C17 or C11, respectively, of the steroid molecule ($\rightarrow$ **A**, top left panel for numerical order). Hydroxylation at C21 ($\rightarrow$ **A, c**)—as realized in the *glomerular zone* of the adrenal cortex—makes the steroid insensitive to the effects of 17-hydroxylase. As a result, only **mineralocorticoids** like corticosterone and aldosterone (**A, d $\Rightarrow$ e**; see also

p. 182) can be synthesized. Initial hydroxylation at C17 ($\rightarrow$ **A, f** or **g**) results in the synthesis of **glucocorticoids**—realized mainly in the *fascicular zone* of the adrenal cortex ($\rightarrow$ **A, h $\Rightarrow$ j $\Rightarrow$ k**)—and **17-ketosteroids**, steroids with a keto group at C17 ($\rightarrow$ **A, l** and **m**). Glucocorticoids and 17-ketosteroids can therefore be synthesized from 17α-hydroxypregnenolone without the aid of progesterone ($\rightarrow$ **A, n $\Rightarrow$ h $\Rightarrow$ j**).

The **estrogens** ($\rightarrow$ p. 302) *estrone* and *estradiol* can be directly or indirectly synthesized from 17-ketosteroids ($\rightarrow$ **A, o $\Rightarrow$ p**); they are produced indirectly by way of testosterone ($\rightarrow$ **A, q $\Rightarrow$ r $\Rightarrow$ p**). The true active substance of certain target cells for androgens (e.g., in the prostate) is either *dihydrotestosterone* or *estradiol* ; both are synthesized from testosterone ($\rightarrow$ **A,s** and **A,r**, respectively).

17-ketosteroids are synthesized by the gonads (testes and ovaries) and adrenal cortex. Since they are found in the urine, the *metyrapone test* of pituitary function is used to assess the *ACTH reserve* based on urinary 17-ketosteroids levels. ACTH secretion is normally subject to feedback control by glucocorticoids ($\rightarrow$ p. 296). Metyrapone inhibits 11-hydroxylase activity ($\rightarrow$ **A, d** and **j**), which leaves ACTH unsuppressed in healthy subjects. Urinary 17-ketosteroid levels should therefore increase after metyrapone administration. An abnormality of ACTH secretion can be assumed when this does not occur in patients with a healthy adrenal cortex.

Degradation of steroid hormones occurs mainly in the **liver**. Their OH groups are usually linked to *sulfate* or *glucuronic acid* molecules and are ultimately excreted in the bile or urine ($\rightarrow$ pp. 160, 183 and 250). The chief urinary metabolite of the estrogens is *estriol*, while that of the gestagens (mainly progesterone and 17α-hydroxyprogesterone) is *pregnanediol* ($\rightarrow$ p. 304). Pregnanediol levels in urine can be measured to confirm or exclude *pregnancy test* (pregnanediol test). Chronically increased estrogen levels due, for example, to decreased estrogen degradation secondary to liver damage, can lead to breast development (*gynecomastia*) in the male, among other things. For normal estrogen ranges, see table on p. 302.

Plate 11.15 Biosynthesis of Steroid Hormones

A. Biosynthesis of steroid hormones

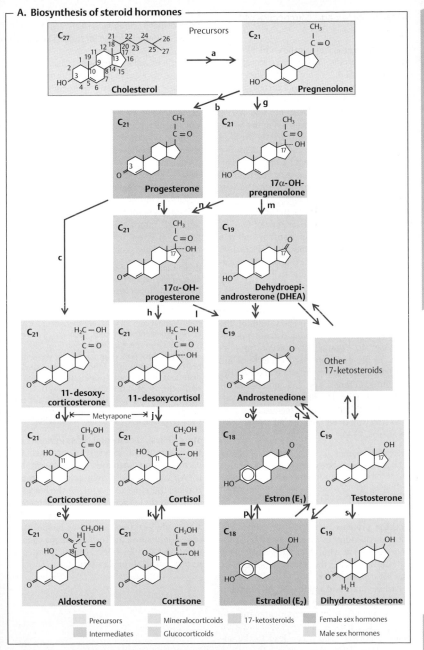

Precursors	Mineralocorticoids	17-ketosteroids	Female sex hormones
Intermediates	Glucocorticoids		Male sex hormones

Adrenal Cortex and Glucocorticoid Synthesis

The **mineralocortico(stero)ids** *aldosterone*, corticosterone and 11-desoxycorticosterone (→ pp. 182ff. and 294) are synthesized in the *glomerular zone* of the **adrenal cortex** (→ **A1**), whereas the **glucocortico(stero)ids** *cortisol* (hydrocortisone) and cortisone (→ p. 294, small quantities) are synthesized in the *fascicular zone* (→ **A2**). **Androgens** are synthesized in the *reticular zone* of the adrenal cortex (→ **A3**). One of the androgens is *dehydroepiandrosterone* (DHEA), which is used (partly in its sulfated form, DHEA-S) to synthesize various sex hormones in other tissues (→ p. 304).

Cortisol transport. Most of the plasma cortisol is bound to *transcortin*, or *cortisol-binding globulin* (CBG), a specific transport protein with a high-affinity binding site for cortisol. Cortisol is released in response to conformational changes of CBG due to inflammation etc.

CRH and **ACTH** regulate cortisol synthesis and secretion (→ **A4, A5**; see also p. 270). ACTH ensures also structural preservation of the adrenal cortex and supplies cortisol precursors, e.g., by forming cholesterol from its esters, by de novo synthesis of cholesterol and by converting it to progesterone and 17α-hydroxyprogesterone (→ pp. 256 and 294). *ACTH secretion* is stimulated by *CRH* and epinephrine and inhibited (negative feedback control) by *cortisol* with or without the aid of CRH (→ **A**; see also p. 273 A).

A **circadian rhythm** of CRH secretion and thus of ACTH and cortisol secretion can be observed. The peak secretion is in the morning (→ **B**, mean values). Continuous hormone conc. sampling at short intervals have shown that ACTH and cortisol are secreted in 2–3-hour *episodes* (→ **B**).

Receptor proteins (→ p. 278) for glucocorticoids can be found in virtually every cell. Glucocorticoids are vital hormones that exert numerous **effects**, the most important of which are listed below.

Carbohydrate and amino acid (AA) metabolism (see also pp. 283 A and 285 C): Cortisol uses AA derived from protein degradation to increase the plasma glucose concentration (*gluconeogenesis*), which can lead to the so-called *steroid diabetes* in extreme cases. Thus, cortisol has a *catabolic* effect (degrades proteins) that results in the increased excretion of urea.

Cardiovascular function: Glucocorticoids increase myocardial contractility and vasoconstriction due to enhancement of catecholamine effects (→ pp. 194 and 214). These are described as *permissive effects* of cortisol. Cortisol increases the synthesis of epinephrine in the adrenal medulla (→ **A6**) and of angiotensinogen in the liver (→ p. 184).

Especially when administered at high doses, glucocorticoids induce **anti-inflammatory and anti-allergic effects** because they stabilize lymphokine synthesis and histamine release (→ p. 100). On the other hand, interleukin-1, interleukin-2 and TNF-α (e.g., in severe infection) leads to increased secretion of CRH and high cortisol conc. (see below).

Renal function: Glucocorticoids delay the *excretion of water* and help to maintain a normal glomerular filtration rate. They can react also with aldosterone receptors but are converted to cortisone by *11β-hydroxysteroid oxidoreductase* in aldosterone target cells. Normal cortisol conc. are therefore ineffective at the aldosterone receptor. High conc., however, have the same effect as aldosterone (→ p. 182).

Gastric function: Glucocorticoids weaken the protective mechanisms of the gastric mucosa. Thus, high-dose glucocorticoids or stress (see below) increase the risk of gastric ulcers (→ p. 242).

Cerebral function: High glucocorticoid conc. change hypothalamic (→ **A**) and electrical brain activity (EEG) and lead to psychic abnormalities.

Stress: Physical or mental stress increases cortisol secretion as a result of increased CRH secretion and increased sympathetic tone (→ **A**). Many of the aforementioned effects of cortisol therefore play a role in the body's response to stress (activation of energy metabolism, increase in cardiac performance, etc.). In severe physical (e.g., sepsis) or mental stress (e.g., depression), the cortisol plasma conc. remains at a very high level (up to 10 times the normal value) throughout the day.

Plate 11.16 Adrenal Cortex and Glucocorticoid Synthesis

A. Adrenal gland

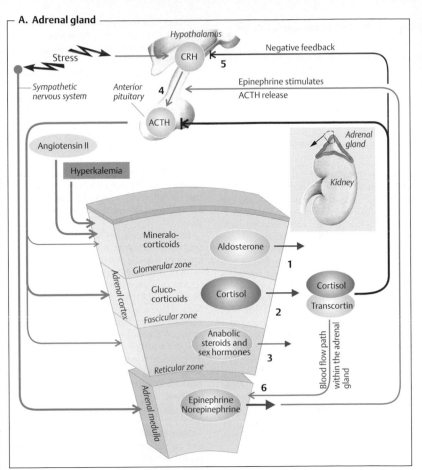

Stress

Hypothalamus

CRH 5

Negative feedback

Sympathetic nervous system

Anterior pituitary

4

Epinephrine stimulates ACTH release

ACTH

Angiotensin II

Hyperkalemia

Adrenal gland

Kidney

Mineralo-corticoids — Aldosterone 1

Glomerular zone

Gluco-corticoids — Cortisol 2

Cortisol

Transcortin

Fascicular zone

Adrenal cortex

Anabolic steroids and sex hormones 3

Reticular zone

Blood flow path within the adrenal gland

Adrenal medulla

Epinephrine Norepinephrine 6

B. Circadian rhythm of ACTH and cortisol secretion

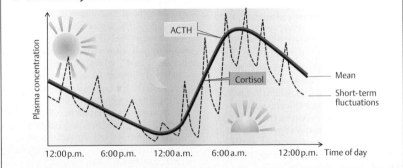

Plasma concentration

ACTH

Cortisol

Mean

Short-term fluctuations

12:00 p.m.   6:00 p.m.   12:00 a.m.   6:00 a.m.   12:00 p.m.   Time of day

Plate 11.16 Adrenal Cortex and Glucocorticoid Synthesis

Oogenesis and the Menstrual Cycle

Oogenesis. The development of the female gametes (ova) extends from the oogonium stage to the primary oocyte stage (in the primordial follicle), starting long before birth. Oogenesis therefore occurs much sooner than the corresponding stages of spermatogenesis ($\rightarrow$ p. 306). The fetal phase of oogenesis is completed by the first week of gestation; these oocytes remain latent until puberty. In the sexually mature female, a fertilizable ovum develops in the graafian follicles approximately every 28 days.

Menstrual cycle. After the start of sexual maturation, a woman starts to secrete the following hormones in a cyclic (approx.) 28-day rhythm ($\rightarrow$ **A1, A2**). *Gonadoliberin* (= *gonadotropin-releasing hormone*, Gn-RH) and *dopamine* (PIH) are secreted by the hypothalamus. *Follicle-stimulating hormone* (FSH), *luteinizing hormone* (LH) and *prolactin* (PRL) are released by the anterior pituitary. *Progesterone, estrogens* (chiefly estradiol, E_2) and *inhibin* are secreted by the ovaries. Gn-RH controls the *pulsatile secretion* of FSH and LH ($\rightarrow$ p. 300), which in turn regulate the secretion of estradiol and progesterone. The female sex functions are controlled by the periodic release of hormones, the purpose of which is to produce a fertilizable egg in the ovaries each month ($\rightarrow$ **A4**) and produce an environment suitable for sperm reception (*fertilization*) and implantation of the fertilized ovum (*nidation*) ($\rightarrow$ **A5**). This cyclic activity is reflected by the monthly *menses* (*menstruation*) which, by definition, marks the start of the menstrual cycle.

Girls in Central Europe usually have their first menstrual period (**menarche**) around the age of 13. By about age 40, the cycle becomes increasingly irregular over a period of up to 10 years (**climacteric**) as the end of the reproductive period nears. The last menses (**menopause**) generally occurs around the age of 48–52.

A menstrual cycle can last 21–35 days. The second half of the cycle (**luteal phase** = secretory phase) usually lasts 14 days, while the first half (**follicular phase** = proliferative phase) lasts 7–21 days. **Ovulation** separates the two phases ($\rightarrow$ **A**). If the cycle length varies by more than 2–3 days, ovulation generally does not occur. Such *anovulatory cycles* account for 20% of all cycles in healthy females.

In addition to general changes in the body and mood, the following changes occur in the ovaries, uterus and cervix during the menstrual cycle ($\rightarrow$ **A**):

Day 1: Start of menstruation (lasting about 2–6 days).

Days 1–14 (variable, see above): The **follicular phase** starts on the first day of menstruation. The endometrium thickens to become prepared for the implantation of the fertilized ovum during the luteal phase ($\rightarrow$ **A5**), and about 20 ovarian follicles mature under the influence of FSH. One of these becomes the *dominant follicle*, which produces increasing quantities of *estrogens* ($\rightarrow$ **A4** and p. 300). The small cervical os is blocked by a viscous mucous plug.

Day 14 (variable, see above): **Ovulation**. The amount of *estrogens* produced by the follicle increases rapidly between day 12 and 13 ($\rightarrow$ **A2**). The increased secretion of LH in response to higher levels of estrogen leads to ovulation ($\rightarrow$ **A1, A4**; see also p. 300). The *basal body temperature* (measured on an empty stomach before rising in the morning) rises about 0.5°C about 1–2 days later and remains elevated until the end of the cycle ($\rightarrow$ **A3**). This temperature rise generally indicates that ovulation has occurred. During ovulation, the *cervical mucus* is less viscous (it can be stretched into long threads—*spinnbarkeit*) and the cervical os opens slightly to allow the sperm to enter.

Days 14–28: The **luteal phase** is characterized by the development of a *corpus luteum* ($\rightarrow$ **A4**), which secretes *progesterone*, ($\rightarrow$ **A2**); an increase in mucoid secretion from the uterine glands also occurs ($\rightarrow$ **A5**). The endometrium is most responsive to progesterone around the 22nd day of the cycle, which is when *nidation* should occur if the ovum has been fertilized. Otherwise, progesterone and estrogens now inhibit Gn-RH secretion ($\rightarrow$ p. 300), resulting in degeneration of the corpus luteum. The subsequent rapid decrease in the plasma concentrations of estrogens and progesterone ($\rightarrow$ **A2**) results in constriction of endometrial blood vessels and ischemia. This ultimately leads to the breakdown and discharge of the uterine lining and to bleeding, i.e., menstruation ($\rightarrow$ **A5**).

Plate 11.17 Oogenesis and the Menstrual Cycle

A. Menstrual cycle

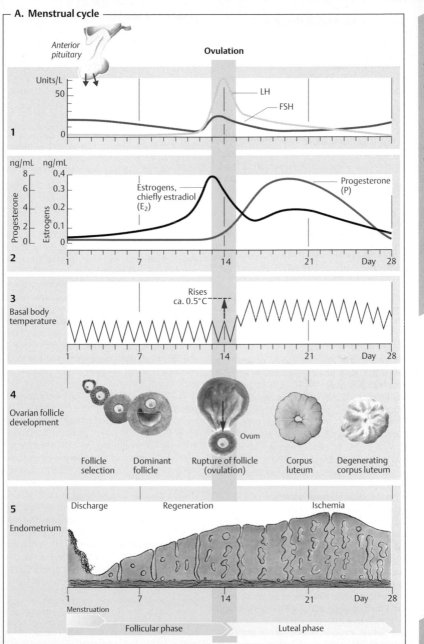

Anterior pituitary

Ovulation

1

Units/L

50

0

LH

FSH

2

ng/mL Progesterone

8
6
4
2
0

ng/mL Estrogens

0.4
0.3
0.2
0.1
0

Estrogens, chiefly estradiol (E₂)

Progesterone (P)

1   7   14   21   Day   28

3 Basal body temperature

Rises ca. 0.5°C

1   7   14   21   Day   28

4 Ovarian follicle development

Ovum

Follicle selection

Dominant follicle

Rupture of follicle (ovulation)

Corpus luteum

Degenerating corpus luteum

5 Endometrium

Discharge   Regeneration   Ischemia

1   7   14   21   Day   28

Menstruation

Follicular phase   Luteal phase

Hormonal Control of the Menstrual Cycle

In sexually mature women, **gonadoliberin** or **gonadotropin-releasing hormone (Gn-RH)** is secreted in one-minute pulses every 60–90 min in response to signals from various neurotransmitters. This, in turn, induces the pulsatile secretion of FSH and LH from the anterior pituitary. If the rhythm of Gn-RH secretion is much faster or continuous, less FSH and LH will be secreted, which can result in infertility. The LH : FSH secretion ratio changes during the course of the menstrual cycle. Their release must be therefore subject to additional factors besides Gn-RH.

The secretion of LH and FSH is, for example, subject to **central nervous effects** (psychogenic factors, stress) mediated by various **transmitters** circulating in the portal blood in the hypothalamic region, e.g., norepinephrine (NE) and neuropeptide Y (NPY) as well as by ovarian hormones, i.e., by *estrogens* (estrone, estradiol, estriol, etc.), *progesterone* and *inhibin*. Ovarian hormones affect Gn-RH secretion indirectly by stimulating central nerve cells that activate Gn-RH-secreting neurons by way of neurotransmitters such as norepinephrine and NPY and inhibit Gn-RH secretion by way of GABA and opioids.

FSH production again increases toward the end of the luteal phase ($\rightarrow$ p. 299, A1). In the **early follicular phase** ($\rightarrow$ **A1**), FSH induces the proliferation of the *stratum granulosum* in about 20 follicles and stimulates the secretion of *aromatase* in their *granulosa cells*. Aromatase catalyzes the conversion of the androgens *testosterone* and *androstenedione* to estradiol (E_2) and estrone (E_1) ($\rightarrow$ p. 295 A, steps *r* and *o*). Estrogens are synthesized in *theca cells* and absorbed by granulosa cells. Although relatively small amounts of LH are secreted ($\rightarrow$ **A1** and p. 299 A1), this is enough to activate theca cell-based enzymes (17β-hydroxysteroid dehydrogenase and C17/C20-lyase) that help to produce the androgens needed for estrogen synthesis. The follicle-based estrogens increase their own FSH receptor density. The follicle with the highest estrogen content is therefore the most sensitive to FSH. This loop has a self-amplifying effect, and the follicle in question is selected as the *dominant follicle* around the 6th day of the cycle ($\rightarrow$ **A2**). In the **mid-follicular phase**, estrogens restrict FSH and LH secretion (via negative feedback control and with the aid of *inhibin*; $\rightarrow$ **A2**) but later stimulate LH receptor production in granulosa cells. These cells now also start to produce progesterone (start of *luteinization*), which is absorbed by the theca cells ($\rightarrow$ **A3**) and used as precursor for further increase in androgen synthesis ($\rightarrow$ p. 295 A, steps *f* and *l*).

Inhibin and estrogens secreted by the dominant follicle increasingly inhibit FSH secretion, thereby decreasing the estrogen production in other follicles. This leads to an androgen build-up in and **apoptosis of the unselected follicles**.

Increasing quantities of LH and FSH are released in the **late follicular phase** ($\rightarrow$ **A3**), causing a sharp rise in their plasma concentrations. The **FSH peak** occurring around day 13 of the cycle induces the *first meiotic division* of the ovum. Estrogens increase the LH secretion (mainly via the hypothalamus), resulting in the increased production of androgens and estrogens (*positive feedback*) and a rapid rise in the LH conc. (*LH surge*). The **LH peak** occurs around **day 14** ($\rightarrow$ **A2**). The follicle ruptures and discharges its ovum about 10 hours later (**ovulation**). Ovulation does not take place if the LH surge does not occur or is too slow. Pregnancy is not possible in the absence of ovulation.

Luteal phase ($\rightarrow$ **A4**). LH, FSH and estrogens transform the ovarian follicle into a *corpus luteum*. It actively secretes large quantities of **progesterone** (progestational hormone), marking the beginning of the luteal phase ($\rightarrow$ **A**). Estrogens and progesterone now *inhibit* the secretion of FSH and LH directly and indirectly (e.g., through inhibition of Gn-RH; see above), causing a rapid drop in their plasma conc. This *negative feedback* leads to a marked drop in the plasma conc. of estrogens and progesterone towards the end of the menstrual cycle (approx. day 26), thereby triggering the menses ($\rightarrow$ p. 299, A2). FSH secretion starts to rise just before the start of menstruation ($\rightarrow$ **A4**).

Combined administration of estrogens and gestagens during the first half of the menstrual cycle prevents ovulation. Since ovulation does not occur, pregnancy cannot take place. Most **contraceptives** work according to this principle.

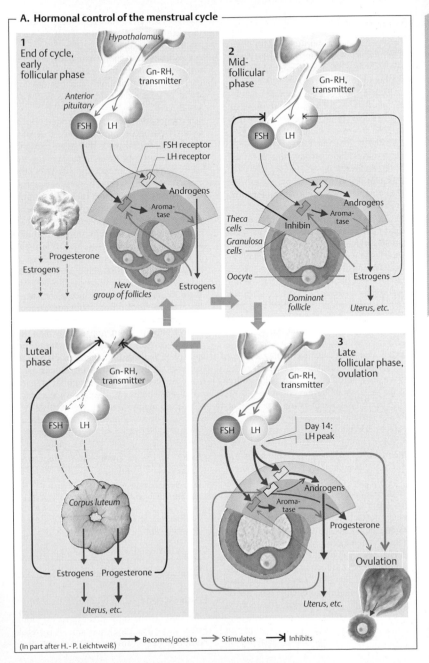

A. Hormonal control of the menstrual cycle

1
End of cycle, early follicular phase

Hypothalamus

Gn-RH, transmitter

Anterior pituitary

FSH   LH

FSH receptor
LH receptor

Androgens

Aroma-tase

Progesterone

Estrogens

New group of follicles

Estrogens

2
Mid-follicular phase

Gn-RH, transmitter

FSH   LH

Androgens

Inhibin

Aroma-tase

Theca cells
Granulosa cells

Oocyte

Dominant follicle

Estrogens

Uterus, etc.

4
Luteal phase

Gn-RH, transmitter

FSH   LH

Corpus luteum

Estrogens   Progesterone

Uterus, etc.

3
Late follicular phase, ovulation

Gn-RH, transmitter

FSH   LH

Day 14:
LH peak

Androgens

Aroma-tase

Progesterone

Ovulation

Uterus, etc.

(In part after H.-P. Leichtweiß)   → Becomes/goes to   → Stimulates   → Inhibits

Plate 11.18   Hormonal Control of the Menstrual Cycle

Estrogens

Estrogens are steroid hormones with 18 carbon atoms. Estrogens are primarily synthesized from the 17-ketosteroid *androstenedione*, but testosterone can also be a precursor (→ p. 295 A). The *ovaries*, (granulosa and theca cells), *placenta* (→ p. 304), *adrenal cortex*, and in *Leydig's cells* (interstitial cells) of the *testes* (→ p. 306) are the physiological **sites of estrogen synthesis.**. In some target cells for testosterone, it must first be converted to estradiol to become active.

Estradiol (E₂) is the most potent estrogen (E). The potencies of **estrone (Eᵢ)** and **estriol (E₃)** are relatively low in comparison (E₂ : E₁ : E₃ = 10 : 5 : 1). Most estrogens (and testosterone) circulating in the blood are bound to *sex hormone-binding globulin (SHBG)*. Estriol (E₃) is the chief degradation product of estradiol (E₂).

Oral administration of estradiol has virtually no effect because almost all of the hormone is removed from the bloodstream during its first pass through the liver. Therefore, other estrogens (with a different chemical structure) must be used for effective **oral estrogen therapy.**

Actions. Although estrogens play a role in the development of female sex characteristics, they are not nearly as important as the androgens for male sexual development (→ p. 306). The preparatory action of estrogen is often required for optimal progesterone effects (e.g., in the uterus; see below). Other important effects of estrogens in human females are as follows.

◆ **Menstrual cycle.** Estrogens accelerate maturation of the ovarian follicle during the menstrual cycle (→ p. 298 and table). In the *uterus*, estrogen promotes the proliferation (thickening) of the endometrium and increases uterine muscle contraction. In the *vagina*, estrogen thickens the mucosal lining, leading to the increased discharge of glycogen-containing epithelial cells. The liberated glycogen is used for an increased production of lactic acid by Döderlein's bacillus. This lowers the vaginal pH to 3.5–5.5, thereby reducing the risk of vaginal infection. In the *cervix*, the mucous plug sealing the cervical os functions as a barrier that prevents sperm from entering the uterus. Estrogens change the consistency of

the cervical mucus, making it more conducive to sperm penetration and survival, especially around the time of ovulation.

◆ **Fertilization.** In the female body, estrogens prepare the sperm to penetrate and fertilize the ovum (*capacitation*) and regulate the speed at which the ovum travels in the fallopian tube.

◆ **Extragonadal effects of estrogen.** During *puberty*, estrogens stimulate breast development, induces changes in the vagina and in the distribution of subcutaneous fat, and (together with androgens) stimulate the growth of pubic and axillary hair. Since estrogens increase the coagulability of the blood, the administration of estrogens (e.g., in contraceptives) increases the risk of thrombosis and leads renal salt and water retention. Estrogens slow longitudinal bone growth, accelerate epiphyseal closure (in men and women) and increase osteoblast activity. Estrogen deficiencies in menopause consequently lead to the loss of bone mass (*osteoporosis*). Estrogens induce a decrease in LDL and a rise in VLDL and HDL concentrations (→ p. 254ff.), which is why arteriosclerosis is less common in premenopausal women than in men. Estrogen also makes the *skin* thinner and softer, reduces the sebaceous glands, and increases fat deposits in subcutaneous tissue. Lastly, estrogen influences a number of *central nervous functions*, e.g., sexual response, social behavior, and mood.

Plasma concentrations of estradiol and progesterone (ng/mL)

Phase	Estradiol	Progesteron
Women		
Early follicular phase	0.06	0.3
Mid- and late follicular phase	0.1 ⇒ 0.4	1.0
Ovulation	0.4	2.0
Mid-luteal phase	0.2	8–16
Pregnancy	7–14	40 ⇒ 130
Day 1 after parturition		20
Men	0.05	0.3

Progesterone

Progesterone, the most potent progestational (pregnancy-sustaining) hormone, is a steroid hormone (21 C atoms) synthesized from cholesterol via pregnenolone ($\rightarrow$ p. 295). It is produced in the corpus luteum, ovarian follicles and placenta ($\rightarrow$ p. 304) of the female, and in the adrenal cortex of males and females. Like cortisol, most circulating progesterone is bound to cortisol-binding globulin (CBG = transcortin). Like estradiol (E_2), most progesterone is broken down during its first pass through the liver, so oral doses of progesterone are almost completely ineffective. *Pregnanediol* is the most important degradation product of progesterone.

Actions of progesterone. The *main functions* of progesterone are to prepare the female genital tract for implantation and maturation of the fertilized ovum and to **sustain pregnancy** ($\rightarrow$ see table). Progesterone counteracts many of the effects induced by estrogens, but various effects of progesterone depend on the preparatory activity or simultaneous action of estrogens. During the follicular phase, for example, estrogens increases the density of progesterone receptors, while simultaneous estrogen activity is needed to induce mammary growth (see below).

◆ The **uterus** is the chief target organ of progesterone. Once estrogen induces endometrial thickening, progesterone stimulates growth of the uterine muscle (myometrium), restructures the endometrial glands ($\rightarrow$ p. 298), alters the blood supply to the endometrium, and changes the glycogen content. This represents the transformation from a proliferative endometrium to a secretory endometrium, with a peak occurring around day 22 of the cycle. Progesterone later plays an important role in the potential implantation (nidation) of the fertilized ovum because it reduces myometrial activity (important during pregnancy), narrows the **cervical os**, and changes the consistency of the cervical mucous plug so that it becomes virtually impregnable to sperm.

◆ Progesterone inhibits the release of LH during the luteal phase. The administration of gestagens like progesterone during the follicular phase *inhibits ovulation*. Together with its effects on the cervix (see above) and its inhibitory effect on capacitation ($\rightarrow$ p. 302), progesterone can therefore have a contraceptive effect ("**mini pill**").

◆ High levels of progesterone have an anesthetic effect on the **central nervous system**. Progesterone also increases the susceptibility to epileptic fits and exerts thermogenic action, i.e., it raises the basal body temperature ($\rightarrow$ p. 298). In addition, a decrease in the progesterone concentration is also believed to be responsible for the mood changes and depression observed before menstruation (*premenstrual syndrome, PMS*) and after pregnancy (*postpartum depression*).

◆ In the **kidneys**, progesterone slightly inhibits the effects aldosterone, thereby inducing increased NaCl excretion.

Prolactin and Oxytocin

The secretion of **prolactin (PRL)** is inhibited by prolactin-inhibiting hormone (**PIH = dopamine**) and stimulated by thyroliberin (**TRH**) ($\rightarrow$ p. 270). Prolactin increases the hypothalamic secretion of PIH in both men and women (negative feedback control). Conversely, estradiol (E_2) and progesterone inhibit PIH secretion (indirectly via transmitters, as observed with Gn-RH; see above). Consequently, prolactin secretion rises significantly during the second half of the menstrual cycle and during pregnancy. Prolactin (together with estrogens, progesterone, glucocorticoids and insulin) stimulate *breast enlargement* during pregnancy and *lactogenesis* after parturition. In breast-feeding, stimulation of the nerve endings in the nipples by the suckling infant stimulates the secretion of prolactin (**lactation reflex**). This also increases release of oxytocin which triggers *milk ejection* and increases uterine contractions, thereby increasing lochia discharge after birth. When the mother stops breast-feeding, the prolactin levels drop, leading to the rapid stoppage of milk production.

Hyperprolactinemia. Stress and certain drugs inhibit the secretion of PIH, causing an increase in prolactin secretion. *Hypothyroidism* ($\rightarrow$ p. 288) can also lead to hyperprolactinemia, because the associated increase in TRH stimulates the release of prolactin. Hyperprolactinemia inhibits ovulation and leads to *galactorrhea*, i.e., the secretion of milk irrespective of pregnancy. Some women utilize the anti-ovulatory effect of nursing as a natural method of birth control, which is often but not always effective.

Hormonal Control of Pregnancy and Birth

Beside its other functions, the **placenta** produces most of the hormones needed during pregnancy ($\rightarrow$ p. 220). *Ovarian hormones* also play a role, especially at the start of pregnancy ($\rightarrow$ **A**).

Placental hormones. The primary hormones produced by the placenta are *human chorionic gonadotropin* (**hCG**), *corticotropin-releasing hormone* (**CRH**), *estrogens, progesterone, human placental lactogen* (**hPL**), and *proopiomelanocortin* (POMC; $\rightarrow$ p. 280). hCG is the predominant hormone during the first trimester of pregnancy (3-month period calculated from the beginning of the last menses). Maternal conc. of hPL and CRH-controlled estrogens rise sharply during the third trimester ($\rightarrow$ **B**). Placental hormones are distributed to mother and fetus. Because of the close connection between maternal, fetal and placental hormone synthesis, they are jointly referred to as the **fetoplacental unit** ($\rightarrow$ **A**).

Human chorionic gonadotropin (hCG) (a) stimulates the synthesis of steroids like DHEA and DHEA-S by the fetal adrenal cortex (see below); (b) suppresses follicle maturation in the maternal ovaries, and (c) maintains the production of progesterone and estrogen in the corpus luteum ($\rightarrow$ **A1**) until the 6th week of gestation, i.e., until the placenta is able to produce sufficient quantities of the hormones.

Most **pregnancy tests** are based on the fact that hCG is detectable in the urine about 6–8 days after conception. Since the levels of estrogen and progesterone greatly increase during pregnancy (see table on p. 302), larger quantities of these hormones and their metabolites *estriol* and *pregnanediol* are excreted in the urine. Therefore, their conc. can also be measured to test for pregnancy.

In contrast to other endocrine organs, the placenta has to receive the appropriate precursors (cholesterol or androgens, $\rightarrow$ p. 294) from the maternal and fetal adrenal cortex, respectively, before it can synthesize progesterone and estrogen ($\rightarrow$ **A2**). The **fetal adrenal cortex (FAC)** is sometimes larger than the kidneys and consists of a *fetal zone* and an *adult zone*. The placenta takes up cholesterol and pregnenolone and uses them to synthesize progesterone. It is transported to the fetal zone of the FAC, where it is converted to *dehydroepiandrosterone* (**DHEA**) and *dehydroepiandrosterone sulfate* (**DHEA-S**). DHEA and DHEA-S pass to the **placenta**, where they are used for *estrogen synthesis*. Progesterone is converted to *testosterone* in the *testes* of the male fetus.

Human placental lactogen (hPL = human chorionic somatomammotropin, HCS) levels rise steadily during pregnancy. Like prolactin ($\rightarrow$ p. 303), hPL stimulates mammary enlargement and lactogenesis in particular and, like GH ($\rightarrow$ p. 280), stimulates physical growth and development in general. hPL also seems to increase maternal plasma glucose conc.

Corticotropin-releasing hormone (CRH) secreted by the *placenta* seems to play a key role in the **hormonal regulation of birth**. The plasma levels of maternal CRH increase exponentially from the 12th week of gestation on. This rise is more rapid in premature births and slower in post-term births. In other words, the rate at which the CRH concentration rises seems to determine the duration of the pregnancy. Placental CRH stimulates the release of ACTH by the fetal pituitary, resulting in increased cortisol production in the adult zone of FAC; this again stimulates the release of CRH (positive feedback). CRH also stimulates lung development and the production of DHEA and DHEA-S in the fetal zone of FAC.

The maternal estrogen conc. rises sharply towards the **end of the pregnancy**, thereby counteracting the actions of progesterone, including its pregnancy-sustaining effect. Estrogens induce oxytocin receptors ($\rightarrow$ p. 303), α_1-adrenoceptors ($\rightarrow$ p. 84ff.), and gap junctions in the uterine musculature ($\rightarrow$ p. 16ff.), and uterine cells are depolarized. All these effects increase the responsiveness of the uterine musculature. The simultaneous increase in progesterone synthesis triggers the production of collagenases that soften the taut cervix. Stretch receptors in the uterus respond to the increase in size and movement of the fetus. Nerve fibers relay these signals to the hypothalamus, which responds by secreting larger quantities of oxytocin which, in turn, increases uterine contractions (positive feedback). The gap junctions conduct the spontaneous impulses from individual pacemaker cells in the fundus across the entire myometrium at a rate of approximately 2 cm/s ($\rightarrow$ p. 70).

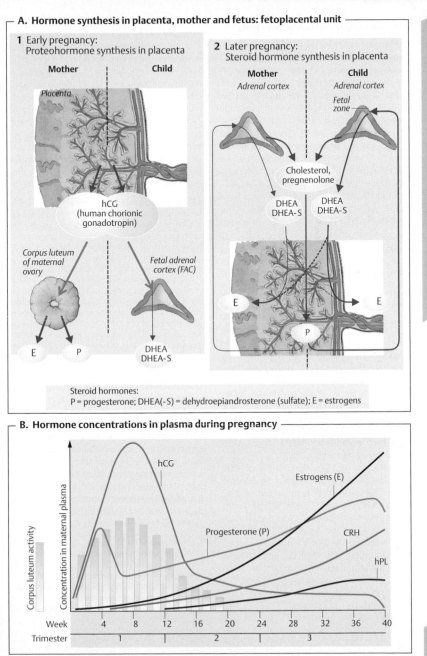

Plate 11.19   Hormonal Control of Pregnancy and Birth

A. Hormone synthesis in placenta, mother and fetus: fetoplacental unit

1 Early pregnancy:
Proteohormone synthesis in placenta

Mother | Child

Placenta

hCG
(human chorionic
gonadotropin)

*Corpus luteum
of maternal
ovary*

*Fetal adrenal
cortex (FAC)*

E   P

DHEA
DHEA-S

2 Later pregnancy:
Steroid hormone synthesis in placenta

Mother
Adrenal cortex | Child
Adrenal cortex

*Fetal
zone*

Cholesterol,
pregnenolone

DHEA
DHEA-S   DHEA
DHEA-S

E   E

P

Steroid hormones:
P = progesterone; DHEA(-S) = dehydroepiandrosterone (sulfate); E = estrogens

B. Hormone concentrations in plasma during pregnancy

hCG

Estrogens (E)

Progesterone (P)

CRH

hPL

Corpus luteum activity

Concentration in maternal plasma

Week   4   8   12   16   20   24   28   32   36   40
Trimester   1   2   3

Androgens and Testicular Function

Androgens (male sex hormones) are steroid hormones with 19 C atoms. This group includes potent hormones like *testosterone* (**T**) and *5α-dihydrotestosterone* (**DHT**) and less potent *17-ketosteroids* (**17-KS**) such as DHEA ($\rightarrow$ p. 294). In males, up to 95% of testosterone is synthesized by the *testes* ($\rightarrow$ **A2**) and 5% by the *adrenal cortex* ($\rightarrow$ **A1**). The *ovaries* and adrenal cortex synthesize testosterone in females. The plasma testosterone conc. in males is about 15 times higher than in females, but decreases with age. Up to 98% of testosterone circulating in blood is bound to plasma proteins (albumin and sex hormone-binding globulin, SHBG; $\rightarrow$ **A2**).

The testes secrete also small quantities of **DHT** and **estradiol** (**E₂**). Larger quantities of DHT (via *5-α-reductase*) and estradiol are synthesized from testosterone (via *aromatase*) by their respective target cells. A portion of this supply is released into the plasma. DHT and testosterone bind to the same intracellular receptor. Estradiol influences many functions in the male, e.g., epiphyseal cartilage and ejaculate formation and pituitary and hypothalamic activity.

Testosterone secretion is **regulated** by luteinizing hormone (= LH, also called ICSH, $\rightarrow$ p. 269), the pulsatile secretion of which is controlled by Gn-RH at 1.5- to 2-hourly intervals, as in the female. LH stimulates the release of testosterone from Leydig's cells (interstitial cells) in the testes ($\rightarrow$ **A2**), whereas testosterone and estradiol inhibit LH and Gn-RH secretion (negative feedback).

Gn-RH also induces the release of **FSH**, which stimulates the secretion of **inhibin** and induces the expression of *androgen-binding protein* (**ABP**) in Sertoli cells of the testes ($\rightarrow$ **A3**). Testosterone cannot induce spermatogenesis without the help of ABP (see below). FSH also induces the formation of LH receptors in the interstitial cells of Leydig. Testosterone, DHT, estradiol and inhibin inhibit the secretion of FSH (negative feedback; $\rightarrow$ **A**). **Activin**, the physiological significance of which is still unclear, inhibits FSH secretion.

Apart from the important **effects of testosterone** on male sexual differentiation, spermatogenesis and sperm growth as well as on the functions of the genitalia, prostate and seminal vesicle (see below), testosterone also induces the *secondary sex characteristics* that occur in males around the time of puberty, i.e., body hair distribution, physique, laryngeal size (voice change), acne, etc. In addition, testosterone is necessary for normal sex drive (*libido*), procreative capacity (*fertility*) and coital capacity (*potentia coeundi*) in the male. Testosterone also stimulates *hematopoiesis* and has *anabolic properties*, leading to increased muscle mass in males. It also has central nervous effects and can influence behavior—e.g., cause aggressiveness.

Sexual development and differentiation. The *genetic sex* ($\rightarrow$ **B**) determines the development of the sex-specific *gonads* (gamete-producing glands). The germ cells (*spermatogonia*; see below) then migrate into the gonads. The *somatic sex* is female when the subsequent somatic sex development and sex differentiation occurs in the absence of testosterone ($\rightarrow$ **C**). Male development requires the presence of testosterone in both steps ($\rightarrow$ **C**) with or without the aid of additional factors (e.g., calcitonin gene-related peptide, CGRP?) in certain stages of development (e.g., descent of testes into scrotum). High conc. of testosterone, either natural or synthetic (anabolic steroids), lead to masculinization (*virilization*) of the female ($\rightarrow$ **C**).

Testicular function. *Spermatogenesis* occurs in several stages in the testes (target organ of testosterone) and produces *sperm* (*spermatozoa*) ($\rightarrow$ **A3**). Sperm are produced in the *seminiferous tubules* (total length, ca. 300 m), the epithelium of which consists of *germ cells* and *Sertoli cells* that support and nourish the spermatogenic cells. The seminiferous tubules are strictly separated from other testicular tissues by a *blood–testis barrier*. The testosterone required for sperm maturation and semen production ($\rightarrow$ p. 308) must be bound to androgen-binding protein (**ABP**) to cross the barrier.

Spermatogonia ($\rightarrow$ **B**) are primitive sex cells. At puberty, a spermatogonium divides mitotically to form two daughter cells. One of these is kept as a lifetime *stem cell* reservoir (in contrast to oogonia in the female; $\rightarrow$ p. 298). The other undergoes several divisions to form a **primary spermatocyte**. It undergoes a *first meiotic division* (MD1) to produce two **secondary spermatocytes**, each of which undergoes a *second meiotic division* (MD2), producing a total of four **spermatids**, which ultimately differentiate into *spermatozoa*. After MD1, the spermatocytes have a single (haploid) set of chromosomes.

Plate 11.20 Androgens and Testicular Function

307

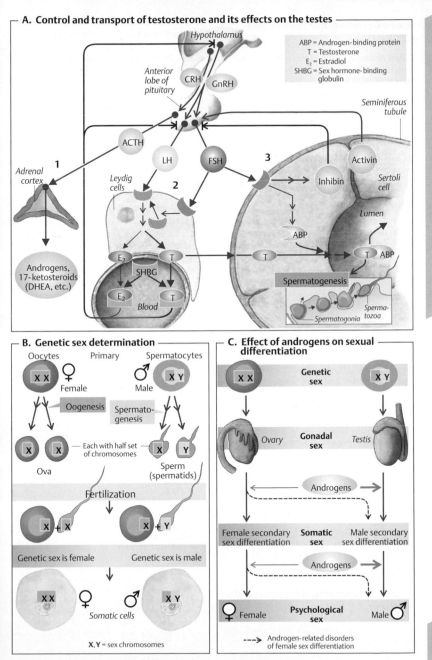

A. Control and transport of testosterone and its effects on the testes

ABP = Androgen-binding protein
T = Testosterone
E₂ = Estradiol
SHBG = Sex hormone-binding globulin

Hypothalamus

Anterior lobe of pituitary

CRH   GnRH

Seminiferous tubule

ACTH   LH   FSH   3   Activin

Adrenal cortex   1   Leydig cells   2   Inhibin   Sertoli cell

Androgens, 17-ketosteroids (DHEA, etc.)

Lumen

ABP

E₂   T   T   T   ABP

SHBG   Spermatogenesis

E₂   T

Blood   Spermatogonia   Spermatozoa

B. Genetic sex determination

Oocytes   Primary   Spermatocytes

X X ♀   X Y ♂

Female   Male

Oogenesis   Spermatogenesis

X   X   Each with half set of chromosomes   X   Y

Ova   Sperm (spermatids)

Fertilization

X + X   X + Y

Genetic sex is female   Genetic sex is male

X X ♀   X Y ♂

Somatic cells

X, Y = sex chromosomes

C. Effect of androgens on sexual differentiation

X X   Genetic sex   X Y

Ovary   Gonadal sex   Testis

Androgens

Female secondary sex differentiation   Somatic sex   Male secondary sex differentiation

Androgens

♀ Female   Psychological sex   Male ♂

- - -> Androgen-related disorders of female sex differentiation

Sexual Response, Intercourse and Fertilization

Sexual response in the male (→ **A1**). Impulses from tactile receptors on the skin in the genital region (especially the glans penis) and other parts of the body (*erogenous areas*) are transmitted to the *erection center* in the sacral spinal cord (S2–S4), which conducts them to parasympathetic neurons of the pelvic splanchnic nerves, thereby triggering **sexual arousal**. Sexual arousal is decisively influenced by stimulatory or inhibitory *impulses from the brain* triggered by sensual perceptions, imagination and other factors. Via nitric oxide (→ p. 278), efferent impulses lead to dilatation of deep penile artery branches (helicine arteries) in the erectile body (corpus cavernosum), while the veins are compressed to restrict the drainage of blood. The resulting high pressure (> 1000 mmHg) in the erectile body causes the penis to stiffen and rise (**erection**). The *ejaculatory center* in the spinal cord (L2–L3) is activated when arousal reaches a certain threshold (→ **A2**). Immediately prior to ejaculation, efferent sympathetic impulses trigger the partial evacuation of the prostate gland and the **emission** of semen from the *vas deferens* to the posterior part of the urethra. This triggers the **ejaculation reflex** and is accompanied by **orgasm**, the apex of sexual excitement. The effects of orgasm can be felt throughout the entire body, which is reflected by perspiration and an increase in respiratory rate, heart rate, blood pressure, and skeletal muscle tone. During ejaculation, the internal sphincter muscle closes off the urinary bladder while the vas deferens, seminal vesicles and bulbocavernous and ischiocavernous muscles contract rhythmically to propel the semen out of the urethra.

Semen. The fluid expelled during ejaculation (2–6 mL) contains 35–200 million sperm in a nutrient fluid (*seminal plasma*) composed of various substances, such as prostaglandins (from the prostate) that stimulate uterine contraction. Once semen enters the vagina during **intercourse**, the alkaline seminal plasma increase the vaginal pH to increase sperm motility. At least one sperm cell must reach the ovum for fertilization to occur.

Sexual response in the female (→ **A2**). Due to impulses similar to those in the male, the erectile tissues of the clitoris and vestibule of the vagina engorge with blood during the erection phase. Sexual arousal triggers the release of secretions from glands in the labia minora and transudates from the vaginal wall, both of which lubricate the vagina, and the nipples become erect. On continued stimulation, afferent impulses are transmitted to the lumbar spinal cord, where sympathetic impulses trigger **orgasm** (**climax**). The vaginal walls contract rhythmically (*orgasmic cuff*), the vagina lengthens and widens, and the uterus becomes erect, thereby creating a space for the semen. The cervical os also widens and remains open for about a half an hour after orgasm. Uterine contractions begin shortly after orgasm (and are probably induced locally by oxytocin). Although the accompanying physical reactions are similar to those in the male (see above), there is a wide range of variation in the orgasmic phase of the female. Erection and orgasm are not essential for conception.

Fertilization. The fusion of sperm and egg usually occurs in the *ampulla* of the fallopian tube. Only a small percentage of the sperm expelled during ejaculation (1000–10 000 out of 10^7 to 10^8 sperm) reach the fallopian tubes (*sperm ascension*). To do so, the sperm must penetrate the mucous plug sealing the cervix, which also acts as a sperm reservoir for a few days. In the time required for them to reach the ampullary portion of the fallopian tube (about 5 hours), the sperm must undergo certain changes to be able to fertilize an ovum; this is referred to as **capacitation** (→ p. 302).

After ovulation (→ p. 298ff.) the ovum enters the tube to the uterus (*oviduct*) via the abdominal cavity. When a sperm makes contact with the egg (via chemotaxis), species-specific sperm-binding receptors on the ovum are exposed and the proteolytic enzyme *acrosin* is thereby activated (**acrosomal reaction**). Acrosin allows the sperm to penetrate the cells surrounding the egg (*corona radiata*). The sperm bind to receptors on the envelope surrounding the ovum (*zona pellucida*) and enters the egg. The membranes of both cells then *fuse*. The ovum now undergoes a *second meiotic division*, which concludes the act of *fertilization*. Rapid proteolytic changes in the receptors on the ovum (**zona pellucida reaction**) prevent other sperm from entering the egg. Fertilization usually takes place on the first day after intercourse and is only possible within 24 hours after ovulation.

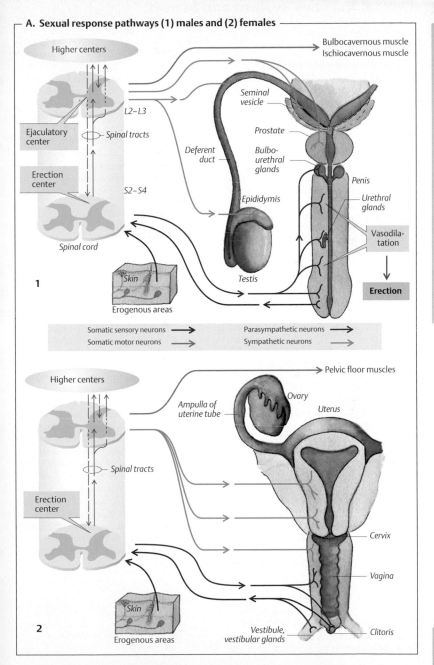

A. Sexual response pathways (1) males and (2) females

1

Higher centers

Ejaculatory center

Erection center

L2–L3

Spinal tracts

S2–S4

Spinal cord

Skin

Erogenous areas

Bulbocavernous muscle
Ischiocavernous muscle

Seminal vesicle

Prostate

Bulbo-urethral glands

Deferent duct

Penis

Epididymis

Urethral glands

Vasodilatation

Testis

Erection

| Somatic sensory neurons → | Parasympathetic neurons → |
| Somatic motor neurons → | Sympathetic neurons → |

2

Higher centers

Erection center

Spinal tracts

Skin

Erogenous areas

Pelvic floor muscles

Ampulla of uterine tube

Ovary

Uterus

Cervix

Vagina

Vestibule, vestibular glands

Clitoris

Plate 11.21   Sexual Response, Fertilization

Central Nervous System

The *brain* and *spinal cord* make up the *central nervous system* (**CNS**) (→ **A**). The spinal cord is divided into similar *segments*, but is 30% shorter than the spinal column. The **spinal nerves** exit the spinal canal at the level of their respective vertebrae and contains the afferent somatic and visceral fibers of the *dorsal root*, which project to the spinal cord, and the efferent somatic (and partly autonomic) fibers of the *anterior root*, which project to the periphery. Thus, a nerve is a bundle of nerve fibers that has different functions and conducts impulses in different directions (→ p. 42).

Spinal cord (→ **A**). Viewed in cross-section, the spinal cord has a dark, butterfly-shaped inner area (*gray matter*) surrounded by a lighter outer area (*white matter*). The four wings of the gray matter are called *horns* (cross-section) or columns (longitudinal section). The *anterior horn* contains motoneurons (projecting to the muscles), the *posterior horn* contains interneurons. The cell bodies of most afferent fibers lie within the **spinal ganglion** outside the spinal cord. The white matter contains the axons of ascending and descending tracts.

Brain (→ **D**). The main parts of the brain are the *medulla oblongata* (→ **D7**) *pons* (→ **D6**), *mesencephalon* (→ **D5**), *cerebellum* (→ **E**), *diencephalon* and *telencephalon* (→ **E**). The medulla, pons and mesencephalon are collectively called the **brain stem**. It is structurally similar to the spinal cord but also contains cell bodies (nuclei) of *cranial nerves*, neurons controlling *respiration* and *circulation* (→ pp. 132 and 212ff.) etc. The **cerebellum** is an important control center for motor function (→ p. 326ff.).

Diencephalon. The *thalamus* (→ **C6**) of the diencephalon functions as a relay station for most afferents, e.g., from the eyes, ears and skin as well as from other parts of the brain. The *hypothalamus* (→ **C9**) is a higher autonomic center (→ p. 330), but it also plays a dominant role in endocrine function (→ p. 266ff.) as it controls the release of hormones from the adjacent *hypophysis* (→ **D4**).

The **telencephalon** consists of the cortex and nuclei important for motor function, the **basal ganglia**, i.e. *caudate nucleus* (→ **C5**), *putamen* (→ **C7**), *globus pallidus* (→ **C8**), and parts of the *amygdala* (→ **C10**). The amygdaloid nucleus and *cingulate gyrus* (→ **D2**) belong to the **limbic system** (→ p. 330). The **cerebral cortex** consists of four *lobes* divided by fissures (sulci), e.g., the *central sulcus* (→ **D1, E**) and *lateral sulcus* (→ **C3, E**). According to *Brodmann's map*, the cerebral cortex is divided into histologically distinct regions (→ **E**, italic letters) that generally have different functions (→ **E**). The *hemispheres* of the brain are closely connected by nerve fibers of the *corpus callosum* (→ **C1, D3**).

Cerebrospinal Fluid

The brain is surrounded by external and internal *cerebrospinal fluid (CSF) spaces* (→ **B**). The internal CSF spaces are called *ventricles*. The two lateral ventricles, I and II, (→ **B, C2**) are connected to the IIIrd and IVth ventricle and to the central canal of the spinal cord (→ **B**). Approximately 650 mL of CSF forms in the *choroid plexus* (→ **B, C4**) and drains through the *arachnoid villi* each day (→ **B**). Lesions that obstruct the drainage of CSF (e.g., brain tumors) result in cerebral compression; in children, they lead to fluid accumulation (*hydrocephalus*). The *blood–brain barrier* and the *blood–CSF barrier* prevents the passage of most substances except CO_2, O_2, water and lipophilic substances. (As an exception, the *circumventricular organs* of the brain such as the *organum vasculosum laminae terminalis* (OVLT; → p. 280) and the *area postrema* (→ p. 238) have a less tight blood–brain barrier.) Certain substances like glucose and amino acids can cross the blood–brain barrier with the aid of carriers, whereas proteins cannot. The ability or inability of a drug to cross the blood–brain barrier is an important factor in pharmacotherapeutics.

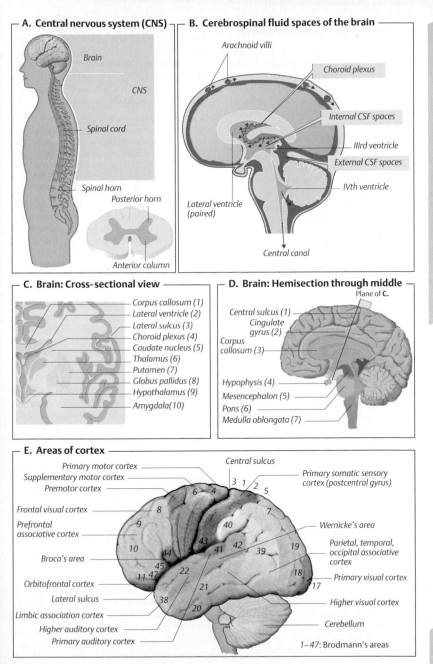

A. Central nervous system (CNS)

Brain

CNS

Spinal cord

Spinal horn

Posterior horn

Anterior column

B. Cerebrospinal fluid spaces of the brain

Arachnoid villi

Choroid plexus

Internal CSF spaces

IIIrd ventricle

External CSF spaces

IVth ventricle

Lateral ventricle (paired)

Central canal

C. Brain: Cross-sectional view

Corpus callosum (1)
Lateral ventricle (2)
Lateral sulcus (3)
Choroid plexus (4)
Caudate nucleus (5)
Thalamus (6)
Putamen (7)
Globus pallidus (8)
Hypothalamus (9)
Amygdala (10)

D. Brain: Hemisection through middle

Plane of **C.**

Central sulcus (1)
Cingulate gyrus (2)
Corpus callosum (3)

Hypophysis (4)
Mesencephalon (5)
Pons (6)
Medulla oblongata (7)

E. Areas of cortex

Central sulcus

Primary motor cortex
Supplementary motor cortex
Premotor cortex

Primary somatic sensory cortex (postcentral gyrus)

Frontal visual cortex

Prefrontal associative cortex

Wernicke's area

Parietal, temporal, occipital associative cortex

Broca's area

Orbitofrontal cortex

Primary visual cortex

Lateral sulcus

Higher visual cortex

Limbic association cortex

Cerebellum

Higher auditory cortex
Primary auditory cortex

1–47: Brodmann's areas

Plate 12.1 Central Nervous System

Stimulus Reception and Processing

With our **senses**, we receive huge quantities of **information** from the surroundings (10^9 bits/ s). Only a small portion of it is consciously perceived (10^1–10^2 bits/s); the rest is either subconsciously processed or not at all. Conversely, we transmit ca. 10^7 bits/s of information to the environment through speech and motor activity, especially facial expression ($\rightarrow$ **A**).

A **bit** (binary digit) is a single unit of information (1 byte = 8 bits). The average page of a book contains roughly 1000 bits, and TV images convey more than 10^6 bits/s.

Stimuli reach the body in *different forms of energy*, e.g., electromagnetic (visual stimuli) or mechanical energy (e.g., tactile stimuli). Various *sensory receptors* or **sensors** for these stimuli are located in the five "classic" sense organs (eye, ear, skin, tongue, nose) at the body surface as well as inside the body (e.g., propriosensors, vestibular organ). (In this book, sensory receptors are called sensors to distinguish them from binding sites for hormones and transmitters.) The sensory system extracts four stimulatory elements: *modality, intensity, duration, and localization.* Each type of sensor is specific for a unique or *adequate stimulus* that evokes specific sensory **modalities** such as sight, sound, touch, vibration, temperature, pain, taste, smell, as well as the body's position and movement, etc. Each modality has several *submodalities*, e.g., taste can be sweet or bitter, etc.

In **secondary sensors** (e.g., gustatory and auditory sensors), sensor and afferent fibers are separated by a synapse, whereas **primary sensors** (e.g., olfactory sensors and nocisensors) have their own afferent fibers.

A stimulus induces a change in *sensor potential* (**transduction**), which results in depolarization of the sensor cell (in most types; $\rightarrow$ **B1**) or hyperpolarization as in retinal sensors. The stronger the stimulus, the greater the *amplitude* of the sensor potential ($\rightarrow$ **C1**). Once the sensor potential exceeds a certain threshold, it is **transformed** into an **action potential**, AP ($\rightarrow$ **B1**; p. 46ff.).

Coding of signals. The stimulus is encoded in *AP frequency* (impulses/s = Hz), i.e., the higher the sensor potential, the higher the AP frequency ($\rightarrow$ **C2**). This information is decoded at the next synapse: The higher the frequency of arriving APs, the higher the *excitatory postsynaptic potential* (EPSP; $\rightarrow$ 50ff.). New APs are fired by the postsynaptic neuron when the EPSP exceeds a certain threshold ($\rightarrow$ **B2**).

Frequency coding of APs is a more reliable way of transmitting information over long distances than amplitude coding because the latter is much more susceptible to change (and falsification of its information content). At the synapse, however, the signal must be amplified or attenuated (by other neurons), which is better achieved by amplitude coding.

Adaptation. At constant stimulation, most sensors adapt, i.e., their potential decreases. The potential of slowly adapting sensors becomes **p**roportional to stimulus intensity (*P sensors* or *tonic sensors*). Fast-adapting sensors respond only at the onset and end of a stimulus. They sense **d**ifferential changes in the stimulus intensity (*D sensors* or *phasic sensors*). *PD sensors* have both characteristics ($\rightarrow$ p. 314).

Central processing. In a first phase, inhibitory and stimulatory impulses conducted to the CNS are integrated—e.g., to increase the *contrast* of stimuli ($\rightarrow$ **D**; see also p. 354). In this case, stimulatory impulses originating from adjacent sensor are attenuated in the process (*lateral inhibition*). In a second step, a **sensory impression** of the stimuli (e.g. "green" or "sweet") takes form in low-level areas of the sensory cortex. This is the first step of subjective sensory physiology. Consciousness is a prerequisite for this process. Sensory impressions are followed by their interpretation. The result of it is called **perception**, which is based on experience and reason, and is subject to individual interpretation. The impression "green," for example, can evoke the perception "There is a tree" or "This is a meadow."

Absolute threshold ($\rightarrow$ pp. 340ff., 352, 362), difference threshold ($\rightarrow$ pp. 340ff., 352, 368), spatial and temporal summation ($\rightarrow$ pp. 52, 352), receptive field ($\rightarrow$ p. 354), habituation and sensitization are other important concepts of sensory physiology. The latter two mechanisms play an important role in learning processes ($\rightarrow$ p. 336).

Plate 12.2 **Stimulus Reception and Processing**

A. Reception, perception and transmission of information

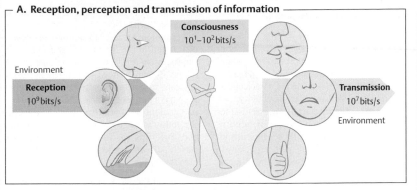

Consciousness
10^1–10^2 bits/s

Environment

Reception
10^9 bits/s

Transmission
10^7 bits/s

Environment

B. Stimulus processing and information coding

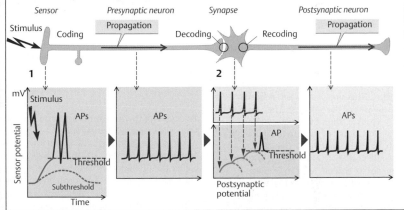

Sensor   *Presynaptic neuron*   *Synapse*   *Postsynaptic neuron*

Stimulus   Coding   Propagation   Decoding   Recoding   Propagation

1   **2**

mV

Stimulus   APs   APs   AP   APs

Sensor potential

Threshold   Threshold

Subthreshold

Postsynaptic
potential

Time

C. Stimulus, sensor and action potential relationships

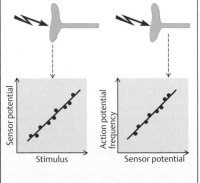

1 Transduction   **2** Transformation

Sensor potential

Stimulus

Action potential frequency

Sensor potential

D. Contrasting

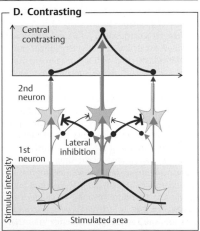

Central
contrasting

2nd
neuron

Lateral
inhibition

1st
neuron

Stimulus intensity

Stimulated area

Sensory Functions of the Skin

Somatovisceral sensibility is the collective term for all sensory input from receptors or *sensors* of the body (as opposed to the sensory organs of the head). It includes the areas of proprioception ($\rightarrow$ p. 316), nociception ($\rightarrow$ p. 318), and skin or surface sensitivity.

The **sense of touch** (taction) is essential for perception of *form, shape*, and *spatial nature* of objects (*stereognosis*). Tactile sensors are located predominantly in the palm, especially in the fingertips, and in the tongue and oral cavity. Stereognostic perception of an object requires that the CNS integrate signals from adjacent receptors into a spatial pattern and coordinate them with **tactile motor function**.

Mechanosensors. *Hairless areas* of the skin contain the following mechanosensors ($\rightarrow$ **A**), which are afferently innervated by myelinated nerve fibers of class II/Aβ ($\rightarrow$ p. 49 C):

◆ The spindle-shaped **Ruffini's corpuscle** ($\rightarrow$ **A3**) partly encapsulates the afferent axon branches. This unit is a *slowly adapting* (SA) pressosensor of the **SA2** type. They are P sensors ($\rightarrow$ p. 312). Thus, the greater the pressure on the skin (depth of indentation or weight of an object), the higher the AP frequency ($\rightarrow$ **B1**).

◆ **Merkel's cells** ($\rightarrow$ **A2**) are in synaptic contact to meniscus-shaped axon terminals. These complexes are pressure-sensitive **SA1** sensors. They are PD sensors (combination of **B1** and **B2**) since their AP frequency is not only dependent on the pressure intensity but also on the rate of its change (dp/dt; $\rightarrow$ p. 312).

◆ **Meissner's corpuscles** ($\rightarrow$ **A1**) are composed of lamellar cell layers between which club-shaped axons terminate. This unit represents a *rapidly adapting* pressure sensor (**RA** *sensor*) that responds only to pressure changes, dp/dt (pure *D sensor* or *velocity sensor*). The RA sensors are specific for *touch* (skin indentation of 10–100 μm) and low-frequency *vibration* (10–100 Hz). **Hair follicle receptors** ($\rightarrow$ **A5**), which respond to bending of the hairs, assume these functions in hairy areas of the skin.

◆ **Pacinian corpuscles** ($\rightarrow$ **A4**) are innervated by a centrally situated axon. They *adapt very rapidly* and therefore respond to changes in pressure change velocity, i.e. to *acceleration* (d^2p/dt^2), and sense high-frequency vibration (100–400 Hz; indentation depths < 3 μm). The AP frequency is proportional to the vibration frequency ($\rightarrow$ **B3**).

Resolution. RA and SA1 sensors are densely distributed in the mouth, lips and fingertips, especially in the index and middle finger (about 100/cm^2). They can distinguish closely adjacent stimuli as separate, i.e., each afferent axon has a *narrow receptive field*. Since the signals do not converge as they travel to the CNS, the ability of these sensors in the mouth, lips and fingertips to distinguish between two closely adjacent tactile stimuli, i.e. their *resolution*, is very high.

The **spatial threshold for two-point discrimination**, i.e., the distance at which two simultaneous stimuli can be perceived as separate, is used as a measure of tactile resolution. The spatial thresholds are roughly 1 mm on the fingers, lips and tip of the tongue, 4 mm on the palm of the hand, 15 mm on the arm, and over 60 mm on the back.

SA2 receptors and **pacinian corpuscles** have a broad receptive field (the exact function of SA2 receptors is not known). Pacinian corpuscles are therefore well adapted to detect vibrations, e.g., earth tremors.

Two types of **thermosensors** are located in the skin: *cold sensors* for temperatures < 36 °C and *warm sensors* for those > 36 °C. The lower the temperature (in the 20–36 °C range), the higher the AP frequency of the cold receptors. The reverse applies to warm receptors in the 36–43 °C range ($\rightarrow$ **C**). Temperatures ranging from 20° to 40 °C are subject to *rapid adaptation of thermosensation* (PD characteristics). Water warmed, for example, to 25 °C initially feels cold. More extreme temperatures, on the other hand, are persistently perceived as cold or hot (this helps to maintain a constant core temperature and prevent skin damage). The density of these cold and warm sensors in most skin areas is low as compared to the much higher densities in the mouth and lips. (That is why the lips or cheeks are used for temperature testing.)

Different sensors are responsible for thermoception at temperatures exceeding 45 °C. These **heat sensors** are also used for the perception of pungent substances such as *capsaicin*, the active constituent of hot chili peppers. Stimulation of VR1 receptors (vanilloid receptor type 1) for capsaicin mediates the opening of cation channels in nociceptive nerve endings, which leads to their depolarization.

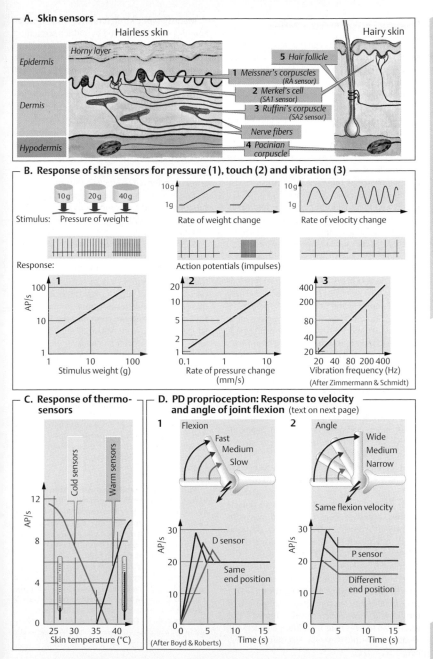

A. Skin sensors

Hairless skin | Hairy skin

Epidermis — Horny layer

5 Hair follicle

1 Meissner's corpuscles (RA sensor)

2 Merkel's cell (SA1 sensor)

Dermis

3 Ruffini's corpuscle (SA2 sensor)

Nerve fibers

Hypodermis — **4** Pacinian corpuscle

B. Response of skin sensors for pressure (1), touch (2) and vibration (3)

Stimulus: Pressure of weight | Rate of weight change | Rate of velocity change

Response: | Action potentials (impulses)

(After Zimmermann & Schmidt)

C. Response of thermo-sensors

Cold sensors | Warm sensors

AP/s — Skin temperature (°C)

D. PD proprioception: Response to velocity and angle of joint flexion (text on next page)

1 Flexion — Fast / Medium / Slow

D sensor | Same end position

2 Angle — Wide / Medium / Narrow

Same flexion velocity

P sensor | Different end position

(After Boyd & Roberts) | Time (s)

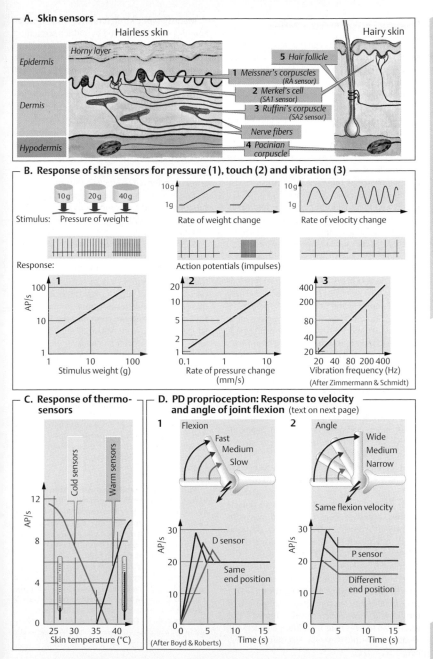

Plate 12.3 Sensory Functions of the Skin

315

Proprioception, Stretch Reflex

Proprioception is the mechanism by which we sense the *strength* that our muscles develop as well as the *position* and *movement* of our body and limbs. The *vestibular organ* ($\rightarrow$ p. 342) and *cutaneous mechanosensors* ($\rightarrow$ p. 314) assist the *propriosensors* in muscle spindles, joints and tendons. Sensors of *Golgi tendon organs* are located near muscle–tendon junctions.

Muscle spindles ($\rightarrow$ **A1**) contain intensity (P) and differential (D) sensors for monitoring of **joint position** and **movement**. The *velocity of position change* is reflected by a transient rise in impulse frequency (D sensor; $\rightarrow$ p. 315 D1, spike), and the final joint *position* is expressed as a constant impulse frequency (P-sensor, $\rightarrow$ p. 315 D2, plateau). Muscle spindles function to **regulate muscle length**. They lie parallel to the skeletal muscle fibers (*extrafusal muscle fibers*) and contain their own muscle fibers (*intrafusal muscle fibers*). There are two types of intrafusal muscle fibers: (1) *nuclear chain fibers* (P sensors) and (2) *nuclear bag fibers* (D sensors). The endings of *type Ia* afferent neurons coil around both types, whereas *type II* neurons wind around the nuclear chain fibers only (neuron types described on p. 49 C). These *annulospiral endings* detect longitudinal stretching of intrafusal muscle fibers and report their length (type Ia and II afferents) and changes in length (Ia afferents) to the spinal cord. The efferent γ *motoneurons* (fusimotor fibers) innervate both intrafusal fiber types, allowing variation of their length and stretch sensitivity ($\rightarrow$ **A1, B1**).

Golgi tendon organs ($\rightarrow$ **A2**) are arranged in series with the muscle and respond to the contraction of only a few motor units or more. Their primary function is to **regulate muscle tension**. Impulses from Golgi tendon organs (conveyed by *type Ib* afferents), the skin and joints, and muscle spindles (some of which are type Ia and II afferent fibers), as well as descending impulses, are jointly integrated in *type Ib interneurons* of the spinal cord; this is referred to as **multimodal integration** ($\rightarrow$ **D2**). Type Ib interneurons inhibit α motoneurons of the muscle from which the Ib afferent input originated (*autogenous inhibition*) and activate antagonistic muscles via excitatory interneurons ($\rightarrow$ **D5**).

Monosynaptic stretch reflex ($\rightarrow$ **C**). Muscles spindles are also affected by sudden stretching of a skeletal muscle, e.g. due to a tap on the tendon attaching it. Stretching of the muscle spindles triggers the activation of type Ia afferent impulses ($\rightarrow$ **B2, C**), which enter the spinal cord via the dorsal root and terminate in the ventral horn at the α motoneurons of the same muscle. This type Ia afferent input therefore induces contraction of the same muscle by only one synaptic connection. The reflex time for this monosynaptic stretch reflex is therefore very short (ca. 30 ms). This is classified as a *proprioceptive reflex,* since the stimulation and response arise in the same organ. The monosynaptic stretch reflex **functions** to rapidly correct "involuntary" changes in muscle length and joint position.

Supraspinal activation ($\rightarrow$ **B3**). Voluntary muscle contractions are characterized by *co-activation* of α and γ neurons. The latter adjust the muscle spindles (length sensors) to a certain set-point of length. Any deviations from this set-point due, for example, to unexpected shifting of weight, are compensated for by readjusting the α-innervation (*load compensation reflex*). Expected changes in muscle length, especially during complex movements, can also be more precisely controlled by (centrally regulated) γ fiber activity by increasing the preload and stretch sensitivity of the intrafusal muscle fibers (*fusimotor set*).

Hoffmann's reflex can be also used to test the stretch reflex pathway. This can be done by positioning electrodes on the skin over (mixed) muscle nerves and subsequently recording the muscle contraction induced by electrical stimuli of different intensity.

Polysynaptic circuits, also arising from type II afferents complement the stretch reflex. If a stretch reflex (e.g., knee-jerk reflex) occurs in an extensor muscle, the α motoneurons of the antagonistic flexor muscle must be inhibited via *inhibitory Ia interneurons* to achieve efficient extension ($\rightarrow$ **D1**).

Deactivation of stretch reflex is achieved by inhibiting muscle contraction as follows: 1) The muscle spindles relax, thereby allowing the deactivation of type Ia fibers; 2) the Golgi tendon organs inhibit the α motoneurons via type Ib interneurons ($\rightarrow$ **D2**); 3) the α motoneurons are inhibited by the interneurons (*Renshaw cells*; $\rightarrow$ **D4**) that they themselves stimulated via axon collaterals (*recurrent inhibition*; $\rightarrow$ **D3**; p. 321 C1).

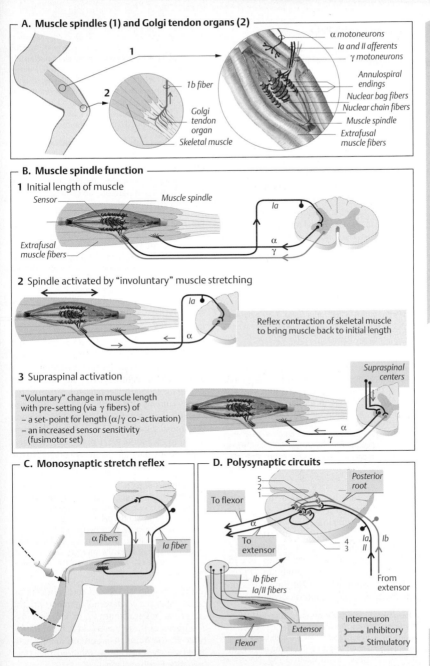

A. Muscle spindles (1) and Golgi tendon organs (2)

1

2

1b fiber
Golgi tendon organ
Skeletal muscle

α motoneurons
Ia and II afferents
γ motoneurons
Annulospiral endings
Nuclear bag fibers
Nuclear chain fibers
Muscle spindle
Extrafusal muscle fibers

B. Muscle spindle function

1 Initial length of muscle

Sensor
Muscle spindle
Extrafusal muscle fibers

Ia
α
γ

2 Spindle activated by "involuntary" muscle stretching

Ia
α

Reflex contraction of skeletal muscle to bring muscle back to initial length

3 Supraspinal activation

"Voluntary" change in muscle length with pre-setting (via γ fibers) of
– a set-point for length (α/γ co-activation)
– an increased sensor sensitivity (fusimotor set)

Supraspinal centers

α
γ

C. Monosynaptic stretch reflex

α fibers
Ia fiber

D. Polysynaptic circuits

5
2
1

Posterior root

To flexor

α

To extensor

4
3

Ia
II

Ib

From extensor

Ib fiber
Ia/II fibers

Extensor
Flexor

Interneuron
Inhibitory
Stimulatory

Plate 12.4 Proprioception, Stretch Reflex

Nociception and Pain

Pain is an unpleasant sensory experience associated with discomfort. It is protective insofar as it signals that the body is being threatened by an injury (*noxa*). *Nociception* is the perception of noxae via *nocisensors*, neural conduction and central processing. The *pain* that is ultimately felt is a subjective experience. Pain can also occur without stimulation of nocisensors, and excitation of nocisensors does not always evoke pain.

All body tissues except the brain and liver contain sensors for pain, i.e., **nocisensors** or *nociceptors* (→ **A**). Nocisensors are bead-like endings of peripheral axons, the somata of which are located in dorsal root ganglia and in nuclei of the trigeminal nerve. Most of these fibers are slowly conducting C fibers (< 1 m/s); the rest are myelinated Aδ fibers (5–30 m/s; fiber types described on p. 49 C).

When an injury occurs, one first senses sharp "fast pain" (Aδ fibers) before feeling the dull "slow pain" (C fibers), which is felt longer and over a broader area. Since nocisensors do not adapt, the pain can last for days. Sensitization can even lower the stimulus threshold.

Most nocisensors are **polymodal sensors** (C fibers) activated by mechanical stimuli, chemical mediators of inflammation, and high-intensity heat or cold stimuli. **Unimodal nociceptors**, the less common type, consist of *thermal nocisensors* (Aδ fibers), *mechanical nocisensors* (Aδ fibers), and "*dormant nocisensors*." Thermal nocisensors are activated by extremely hot (> 45 °C) or cold (< 5 °C) stimuli (→ p. 314). Dormant nocisensors are chiefly located in internal organs and are "awakened" after prolonged exposure (sensitization) to a stimulus, e.g., inflammation.

Nocisensors can be inhibited by opioids (**desensitization**) and stimulated by prostaglandin E_2 or bradykinin, which is released in response to inflammation (**sensitization**; → **A**). Endogenous opioids (e.g., dynorphin, enkephalin, endorphin) and exogenous opioids (e.g., morphium) as well as inhibitors of prostaglandin synthesis (→ p. 269) are therefore able to alleviate pain (*analgesic action*).

Inflammatory sensitization (e.g., sunburn) lowers the threshold for noxious stimuli, leading to excessive sensitivity (hyperalgesia) and additional pain resulting from non-noxious stimuli to the skin (allodynia), e.g., touch or warm water (37 °C). Once the nocisensors are stimulated, they start to release *neuropeptides* such as substance P or CGRP (calcitonin gene-related peptide) that cause inflammation of the surrounding vessels (**neurogenic inflammation**).

Projected pain. Damage to nociceptive fibers causes pain (neurogenic or neuropathic) that is often projected to and perceived as arising from the periphery. A prolapsed disk compressing a spinal nerve can, for example, cause leg pain. Nociceptive fibers can be blocked by cold or local anesthesia.

Nociceptive tracts (→ **C1**). The central axons of nociceptive somatic neurons and nociceptive afferents of internal organs end on neurons of the dorsal horn of the spinal cord. In many cases, they terminate on the same neurons as the skin afferents.

Referred pain (→ **B**). Convergence of somatic and visceral nociceptive afferents is probably the main cause of referred pain. In this type of pain, noxious visceral stimuli cause a perception of pain in certain skin areas called **Head's zones**. That for the heart, for example, is located mainly in the chest region. Myocardial ischemia is therefore perceived as pain on the surface of the chest wall (angina pectoris) and often also in the lower arm and upper abdominal region.

In the spinal cord, the neuroceptive afferents cross to the opposite side (decussation) and are conducted in the **tracts of the anterolateral funiculus**—mainly in the spinothalamic tract—and continue centrally via the brain stem where they join nociceptive afferents from the head (mainly trigeminal nerve) to the **thalamus** (→ **C1**). From the ventrolateral thalamus, sensory aspects of pain are projected to S1 and S2 areas of the **cortex**. Tracts from the medial thalamic nuclei project to the limbic system and other centers.

Components of pain. Pain has a *sensory component* including the conscious perception of site, duration and intensity of pain; a *motor component* (e.g., defensive posture and withdrawal reflex; → p. 320), an *autonomic component* (e.g., tachycardia), and an *affective component* (e.g., aversion). In addition, **pain assessments** based on the memory of a previous pain experience can lead to **pain-related behavior** (e.g., moaning).

In the thalamus and spinal cord, nociception can be **inhibited via descending tracts** with the aid of various transmitters (mainly opioids). The nuclei of these tracts (→ **C2**, blue) are located in the *brain stem* and are mainly activated via the nociceptive spinoreticular tract (negative feedback loop).

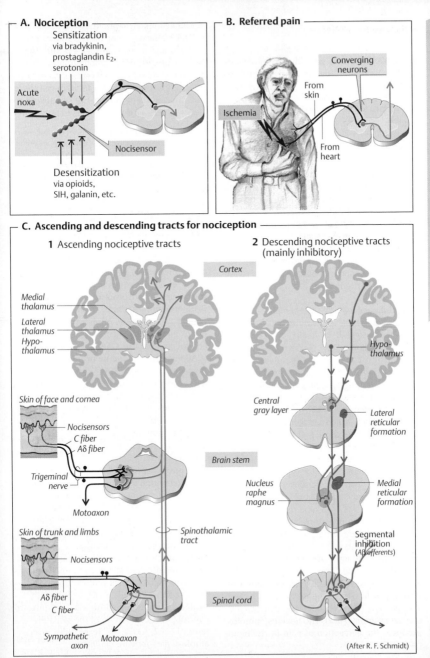

Plate 12.5 Nociception and Pain

A. Nociception

Sensitization
via bradykinin,
prostaglandin E_2,
serotonin

Acute
noxa

Nocisensor

Desensitization
via opioids,
SIH, galanin, etc.

B. Referred pain

Converging
neurons

From
skin

Ischemia

From
heart

C. Ascending and descending tracts for nociception

1 Ascending nociceptive tracts

2 Descending nociceptive tracts
(mainly inhibitory)

Cortex

Medial
thalamus

Lateral
thalamus

Hypo-
thalamus

Hypo-
thalamus

Skin of face and cornea

Nocisensors
C fiber
Aδ fiber

Central
gray layer

Lateral
reticular
formation

Trigeminal
nerve

Brain stem

Nucleus
raphe
magnus

Medial
reticular
formation

Motoaxon

Skin of trunk and limbs

Nocisensors

Spinothalamic
tract

Segmental
inhibition
(Aβ afferents)

Aδ fiber
C fiber

Spinal cord

Sympathetic
axon

Motoaxon

(After R. F. Schmidt)

Polysynaptic Reflexes

Unlike proprioceptive reflexes (→ p. 316), polysynaptic reflexes are activated by sensors that are spatially separate from the effector organ. This type of reflex is called *polysynaptic,* since the reflex arc involves many synapses in series. This results in a relatively *long reflex time.* The intensity of the response is dependent on the duration and intensity of stimulus, which is temporally and spatially summated in the CNS (→ p. 52). Example: itching sensation in nose _⇒ sneezing. The response spreads when the stimulus intensity increases (e.g., coughing ⇒ choking cough). *Protective reflexes* (e.g., withdrawal reflex, corneal and lacrimal reflexes, coughing and sneezing), *nutrition reflexes* (e.g., swallowing, sucking reflexes), *locomotor reflexes,* and the various *autonomic reflexes* are polysynaptic reflexes. Certain reflexes, e.g., plantar reflex, cremasteric reflex and abdominal reflex, are used as diagnostic tests.

Withdrawal reflex (→ **A**). *Example:* A painful stimulus in the sole of the right foot (e.g., stepping on a tack) leads to flexion of all joints of that leg (**flexion reflex**). Nociceptive afferents (→ p. 318) are conducted via stimulatory interneurons (→ **A1**) in the spinal cord to motoneurons of ipsilateral *flexors* and via inhibitory interneurons (→ **A2**) to motoneurons of ipsilateral *extensors* (→ **A3**), leading to their relaxation; this is called *antagonistic inhibition.* One part of the response is the **crossed extensor reflex**, which promotes the withdrawal from the injurious stimulus by increasing the distance between the nociceptive stimulus (e.g. the tack) and the nocisensor and helps to support the body. It consists of contraction of extensor muscles (→ **A5**) and relaxation of the flexor muscles in the contralateral leg (→ **A4, A6**). Nociceptive afferents are also conducted to other segments of the spinal cord (ascending and descending; → **A7, A8**) because different extensors and flexors are innervated by different segments. A noxious stimulus can also trigger flexion of the ipsilateral arm and extension of the contralateral arm (*double crossed extensor reflex*). The noxious stimulus produces the perception of pain in the brain (→ p. 316).

Unlike monosynaptic stretch reflexes, polysynaptic reflexes occur through the *co-activation of α and γ motoneurons* (→ p. 316). The *reflex excitability* of α motoneurons is largely **controlled by supraspinal centers** via multiple interneurons (→ p. 324). The brain can therefore shorten the reflex time of spinal cord reflexes when a noxious stimulus is anticipated.

Supraspinal lesions or interruption of descending tracts (e.g., in paraplegics) can lead to exaggeration of reflexes (**hyperreflexia**) and stereotypic reflexes. The absence of reflexes (**areflexia**) corresponds to specific disorders of the spinal cord or peripheral nerve.

Synaptic Inhibition

GABA (γ-aminobutyric acid) and glycine (→ p. 55f.) function as inhibitory transmitters in the spinal cord. **Presynaptic inhibition** (→ **B**) occurs frequently in the CNS, for example, at synapses between type Ia afferents and α motoneurons, and involves *axoaxonic synapses* of GABAergic interneurons at presynaptic nerve endings. GABA exerts inhibitory effects at the nerve endings by increasing the membrane conductance to Cl^- (GABA$_A$ receptors) and K^+ (GABA$_B$ receptors) and by decreasing the conductance to Ca^{2+} (GABA$_B$ receptors). This decreases the release of transmitters from the nerve ending of the target neuron (→ **B2**), thereby lowering the amplitude of its postsynaptic EPSP (→ p. 50). The *purpose* of presynaptic inhibition is to reduce certain influences on the motoneuron without reducing the overall excitability of the cell.

In **postsynaptic inhibition** (→ **C**), an *inhibitory interneuron* increases the membrane conductance of the postsynaptic neuron to Cl^- or K^+, especially near the axon hillock, thereby short-circuiting the depolarizing electrical currents from excitatory EPSPs (→ p. 54 D).

The interneuron responsible for postsynaptic inhibition is either activated by feedback from axonal collaterals of the target neurons (**recurrent inhibition** of motoneurons via glycinergic Renshaw cells; → **C1**) or is directly activated by another neuron via feed-forward control (→ **C2**). Inhibition of the ipsilateral extensor (→ **A2, A3**) in the flexor reflex is an example of *feed-forward inhibition.*

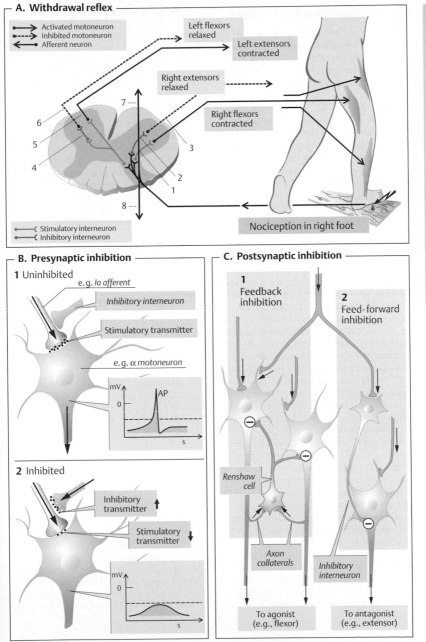

Plate 12.6 Polysynaptic Reflexes

A. Withdrawal reflex

Activated motoneuron
Inhibited motoneuron
Afferent neuron

Left flexors relaxed
Left extensors contracted

Right extensors relaxed

Right flexors contracted

7
6
5
4
3
2
1
8

Stimulatory interneuron
Inhibitory interneuron

Nociception in right foot

B. Presynaptic inhibition

1 Uninhibited

e.g. Ia afferent
Inhibitory interneuron
Stimulatory transmitter
e.g. α motoneuron

mV
0
AP
s

2 Inhibited

Inhibitory transmitter ↑
Stimulatory transmitter ↓

mV
0
s

C. Postsynaptic inhibition

1 Feedback inhibition

2 Feed-forward inhibition

Renshaw cell

Axon collaterals

Inhibitory interneuron

To agonist (e.g., flexor)

To antagonist (e.g., extensor)

Central Conduction of Sensory Input

The **posterior funiculus–lemniscus system** (→ **C**, green) is the principal route by which the somatosensory cortex S1 (postcentral gyrus) receives sensory input from skin sensors and propriosensors. Messages from the skin (*superficial sensibility*) and locomotor system (*proprioceptive sensibility*) reach the spinal cord via the *dorsal roots*. Part of these primarily afferent fibers project in *tracts of the posterior funiculus* without synapses in the *posterior funicular nuclei* of the caudal medulla oblongata (nuclei cuneatus and gracilis). The tracts of the posterior funiculi exhibit a somatotopic arrangement, i.e., the further cranial the origin of the fibers the more lateral their location. At the *medial lemniscus*, the secondary afferent somatosensory fibers cross to the contralateral side (decussate) and continue to the posterolateral ventral nucleus (PLVN) of the **thalamus**, where they are also somatotopically arranged. The secondary afferent trigeminal fibers (lemniscus trigeminalis) end in the *posteromedial ventral nucleus* (PMVN) of the thalamus. The tertiary afferent somatosensory fibers end at the quaternary somatosensory neurons in the **somatosensory cortex S1**. The main **function** of the posterior funiculus–lemniscus pathway is to relay information about tactile stimuli (pressure, touch, vibration) and joint position and movement (proprioception) to the brain cortex via its predominantly rapidly conducting fibers with a high degree of spatial and temporal resolution.

As in the motor cortex (→ p. 325 B), each body part is assigned to a corresponding projection area in the **somatosensory cortex S1** (→ **A**) following a *somatotopic arrangement* (→ **B**). Three features of the organization of S1 are (1) that one hemisphere of the brain receives the information from the contralateral side of the body (tracts decussate in the medial lemniscus; → **C**); (2) that most neurons in S1 receive afferent signals from tactile sensors in the fingers and mouth (→ p. 314); and (3) that the afferent signals are processed in columns of the cortex (→ p. 333 A) that are activated by specific types of stimuli (e.g., touch).

Anterolateral spinothalamic pathway (→ **C**; violet). Afferent signals from nocisensors, thermosensors, and the second part of pressure and touch afferent neurons are already relayed (partly via interneurons) at various levels of the *spinal cord*. The secondary neurons cross to the opposite side at the corresponding segment of the spinal cord, form the lateral and ventral *spinothalamic tract* in the anterolateral funiculus, and project to the thalamus.

Descending tracts (from the cortex) can inhibit the flow of sensory input to the cortex at all relay stations (spinal cord, medulla oblongata, thalamus). The main function of these tracts is to modify the receptive field and adjust stimulus thresholds. When impulses from different sources are conducted in a common afferent, they also help to suppress unimportant sensory input and selectively process more important and interesting sensory modalities and stimuli (e.g., eavesdropping).

Hemiplegia. (→ **D**) *Brown–Séquard syndrome* occurs due to hemisection of the spinal cord, resulting in ipsilateral paralysis and loss of various functions below the lesion. The injured side exhibits *motor paralysis* (initially flaccid, later spastic) and loss of tactile sensation (e.g., impaired two-point discrimination, → p. 314). An additional loss of pain and temperature sensation occurs on the contralateral side (*dissociated paralysis*).

Reticular activating system. (→ **E**) The sensory input described above as well as the input from the sensory organs are specific, whereas the reticular activating system (RAS) is an *unspecific system*. The RAS is a complex processing and integrating system of cells of the *reticular formation* of the brainstem. These cells receive *sensory input* from all sensory organs and ascending spinal cord pathways (e.g., eyes, ears, surface sensitivity, nociception), basal ganglia, etc. Cholinergic and adrenergic output from the RAS is conducted along descending pathways to the spinal cord and along ascending "unspecific" thalamic nuclei and "unspecific" thalamocortical tracts to almost all cortical regions (→ p. 333 A), the limbic system and the hypothalamus. The ascending RAS or **ARAS** controls the state of consciousness and the degree of wakefulness (*arousal activity*).

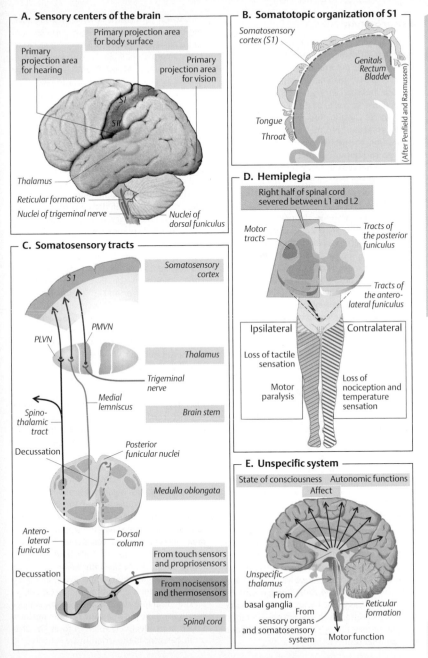

A. Sensory centers of the brain

Primary projection area for body surface

Primary projection area for hearing

Primary projection area for vision

SI

SII

Thalamus

Reticular formation

Nuclei of trigeminal nerve

Nuclei of dorsal funiculus

B. Somatotopic organization of S1

Somatosensory cortex (S1)

Genitals
Rectum
Bladder

Tongue

Throat

(After Penfield and Rasmussen)

C. Somatosensory tracts

Somatosensory cortex

S1

PLVN

PMVN

Thalamus

Trigeminal nerve

Medial lemniscus

Brain stem

Spino-thalamic tract

Decussation

Posterior funicular nuclei

Medulla oblongata

Antero-lateral funiculus

Dorsal column

Decussation

From touch sensors and propriosensors

From nocisensors and thermosensors

Spinal cord

D. Hemiplegia

Right half of spinal cord severed between L1 and L2

Motor tracts

Tracts of the posterior funiculus

Tracts of the antero-lateral funiculus

Ipsilateral

Contralateral

Loss of tactile sensation

Motor paralysis

Loss of nociception and temperature sensation

E. Unspecific system

State of consciousness   Autonomic functions

Affect

Unspecific thalamus

From basal ganglia

From sensory organs and somatosensory system

Reticular formation

Motor function

Plate 12.7   Central Conduction of Sensory Input

323

Motor System

Coordinated muscular movements (walking, grasping, throwing, etc.) are functionally dependent on the *postural motor system*, which is responsible for maintaining upright posture, balance, and spatial integration of body movement. Since control of postural motor function and muscle coordination requires the simultaneous and uninterrupted flow of sensory impulses from the periphery, this is also referred to as **sensorimotor function**.

α **motoneurons** in the anterior horn of the spinal cord and in cranial nerve nuclei are the terminal tracts for skeletal muscle activation. Only certain parts of the corticospinal tract and type Ia afferents connect to α motoneurons monosynaptically. Other afferents from the periphery (propriosensors, nocisensors, mechanosensors), other spinal cord segments, the motor cortex, cerebellum, and motor centers of the brain stem connect to α motoneurons via hundreds of inhibitory and stimulatory interneurons per motoneuron.

Voluntary motor function. Voluntary movement requires a series of actions: decision to move ⇒ programming (recall of stored subprograms) ⇒ command to move ⇒ execution of movement (→ **A1–4**). Feedback from afferents (re-afferents) from motor subsystems and information from the periphery is constantly integrated in the process. This allows for adjustments before and while executing voluntary movement.

The neuronal activity associated with the first two phases of voluntary movement activates numerous motor areas of the cortex. This electrical brain activity is reflected as a negative **cortical expectancy potential**, which can best be measured in association areas and the vertex. The more complex the movement, the higher the expectancy potential and the earlier its onset (roughly 0.3–3 s).

The **motor cortex** consists of three main areas (→ **C**, top; → see p. 311 E for area numbers): (a) *primary motor area*, M1 (area 4), (b) *premotor area*, PMA (lateral area 6); and (c) *supplementary motor area*, SMA (medial area 6). The motor areas of the cortex exhibit somatotopic organization with respect to the target muscles of their fibers (shown for M1 in **B**) and their mutual connections.

Cortical afferents. The cortex receives motor input from (a) the *body periphery* (via thalamus ⇒ S1 [→ p. 323 A] ⇒ sensory association cortex ⇒ PMA); (b) the *basal ganglia* (via thalamus ⇒ M1, PMA, SMA [→ **A2**] ⇒ prefrontal association cortex); (c) the *cerebellum* (via thalamus ⇒ M1, PMA; → **A2**); and (d) sensory and posterior parietal areas of the *cortex* (areas 1–3 and 5–7, respectively).

Cortical efferents. (→ **C, D, E, F**) Motor output from the cortex is mainly projected to (a) the spinal cord, (b) subcortical motor centers (see below and p. 328), and (c) the contralateral cortex via commissural pathways.

The **pyramidal tract** includes the *corticospinal tract* and part of the *corticobulbar tract*. Over 90% of the pyramidal tract consists of thin fibers, but little is known about their function. The thick, rapidly conducting corticospinal tract (→ **C**) project to the spinal cord from areas 4 and 6 and from areas 1–3 of the sensory cortex. Some of the fibers connect monosynaptically to α and γ motoneurons responsible for finger movement (precision grasping). The majority synapse with interneurons of the spinal cord, where they influence input from peripheral afferents as well as motor output (via Renshaw's cells) and thereby spinal reflexes.

Function of the Basal Ganglia

Circuitry. The basal ganglia are part of multiple parallel **corticocortical signal loops**. *Associative loops* arising in the frontal and limbic cortex play a role in mental activities such as assessment of sensory information, adaptation of behavior to emotional context, motivation, and long-term action planning. The function of the *skeletomotor* and *oculomotor loops* (see below) is to coordinate and control the velocity of movement sequences. Efferent projections of the basal ganglia *control thalamocortical signal conduction* by (a) attenuating the inhibition (*disinhibiting effect*, direct mode) of the *thalamic motor nuclei* and the *superior colliculus*, respectively, or (b) by intensifying their inhibition (indirect mode).

The principal **input** to the basal ganglia comes from the putamen and caudate nucleus, which are collectively referred to as the *striatum*. Neurons of the striatum are activated by

▶

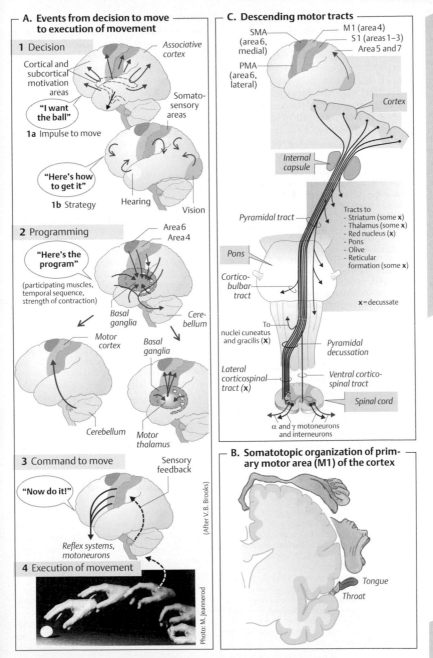

A. Events from decision to move to execution of movement

1 Decision

Cortical and subcortical motivation areas

Associative cortex

Somatosensory areas

"I want the ball"

1a Impulse to move

"Here's how to get it"

1b Strategy

Hearing

Vision

2 Programming

Area 6
Area 4

"Here's the program"

(participating muscles, temporal sequence, strength of contraction)

Basal ganglia

Cerebellum

Motor cortex

Basal ganglia

Cerebellum

Motor thalamus

3 Command to move

Sensory feedback

"Now do it!"

Reflex systems, motoneurons

4 Execution of movement

(After V. B. Brooks)

Photo: M. Jeannerod

C. Descending motor tracts

SMA (area 6, medial)

M1 (area 4)
S1 (areas 1–3)
Area 5 and 7

PMA (area 6, lateral)

Cortex

Internal capsule

Pyramidal tract

Tracts to
- Striatum (some **x**)
- Thalamus (some **x**)
- Red nucleus (**x**)
- Pons
- Olive
- Reticular formation (some **x**)

Pons

Cortico-bulbar tract

x = decussate

To nuclei cuneatus and gracilis (**x**)

Pyramidal decussation

Lateral corticospinal tract (**x**)

Ventral cortico-spinal tract

Spinal cord

α and γ motoneurons and interneurons

B. Somatotopic organization of primary motor area (M1) of the cortex

Tongue

Throat

Plate 12.8 **Motor System I**

325

tracts from the entire **cortex** and use *glutamate* as their transmitter (→ **D**). Once activated, neurons of the striatum release an inhibitory transmitter (GABA) and a co-transmitter— either *substance P* (SP) or *enkephalin* (Enk., → **D**; → p. 55). The principal **output** of the basal ganglia runs through the *pars reticularis of the substantia nigra* (**SNr**) and the *pars interna of the globus pallidus* (**GPi**), both of which are inhibited by SP/GABAergic neurons of the striatum (→ **D**).

Both SNr and GPi inhibit (by GABA) the ventrolateral thalamus with a high level of spontaneous activity. Activation of the striatum therefore leads to **disinhibition of the thalamus** by this **direct pathway**. If, however, enkephalin/GABA-releasing neurons of the striatum are activated, then they inhibit the *pars externa of the globus pallidus* (**GPe**) which, in turn, inhibits (by GABA) the subthalamic nucleus. The subthalamic nucleus induces glutamatergic activation of SNr and GPi. The ultimate effect of this **indirect pathway** is **increased thalamic inhibition**. Since the thalamus projects to motor and prefrontal cortex, a corticothalamocortical loop that influences skeletal muscle movement (**skeletomotor loop**) via the putamen runs through the basal ganglia. An *oculomotor loop* projects through the caudate nucleus, pars reticularis and superior colliculus and is involved in the control of eye movement (→ pp. 342, 360). *Descending tracts* from the SNr project to the tectum and nucleus pedunculus pontinus.

The fact that the *pars compacta of the substantia nigra* (**SNc**) showers the entire striatum with **dopamine** (dopaminergic neurons) is of pathophysiological importance (→ **D**). On the one hand, dopamine binds to D1 receptors (rising cAMP levels), thereby activating SP/GABAergic neurons of the striatum; this is the direct route (see above). On the other hand, dopamine also reacts with D2 receptors (decreasing cAMP levels), thereby inhibiting enkephalin/GABAergic neurons; this is the indirect route. These effects of dopamine are essential for normal striatum function. Degeneration of more than 70% of the dopaminergic neurons of the pars compacta results in *excessive inhibition* of the motor areas of the thalamus, thereby impairing voluntary motor function. This occurs in **Parkinson's disease** and can be due genetic predisposition, trauma (e.g., boxing), cerebral infection and other causes. The characteristic **symptoms** of disease include poverty of movement (*akinesia*), slowness of movement (*bradykinesia*), a festinating gait, small handwriting (*micrography*), masklike facial expression, muscular hypertonia (*rigor*), bent posture, and a *tremor* of resting muscles ("money-counting" movement of thumb and fingers).

Function of the Cerebellum

The cerebellum contains as many neurons as the rest of the brain combined. It is an important **control center for motor function** that has afferent and efferent connections to the cortex and periphery (→ **F, top** panel). The cerebellum is involved in the planning, execution and control of movement and is responsible for motor adaptation to new movement sequences (*motor learning*). It is also cooperates with higher centers to control attention, etc.

Anatomy (→ **F, top**). The *archeocerebellum* (flocculonodular lobe) and *paleocerebellum* (pyramids, uvula, paraflocculus and parts of the anterior lobe) are the phylogenetically older parts of the cerebellum. These structures and the pars intermedia form the **median cerebellum**. The *neocerebellum* (posterior lobe of the body of the cerebellum) is the phylogenetically younger part of the cerebellum and forms the **lateral cerebellum**. Based on the origin of their principal efferents, the archicerebellum and vermis are sometimes referred to as the *vestibulocerebellum*, the paleocerebellum as the *spinocerebellum*, and the neocerebellum as the *pontocerebellum*. The cerebellar cortex is the folded (fissured) superficial gray matter of the cerebellum consisting of an outer molecular layer of Purkinje cell dendrites and their afferents, a middle layer of Purkinje cells (Purkinje somata), and an inner layer of granular cells. The outer surface of the cerebellum exhibits small, parallel convolutions called *folia*.

The **median cerebellum** and pars intermedia of the cerebellum mainly control postural and supportive motor function (→ **F1,2**) and oculomotor function (→ pp. 342 and 360). **Input:** The median cerebellum receives *afference copies* of spinal, vestibular and ocular origin and *efference copies* of descending motor signals to the skeletal muscles. **Output** from the median cerebellum flows through the intracerebellar fastigial, globose, and emboliform nuclei to motor centers of the spinal cord and brain stem and to extracerebellar vestibular nuclei (mainly Deiter's nucleus). These centers control oculomotor function and influence locomotor and postural/supportive motor function via the vestibulospinal tract.

The **lateral cerebellum** (hemispheres) mainly takes part in programmed movement (→ **F3**), but its plasticity also permits motor adaptation and the learning of motor sequences. The hemispheres have two-way

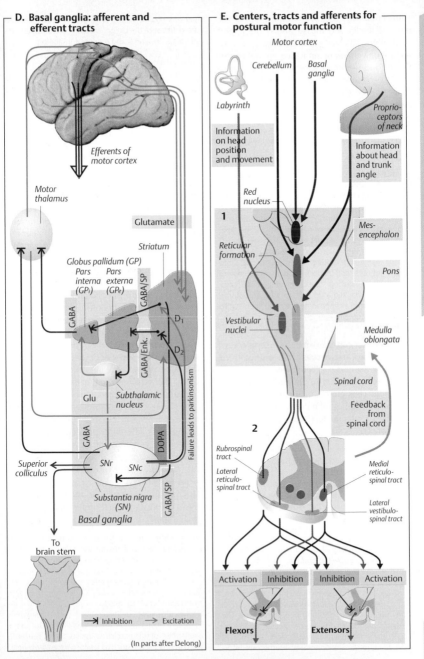

Plate 12.9 Motor System II

D. Basal ganglia: afferent and efferent tracts

Efferents of motor cortex

Motor thalamus

Glutamate

Striatum

Globus pallidum (GP)
Pars interna (GP$_i$) Pars externa (GP$_e$)

GABA/SP

GABA

GABA/Enk.

D$_1$

D$_2$

Failure leads to parkinsonism

Glu

Subthalamic nucleus

GABA

DOPA

Superior colliculus

SNr SNc

Substantia nigra (SN)

GABA/SP

Basal ganglia

To brain stem

→ Inhibition   → Excitation

(In parts after Delong)

E. Centers, tracts and afferents for postural motor function

Motor cortex

Cerebellum Basal ganglia

Labyrinth

Proprioceptors of neck

Information on head position and movement

Information about head and trunk angle

Red nucleus

1

Reticular formation

Mesencephalon

Pons

Vestibular nuclei

Medulla oblongata

Spinal cord

Feedback from spinal cord

2

Rubrospinal tract

Lateral reticulospinal tract

Medial reticulospinal tract

Lateral vestibulospinal tract

Activation Inhibition   Inhibition Activation

Flexors   Extensors

connections to the cortex. **Input**: **a.** Via the pontine nuclei and mossy fibers, the lateral cerebellum receives input from cortical centers for movement planning (e.g., parietal, prefrontal and premotor *association cortex*; sensorimotor and visual areas). **b.** It also receives input from cortical and subcortical motor centers via the inferior olive and climbing fibers (see below). **Output** from the lateral cerebellum projects across motor areas of the thalamus from the dentate nucleus to motor areas of the cortex.

Lesions of the *median cerebellum* lead to disturbances of balance and oculomotor control (vertigo, nausea, pendular nystagmus) and cause trunk and gait ataxia. Lesions of the *lateral cerebellum* lead to disturbances of initiation, coordination and termination of goal-directed movement and impair the rapid reprogramming of diametrically opposing movement (diadochokinesia). The typical patient exhibits tremor when attempting voluntary coordinated movement (*intention tremor*), difficulty in measuring the distances during muscular movement (*dysmetria*), pendular rebound motion after stopping a movement (*rebound phenomenon*), and inability to perform rapid alternating movements (*adiadochokinesia*).

The cerebellar cortex exhibits a uniform neural **ultrastructure and circuitry**. All **output** from the cerebellar cortex is conducted via neurites of approximately 15×10^6 Purkinje cells. These GABAergic cells project to and *inhibit* neurons of the fastigial, emboliform, dentate, and lateral vestibular nuclei (Deiter's nucleus; → **F**, **right** panel).

Input and **circuitry**: Input from the spinal cord (spinocerebellar tracts) is relayed by the *inferior olive* and projected via stimulatory (1 : 15 diverging) **climbing fibers** that terminate on a band of Purkinje cells extending across the folia of the cerebellum, forming the *sagittal excitatory foci*. The climbing fibers use aspartate as their transmitter. Serotoninergic fibers from the raphe nuclei and noradrenergic fibers from the locus caeruleus terminate also on the excitatory foci. **Mossy fibers** (pontine, reticular and spinal afferents) excite the granular cells. Their axons form T-shaped branches (*parallel fibers*). In the molecular layer, they densely converge (ca. $10^5 : 1$) on strips of Purkinje cells that run alongside the folium; these are called *longitudinal excitatory foci*. It is assumed that the climbing fiber system (at the "crossing points" of the perpendicular excitatory foci) amplify the relatively weak signals of mossy fiber afferents to Purkinje cells. Numerous interneurons (Golgi, stellate and basket cells) heighten the contrast of the excitatory pattern on the cerebellar cortex by lateral and recurrent inhibition.

Postural Motor Control

Simple *stretch reflexes* (→ p. 316)) as well as the more complicated *flexor reflexes* and *crossed extensor reflexes* (→ p. 320) are controlled at the level of the **spinal cord**.

Spinal cord transection (paraplegia) leads to an initial loss of peripheral reflexes below the lesion (areflexia, spinal shock), but the reflexes can later be provoked in spite of continued transection.

The spinal reflexes are mainly subordinate to supraspinal centers (→ **E**). Postural motor function is chiefly controlled by motor centers of the brain stem (→ **E1**), i.e., the *red nucleus,* vestibular nuclei (mainly *lateral vestibular nucleus*), and parts of the *reticular formation*. These centers function as relay stations that pass along information pertaining to postural and labyrinthine postural reflexes required to maintain *posture* and *balance* (involuntary). Postural reflexes function to regulate muscle tone and eye adaptation movements (→ p. 343 C). **Input** is received from the equilibrium organ (*tonic labyrinthine reflexes*) and from propriosensors in the neck (*tonic neck reflexes*). The same afferents are involved in postural reflexes (labyrinthine and neck reflexes) that help to maintain the body in its normal position. The trunk is first brought to its normal position in response to inflow from neck proprioceptors. Afferents projecting from the cerebellum, cerebral motor cortex (→ **C**), eyes, ears, and olfactory organ as well as skin receptors also influence postural reflexes. *Statokinetic reflexes* also play an important role in the control of body posture and position. They play a role e.g. in startle reflexes and nystagmus (→ p. 360).

Descending tracts to the spinal cord arising from the red nucleus and medullary reticular formation (*rubrospinal* and *lateral reticulospinal tracts*) have a generally inhibitory effect on α and γ motoneurons (→ p. 316) of extensor muscles and an excitatory effect on flexor muscles (→ **E2**). Conversely, the tracts from Deiter's nucleus and the pontine areas of the reticular formation (*vestibulospinal* and *medial reticulospinal tracts*) inhibit the flexors and excite the α and γ fibers of the extensors. Transection of the brain stem below the red nucleus leads to *decerebrate rigid-*

F. Tracts and function of cerebellum

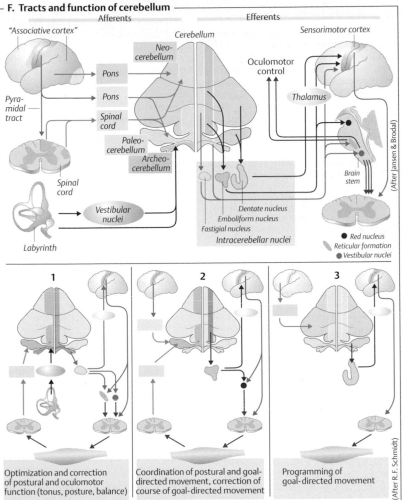

Afferents — **Efferents**

"Associative cortex" — Cerebellum — Sensorimotor cortex

Neo-cerebellum

Oculomotor control

Pons

Pyra-midal tract

Pons

Thalamus

Spinal cord

Paleo-cerebellum

Archeo-cerebellum

Brain stem

Spinal cord

Vestibular nuclei

Dentate nucleus
Emboliform nucleus
Fastigial nucleus
Intracerebellar nuclei

Labyrinth

● Red nucleus
Reticular formation
● Vestibular nuclei

(After Jansen & Brodal)

1 Optimization and correction of postural and oculomotor function (tonus, posture, balance)

2 Coordination of postural and goal-directed movement, correction of course of goal-directed movement

3 Programming of goal-directed movement

(After R. F. Schmidt)

Plate 12.10 Motor System III

---

ity because the extensor effect of Deiter's nucleus predominates.

The **integrating and coordinating function** of the sensorimotor system can be illustrated in two tennis players. When one player serves, the body of the other player moves to meet the ball (*goal-directed movement*) while using the right leg for support and the left arm for balance (*postural motor control*). The player keeps his eye on the ball (*oculomotor control*) and the visual area of the cortex assesses the trajectory and velocity of the ball. The associative cere-bral cortex initiates the movement of returning the ball while taking the ball, net, other side of the court, and position of the opponent into consideration. Positional adjustments may be necessary when returning the ball. Using the movement concept programmed in the cerebellum and basal ganglia, the motor cortex subsequently executes the directed movement of returning the ball. In doing so, the player may "slice" the ball to give it an additional spinning motion (*acquired rapid directed movement*).

Hypothalamus, Limbic System

The **hypothalamus** coordinates all autonomic and most endocrine processes ($\to$ p. 266ff.) and integrates signals for control of internal milieu, sleep–wake cycle, growth, mental/physical development, reproduction and other functions. The hypothalamus receives numerous sensory and humoral signals ($\to$ **A**). Peptide hormones can circumvent the blood–brain barrier by way of the *circumventricular organs* ($\to$ p. 280).

Afferents. *Thermosensors* for control of body temperature ($\to$ p. 224), *osmosensors* for regulation of osmolality and water balance ($\to$ p. 168), and glucose sensors for maintenance of a minimum *glucose concentration* are located within the hypothalamus. Information about the current status of the internal milieu is neuronally projected to the hypothalamus from distant sensors, e.g., thermosensors in the *skin*, osmosensors in the *liver* ($\to$ p. 170), and stretch sensors in the cardiac *atria* ($\to$ p. 214ff.). The hypothalamus/circumventricular organs also contain receptors for various hormones (e.g., cortisol and angiotensin II), some of which form part of control loops for energy metabolism and metabolic homeostasis (e.g., receptors for cortisol, ACTH, CRH, leptin, and CCK). For functions related to *growth* and *reproduction*, the hypothalamus receives hormonal signals from the gonads and input from neuronal afferents that report cervical widening at the beginning of the birth process and breast stimulation (suckling reflexes), among other things.

The **limbic system** ($\to$ **A**) and other areas of the brain influence hypothalamic function. The limbic system controls inborn and acquired *behavior* ("*program selection*") and is the seat of instinctive behavior, emotions and motivation ("inner world"). It controls the expression of emotions conveying important *signals to the environment* (e.g., fear, anger, wrath, discomfort, joy, happiness). Inversely, signals from the environment (e.g., odors) are closely associated to behavior.

The limbic system has *cortical components* (hippocampus, parahippocampal gyrus, cingulate gyrus, parts of olfactory brain) and *subcortical components* (amygdaloid body, septal nuclei, anterior thalamic nucleus). It has reciprocal connections to the lateral hypothalamus (chiefly used for recall of "programs", see below) and to the temporal and frontal cortex. Its connections to the **cortex** are primarily used to perceive and assess signals from the "outer world" and from memories. Processing of both types of input is important for behavior.

Programmed behavior ($\to$ **A**). The lateral hypothalamus has various programs to control lower hormonal, autonomic and motor processes. This is reflected internally by numerous autonomic and hormonal activities, and is reflected outwardly by different types of behavior.

Different programs exist for different behavioral reactions, for example:

◆ **Defensive behavior** ("fight or flight"). This program has somatic (repulsive facial expression and posture, flight or fight behavior), hormonal (epinephrine, cortisol) and autonomic (sympathetic nervous system) components. Its activation results in the release of energy-rich free fatty acids, the inhibition of insulin release, and a decrease in blood flow to the gastrointestinal tract as well as to rises in cardiac output, respiratory rate, and blood flow to the skeletal muscles.

◆ **Physical exercise**. The components of this program are similar to those of defensive behavior.

◆ **Nutritive behavior**, the purpose of which is to ensure an adequate supply, digestion and intake of foods and liquids. This includes searching for food, e.g. in the refrigerator, activation of the parasympathetic system with increased gastrointestinal secretion and motility in response to food intake, postprandial reduction of skeletal muscle activity and similar activities.

◆ **Reproductive behavior**, e.g., courting a partner, neuronal mechanisms of sexual response, hormonal regulation of pregnancy ($\to$ p. 304), etc.

◆ **Thermoregulatory behavior**, which enables us to maintain a relatively constant core temperature ($\to$ p. 224), even in extreme ambient temperatures or at the high level of heat production during strenuous physical work.

Monoaminergic neuron systems contain neurons that release the monoamine neurotransmitters norepinephrine, epinephrine, dopamine, and serotonin. These neuron tracts extend from the brain stem to almost all parts of the brain and play an important role in the overall regulation of behavior. Experimental activation of noradrenergic neurons, for example, led to positive reinforcement (liking, rewards), whereas the serotoninergic neurons are thought to be associated with dislike. A number of psychotropic drugs target monoaminergic neuron systems.

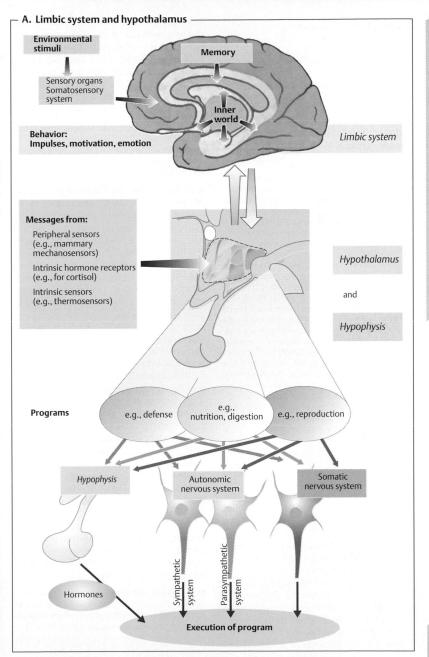

A. Limbic system and hypothalamus

Environmental stimuli

Sensory organs
Somatosensory system

Memory

Inner world

Behavior:
Impulses, motivation, emotion

Limbic system

Messages from:

Peripheral sensors
(e.g., mammary mechanosensors)

Intrinsic hormone receptors
(e.g., for cortisol)

Intrinsic sensors
(e.g., thermosensors)

Hypothalamus

and

Hypophysis

Programs

e.g., defense

e.g., nutrition, digestion

e.g., reproduction

Hypophysis

Autonomic nervous system

Somatic nervous system

Hormones

Sympathetic system

Parasympathetic system

Execution of program

Plate 12.11 Hypothalamus, Limbic System

Cerebral Cortex, Electroencephalogram (EEG)

Proper function of the cerebral cortex is essential for *conscious perception, planning, action,* and *voluntary movement* (→ p. 322ff.).

Cortical ultrastructure and **neuronal circuitry** (→ **A**). The cerebral cortex consists of six layers, I–VI, lying parallel to the brain surface. Vertically, it is divided into columns and modules (diameter 0.05–0.3 mm, depth 1.3–4.5 mm) that extend through all six layers.

Input from specific and unspecific areas of the thalamus terminate mainly on layers IV and on layers I and II, respectively (→ **A3**); those from other areas of the cortex terminate mainly on layer II (→ **A2**). The large and small **pyramidal cells** (→ **A1**) comprise 80% of all cells in the cortex and are located in layers V and III, respectively (glutamate generally serves as the transmitter, e.g., in the striatum; → p. 325 D). The pyramidal cell axons leave the layer VI of their respective columns and are the sole source of **output** from the cortex. Most of the axons project to other areas of the ipsilateral cortex (association fibers) or to areas of the contralateral cortex (commissural fibers) (→ **A2**); only a few extend to the periphery (→ **A4** and p. 325 C). Locally, the pyramidal cells are connected to each other by axon collaterals. The *principal dendrite* of a pyramidal cell projects to the upper layers of its column and has many thorn-like processes (**spines**) where many thalamocortical, commissural and association fibers terminate. The afferent fibers utilize various transmitters, e.g., norepinephrine, dopamine, serotonin, acetylcholine and histamine. Inside the cerebral cortex, information is processed by many morphologically variable **stellate cells** (→ **A1**), some of which have stimulatory effects (VIP, CCK and other peptide transmitters), while others have inhibitory effects (GABA). Dendrites of pyramidal and stellate cells project to neighboring columns, so the columns are connected by thousands of threads. **Plasticity** of pyramidal cell synapses — i.e., the fact that they can be modified in conformity with their activity pattern — is important for the learning process (→ p. 336).

Cortical potentials. Similar to electrocardiography, collective fluctuations of electrical potentials (brain waves) in the cerebral cortex can be recorded by **electroencephalography** using electrodes applied to the skin over the cranium (→ **B**). The EPSPs contribute the most to the electroencephalogram (**EEG**) whereas the share of the relatively low IPSPs (→ p. 50ff.) generated at the synapses of pyramidal cell dendrites is small. Only a portion of the rhythms recorded in the EEG are produced directly in the cortex (α and γ waves in conscious perception; see below). Lower frequency waves from other parts of the brain, e.g. α waves from the thalamus and θ waves from the hippocampus, are "forced on" the cortex (*brain wave entrainment*).

By convention, downward **deflections of the EEG** are positive. Generally speaking, depolarization (excitation) of deeper layers of the cortex and hyperpolarization of superficial layers cause downward deflection (+) and vice versa.

Brain wave types. The electrical activity level of the cortex is mainly determined by the degree of *wakefulness* and can be distinguished based on the amplitude (a) and frequency (f) of the waves (→ **B, C**). α **Waves** (f ≈ 10 Hz; a ≈ 50 μV), which predominate when an adult subject is awake and relaxed (with eyes closed), are generally detected in multiple electrodes (*synchronized activity*). When the eyes are opened, other sensory organs are stimulated, or the subject solves a math problem, the α waves subside (α blockade) and **β waves** appear (f ≈ 20 Hz). The amplitude of β waves is lower than that of α waves, and they are chiefly found in occipital (→ **B**) and parietal regions when they eyes are opened. The frequency and amplitude of β waves varies greatly in the different leads (*desynchronization*). β Waves reflect the increased attention and activity (arousal activity) of the *ascending reticular activating system* (ARAS; → p. 322). **γ Waves** (> 30 Hz) appear during learning activity. Low-frequency **θ waves** appear when drowsiness descends to sleep (sleep stages A/B/C; → **D**); they transform into even slower **δ waves** during deep sleep (→ **C, D**).

The EEG is used to diagnose epilepsy (localized or generalized paroxysmal waves and spikes; → **C**), to assess the degree of brain maturation, monitor anesthesia, and to determine brain death (*isoelectric EEG*).

Magnetoencephalography (MEG), i.e. recording magnetic signals induced by cortical ion currents, can be combined with the EEG to precisely locate the site of cortical activity (resolution a few mm).

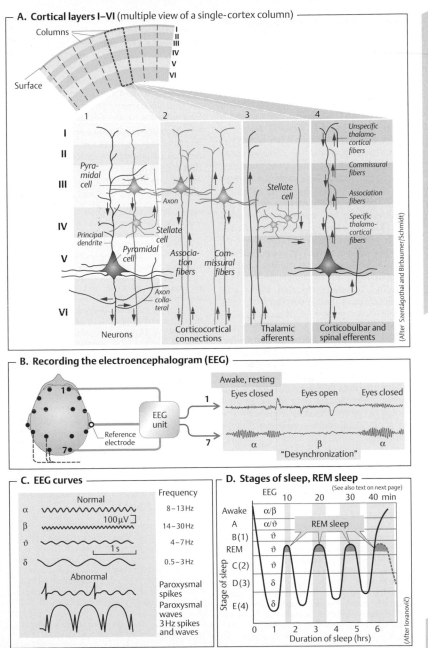

A. Cortical layers I–VI (multiple view of a single-cortex column)

Columns

I
II
III
IV
V
VI

Surface

1   2   3   4

I
II
Pyra-
midal
cell
III   *Axon*   *Stellate cell*   *Unspecific thalamo-cortical fibers*
Commissural fibers
Association fibers
Stellate cell
IV   Principal dendrite   *Specific thalamo-cortical fibers*
Pyramidal cell
V   Associa-tion fibers   Com-missural fibers
Axon colla-teral
VI

Neurons   Corticocortical connections   Thalamic afferents   Corticobulbar and spinal efferents

(After Szentágothai and Birbaumer/Schmidt)

B. Recording the electroencephalogram (EEG)

1
7
Reference electrode

EEG unit

1
7

Awake, resting
Eyes closed   Eyes open   Eyes closed

α   β   α
"Desynchronization"

C. EEG curves

Normal   Frequency
α ∿∿∿∿∿∿∿   8–13 Hz
β ∿∿∿∿∿∿∿ ⌐100µV   14–30 Hz
ϑ ∿∿∿∿∿   4–7 Hz
⌐ 1 s
δ ∿∿   0.5–3 Hz

Abnormal
          Paroxysmal spikes
          Paroxysmal waves
          3 Hz spikes and waves

D. Stages of sleep, REM sleep

(See also text on next page)

EEG   10   20   30   40 min
Awake   α/β
A   α/ϑ   REM sleep
B(1)   ϑ
REM   ϑ
C(2)   ϑ
D(3)   δ
E(4)   δ

Stage of sleep

0   1   2   3   4   5   6
Duration of sleep (hrs)

(After Jovanović)

Plate 12.12   Cerebral Cortex EEG

Sleep–Wake Cycle, Circadian Rhythms

Various **stages of sleep** can be identified in the EEG (→ p. 333 D). When a normal person who is awake, relaxed and has the eyes closed (α waves) starts to fall asleep, the level of consciousness first descends to *sleep phase A* (dozing), where only a few isolated α waves can be detected. Drowsiness further descends to *sleep stage B* (stage 1), where θ waves appear, then to *stage C* (stage 2), where burst of fast waves (sleep spindles) and isolated waves (K complexes) can be recorded, and ultimately to the stages of *deep sleep* (stages D/E = stages 3/4), characterized by the appearance of δ waves. Their amplitude increases while their frequency drops to a minimum in phase E (→ p. 333 D). This phase is therefore referred to as *slow-wave sleep* (**SWS**). The arousal threshold is highest about 1 hour after a person falls asleep. Sleep then becomes less deep and the first episode of *rapid eye movement* (**REM**) occurs. This completes the first *sleep cycle*. During REM sleep, most of the skeletal muscles become atonic (inhibition of motoneurons) while the breathing and heart rates increase. The face and fingers suddenly start to twitch, and penile erection and rapid eye movements occur. All other stages of sleep are collectively referred to as *non-REM sleep* (**NREM**). Sleepers aroused from REM sleep are more often able to describe their dreams than when aroused from NREM sleep. The **sleep cycle** normally lasts about 90 min and is repeated 4–5 times each night (→ p. 333 D). Towards morning, NREM sleep becomes shorter and more even, while the REM episodes increase from ca. 10 min to over 30 min.

Infants sleep longest (about 16 hours/day, 50% REM), 10-year-olds sleep an average 10 hours (20% REM), young adults sleep 7–8 hours a day, and adults over 50 sleep an average 6 hours or so (both 20% REM). The proportion of SWS clearly decreases in favor of stage C (stage 2) sleep.

When a person is **deprived of REM sleep** (awakened during this phase), the duration of the next REM phase increases to compensate for the deficit. The first two to three sleep cycles (**core sleep**) are *essential*. Total **sleep deprivation** leads to death, but the reason is still unclear because too little is known about the physiological role of sleep.

The daily sleep–wake cycle and other **circadian rhythms** (diurnal rhythms) are controlled by *endogenous rhythm generators*. The *central biological clock* (*oscillator*) that times these processes is located in the *suprachiasmatic nucleus* (**SCN**) of the hypothalamus (→ **A**). The endogenous circadian rhythm occurs in cycles of roughly 24–25 hours, but is unadulterated only when a person is completely isolated from the outside influences (e.g., in a windowless basement, dark cave, etc.). External *zeitgebers* (entraining signals) synchronize the biological clock to precise 24-hour cycles. It takes several days to "reset" the biological clock, e.g., after a long journey from east to west (*jet lag*).

Important genetic **"cogwheels"** of the central biological clock of mammals were recently discovered (→ **A1**). Neurons of the SCN contain specific proteins (CLOCK and BMAL1), the PAS domains of which bind to form heterodimers. The resulting CLOCK/BMAL1 complexes enter the cell nuclei, where their promoter sequences (E-box) bind to period (*per*) oscillator genes *per*1, *per*2, and *per*3, thereby activating their transcription. After a latency period, expression of the genes yields the proteins PER1, PER2, and PER3, which jointly function as a trimer to block the effect of CLOCK/BMAL1, thereby completing the negative feedback loop. The mechanism by which this cycle activates subsequent neuronal actions (membrane potentials) is still unclear.

The main **external zeitgeber** for 24-hour synchronization of the sleep–wake cycle is *bright light* (photic entrainment). Light stimuli are directly sensed by a small, *melanopsin*-containing fraction of retinal ganglion cells and conducted to the SCN via the retinohypothalamic tract (→ **A2,3**). The coupled cells of the SCN (→ **A3**) bring about circadian rhythms of hormone secretion, core temperature, and sleep–wake cycles (→ **A5, B**) by various effector systems of the CNS (→ **A4**).

The *zeitgeber* slows or accelerates the rhythm, depending on which phase it is in. Signals from the *zeitgeber* also reaches the **epiphysis** (pineal body, pineal gland) where it inhibits the secretion of **melatonin** which is high at night. Since it exerts its effects mainly on the SCN, administration of melatonin before retiring at night can greatly reduce the time required to "reset" the biological clock. The main reason is that it temporarily "deactivates" the SCN (via MT$_2$ receptors), thereby excluding most nocturnal neuronal input (except light stimuli).

Plate 12.13 Sleep–Wake Cycle, Circadian Rhythms

A. Circadian rhythm generator in suprachiasmatic nucleus (SCN)

1 Genetic feedback loop in SCN cells (oscillator)

Neuron of SCN

CLOCK
PAS motif
BMAL1

PER 1
PER 2
PER 3

Translation

Transcription

RNA

per 1
E-box

per 2
E-box

per 3
E-box

Nucleus   *Cytosol*

Intracellular Cl⁻ concentration **?**

2 Zeitgeber

Bright light

Retina, etc.

Melatonin secretion

3 Coupled oscillators in SCN

Membrane potential

4 Effector systems in CNS

Neocortex
Hypo-thalamus   *Thalamus*

CRH

Sleep-wake rhythm

Core temperature

5
Circadian rhythms of:
– CRH secretion
– Core temperature,
– Sleep-wake cycle (see **B**), etc.

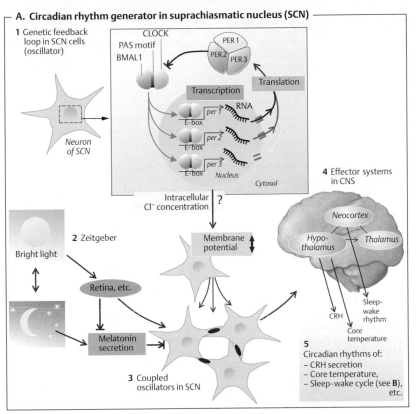

B. Circadian rhythm of sleep-wake cycle

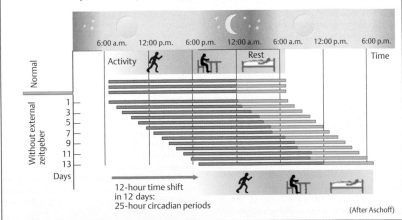

6:00 a.m. 12:00 p.m. 6:00 p.m. 12:00 a.m. 6:00 a.m. 12:00 p.m. 6:00 p.m.

Activity   Rest   Time

Normal

Without external zeitgeber

1
3
5
7
9
11
13
Days

12-hour time shift in 12 days:
25-hour circadian periods

(After Aschoff)

Consciousness, Memory, Language

Consciousness. Selective attention, abstract thinking, the ability to verbalize experiences, the capacity to plan activities based on experience, self-awareness and the concept of values are some of the many characteristics of *consciousness.* Consciousness enables us to deal with difficult environmental conditions (adaptation). Little is known about the brain activity associated with consciousness and controlled attention (LCCS, see below), but we do know that subcortical activation systems such as the reticular formation ($\rightarrow$ p. 322) and corticostriatal systems that inhibit the afferent signals to the cortex in the thalamus ($\rightarrow$ p. 326) play an important role.

Attention. Sensory stimuli arriving in the *sensory memory* are evaluated and compared to the contents of the long-term memory within fractions of a second ($\rightarrow$ **A**). In routine situations such as driving in traffic, these stimuli are unconsciously processed (*automated attention*) and do not interfere with other reaction sequences such as conversation with a passenger. Our *conscious, selective (controlled) attention* is stimulated by novel or ambiguous stimuli, the reaction to which (e.g., the setting of priorities) is controlled by vast parts of the brain called the *limited capacity control system* (**LCCS**). Since our capacity for selective attention is limited, it normally is utilized only in stress situations.

The **implicit memory** (procedural memory) stores skill-related information and information necessary for associative learning (conditioning of conditional reflexes; $\rightarrow$ p. 236) and non-associative learning (habituation and sensitization of reflex pathways). This type of unconscious memory involves the basal ganglia, cerebellum, motor cortex, amygdaloid body (emotional reactions) and other structures of the brain.

The **explicit memory** (declarative/knowledge memory) stores facts (semantic knowledge) and experiences (episodic knowledge, especially when experienced by selective attention) and *consciously* renders the data. Storage of information processed in the uni- and polymodal association fields is the responsibility of the temporal lobe system (hippocampus, perirhinal, entorhinal and parahippocampal cortex, etc.). It establishes the temporal and spatial *context* surrounding an experience and recurrently stores the information back into the *spines* of cortical dendrites in the association areas ($\rightarrow$ p. 322). The recurrence of a portion of the experience then suffices to recall the contents of the memory.

Explicit learning ($\rightarrow$ **A**) starts in the *sensory memory*, which holds the sensory impression automatically for less than 1 s. A small fraction of the information reaches the *primary memory* (*short-term memory*), which can retain about 7 units of information (e.g., groups of numbers) for a few seconds. In most cases, the information is also verbalized. *Long-term storage* of information in the *secondary memory* (*long-term memory*) is achieved by repetition (*consolidation*). The *tertiary memory* is the place where frequently repeated impressions are stored (e.g., reading, writing, one's own name); these things are never forgotten, and can be quickly recalled throughout one's lifetime. *Impulses circulating* in neuronal tracts are presumed to be the physiological correlative for short-term (primary) memory, whereas *biochemical mechanisms* are mainly responsible for long-term memory. Learning leads to long-term genomic changes. In addition, frequently repeated stimulation can lead to *long-term potentiation* (**LTP**) of synaptic connections that lasts for several hours to several days. The spines of dendrites in the cortex play an important role in LTP.

Mechanisms for LTP. Once receptors for AMPA (α-amino-3-hydroxy-5-methyl-4-isoxazolepropionic acid) are activated by the presynaptic release of glutamate ($\rightarrow$ p. 55 F), influxing Na^+ depolarizes the postsynaptic membrane. Receptors for NMDA (*N*-methyl-D-aspartic acid) are also activated, but the Ca^{2+} channels of the NMDA receptors are blocked by Mg^{2+}, thereby inhibiting the influx of Ca^{2+} until the Mg^{2+} block is relieved by depolarization. The cytosolic Ca^{2+} concentration $[Ca^{2+}]_i$ then rises. If this is repeated often enough, calmodulin mediates the autophosphorylation of CaM kinase II ($\rightarrow$ p. 36), which persists even after the $[Ca^{2+}]_i$ falls back to normal. CaM kinase II phosphorylates AMPA receptors (increases their conductivity) and promotes their insertion into the postsynaptic membrane, thereby enhancing synaptic transmission over longer periods of time (LTP).

A. Storage of information in the brain (explicit memory)

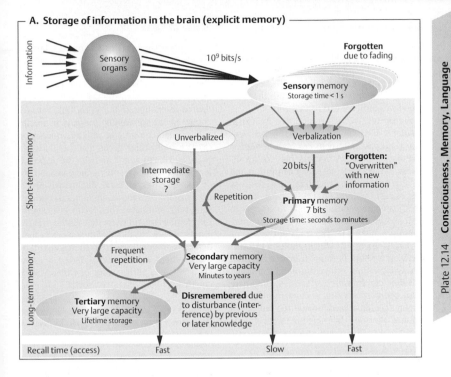

Information

10^9 bits/s

Sensory organs

Forgotten due to fading

Sensory memory
Storage time < 1 s

Short-term memory

Unverbalized

Verbalization

Intermediate storage ?

20 bits/s

Forgotten: "Overwritten" with new information

Repetition

Primary memory
7 bits
Storage time: seconds to minutes

Long-term memory

Frequent repetition

Secondary memory
Very large capacity
Minutes to years

Tertiary memory
Very large capacity
Lifetime storage

Disremembered due to disturbance (interference) by previous or later knowledge

Recall time (access)   Fast   Slow   Fast

Amnesia (memory loss). *Retrograde amnesia* (loss of memories of past events) is characterized by the loss of primary memory and (temporary) difficulty in recalling information from the secondary memory due to various causes (concussion, electroshock, etc.). *Anterograde amnesia* (inability to form new memories) is characterized by the inability to transfer new information from the primary memory to the secondary memory (*Korsakoff's syndrome*).

Language is a mode of *communication* used (1) to receive information through visual and aural channels (and through tactile channels in the blind) and (2) to transmit information in written and spoken form (see also p. 370). Language is also needed to form and verbalize *concepts and strategies* based on consciously processed sensory input. Memories can therefore be stored efficiently. The centers for formation and processing of concepts and language are unevenly distributed in the cerebral hemispheres. The left hemisphere is usually the main center of speech in right-handed individuals *("dominant" hemisphere*, large planum temporale), whereas the right hemisphere is dominant in 30–40% of all left-handers. The *non-dominant hemisphere* is important for word recognition, sentence melody, and numerous nonverbal capacities (e.g., music, spatial thinking, face recognition).

This can be illustrated using the example of patients in whom the two hemispheres are surgically disconnected due to conditions such as otherwise untreatable, severe epilepsy. If such a **split-brain patient** touches an object with the right hand (reported to the left hemisphere), he can name the object. If, however, he touches the object with the left hand (right hemisphere), he cannot name the object but can point to a picture of it. Since complete separation of the two hemispheres also causes many other severe disturbances, this type of surgery is used only in patients with otherwise unmanageable, extremely severe seizures.

Plate 12.14   Consciousness, Memory, Language

Glia

The central nervous system contains around 10^{11} nerve cells and 10 times as many **glia cells** such as *oligodendrocytes, astrocytes, ependymal cells*, and *microglia* ($\rightarrow$ **A**). **Oligodendrocytes** (ODC) form the myelin sheath that surrounds axons of the CNS ($\rightarrow$ **A**).

Astrocytes (AC) are responsible for **extracellular K^+ and H^+ homeostasis** in the CNS. Neurons release K^+ in response to high-frequency stimulation ($\rightarrow$ **B**). Astrocytes prevent an increase in the interstitial K^+ concentration and thus an undesirable depolarization of neurons (see Nernst equation, Eq. 1.18, p. 32) by taking up K^+, and intervene in a similar manner with H^+ ions. Since AC are connected by *gap junctions* ($\rightarrow$ p. 16ff.), they can transfer their K^+ or H^+ load to nearby AC ($\rightarrow$ **B**). In addition to forming a *barrier* that prevents transmitters from one synapse from being absorbed by another, AC also *take transmitters up*, e.g. glutamate (Glu). Intracellular Glu is converted to glutamine ($GluNH_2$), then transported out of the cell and taken up by the nerve cells, which convert it back to Glu (**transmitter recycling**; $\rightarrow$ **B**).

Some AC have **receptors for transmitters** such as Glu, which triggers a Ca^{2+} wave from one AC to another. Astrocytes are also able to modify the Ca^{2+} concentration in the neuronal cytosol so that the two cell types can "communicate" with each other. AC also mediate the transport of materials between capillaries and neurons and play an important part in *energy homeostasis* of the neurons by mediating glycogen synthesis and breakdown.

During **embryonal development**, the long processes of AC serve as *guiding structures* that help undifferentiated nerve cells migrate to their target areas. Glia cells also play an important role in CNS development by helping to control gene expression in nerve cell clusters with or without the aid of **growth factors** such as NGF (nerve growth factor), BDGF (brain-derived growth factor), and GDNF (glial cell line-derived neurotropic factor). GDNF also serves as a trophic factor for all mature neurons. Cell division of glia cells can lead to in scarring (*epileptic foci*) and tumor formation (*glioma*).

Immunocompetent **microglia** ($\rightarrow$ **A**) assume many functions of macrophages outside the CNS when CNS injuries or infections occur ($\rightarrow$ p. 94ff.). **Ependymal cells** line internal hollow cavities of the CNS ($\rightarrow$ **A**).

Sense of Taste

Gustatory pathways. The *taste buds* ($\rightarrow$ **D**) consist of clusters of 50–100 *secondary sensory cells* on the tongue (renewed in 2-week cycles); humans have around 5000 taste buds. Sensory stimuli from the taste buds are conducted to endings of the VIIth, IXth and Xth cranial nerves, relayed by the *nucleus tractus solitarii*, and converge at a high frequency on (a) the postcentral gyrus via the thalamus ($\rightarrow$ p. 323 B, "tongue") and (b) the hypothalamus and limbic system via the pons ($\rightarrow$ **C**).

The **qualities of taste** distinguishable in humans are conventionally defined as *sweet, sour, salty*, and *bitter*. The specific taste sensor cells for these qualities are distributed over the whole tongue but differ with respect to their densities. *Umami*, the sensation caused by monosodium-L-glutamate (MSG), is now classified as a fifth quality of taste. MSG is chiefly found in protein-rich foods.

Taste sensor cells distinguish the types of taste as follows: **Salty**: Cations (Na^+, K^+, etc.) taste salty, but the presence of anions also plays a role. E.g., Na^+ enters the taste sensor cell via Na^+ channels and depolarizes the cell. **Sour**: H^+ ions lead to a more frequent closure of K^+ channels, which also has a depolarizing effect. **Bitter**: A family of > 50 genes codes for an battery of bitter sensors. A number of sensory proteins specific for a particular substance are expressed in a single taste sensor cell, making it sensitive to different bitter tastes. The sensory input is relayed by the G-protein α-*gustducin*. No nuances but only the overall warning signal "bitter" is perceived. **Umami**: Certain taste sensor contain a metabotropic glutamate receptor, mGluR4, the stimulation of which leads to a drop in cAMP conc.

Taste thresholds. The threshold (mol/L) for recognition of taste stimuli applied to the tongue is roughly 10^{-5} for quinine sulfate and saccharin, 10^{-3} for HCl, and 10^{-2} for sucrose and NaCl. The relative *intensity differential threshold* $\Delta I/I$ ($\rightarrow$ p. 352) is about 0.20. The concentration of the gustatory stimulus determines whether its taste will be perceived as pleasant or unpleasant ($\rightarrow$ **E**). For the *adaptation* of the sense of taste, see p. 341 C.

Function of taste. The sense of taste has a protective function as spoiled or bitter-tasting food (low taste threshold) is often poisonous. Tasting substances also stimulate the secretion of saliva and gastric juices ($\rightarrow$ pp. 236, 242).

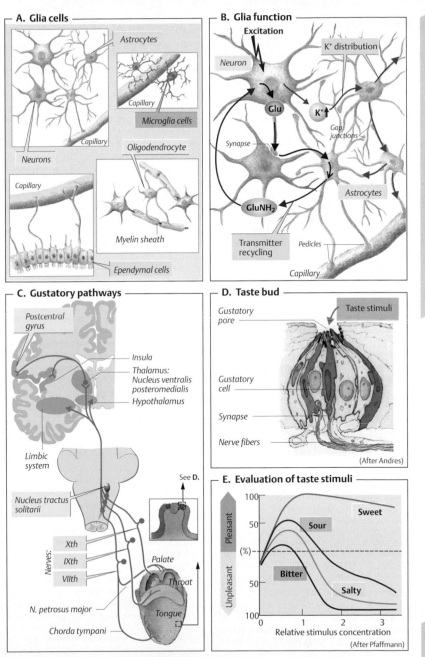

Plate 12.15 Glia, Sense of Taste

A. Glia cells

Astrocytes

Capillary

Microglia cells

Neurons

Capillary

Oligodendrocyte

Capillary

Myelin sheath

Ependymal cells

B. Glia function

Excitation

Neuron

K⁺ distribution

Glu

K⁺

Synapse

Gap junctions

GluNH₂

Astrocytes

Transmitter recycling

Pedicles

Capillary

C. Gustatory pathways

Postcentral gyrus

Insula

Thalamus: Nucleus ventralis posteromedialis

Hypothalamus

Limbic system

Nucleus tractus solitarii

See D.

Nerves:

Xth

IXth

VIIth

Palate

Throat

N. petrosus major

Tongue

Chorda tympani

D. Taste bud

Gustatory pore

Taste stimuli

Gustatory cell

Synapse

Nerve fibers

(After Andres)

E. Evaluation of taste stimuli

Pleasant

Unpleasant

(%)

Sweet

Sour

Bitter

Salty

Relative stimulus concentration

(After Pfaffmann)

339

Sense of Smell

The neuroepithelium of the *olfactory region* contains ca. 10^7 *primary* **olfactory sensor cells** ($\rightarrow$ **A1**) which are bipolar neurons. Their dendrites branch to form 5–20 mucus-covered cilia, whereas the axons extend centrally in bundles called *fila olfactoria* ($\rightarrow$ **A1,2**). Olfactory neurons are replenished by basal cell division in 30–60-day cycles. *Free nerve endings* (trigeminal nerve) in the nasal mucosa also react to certain aggressive odors (e.g., acid or ammonia vapors).

Olfactory sensors. Odorant molecules (M_r 15–300) are transported by the inhaled air to the olfactory region, where they first dissolve in the mucous lining before combining with **receptor proteins** in the *cilial membrane*. These are coded by a huge family of genes (500–750 genes distributed in most chromosomes), whereby probably one olfactory sensor cell only expresses one of these genes. Since only a part of the sequence of about 40% of these genes is expressed, humans have roughly 200–400 different **sensor cell types**. Olfactory receptors couple with G_s-proteins (G_{olf} proteins; $\rightarrow$ **B** and p. 274ff.) that increase the conductivity of the sensor cell membrane to cations, thereby increasing the influx of Na^+ and Ca^{2+} and thus depolarizing the cell.

Sensor specificity ($\rightarrow$ **A3**). Olfactory sensor cells recognize a very specific structural feature of the odorant molecules they are sensitive to. The cloned receptor 17 of the rat, for example, reacts with the aldehyde *n*-octanal but not to octanol, octanoic acid, or aldehydes which have two methyl groups more or less than *n*-octanal. In the case of aromatic compounds, one sensor recognizes whether the compound is *ortho*, *meta* or *para*-substituted, while another detects the length of the substituent regardless of where it is located on the ring. The different molecular moieties of an odorant molecule therefore activate different types of sensors ($\rightarrow$ **A3, top** right). Jasmine leaves and wine contain several dozens and hundreds of odorants, respectively, so their overall scent is a more complex perception (integrated in the rhinencephalon).

Olfactory pathway ($\rightarrow$ **A2**). Axons of (ca. 10^3) same-type sensors distributed over the olfactory epithelium synapse to dendrites of their respective *mitral cells* (MC) and *bristle cells* (BC) within the glomeruli olfactorii of the olfactory bulb. The glomeruli therefore function as convergence centers that integrate and relay signals from the same sensor type. Their respective sensor protein also determines which glomerulus newly formed sensor axons will connect to. *Periglomerular cells* and *granular cells* connect and inhibit mitral and bristle cells ($\rightarrow$ **A2**). Mitral cells act on the same *reciprocal synapses* ($\rightarrow$ **A**, "+/−") in reverse direction to activate the periglomerular cells and granular cells which, on the other hand, are inhibited by efferents from the primary olfactory cortex and contralateral anterior olfactory nucleus ($\rightarrow$ **A2**, violet tracts). These connections enable the cells to inhibit themselves or nearby cells (contrast), or they can be disinhibited by higher centers. The signals of the *axons of mitral cells* (1) reach the anterior olfactory nucleus. Its neurons cross over (in the anterior commissure) to the mitral cells of the contralateral bulb and (2) form the *olfactory tract* projecting to the *primary olfactory cortex* (prepiriform cortex, tuberculum olfactorium, nucleus corticalis amygdalae). The olfactory input processed there is relayed to the hypothalamus, limbic system (see also p. 330), and reticular formation; it is also relayed to the neocortex (*insula, orbitofrontal area*) either directly or by way of the thalamus.

Thresholds. It takes only 4×10^{-15} g of methylmercaptan (in garlic) per liter of air to trigger the vague sensation of smell (*perception* or *absolute threshold*). The odor can be properly identified when 2×10^{-13} g/L is present (*identification threshold*). Such thresholds are affected by air temperature and humidity; those for other substances can be 10^{10} times higher. The relative intensity differential threshold $\Delta I/I$ (0.25) is relatively high ($\rightarrow$ p. 352). *Adaptation* to smell is sensor-dependent (desensitization) and neuronal ($\rightarrow$ **C**).

The sense of smell has various **functions**. Pleasant smells trigger the secretion of saliva and gastric juices, whereas unpleasant smells warn of potentially spoiled food. Body odor permits hygiene control (sweat, excrement), conveys social information (e.g., family, enemy; $\rightarrow$ p. 330), and influences sexual behavior. Other aromas influence the emotional state.

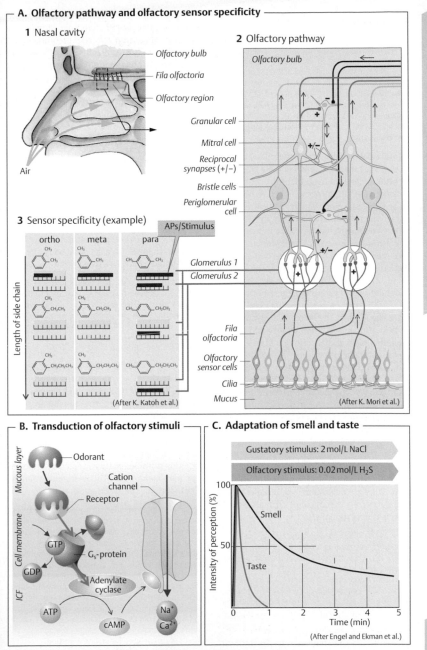

A. Olfactory pathway and olfactory sensor specificity

1 Nasal cavity

Olfactory bulb
Fila olfactoria
Olfactory region

Air

2 Olfactory pathway

Olfactory bulb

Granular cell
Mitral cell
Reciprocal synapses (+/−)
Bristle cells
Periglomerular cell

APs/Stimulus

Glomerulus 1
Glomerulus 2

Fila olfactoria
Olfactory sensor cells
Cilia
Mucus

(After K. Mori et al.)

3 Sensor specificity (example)

ortho   meta   para

Length of side chain

CH_3 CH_3
CH_3 CH_3

CH_3 CH_3
CH_2CH_3 CH_2CH_3

CH_3 CH_3
$CH_2CH_2CH_3$ $CH_2CH_2CH_3$

(After K. Katoh et al.)

B. Transduction of olfactory stimuli

Mucous layer — Odorant
— Receptor

Cell membrane

Cation channel

GTP
GDP
G_s-protein
Adenylate cyclase
ATP
cAMP
Na^+
Ca^{2+}

ICF

C. Adaptation of smell and taste

Gustatory stimulus: 2 mol/L NaCl
Olfactory stimulus: 0.02 mol/L H₂S

Intensity of perception (%)

100

Smell

50

Taste

0   1   2   3   4   5
Time (min)

(After Engel and Ekman et al.)

Plate 12.16 Sense of Smell

Sense of Balance

Anatomy. Each of the three **semicircular canals** (→ **A1**) is located in a plane about at right angles to the others. The ampulla of each canal contains a ridge-like structure called the *crista ampullaris* (→ **A2**). It contains *hair cells* (secondary sensory cells), the *cilia* of which (→ **A3**) project into a gelatinous membrane called the *cupula* (→ **A2**). Each hair cell has a long *kinocilium* and ca. 80 *stereocilia* of variable length. Their tips are connected to longer adjacent cilia via the "*tip links*" (→ **A3**).

Semicircular canals. When the cilia are in a resting state, the hair cells release a transmitter (glutamate) that triggers the firing of action potentials (AP) in the nerve fibers of the vestibular ganglion. When the head is turned, the semicircular canal automatically moves with it, but *endolymph* in the canal moves more sluggishly due to inertia. A brief *pressure difference* thus develops between the two sides of the cupula. The resultant vaulting of the cupula causes the stereocilia to bend (→ **A2**) and shear against each other, thereby changing the cation conductance of the hair cell membrane. Bending of the stereocilia towards the kinocilium increases conductivity and allows the influx of K^+ and Na^+ along a high electrochemical gradient between the endolymph and hair cell interior (see also pp. 366 and 369 C). Thus, the hair cell becomes depolarized, Ca^{2+} channels open, more glutamate is released, and the AP frequency increases. The reverse occurs when the cilia bend in the other direction (away from the kinocilium). The semicircular canals **function** to detect *angular (rotational) accelerations* of the head in all planes (rotation, nodding, tilting sideways). Since normal head movements take less than $0.3\,s$ (acceleration $\Rightarrow$ deceleration), stimulation of the semicircular canals usually reflects the rotational *velocity*.

The pressure difference across the cupula disappears when the **body rotates for longer periods** of time. Deceleration of the rotation causes a pressure gradient in the opposite direction. When bending of the cilia increased the AP frequency at the start of rotation, it decreases during deceleration and vice versa. Abrupt cessation of the rotation leads to **vertigo** and **nystagmus** (see below).

The **saccule** and **utricle** contain **maculae** (→ **A1, A4**) with cilia that project into a gelatinous

membrane (→ **A4**) with high density (≈ 3.0) calcite crystals called *statoconia, statoliths* or *otoliths*. They displace the membrane and thereby bend the embedded cilia (→ **A4**) due to changes of the direction of *gravity*, e.g. when the head position deviates from the perpendicular axis. The maculae respond also to other *linear (translational) accelerations or decelerations, e.g.* of a car or an elevator.

Central connections. The bipolar neurons of the vestibular ganglion synapse with the *vestibular nuclei* (→ **A, B**). Important tracts extend from there to the contralateral side and to *ocular muscle nuclei, cerebellum* (→ p. 326), *motoneurons* of the skeletal muscles, and to the *postcentral gyrus* (conscious spatial orientation). **Vestibular reflexes** (a) maintain the balance of the body (*postural motor function,* → p. 328) and (b) keep the visual field in focus despite changes in head and body position (*oculomotor control,* → **B** and p. 360).

Example (→ **C**): If a support holding a test subject is tilted, the activated vestibular organ prompts the subject to extend the arm and thigh on the declining side and to bend the arm on the inclining side to maintain balance (→ **C2**). The patient with an impaired equilibrium organ fails to respond appropriately and topples over (→ **C3**).

Since the vestibular organ cannot determine whether the head alone or the entire body moves (*sense of movement*) or changed position (postural *sense*), the vestibular nuclei must also receive and process visual information and that from propriosensors in the neck muscles. Efferent fibers project bilaterally to the eye muscle nuclei, and any change in head position is immediately corrected by opposing *eye movement* (→ **B**). This *vestibulo-ocular reflex* maintains **spatial orientation**.

Vestibular organ function can be assessed by testing oculomotor control. *Secondary* or *postrotatory* **nystagmus** occurs after abrupt cessation of prolonged rotation of the head around the vertical axis (e.g., in an office chair) due to activation of the horizontal semicircular canals. It is characterized by slow horizontal movement of the eyes in the direction of rotation and rapid return movement. Rightward rotation leads to left nystagmus and vice versa (→ p. 360). Caloric stimulation of the horizontal semicircular canal by instilling cold (30 °C) or warm water (44 °C) in the auditory canal leads to *caloric nystagmus*. This method can be used for unilateral testing.

Plate 12.17 Sense of Balance

A. Equilibrium organ (vestibular organ)

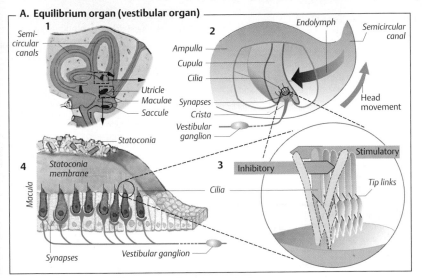

1
Semi-circular canals
Utricle
Maculae
Saccule

2
Endolymph
Semicircular canal
Ampulla
Cupula
Cilia
Synapses
Crista
Vestibular ganglion
Head movement

4
Statoconia
Statoconia membrane
Macula
Cilia
Synapses
Vestibular ganglion

3
Inhibitory
Stimulatory
Tip links

B. Vestibular organ: effects on oculomotor control

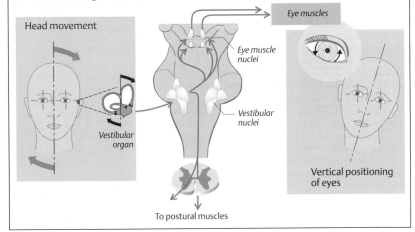

Head movement
Eye muscles
Eye muscle nuclei
Vestibular organ
Vestibular nuclei
Vertical positioning of eyes
To postural muscles

C. Vestibular organ: effects on postural motor control

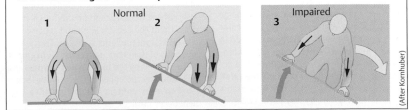

Normal
1 **2**
Impaired
3

(After Kornhuber)

Eye Structure, Tear Fluid, Aqueous Humor

Light entering the eye must pass through the *cornea, aqueous humor, lens* and *vitreous body*, which are collectively called the **optical apparatus,** before reaching the *retina* and its light-sensitive photosensors ($\rightarrow$ **A**). This produces a *reduced and inverse image* of the visual field on the retina. All parts of the apparatus must be transparent and have a stable shape and smooth surface to produce an undistorted image, which is the main purpose of **tear fluid** in case of the cornea. Tears are secreted by *lacrimal glands* located in the top outer portion of orbit and their mode of production is similar to that of saliva ($\rightarrow$ p. 236). Tears are distributed by reflex blinking and then pass through the lacrimal puncta and lacrimal canaliculi (or ducts) of the upper and lower eyelid into the lacrimal sac and finally drain into the nasal sinuses by way of the nasolacrimal duct. Tear fluid improves the optical characteristics of the cornea by smoothing uneven surfaces, washing away dust, protecting it from caustic vapors and chemicals, and protects it from drying out. Tears lubricate the eyelid movement and contain lysozyme and immunoglobulin A ($\rightarrow$ pp. 96ff. and 232), which help ward off infections. In addition, tears are a well known mode of expressing emotions.

The entry of light into the eye is regulated by the **iris** ($\rightarrow$ **A**; p. 353 C1), which contains annular and radial smooth muscle fibers. Cholinergic activation of the sphincter muscle of pupil leads to pupil contraction (*miosis*), and adrenergic activation of the dilator muscle of pupil results in pupil dilatation (*mydriasis*).

The **bulbus** (eyeball) maintains its shape due to its tough outer coat or **sclera** ($\rightarrow$ **C**) and **intraocular pressure** which is normally 10–21 mmHg above the atmospheric pressure. The drainage of **aqueous humor** must balance its production to maintain a constant ocular pressure ($\rightarrow$ **C**). Aqueous humor is produced in the *ciliary process* of the posterior ocular chamber with the aid of carbonic anhydrase and active ion transport. It flows through the pupil into the anterior ocular chamber and drains into the venous system by way of the trabecular meshwork and Schlemm's canal. Aqueous humor is renewed once every hour or so.

Glaucoma. Obstruction of humor drainage can occur due to chronic obliteration of the trabecular meshwork (open-angle glaucoma) or due to acute block of the anterior angle (angle-closure glaucoma) leading to elevated intraocular pressure, pain, retinal damage, and blindness. Drugs that decrease humor production (e.g. carbonic anhydrase inhibitors) and induce meiosis are used to treat glaucoma.

The **lens** is held in place by the *ciliary zonules* ($\rightarrow$ **C**). When the eye adjusts for *far vision*, the zonules are stretched and the lens becomes flatter, especially its anterior surface ($\rightarrow$ **D**, top). When looking at nearby objects (*near vision*), the zonules are relaxed due to contraction of the ciliary muscle, and the lens reassumes its original shape due to its elasticity ($\rightarrow$ **D** , bottom, and p. 346).

The **retina** lines the interior surface of the bulbus except the anterior surface and the site where the *optical nerve* ($\rightarrow$ **A**) exits the bulbus via the *optic papilla* ($\rightarrow$ **A**). The *fovea centralis* ($\rightarrow$ **A**) forms a slight depression across from the pupillary opening. The retina consists of several layers, named from inside out as follows ($\rightarrow$ **E**): pigmented epithelium, photosensors (rods and cones), Cajal's horizontal cells, bipolar cells, amacrine cells, and ganglion cells. The central processes of the ganglion cells ($n \approx 10^6$) exit the bulbus as the optical nerve (retinal circuitry; $\rightarrow$ p. 355ff.).

Photosensors. Retinal rods and cones have a light-sensitive *outer segment*, which is connected to a inner segment by a thin connecting part ($\rightarrow$ p. 349 C1). The inner segment contains the normal cell organelles and establishes synaptic contact with the neighboring cells. The outer segment of the rod cells contains ca. 800 membranous disks, and the plasma membrane of the outer segment of the cones is folded. *Visual pigments* are stored in these disks and folds ($\rightarrow$ p. 348). The outer segment is continuously regenerated; the old membranous disks at the tip of the cell are shed and replaced by new disks from the inner segment. The phagocytic cells of the pigmented epithelium engulf the disks shed by the rods in the morning, and those shed by the cones in the evening. Some ganglion cells contain a light-sensitive pigment ($\rightarrow$ p. 334).

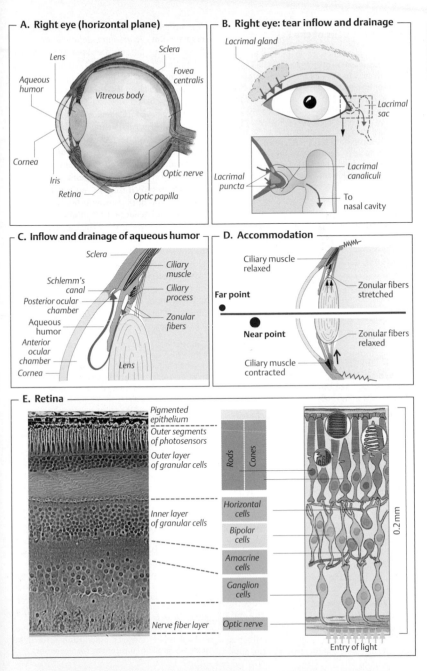

Plate 12.18 Eye Structure, Tear Fluid, Aqueous Humor

A. Right eye (horizontal plane)

Lens
Aqueous humor
Sclera
Fovea centralis
Vitreous body
Cornea
Iris
Retina
Optic nerve
Optic papilla

B. Right eye: tear inflow and drainage

Lacrimal gland
Lacrimal sac
Lacrimal puncta
Lacrimal canaliculi
To nasal cavity

C. Inflow and drainage of aqueous humor

Sclera
Schlemm's canal
Posterior ocular chamber
Aqueous humor
Anterior ocular chamber
Cornea
Ciliary muscle
Ciliary process
Zonular fibers
Lens

D. Accommodation

Ciliary muscle relaxed
Zonular fibers stretched
Far point
Near point
Zonular fibers relaxed
Ciliary muscle contracted

E. Retina

Pigmented epithelium
Outer segments of photosensors
Outer layer of granular cells
Rods
Cones
Inner layer of granular cells
Horizontal cells
Bipolar cells
Amacrine cells
Ganglion cells
Nerve fiber layer
Optic nerve
0.2 mm
Entry of light

Optical Apparatus of the Eye

Physics. The production of an **optical image** is based on the refraction of light rays crossing a spherical interface between air and another medium. Such a **simple optical system** illustrated in plate **A** has an *anterior focal point* (F_a) in air, a *posterior focal point* (F_p), a *principal point* (P), and a *nodal point* (N). Light rays from a distant point (∞) can be regarded as parallel. If they enter the system parallel to its *optical axis*, they will converge at F_p ($\rightarrow$ **A1**, red dot). If they enter at an angle to the axis, then they will form an image beside F_p but in the same *focal plane* ($\rightarrow$ **A1**, violet dot). Light rays from a nearby point do not enter the system in parallel and form an image *behind* the focal plane ($\rightarrow$ **A2**, green and brown dots).

The **optical apparatus** of the eye ($\rightarrow$ p. 344) consists of multiple interfaces and media, and is therefore a *complex optical system*. It can, however, be treated as *a simple optical system*. Light rays from a focused object (O) pass through N and diverge at angle α until they reach the retina and form an image (I) there ($\rightarrow$ **A2**).

Two points separated by a distance of 1.5 mm and located 5 m away from the eye (tan α = 1.5/5000; α = 0.0175 degrees $\approx$ 1′) will therefore be brought into focus 5 μm apart on the retina. In a person with normal vision ($\rightarrow$ p. 348), these two points can be distinguished as separate because 5 μm corresponds to the diameter of three cones in the fovea (two are stimulated, the one in between is not).

Accommodation. When the eyes are adjusted for **far vision**, parallel light rays from a distant point meet at F_p ($\rightarrow$ **B1**, red dot). Since the retina is also located at F_p, the distant point is clearly imaged there. The eye adjusted for far vision will not form a clear image of a nearby point (the light rays meet behind the retina, $\rightarrow$ **B1**, green dot) until the eye has adjusted for **near vision**. In other words, the curvature of the lens (and its refractive power) increases and the image of the nearby point moves to the retinal plane ($\rightarrow$ **B2**, green dot). Now, the distant point cannot not be sharply imaged since F_p does not lie in the retinal plane any more ($\rightarrow$ **B2**).

The **refractive power** around the edge of the optical apparatus is higher than near the optical axis. This *spherical aberration* can be minimized by narrowing the pupils. The refractive power of the eye is the reciprocal of the ante-

rior *focal length* in meters, and is measured in *diopters* (**dpt**). In *accommodation for far vision*, focal length = anterior focal point (F_a)—principal point (P) = 0.017 m ($\rightarrow$ **B1**). Thus, the corresponding refractive power is 1/0.017 = 58.8 dpt, which is mainly attributable to refraction at the air–cornea interface (43 dpt). In maximum *accommodation for near vision* in a young person with normal vision (emmetropia), the refractive power increases by around 10–14 dpt. This increase is called **range of accommodation** and is calculated as 1/near point – 1/far point [m^{-1} = dpt]. The **near point** is the closest distance to which the eye can accommodate; that of a young person with normal vision is 0.07–0.1 m. The **far point** is infinity (∞) in subjects with normal vision. The range of accommodation to a near point of 0.1 m is therefore 10 dpt since $1/\infty$ = 0. It decreases as we grow older (to 1–3.5 dpt in 50-year-olds) due to the loss of elasticity of the lens. This visual impairment of aging, called **presbyopia** ($\rightarrow$ **C1–3**), normally does not affect far vision, but convex lenses are generally required for near vision, e.g., reading.

Cataract causes opacity of the lens of one or both eyes. When surgically treated, convex lenses (glasses or artificial intraocular lenses) of at least + 15 dpt must be used to correct the vision.

In **myopia** (near-sightedness), rays of light entering the eye parallel to the optical axis are brought to focus in front of the retina because the eyeball is too long ($\rightarrow$ **C4**). Distant objects are therefore seen as blurred because the far point is displaced towards the eyes ($\rightarrow$ **C5**). Myopia is corrected by *concave lenses* (negative dpt) that disperse the parallel light rays to the corresponding extent ($\rightarrow$ **C6**). *Example:* When the far point = 0.5 m, a lens of [– 1/0.5] = [– 2 dpt] will be required for correction ($\rightarrow$ **C7**). In **hyperopia** (far-sightedness), on the other hand, the eyeball is too short. Since the accommodation mechanism for near vision must then be already used to focus distant objects ($\rightarrow$ **C8**), the range of accommodation no longer suffices to clearly focus nearby objects ($\rightarrow$ **C9**). Hyperopia is corrected by *convex lenses* (+ dpt) ($\rightarrow$ **C10–11**).

Astigmatism. In regular astigmatism, the corneal surface is more curved in one plane (usually the vertical: astigmatism with the rule) than the other, creating a difference in refraction between the two planes. A point source of light is therefore seen as a line or oval. Regular astigmatism is corrected by *cylindrical lenses*. Irregular astigmatism (caused by scars, etc.) can be corrected by contact lenses.

Plate 12.19 Optical Apparatus of the Eye

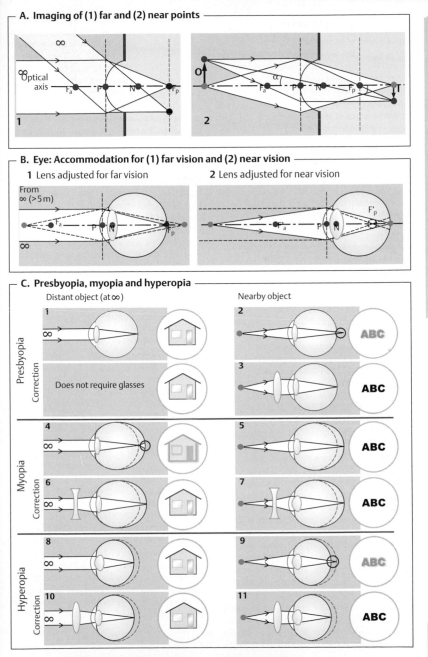

A. Imaging of (1) far and (2) near points

∞
∞
Optical axis
F_a P N F_p
O
α
F_a P N F_p
T
1
2

B. Eye: Accommodation for (1) far vision and (2) near vision

1 Lens adjusted for far vision

From ∞ (>5 m)
F_a P N F_p
∞

2 Lens adjusted for near vision

F'_a P N F'_p

C. Presbyopia, myopia and hyperopia

Distant object (at ∞)   Nearby object

Presbyopia

1 ∞

2 ABC

Correction

3 Does not require glasses   ABC

Myopia

4 ∞

5 ABC

Correction

6 ∞

7 ABC

Hyperopia

8 ∞

9 ABC

Correction

10 ∞

11 ABC

Visual Acuity, Photosensors

Visual acuity is an important measure of eye function. Under *well-lighted conditions*, the normal eye should be able to distinguish two points as separate when the light rays emitted by the point objects converge at an angle (α) of 1 min (1/60 degree) ($\rightarrow$ **A** and p. 346). Visual acuity is calculated as $1/\alpha$ (min^{-1}), and is 1/1 in subjects with normal vision.

Visual acuity testing is generally performed using charts with letters or other optotypes (e.g., Landolt rings) of various sizes used to simulate different distances to the test subject. The letters or rings are usually displayed at a distance of 5 m ($\rightarrow$ **A**). Visual acuity is normal (1/1) if the patient recognizes letters or ring openings seen at an angle of 1 min from a distance of 5 m. *Example*: It should be possible to identify the direction of the opening of the middle ring from a distance of 5 m and that of the left ring from a distance of 8.5 m ($\rightarrow$ **A**). If only the opening of the left ring can be localized from the test distance of 5 m, the visual acuity is 5/8.5 = 0.59.

Photosensors or **photoreceptors**. The light-sensitive sensors of the eye consist of approximately $6 \cdot 10^6$ **rods** and 20 times as many **cones** ($\rightarrow$ p. 345 E) distributed at variable densities throughout the retina ($\rightarrow$ **B1**). (Certain ganglion cells also contain a light-sensitive pigment; $\rightarrow$ p. 334). The fovea centralis is exclusively filled with cones, and their density rapidly decreases towards the periphery. Rods predominate 20–30 degrees away from the fovea centralis. Approaching the periphery of the retina, the density of the rods decreases continuously from $1.5 \times 10^5/mm^2$ (maximum) to about one-third this value. No photosensors are present on the *optic disk*, which is therefore referred to as the **blind spot** in the visual field.

Clear visualization of an object in daylight requires that the gaze be fixed on it, i.e., that an image of the object is produced in the *fovea centralis*. Sudden motion in the periphery of the visual field triggers a **reflex saccade** ($\rightarrow$ p. 360), which shifts the image of the object into the fovea centralis. Thereby, the retinal area with the highest visual acuity is selected ($\rightarrow$ **B2**, yellow peak), which lies 5 degrees temporal to the optical axis. Visual acuity decreases rapidly when moving outward from the fovea ($\rightarrow$ **B2**, yellow field), reflecting the decreasing density of cone distribution ($\rightarrow$ **B1**,

red curve). In a *dark-adapted eye*, on the other hand, the sensitivity of the retina ($\rightarrow$ **B2**, blue curve) is completely dependent on the rod distribution ($\rightarrow$ **B1**, purple curve). The color-sensitive cones are therefore used for visual perception in daylight or good lighting (*day vision, photopic vision*), while the black and white-sensitive cones are used to visualize objects in darkness (*dim-light vision, night vision, scotopic vision*). The high light sensitivity in night vision is associated with a high loss of visual acuity ($\rightarrow$ p. 354).

Photosensor Function

Light-absorbing *visual pigments* and a variety of *enzymes* and *transmitters* in retinal rods and cones ($\rightarrow$ **C1**) mediate the conversion of light stimuli into electrical stimuli; this is called **photoelectric transduction**. The membranous disks of the retinal rods contain *rhodopsin* ($\rightarrow$ **C2**), a photosensitive purple-red chromoprotein (*visual purple*). Rhodopsin consists of the integral membrane protein *opsin* and the aldehyde *11-cis-retinal*. The latter is bound to a lysine residue of opsin which is embedded in this protein; it is stably kept in place by weak interactions with two other amino acid residues. *Photic stimuli* trigger a primary *photochemical reaction* in rhodopsin (duration, $2 \cdot 10^{-14}$ s) in which 11-*cis*-retinal is converted to all-*trans*-retinal ($\rightarrow$ **C3**). Even without continued photic stimulation, the reaction yields bathorhodopsin, the intermediates lumirhodopsin and metarhodopsin I, and finally metarhodopsin II within roughly 10^{-3} s ($\rightarrow$ **D1**).

Metarhodopsin II (MR-II) reacts with a G$_s$-protein ($\rightarrow$ p. 274) called *transducin* (G$_t$-protein), which breaks down into α_s and $\beta\gamma$ subunits once GDP has been replaced by GTP ($\rightarrow$ **D1**). Activated α_s-GTP now binds the inhibitory subunit of cGMP phosphodiesterase (I$_{PDE}$) ($\rightarrow$ **D2**). The consequently disinhibited phosphodiesterase (PDE) then lowers the cytosolic concentration of cyclic guanosine monophosphate (cGMP). The activation of a single retinal rhodopsin molecule by a quantum of light can induce the hydrolysis of up to 10^6 cGMP molecules per second. The reaction cascade therefore has tremendous amplifying power.

▶

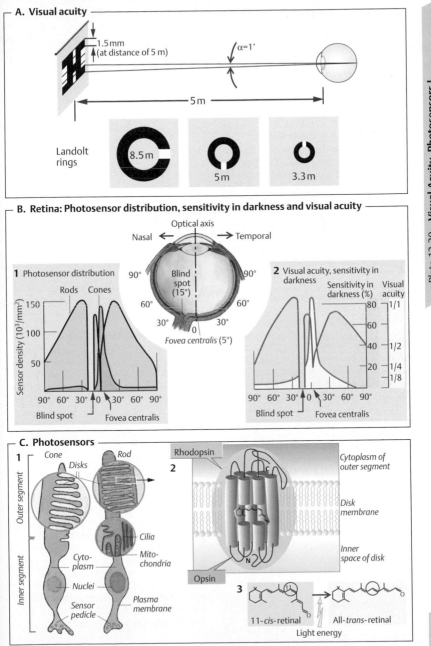

Plate 12.20 Visual Acuity, Photosensors I

A. Visual acuity

1.5 mm (at distance of 5 m)

α=1'

5 m

Landolt rings

8.5 m

5 m

3.3 m

B. Retina: Photosensor distribution, sensitivity in darkness and visual acuity

Optical axis

Nasal ← → Temporal

Blind spot (15°)

Fovea centralis (5°)

1 Photosensor distribution

Rods   Cones

Sensor density (10³/mm²)

150

100

50

90° 60° 30° 0 30° 60° 90°

Blind spot   Fovea centralis

2 Visual acuity, sensitivity in darkness

Sensitivity in darkness (%)   Visual acuity

80   1/1

60

40   1/2

20   1/4
      1/8

90° 60° 30° 0 30° 60° 90°

Blind spot   Fovea centralis

C. Photosensors

1

Cone   Rod

Disks

Outer segment

Inner segment

Cilia

Cytoplasm

Mito-chondria

Nuclei

Plasma membrane

Sensor pedicle

2   Rhodopsin

Cytoplasm of outer segment

Disk membrane

Inner space of disk

Opsin

N

3

11-cis-retinal → All-trans-retinal

Light energy

In **darkness** ($\rightarrow$ **D, left**), cGMP is bound to cation channels (Na^+, Ca^{2+}) in the outer segment of the photosensor, thereby keeping them open. Na^+ and Ca^{2+} can therefore enter the cell and depolarize it to about $-40\,mV$ ($\rightarrow$ **D3, D4**). This darkness-induced influx into the outer segment is associated with the efflux of K^+ from the inner segment of the sensor. The Ca^{2+} entering the outer segment is immediately transported out of the cell by a $3\,Na^+/Ca^{2+}$ exchanger ($\rightarrow$ p. 36), so the cytosolic Ca^{2+} concentration $[Ca^{2+}]_i$ remains constant at ca. 350–500 nmol/L in darkness ($\rightarrow$ **D6**). If the cytosolic cGMP concentration decreases in response to a **light stimulus** ($\rightarrow$ **D2**), cGMP dissociates from the cation channels, allowing them to close. The photosensor then hyperpolarizes to ca. $-70\,mV$ (*sensor potential:* $\rightarrow$ **D, right**). This *inhibits* the release of *glutamate* (transmitter) at the sensor pedicle ($\rightarrow$ **D5**), which subsequently causes changes in the membrane potential in downstream retinal neurons ($\rightarrow$ p. 354).

Deactivation of Photic Reactions and Regeneration Cycles

◆ **Rhodopsin** ($\rightarrow$ **E2**). *Rhodopsin kinase* (RK) competes with transducin for bindings sites on *metarhodopsin II* (MR-II); the concentration of transducin is 100 times higher ($\rightarrow$ **E2**, right). Binding of RK to MR-II leads to phosphorylation of MR-II. As a result, its affinity to transducin decreases while its affinity to another protein, *arrestin*, rises. Arrestin blocks the binding of further transducin molecules to MR-II. All-*trans*-retinal detaches from opsin, which is subsequently dephosphorylated and re-loaded with 11-*cis*-retinal.

◆ **All-trans-retinal** ($\rightarrow$ **E1**) is transported out of the photosensor and into the pigmented epithelium, where it is reduced to all-*trans*-retinol, esterified and ultimately restored to 11-*cis*-retinal. After returning into the photosensor, it binds to opsin again, thereby completing the cycle ($\rightarrow$ **E2**).

◆ **Note:** Retinol is vitamin A_1. A chronic deficiency of vitamin A_1 or its precursors (carotinoids) leads to impaired rhodopsin production and, ultimately, to *night blindness* ($\rightarrow$ p. 352).

◆ **Transducin** ($\rightarrow$ **E3**). Since the GTPase activity of α_s-GTP breaks down GTP into GDP + P_i, the molecule deactivates itself. The α_s-GTP molecule and $\beta\gamma$ subunit then reunite to transducin. GAP (GTPase-activating protein) accelerates the regeneration of transducin. *Phosducin*, another protein, is phosphorylated in the dark ($\rightarrow$ **D6**) and dephosphorylated in light ($\rightarrow$ **D7**). The latter form binds to the $\beta\gamma$ subunit ($\rightarrow$ **D7, E3**), thereby blocking the regeneration of transducin. This plays a role in light adaptation (see below).

◆ **Phosphodiesterase (PDE).** In the course of transducin regeneration, the inhibitory subunit of cGMP phosphodiesterase (I_{PDE}) is released again and PDE is thus inactivated.

◆ **cGMP.** Since the $3\,Na^+/Ca^{2+}$ exchanger still functions even after photostimulation-induced closure of Ca^{2+} channels, the $[Ca^{2+}]_i$ starts to decrease. When a threshold of ca. 100 nmol/L is reached, the Ca^{2+}-binding protein GCAP (guanylyl cyclase-activating protein) loses its $4\,Ca^{2+}$ ions and stimulates guanylyl cyclase, thereby accelerating cGMP synthesis. Thus, the cGMP concentration rises, the cation channels re-open, and the sensor is ready to receive a new light stimulus. This Ca^{2+} cycle therefore mediates a negative feedback loop for cGMP production.

Ca^{2+} Ions and Adaptation (see also p. 352)

In the dark, the $[Ca^{2+}]_i$ is high, and calmodulin-bound Ca^{2+} ($\rightarrow$ p. 36) stimulates the phosphorylation of phosducin with the aid of cAMP and phosphokinase A ($\rightarrow$ **D6**). In light, the $[Ca^{2+}]_i$ is low; phosducin is dephosphorylated and rapid regeneration of transducin is not possible ($\rightarrow$ **D7, E3**). Moreover, Ca^{2+} accelerates the phosphorylation of MR-II in light with the aid of another Ca^{2+} binding protein, *recoverin* ($\rightarrow$ **E2**). Ca^{2+} is therefore essential for the **adaptation of photosensors** ($\rightarrow$ p. 352).

Although they contain similar enzymes and transmittes, the photosensitivity of the **cones** is about 100 times less than that of the rods. Thus, the cones are unable to detect a single quantum of light, presumably because photic reactions in the cones are deactivated too quickly. Compared to rhodopsin in the retinal rods, the three visual pigments in the three types of cones (11-*cis*-retinal with different opsin fractions) only absorb light in a narrow wavelength range ($\rightarrow$ p. 357 E), which is a prerequisite for color vision ($\rightarrow$ p. 356).

Plate 12.21 Visual Acuity, Photosensors II

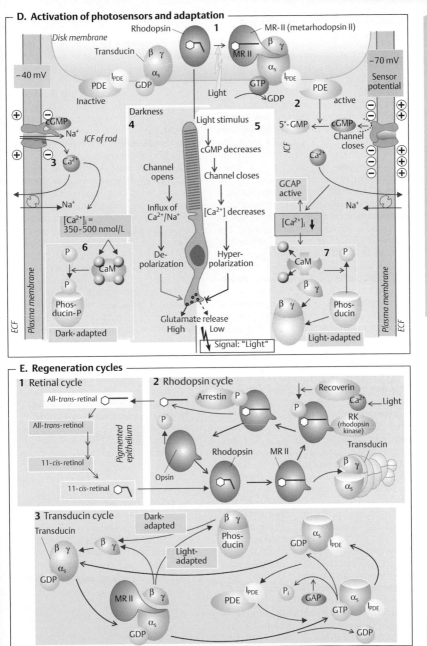

D. Activation of photosensors and adaptation

Disk membrane

Rhodopsin **1**   MR-II (metarhodopsin II)

Transducin

β γ

MR II

α_s   I_PDE

−40 mV   PDE   I_PDE

Inactive   GDP

Light

GTP → GDP

2   PDE   active

Darkness **4**   Light stimulus **5**

cGMP

Na⁺   ICF of rod

3   Ca²⁺

Na⁺

[Ca²⁺]_i = 350–500 nmol/L

−70 mV   Sensor potential

5′-GMP ← cGMP   Channel closes

Ca²⁺

GCAP active

[Ca²⁺]_i ↓

cGMP decreases

Channel closes

[Ca²⁺] decreases

Channel opens

Influx of Ca²⁺/Na⁺

De-polarization

Hyper-polarization

Na⁺

6   P   CaM   P

Phos-ducin-P

Dark-adapted

Plasma membrane   ECF

Glutamate release   High   Low

Signal: "Light"

7   P   CaM   P

β γ   β γ   Phos-ducin

Light-adapted

Plasma membrane   ECF

E. Regeneration cycles

1 Retinal cycle   **2 Rhodopsin cycle**

Recoverin

All-trans-retinal   Arrestin   P   P   Ca²⁺ ← Light

All-trans-retinol   P   RK (rhodopsin kinase)

Pigmented epithelium   Opsin   Rhodopsin   MR II   Transducin

11-cis-retinol   β γ   α_s

11-cis-retinal

3 Transducin cycle

Dark-adapted

Transducin   β γ

β γ   Phos-ducin   α_s   GDP   I_PDE

β γ   Light-adapted

α_s   GDP

MR II   β γ   PDE   I_PDE   P_i ← GAP   α_s   GTP   I_PDE

α_s   GDP   GDP

351

Adaptation of the Eye to Different Light Intensities

The eye is able to perceive a wide range of light intensities, from the extremely low intensity of a small star to the extremely strong intensity of the sun glaring on a glacier. The ability of the eye to process such a wide range of luminance $(1 : 10^{11})$ by adjusting to the prevailing light intensity is called **adaptation**. When going from normal daylight into a darkened room, the room will first appear black because its luminance value (measured in $cd \cdot m^{-2}$) is lower than the current ocular threshold. As the stimulus threshold decreases over the next few minutes, the furniture in the room gradually becomes identifiable. A longer period of adaptation is required to visualize stars. The maximum level of adaptation is reached in about 30 min ($\rightarrow$ **A**). The minimum light intensity that can just be detected after maximum dark adaptation is the **absolute visual threshold**, which is defined as 1 in **A** and **B**.

The **retinal adaptation curve** exhibits a (Kohlrausch) break at roughly 2000 × the absolute threshold ($\rightarrow$ **A**, blue curve). This corresponds to the point where the excitation threshold of the cones is reached (threshold for day vision). The remainder of the curve is governed by the somewhat slower adaptation of the rods ($\rightarrow$ **A**, violet curve). The isolated rod adaptation curve can be recorded in patients with complete color blindness (*rod monochromatism*), and the isolated cone adaptation curve can be observed in night blindness (hemeralopia, $\rightarrow$ p. 350).

Differential threshold (or difference limen). The ability of the eye to distinguish the difference between two similar photic stimuli is an important prerequisite for proper eyesight. At the lowest limit of discriminative sensibility for two light intensities I and I', the *absolute differential threshold* (Δ I) is defined as I minus I'. The *relative differential threshold* is calculated as Δ I/I, and remains relatively constant in the median stimulus range (*Weber's rule*). Under optimal lighting conditions (approx. 10^9 times the absolute threshold; $\rightarrow$ **B**), Δ I/I is very small (0.01). The relative differential threshold rises greatly in dark adaptation, but also rises in response to extremely bright light. Sunglasses decrease the differential threshold in the latter case.

The **mechanisms for adaptation** of the eye are as follows ($\rightarrow$ **C**):

◆ **Pupil reflex** ($\rightarrow$ **C1**). Through reflexive responses to light exposure of the retina ($\rightarrow$ p. 359), the pupils can adjust the quantity of light entering the retina by a factor of 16. Thus, the pupils are larger in darkness than in daylight. The main function of the pupil reflex is to ensure rapid adaptation to sudden changes in light intensity.

◆ **Chemical stimuli** ($\rightarrow$ **C2**) help to adjust the sensitivity of photosensors to the prevailing light conditions. Large quantities of light lead to a prolonged decrease in the receptor's cytosolic Ca^{2+} concentration. This in conjunction with the activity of *recoverin* and *phosducin* reduces the availability of rhodopsin ($\rightarrow$ p. 348ff.). It therefore decreases the probability that a rhodopsin molecule will be struck by an incoming light ray (photon) or that a metarhodopsin II molecule will come in contact with a transducin molecule. When the light intensity is low, large concentrations of rhodopsin and transducin are available and the photosensors become very light-sensitive.

◆ **Spatial summation** ($\rightarrow$ **C3**). Variation of retinal surface area (number of photosensors) exciting an optic nerve fiber causes a form of spatial summation that increases with darkness and decreases with brightness ($\rightarrow$ p. 354).

◆ **Temporal summation** ($\rightarrow$ **C4**). Brief subthreshold stimuli can be raised above threshold by increasing the duration of stimulation (by staring at an object) long enough to trigger an action potential (AP). Thereby, the product of stimulus intensity times stimulus duration remains constant.

Successive contrast occurs due to "local adaptation." When a subject stares at the center of the black-and-white pattern ($\rightarrow$ **D**) for about 20 s and suddenly shifts the focus to the white circle, the previously dark areas appear to be brighter than the surroundings due to sensitization of the corresponding areas of the retina.

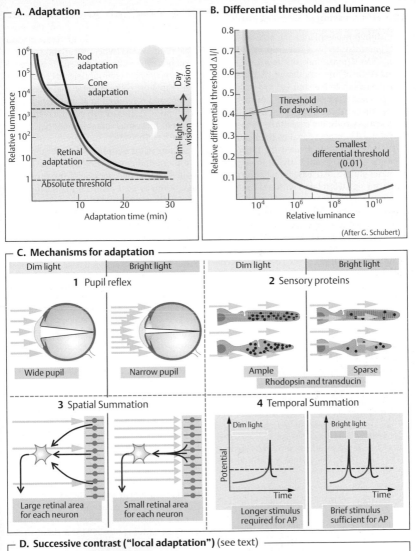

A. Adaptation

Relative luminance (vertical axis, 10^6 to 1)

Rod adaptation

Cone adaptation

Day vision

Dim-light vision

Retinal adaptation

Absolute threshold

Adaptation time (min): 10, 20, 30

B. Differential threshold and luminance

Relative differential threshold $\Delta I/I$ (0.1 to 0.8)

Threshold for day vision

Smallest differential threshold (0.01)

Relative luminance: 10^4, 10^6, 10^8, 10^{10}

(After G. Schubert)

C. Mechanisms for adaptation

1 Pupil reflex

| Dim light | Bright light |
| Wide pupil | Narrow pupil |

2 Sensory proteins

| Dim light | Bright light |
| Ample | Sparse |

Rhodopsin and transducin

3 Spatial Summation

Large retinal area for each neuron

Small retinal area for each neuron

4 Temporal Summation

Dim light — Potential / Time

Bright light — Potential / Time

Longer stimulus required for AP

Brief stimulus sufficient for AP

D. Successive contrast ("local adaptation") (see text)

Plate 12.22 Adaptation to Light Intensity

Retinal Processing of Visual Stimuli

Light stimuli *hyper*polarize the **sensor potential** of photosensors ($\rightarrow$ **A**, left) from ca. – 40 mV to ca. – 70 mV (maximum) due to a decrease in conductance of the membrane of the outer sensor segment to Na^+ and Ca^{2+} ($\rightarrow$ p. 348ff.). The potential rises and falls much more sharply in the cones than in the rods. As in other sensory cells, the magnitude of the sensor potential is proportional to the logarithm of stimulus intensity divided by threshold-intensity (*Fechner's law*). Hyperpolarization decreases glutamate release from the receptor. When this signal is relayed within the retina, a distinction is made between "direct" signal flow for photopic vision and "lateral" signal flow for scotopic vision (see below). Action potentials (APs) can only be generated in ganglion cells ($\rightarrow$ **A**, right), but stimulus-dependent amplitude changes of the potentials occur in the other retinal neurons ($\rightarrow$ **A**, center). These are conducted electrotonically across the short spaces in the retina ($\rightarrow$ p. 48ff.).

Direct signal flow from cones to bipolar cells is conducted via ON or OFF bipolar cells. Photostimulation leads to depolarization of ON bipolar cells (signal inversion) and activation of their respective ON ganglion cells ($\rightarrow$ **A**). OFF bipolar cells, on the other hand, are hyperpolarized by photostimulation, which has an inhibitory effect on their OFF ganglion cells. "Lateral" signal flow can occur via the following pathway: rod $\Rightarrow$ rod–bipolar cell $\Rightarrow$ rod–amacrine cell $\Rightarrow$ ON or OFF bipolar cell $\Rightarrow$ ON or OFF ganglion cell. Both rod–bipolar cells and rod–amacrine cells are depolarized in response to light. Rod–amacrine cells inhibit OFF bipolar cells via a chemical synapse and stimulate ON bipolar cells via an electrical synapse ($\rightarrow$ p. 50).

A light stimulus triggers the firing of an AP in ON ganglion cells ($\rightarrow$ **A**, right). The AP frequency increases with the sensor potential amplitude. The APs of ON ganglion cells can be measured using microelectrodes. This data can be used to identify the retinal region in which the stimulatory and inhibitory effects on AP frequency originate. This region is called the **receptive field (RF)** of the ganglion cell. Retinal ganglion cells have concentric RFs comprising a central zone and a ringlike peripheral zone distinguishable during light adaptation ($\rightarrow$ **B**). Photic stimulation of the center increases the

AP frequency of ON ganglion cells ($\rightarrow$ **B1**). Stimulation of the periphery, on the other hand, leads to a decrease in AP frequency, but excitation occurs when the light source is switched off ($\rightarrow$ **B2**). This type of RF is referred to as an *ON field* (central field ON). The RF of OFF ganglion cells exhibits the reverse response and is referred to as an *OFF field* (central field OFF). *Horizontal cells* are responsible for the functional organization of the RFs ($\rightarrow$ p. 344). They invert the impulses from photosensors in the periphery of the RF and transmit them to the sensors of the center. The opposing central and peripheral responses lead to a **stimulus contrast**. At a light–dark interface, for example, the dark side appears darker and the light side brighter. If the entire RF is exposed to light, the impulses from the center usually predominate.

Simultaneous contrast. A solid gray circle appears darker in light surroundings than in dark surroundings ($\rightarrow$ **C**, left). When a subject focuses on a black-and-white grid ($\rightarrow$ **C**, right), the white grid lines appear to be darker at the cross-sections, black grid lines appear lighter because of reduced contrast in these areas. This effect can be attributed to a variable sum of stimuli within the RFs ($\rightarrow$ **C**, center).

During **dark adaptation**, the center of the RFs increases in size at the expense of the periphery, which ultimately disappears. This leads to an increase in spatial summation ($\rightarrow$ p. 353 C3), but to a simultaneous decrease in stimulus contrast and thus to a lower visual acuity ($\rightarrow$ p. 349 B2).

Color opponency. Red and green light (or blue and yellow light) have opposing effects in the RFs of β ganglion cells ($\rightarrow$ p. 358) and more centrally located cells of the optic tract ($\rightarrow$ p. 357 E). These effects are explained by **Hering's opponent colors theory** and ensure contrast (increase color saturation; $\rightarrow$ p. 356) in color vision. When a subject focuses on a color test pattern ($\rightarrow$ p. 359 C) for about 30 min and then shifts the gaze to a neutral background, the complementary colors will be seen (*color successive contrast*).

RFs of **higher centers** of the optic tract (V1, V2; $\rightarrow$ p. 358) can also be identified, but their characteristics change. Shape (striate or angular), length, axial direction and direction of movement of the photic stimuli play important roles.

A. Potentials of photosensor, ON-bipolar and ON-ganglion cells

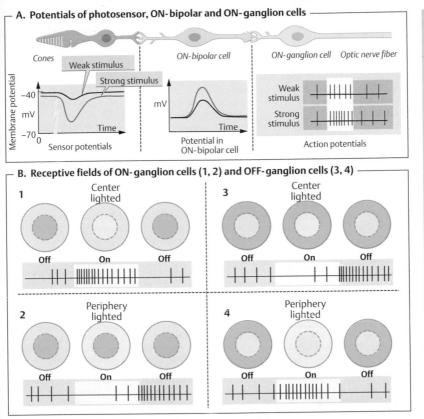

Cones

Weak stimulus

Strong stimulus

Membrane potential

mV

−40

−70

0

Time

Sensor potentials

ON-bipolar cell

mV

Time

Potential in ON-bipolar cell

ON-ganglion cell   Optic nerve fiber

Weak stimulus

Strong stimulus

Action potentials

B. Receptive fields of ON-ganglion cells (1, 2) and OFF-ganglion cells (3, 4)

1   Center lighted

Off   On   Off

2   Periphery lighted

Off   On   Off

3   Center lighted

Off   On   Off

4   Periphery lighted

Off   On   Off

C. Receptive field-related contrast (ON-ganglion cells)

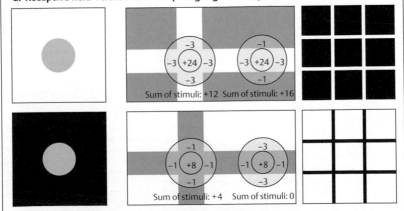

−3
−3 (+24) −3
−3
Sum of stimuli: +12

−1
−3 (+24) −3
−1
Sum of stimuli: +16

−1
−1 (+8) −1
−1
Sum of stimuli: +4

−3
−1 (+8) −1
−3
Sum of stimuli: 0

Plate 12.23   Retinal Processing of Visual Stimuli

Color Vision

White light passing through a prism is split into a color spectrum ranging from red to violet (colors of the rainbow). *Red* is perceived at a wavelength (λ) of 650–700 nm, and violet at around 400–420 nm ($\rightarrow$ **A**). The eye is sensitive to waves in this λ range. Perception of white light does not require the presence of all colors of the visible spectrum. It is sufficient to have the additive effects (mixing) of only two **complementary colors**, e.g., orange (612 nm) and blue light (490 nm).

A **color triangle** ($\rightarrow$ **B**) or similar test panel can be used to illustrate this effect. The two upper limbs of the triangle show the visible spectrum, and a white dot is located inside the triangle. All straight lines passing through this dot intersect the upper limbs of the triangle at two complementary wavelengths (e.g., 612 and 490 nm). **Additive color mixing** ($\rightarrow$ **C**): The color yellow is obtained by mixing roughly equal parts of red and green. Orange is produced by using a higher red fraction, and yellowish green is obtained with a higher green fraction. These colors lie between red and green on the limbs of the color triangle. Similar rules apply when mixing green and violet ($\rightarrow$ **B** and **C**). The combination of red with violet yields a shade of purple not contained in the spectrum ($\rightarrow$ **B**). This means that all colors, including white, can be produced by varying the proportions of three colors—e.g. red (700 nm), green (546 nm) and blue (435 nm) because every possible pair of complementary colors can be obtained by mixing these three colors of the spectrum.

Subtractive color mixing is based on the opposite principle. This technique is applied when color paints and camera filters are used. Yellow paints or filters absorb ("subtract") the blue fraction of white light, leaving the complementary color yellow.

Light absorption. Photosensors must be able to absorb light to be photosensitive. **Rods** ($\rightarrow$ p. 348) contain **rhodopsin**, which is responsible for (achromatic) night vision. Rhodopsin absorbs light at wavelengths of ca. 400–600 nm; the maximum absorption value (λ_{max}) is 500 nm ($\rightarrow$ **E1**). Relatively speaking, greenish blue light therefore appears brightest and red appears darkest at night. Wearing red glasses in daylight therefore leaves the rods adapted for darkness. Three types of color-sensitive **cones** are responsible for (chromatic) day vision ($\rightarrow$ **E1**): (1) *S cones*, which absorb short-wave (S) blue-violet light (λ_{max} =

420 nm); (2) *M cones*, which absorb medium-wave (M) blue-green to yellow light (λ_{max} = 535 nm), and (3) *L cones*, which absorb long-wave (L) yellow to red light (λ_{max} = 565 nm). (The physiological sensitivity curves shown in **E1** make allowances for light absorbed by the lens.) *Ultraviolet rays* (λ_{max} < 400 nm) and *infrared rays* (λ_{max} > 700 nm) are not visible.

Sensory information relayed by the three types of cones (peripheral application of the *trichromatic theory of color vision*) and transduction of these visual impulses to *brightness* and *opponent color channels* ($\rightarrow$ **E2** and p. 354) in the retina and lateral geniculate body (LGB) enables the visual cortex ($\rightarrow$ p. 358) to recognize different **types of colors**. The human eye can distinguish 200 shades of color and different degrees of color saturation. The absolute *differential threshold for color vision* is 1–2 nm ($\rightarrow$ **D**, "normal").

Color perception is more complex. White paper, for example, will look white in white light (sunlight), yellow light (light bulb) and red light. We also do not perceive the different shades of color in a house that is partially illuminated by sunlight and partially in the shade. This **color constancy** is the result of retinal and central processing of the retinal signal.

There is a similar **constancy of size and shape**: Although someone standing 200 meters away makes a much smaller image on the retina that at 2 meters' distance, we still recognize him or her as a person of normal height, and although a square table may appear rhomboid in shape when viewed from the side, we can still tell that it is square.

Color blindness occurs in 9% of all men and in 0.5% of women. The ability to distinguish certain colors is impaired or absent in these individuals, i.e., they have a *high differential threshold for color* ($\rightarrow$ **D**). Various types of color blindness are distinguished: *protanopia* (red blindness), *deuteranopia* (green blindness), and *tritanopia* (blue-violet blindness). Protanomaly, deuteranomaly and tritanomaly are characterized by decreased sensitivity of the cones to colored, green and blue, respectively. Color vision is tested using *color perception charts* or an *anomaloscope*. With the latter, the subject has to mix two color beams (e.g., red and green) with adjustable intensities until their additive mixture matches a specific shade of color (e.g. yellow, $\rightarrow$ **C**) presented for comparison. A protanomal subject needs a too high red intensity, a deuteranomal person a too high green intensity. Protanopes perceive all colors with wavelengths over approx. 520 nm as yellow.

Plate 12.24   Color Vision

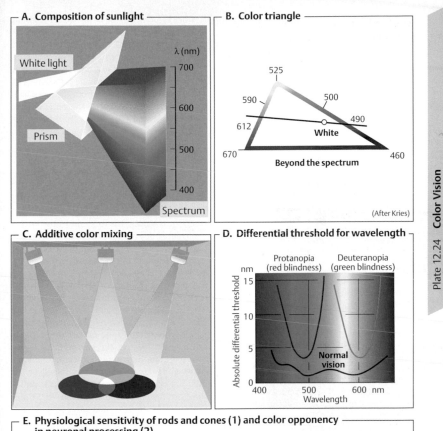

A. Composition of sunlight

White light

Prism

Spectrum

λ (nm)
- 700
- 600
- 500
- 400

B. Color triangle

525

590   500

612   490

White

670   460

Beyond the spectrum

(After Kries)

C. Additive color mixing

D. Differential threshold for wavelength

Protanopia (red blindness)   Deuteranopia (green blindness)

nm

Absolute differential threshold

15

10

5

Normal vision

400   500   600   nm

Wavelength

E. Physiological sensitivity of rods and cones (1) and color opponency in neuronal processing (2)

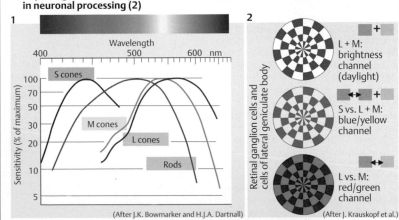

1

Wavelength

400   500   600   nm

Sensitivity (% of maximum)

100
70
50
30
20

10

5

S cones

M cones

L cones

Rods

(After J.K. Bowmaker and H.J.A. Dartnall)

2

Retinal ganglion cells and cells of lateral geniculate body

L + M: brightness channel (daylight)

S vs. L + M: blue/yellow channel

L vs. M: red/green channel

(After J. Krauskopf et al.)

357

Visual Field, Visual Pathway, Central Processing of Visual Stimuli

The **visual field** ($\rightarrow$ A) is the area visualized by the *immobile* eye with the head fixed.

The visual field is examined by **perimetry**. The subject's eye is positioned in the center of the perimeter, which is a hollow hemispherical instrument. The subject must then report when laterally flashed points of light appear in or disappear from the visual field. An area of lost vision within the visual field is a **scotoma**. Lesions of the retina, or of the central visual pathway can cause scotoma. The **blind spot** ($\rightarrow$ A) is a normal scotoma occurring at 15 degrees temporal and is caused by nasal interruption of the retina by the optic disk ($\rightarrow$ p. 349 B). In binocular vision ($\rightarrow$ p. 361 A), the blind spot of one eye is compensated for by the other. The visual field for color stimuli is smaller than that for light–dark stimuli. If, for example, a red object is slowly moved into the visual field, the movement will be identified more quickly than the color of the object.

The retina contains more than 10^8 photosensors connected by retinal neurons ($\rightarrow$ p. 354) to ca. 10^6 retinal ganglion cells. Their axons form the **optic nerve**. The convergence of so many sensors on only a few neurons is particularly evident in the retinal periphery (1000 : 1). In the fovea centralis, however, only one or a few cones are connected to one neuron. Due to the low convergence of impulses from the fovea, there is a high level of visual acuity with a low level of light sensitivity, whereas the high convergence of signals from the periphery produces the reverse effect (*cf. spatial summation*; $\rightarrow$ p. 353 C3).

Ganglion cells. Three types of ganglion cells can be found in the retina: (1) 10% are *M* (or α or Y) **cells** of the *magnocellular system*; their fast-conducting axons emit short phasic impulses in response to light and are very sensitive to *movement*; (2) 80% are the *P* (or β or X) **cells** *of the parvicellular system*; their thin axons have small receptive fields (high spatial resolution), persistently react to constant light (tonic reaction), and therefore permit *pattern and color analysis*. Both types have equal densities of ON and OFF cells ($\rightarrow$ p. 354). (3) 10% are γ *(or W) cells* of the *coniocellular system*; their very thin axons project to the mesencephalon and regulate pupil diameter (see below) and reflex saccades ($\rightarrow$ pp. 348, 360).

Objects located in the nasal half of the visual field of each eye ($\rightarrow$ B, blue and green) are imaged in the temporal half of each retina and vice versa. Along the **visual pathway**, fibers of the **optic nerve** from the temporal half of each retina remain on the same side ($\rightarrow$ B, blue and green), whereas the fibers from the nasal half of each retina decussate at the **optic chiasm** ($\rightarrow$ B, orange and red). Fibers from the fovea centralis are present on both sides.

Lesions of the left optic *nerve* for instance cause deficits in the entire left visual field ($\rightarrow$ B, a), whereas lesions of the left optic *tract* produce deficits in the right halves of both visual fields ($\rightarrow$ B, c). Damage to the median optic *chiasm* results in bilateral temporal deficits, i.e., *bitemporal hemianopia* ($\rightarrow$ B, b).

Fibers of the optic tract extend to the **lateral geniculate body** ($\rightarrow$ B) of the thalamus, the six layers of which are organized in a retinotopic manner. Axons of the ipsilateral eye terminate on layers 2, 3 and 5, while those of the contralateral eye terminate on layers 1, 4 and 6. The M cell axons communicate with cells of *magnocellular layers* 1 and 2, which serve as a relay station for *motion-related stimuli* that are rapidly conducted to the motor cortex. The P cell axons project to the *parvocellular layers* 3–6, the main function of which is to *process colors and shapes*. The neurons of all layers then project further through the **optic radiation** (arranged also retinotopically) to the *primary visual cortex* (V_1) and, after decussating, to further areas of the visual cortex (V_{2-5}) including pathways to the parietal and temporal cortex. Magnocellular input reaches the parietal cortex also via the superior colliculi (see below) and the pulvinar.

The **primary visual cortex** (V_1) is divided depth-wise (x-axis) into six retinotopic layers numbered I to VI ($\rightarrow$ p. 333 A). Cells of the primary visual cortex are arranged as three-dimensional modular **hypercolumns** ($3 \times 1 \times 1$ mm) representing modules for analysis of all sensory information from corresponding areas of both retinas ($\rightarrow$ p. 360). Adjacent hypercolumns represent neighboring retinal regions. Hypercolumns contain *ocular dominance columns* (y-axis), *orientation columns* (z-axis), and *cylinders* (x-axis). The dominance columns receive alternating input from the right and left eye, orientation columns process direction of stimulus movement and cylinders receive information of colors.

Plate 12.25 Visual Field, Visual Pathway

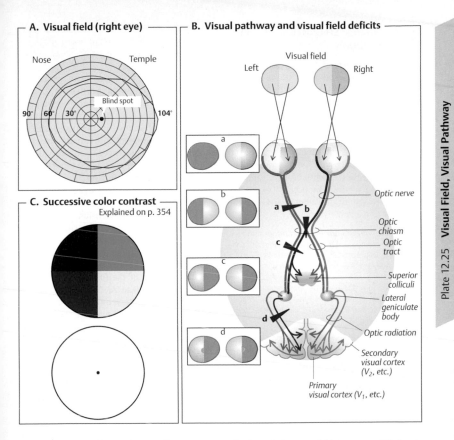

A. Visual field (right eye)

Nose | Temple

Blind spot

90° 60° 30° 104°

C. Successive color contrast
Explained on p. 354

B. Visual pathway and visual field deficits

Visual field
Left | Right

a

b

a | b

c

d

Optic nerve

Optic chiasm

Optic tract

Superior colliculi

Lateral geniculate body

Optic radiation

Secondary visual cortex (V₂, etc.)

Primary visual cortex (V₁, etc.)

Color, high-resolution stationary **shapes**, **movement**, and stereoscopic **depth** are processed in some subcortical visual pathways, and from V₁ onward in *separate information channels*. These individual aspects must be integrated to achieve visual perception. In diurnally active primates like humans, over half of the cortex is involved in processing visual information. On a simplified scale, the parietal cortex analyzes the "where" and involves motor systems, and the temporal cortex takes care of the "what" of visual input comparing it with memory.

Axons of the optic tract (especially those of M and γ cells) also project to **subcortical regions** of the brain such as the *pretectal region*, which regulates the diameter of the pupils (see below); the *superior colliculi* (→ **B**), which are involved in oculomotor function (→ p. 360);

the *hypothalamus*, which is responsible for circadian rhythms (→ p. 334).

The **pupillary reflex** is induced by sudden exposure of the retina to light (→ p. 350). The signal is relayed to the pretectal region; from here, a parasympathetic signal flows via the Edinger–Westphal nucleus, the ciliary ganglion and the oculomotor nerve, and induces narrowing of the pupils (*miosis*) within less than 1 s. Since both pupils respond simultaneously even if the light stimulus is unilateral, this is called a *consensual light response*. Meiosis also occurs when the eyes adjust for near vision (*near-vision response* → p. 360).

The **corneal reflex** protects the eye. An object touching the cornea (afferent: trigeminal nerve) or approaching the eye (afferent: optic nerve) results in reflex closure of the eyelids.

Eye Movements, Stereoscopic Vision, Depth Perception

Conjugated movement of the eyes occurs when the external eye muscles move the eyes in the same direction (e.g., from left to right), whereas *vergence movement* is characterized by opposing (divergent or convergent) eye movement. The axes of the eyes are parallel when gazing into the distance. Fixation of the gaze on a nearby object results in convergence of the visual axes. In addition, the pupil contracts (to increase the depth of focus) and accommodation of the lens occurs (→ p. 346). The three reactions are called **near-vision response** or convergence response.

Strabismus. A greater power of accommodation for near vision is required in hyperopia than in normal vision. Since accommodation is always linked with a convergence impulse, hyperopia is often associated with squinting. If the visual axes wander too far apart, vision in one eye will be suppressed to avoid double vision (diplopia). This type of visual impairment, called *strabismic amblyopia*, can be either temporary or chronic.

Saccades. When scanning the visual field, the eyes make jerky movements when changing the point of fixation, e.g., when scanning a line of print. These quick movements that last 10–80 ms are called *saccades*. Displacement of the image is centrally suppressed during the eye due to *saccadic suppression*. A person looking at both of his or her eyes alternately in a mirror cannot perceive the movement of his or her own eyes, but an independent observer can. The small, darting saccades function to keep an object in focus.

Objects entering the field of vision are reflexively imaged in the fovea centralis (→ p. 348). Slow **pursuit movements** of the eyes function to maintain the gaze on moving objects. **Nystagmus** is characterized by a combination of these slow and rapid (saccade-like) opposing eye movements. The direction of nystagmus (right or left) is classified according to the type of rapid phase, e.g., *secondary nystagmus* (→ p. 342). *Optokinetic nystagmus* occurs when viewing an object passing across the field of vision, e.g., when looking at a tree from inside a moving train. Once the eyes have returned to the normal position (return saccade), a new object can be brought into focus. Damage to the cerebellum or organ of balance (→ p. 342) can result in *pathological nystagmus*.

The brain stem is the main center responsible for **programming** of eye movements. Rapid horizontal (conjugated) movements such as saccades and rapid nystagmus movement are programmed in the pons, whereas vertical and torsion movements are programmed in the mesencephalon. The cerebellum provides the necessary fine tuning (→ p. 326). Neurons in the region of the Edinger–Westphal nucleus are responsible for vergence movements.

In near vision, **depth vision** and **three-dimensional vision** are primarily achieved through the coordinated efforts of both eyes and are therefore limited to the **binocular field of vision** (→ A). If both eyes focus on point A (→ B), an image of the fixation point is projected on both foveae (A_L, A_R), i.e., on the *corresponding areas of the retina*. The same applies for points B and C (→ B) since they both lie on a circle that intersects fixation point A and nodal points N (→ p. 347 B) of the two eyes (**Vieth–Müller horopter**). If there were an imaginary middle eye in which the two retinal regions (in the cortex) precisely overlapped, the retinal sites would correspond to a central point $A_C \triangleq A_L + A_R$ (→ C). Assuming there is a point D outside the horopter (→ C, left), the middle eye would see a double image (D′, D″) instead of point D, where D′ is from the left eye (D_L). If D and A are not too far apart, central processing of the double image creates the perception that D is located behind D, i.e., *depth perception* occurs. A similar effect occurs when a point E (→ C, right) is closer than A; in this case, the E′ image will arise in the right eye (E'_R) and E will be perceived as being closer.

Depth perception from a distance. When viewing objects from great distances or with only one eye, contour overlap, haze, shadows, size differences, etc. are cues for depth perception (→ D). A nearby object moves across the field of vision more quickly than a distant object, e.g., in the case of the sign compared to the wall in plate **D**). In addition, the moon appears to migrate with the moving car, while the mountains disappear from sight.

Plate 12.26   Eye Movements, Depth Perception

A. Binocular visual field

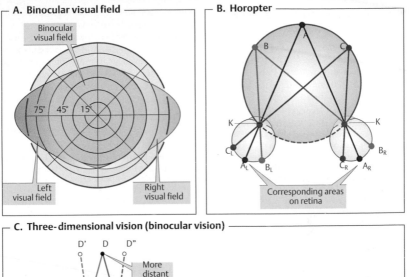

Binocular visual field

75° 45° 15°

Left visual field

Right visual field

B. Horopter

A

B   C

K   K

C_L B_R

A_L B_L C_R A_R

Corresponding areas on retina

C. Three-dimensional vision (binocular vision)

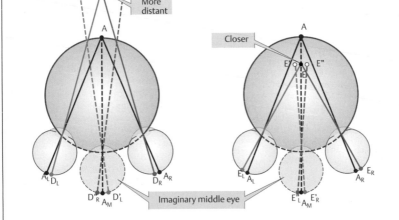

D'   D   D"

More distant

A

$A_L D_L$ $D_R A_R$

$D_R A_M D_L$

Imaginary middle eye

Closer

A

E' E E"

$E_L A_L$ $A_R E_R$

$E_L A_M E_R$

D. Cues for depth vision

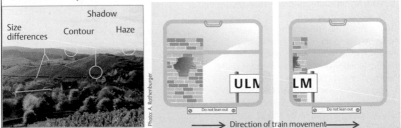

Shadow

Size differences   Contour   Haze

Photo: A. Rothenburger

ULM

LM

Do not lean out

Direction of train movement

Physical Principles of Sound—Sound Stimulus and Perception

Sound waves are the adequate stimulus for the organ of hearing. They arise from a sound source such as a gong (→ **A1**) and are conducted in gases, liquids, and solids. The air is the main carrier of sound.

The air pressure rises and falls rhythmically at the sound source. These pressure waves (sound waves) travel at a characteristic **sound velocity** (**c**) in different materials, e.g., at 332 m/s in air of 0 °C. A graphic recording of sound waves (→ **A1**) will produce waveform curves. The **wavelength** (λ) is the distance between the top of one wave and the identical phase of the succeeding one, and the maximum deviation of pressure from baseline is the **amplitude** (**a**) (→ **A1**). Enlargement (reduction) of wavelength will lower (raise) the tone, whereas a fall (rise) in amplitude will produce a quieter (louder) tone (→ **A1**). The **pitch** of a tone is defined by its **frequency** (**f**), i.e., the number of sound pressure oscillations per unit time. Frequency is measured in **hertz** (Hz = s^{-1}). Frequency, wavelength and the sound velocity are related:

$$f \text{ (Hz)} \cdot \lambda \text{ (m)} = c \text{ (m} \cdot s^{-1}). \qquad [12.1]$$

A **pure tone** has a simple sinus waveform. The tones emanating from most sound sources (e.g., musical instrument, voice) are mixtures of different frequencies and amplitudes that result in complex periodic vibrations referred to as **sound** (→ **A2**). The fundamental (lowest) tone in the complex determines the *pitch* of the sound, and the higher ones determine its *timbre* (overtones). An a^1 (440 Hz) sung by a tenor or played on a harp has a different sound than one produced on an organ or piano. The overlap of two very similar tones produces a distinct effect characterized by a *beat tone* of a much lower frequency (→ **A3**, blue/red).

Audibility limits. Healthy young persons can hear sounds ranging in frequency from 16 to 20 000 Hz. The upper limit of audibility can drop to 5000 Hz due to aging (*presbycusis*). At 1000 Hz, the **absolute auditory threshold** or lowest sound pressure perceived as sound is $3 \cdot 10^{-5}$ Pa. The threshold of sound is frequency-dependent (→ **B**, green curve). The threshold of hearing for a tone rises tremendously when other tones are heard simultaneously. This phenomenon called *masking* is the reason why it is so difficult to carry on a conversation against loud background noise. The ear is overwhelmed by sound pressures over 60 Pa, which corresponds to $2 \cdot 10^6$ times the sound pressure of the limit of audibility at

1000 Hz. Sounds above this level induce the sensation of pain (→ **B**, red curve).

For practical reasons, the *decibel (dB)* is used as a logarithmic measure of the **sound pressure level (SPL)**. Given an arbitrary reference sound pressure of $p_o = 2 \cdot 10^{-5}$ Pa, the sound pressure level (SPL) can be calculated as follows:

$$\text{SPL (dB)} = 20 \cdot \log (p_x/p_o) \qquad [12.2]$$

where p_x is the actual sound pressure. A tenfold increase in the sound pressure therefore means that the SPL rises by 20 dB.

The **sound intensity (I)** is the amount of sound energy passing through a given unit of area per unit of time (W $\cdot$ m^2). The sound intensity is proportional to the *square* of p_x. Therefore, dB values cannot be calculated on a simple linear basis. If, for example, two loudspeakers produce 70 dB each ($p_x = 6.3 \cdot 10^{-2}$ Pa), they do *not* produce 140 dB together, but a mere 73 dB because p_x only increases by a factor of $\sqrt{2}$ when the intensity level doubles. Thus, $\sqrt{2} \cdot 6.3 \cdot 10^{-2}$ Pa has to be inserted for p_x into Eq. 12.2.

Sound waves with different frequencies but equal sound pressures are not *subjectively* perceived as *equally loud*. A 63 Hz tone is only perceived to be as loud as a 20 dB/1000 Hz reference tone if the sound pressure of the 63 Hz tone is 30-fold higher (+ 29 dB). In this case, the sound pressure level of the reference tone (20 dB/1000 Hz) gives the loudness level of the 63 Hz tone in **phon** (20 phon) as at a frequency of 1000 Hz, the phon scale is numerically equals the dB SPL scale (→ **B**). Equal loudness contours or **isophones** can be obtained by plotting the subjective values of equal loudness for test frequencies over the whole audible range (→ **B**, blue curves). The absolute auditory threshold is also an isophone (4 phons; → **B**, green curve). Human hearing is most sensitive in the 2000–5000 Hz range (→ **B**).

Note: Another unit is used to describe how a tone of constant frequency is subjectively perceived as louder or less loud. *Sone* is the unit of this type of loudness, where 1 sone = 40 phons at 1000 Hz. 2 sones equal twice the reference loudness, and 0.5 sone is $^1/_2$ the reference loudness.

The **auditory area** in diagram **B** is limited by the highest and lowest audible frequencies on the one side, and by isophones of the thresholds of hearing and pain on the other. The green area in plate **B** represents the range of frequencies and intensities required for comprehension of ordinary speech (→ **B**).

Plate 12.27 Sound Physics and Thresholds

A. Wavelength, wave amplitude and wave types

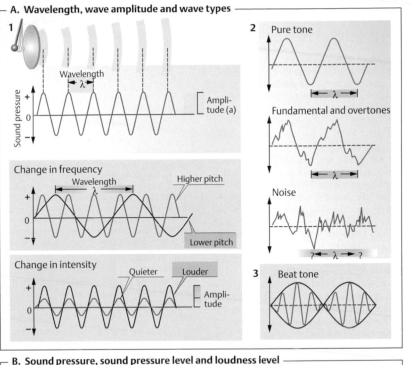

1

Wavelength
λ

Sound pressure

+

−

Amplitude (a)

Change in frequency

Wavelength
λ

Higher pitch

0

+

−

Lower pitch

Change in intensity

Quieter   Louder

0

+

−

Amplitude

2   Pure tone

λ

Fundamental and overtones

λ

Noise

? ← λ → ?

3   Beat tone

B. Sound pressure, sound pressure level and loudness level

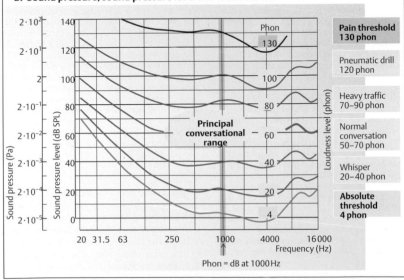

Sound pressure (Pa)

$2\cdot10^2$
$2\cdot10^1$
2
$2\cdot10^{-1}$
$2\cdot10^{-2}$
$2\cdot10^{-3}$
$2\cdot10^{-4}$
$2\cdot10^{-5}$

Sound pressure level (dB SPL)

140
120
100
80
60
40
20
0

Phon

130
100
80
60
40
20
4

Principal conversational range

20 31.5 63   250   1000   4000   16000
Frequency (Hz)

Phon = dB at 1000 Hz

Loudness level (phon)

Pain threshold 130 phon

Pneumatic drill 120 phon

Heavy traffic 70–90 phon

Normal conversation 50–70 phon

Whisper 20–40 phon

Absolute threshold 4 phon

Conduction of Sound, Sound Sensors

Sound waves are transmitted to the organ of hearing via the *external ear* and the *auditory canal*, which terminates at the *tympanic membrane* or *eardrum*. The sound waves are conducted through the air (**air conduction**) and set the eardrum in vibration. These are transmitted via the *auditory ossicles* of the tympanic cavity (*middle ear*) to the membrane of the *oval window* (→ **A 1,2**), where the *internal* or *inner ear* (labyrinth) begins.

In the **middle ear**, the *malleus, incus* and *stapes* conduct the vibrations of the tympanic membrane to the oval window. Their job is to conduct the sound from the low wave resistance/impedance in air to the high resistance in fluid with as little loss of energy as possible. This **impedance transformation** occurs at f < 2400 Hz and is based on a 22-fold pressure amplification (tympanic membrane area/oval window area is 17 : 1, and leverage arm action of the auditory ossicles amplifies force by a factor of 1.3). Impairment of impedance transforming capacity due, e.g., to destruction of the ossicles, causes roughly 20 dB of hearing loss (*conduction deafness*).

Muscles of the middle ear. The middle ear contains two small muscles—the *tensor tympani* (insertion: manubrium of malleus) and the *stapedius* (insertion: stapes)—that can slightly attenuate low-frequency sound. The main functions of the inner ear muscles are to maintain a constant sound intensity level, protect the ear from loud sounds, and to reduce distracting noises produced by the listener.

Bone conduction. Sound sets the skull in vibration, and these bone-borne vibrations are conducted directly to the cochlea. Bone conduction is fairly insignificant for physiological function, but is useful for testing the hearing. In **Weber's test**, a vibrating tuning fork (a[1]) is placed in the middle of the head. A person with normal hearing can determine the location of the tuning fork because of the symmetrical conduction of sound waves. A patient with *unilateral conduction deafness* will perceive the sound as coming from the affected side (lateralization) because of the lack of masking of environmental noises in that ear (bone conduction). A person with *sensorineural deafness*, on the other hand, will perceive the sound as coming from the healthy ear because of sound attenuation in the affected internal ear. In **Rinne's test**, the handle of a tuning fork is placed on one mastoid process (bony process behind the ear) of the patient (bone conduction). If the tone is no longer heard, the tines of the tuning fork are placed in front of the ear (air conduction). Individuals with normal hearing or sensorineural deafness can hear the turning fork in the latter position anew (positive test result), whereas those with conduction deafness cannot (test negative).

The **internal ear** consists of the *equilibrium organ* (→ p. 342) and the **cochlea**, a spiraling bony tube that is 3–4 cm in length. Inside the cochlea is an endolymph-filled duct called the *scala media (cochlear duct)*; the ductus reuniens connects the base of the cochlear duct to the endolymph-filled part of the equilibrium organ. The scala media is accompanied on either side by two perilymph-filled cavities: the *scala vestibuli* and *scala tympani*. These cavities merge at the apex of the cochlea to form the *helicotrema*. The scala vestibuli arises from the oval window, and the scala tympani terminates on the membrane of the round window (→ **A2**). The composition of perilymph is similar to that of plasma water (→ p. 93 C), and the composition of endolymph is similar to that of the cytosol (see below). Perilymph circulates in Corti's tunnel and Nuel's spaces (→ **A4**).

Organ of Corti. The (secondary) sensory cells of the hearing organ consist of approximately 10 000–12 000 external **hair cells** (**HC**s) and 3500 internal hair cells that sit upon the basilar membrane (→ **A4**). Their structure is very similar to that of the vestibular organ (→ p. 342) with the main difference being that the kinocilia are absent or rudimentary.

There are three rows of slender, cylindrical **outer hair cells**, each of which contains approximately 100 **cilia** (actually microvilli) which touch the tectorial membrane. The bases of the hair cells are firmly attached to the basilar membrane by supporting cells, and their cell bodies float in perilymph of Nuel's spaces (→ **A4**). The outer hair cells are principally innervated by efferent, mostly cholinergic neurons from the spiral ganglion (N_M-cholinoceptors; → p. 82). The **inner hair cells** are pear-shaped and completely surrounded by supporting cells. Their cilia project freely into the endolymph. The inner hair cells are arranged in a single row and synapse with over 90% of the afferent fibers of the spiral ganglion. Efferent axons from the nucleus olivaris superior lateralis synapse with the afferent endings.

Sound conduction in the inner ear. The stapes moves against the membrane of the oval window membrane, causing it to vibrate. These are transmitted via the perilymph to the membrane of the round window (→ **A2**). The walls of the endolymph-filled cochlear duct, i.e. Reissner's membrane and the basilar mem-

A. Reception and conduction of sound stimuli

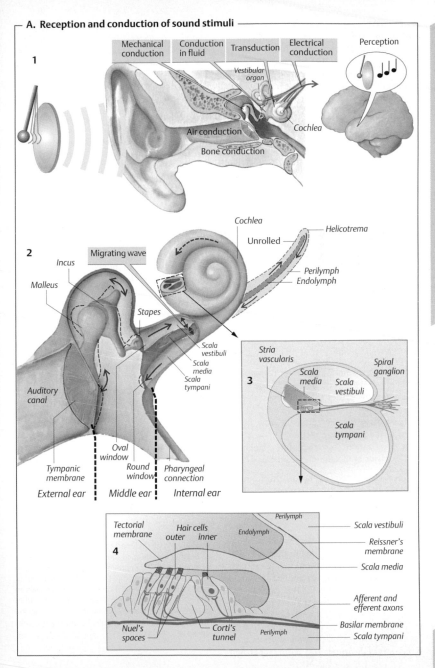

1

Mechanical conduction · Conduction in fluid · Transduction · Electrical conduction · Perception

Vestibular organ
Cochlea
Air conduction
Bone conduction

2

Migrating wave
Incus
Malleus
Stapes
Auditory canal
Scala vestibuli
Scala media
Scala tympani
Tympanic membrane
Oval window
Round window
Pharyngeal connection

Cochlea
Unrolled
Helicotrema
Perilymph
Endolymph

External ear | Middle ear | Internal ear

3

Stria vascularis
Scala media
Scala vestibuli
Spiral ganglion
Scala tympani

4

Tectorial membrane
Hair cells outer · inner
Perilymph
Endolymph
Scala vestibuli
Reissner's membrane
Scala media
Afferent and efferent axons
Basilar membrane
Scala tympani
Nuel's spaces
Corti's tunnel
Perilymph

Plate 12.28 Conduction of Sound, Sound Sensors I

brane ($\rightarrow$ **D1**) give against the pressure wave (**migrating wave**, $\rightarrow$ **B** and C). It can therefore take a "short cut" to reach the round window without crossing the helicotrema. Since the cochlear duct is deformed in waves, Reissner's membrane and the basilar membrane vibrate alternately towards the scala vestibuli and scala tympani ($\rightarrow$ **D1,2**). The velocity and wavelength of the migrating wave that started at the oval window decrease continuously ($\rightarrow$ **B**), while their *amplitude* increases to a *maximum* and then quickly subsides ($\rightarrow$ **B**, envelope curve). (The wave velocity is not equal to the velocity of sound, but is much slower.) The **site** of the maximum excursion of the cochlear duct is characteristic of the wavelength of the stimulating sound. The higher the frequency of the sound, the closer the site is to the stapes ($\rightarrow$ **C**).

Outer hair cells. Vibration of the cochlear duct causes a discrete shearing (of roughly 0.3 nm) of the tectorial membrane against the basilar membrane, causing *bending of the cilia* of the *outer hair cells* ($\rightarrow$ **D3**). This exerts also a shearing force between the rows of cilia of the individual external hair cell. Probably via the "*tip links*" ($\rightarrow$ p. 342), cation channels in the ciliary membranes open (mechanosensitive transduction channels), allowing cations (K^+, Na^+, Ca^{2+}) to enter and depolarize the outer hair cells. This causes the outer hair cells to *shorten* in sync with stimulation ($\rightarrow$ **D3**). The successive shearing force on the cilia bends them in the opposite direction. This leads to hyperpolarization (opening of K^+ channels) and *extension* of the outer hair cells.

The mechanism for this extremely fast **electromotility** (up to 20 kHz or $2 \cdot 10^4$ times per second) is unclear, but it seems to be related to the high turgor of outer hair cells (128 mmHg) and the unusual structure of their cell walls.

These outer hair cell electromotility contributes to the **cochlear amplification** (ca. 100-fold or 40 dB amplification), which occurs before sound waves reach the actual sound sensors, i.e. inner hair cells. This explains the very low threshold within the very narrow location (0.5 nm) and thus within a very small frequency range. The electromotility causes endolymph waves in the subtectorial space which exert *shearing forces on the* **inner hair**

cell *cilia* at the site of maximum reaction to the sound frequency ($\rightarrow$ **D4**), resulting in opening of transduction channels and depolarization of the cells (**sensor potential**). This leads to *transmitter* release (glutamate coupling to AMPA receptors; $\rightarrow$ p. 55 F) by internal hair cells and the subsequent conduction of impulses to the CNS.

Vibrations in the internal ear set off an outward emission of sound. These **evoked otoacoustic emissions** can be measured by placing a microphone in front of the tympanic membrane, e.g., to test internal ear function in infants and other individuals incapable of reporting their hearing sensations.

Inner ear potentials ($\rightarrow$ p. 369 C). On the cilia side, the hair cells border with the endolymph-filled space, which has a potential difference (*endocochlear potential*) of ca. $+80$ to $+110$ mV relative to perilymph ($\rightarrow$ p. 369 C). This potential difference is maintained by an active transport mechanism in the *stria vascularis*. Since the cell potential of outer (inner) hair cells is -70 mV (-40 mV), a potential difference of roughly 150–180 mV (120–150 mV) prevails across the cell membrane occupied by cilia (cell interior negative). Since the K^+ conc. in the endolymph and hair cells are roughly equal (≈ 140 mmol/L), the prevailing K^+ equilibrium potential is ca. 0 mV ($\rightarrow$ p. 32). These high potentials provide the driving forces for the influx not only of Ca^{2+} and Na^+, but also of K^+, prerequisites for provoking the sensor potential.

Hearing tests are performed using an **audiometer**. The patient is presented sounds of various frequencies and routes of conduction (bone, air). The sound pressure is initially set at a level under the threshold of hearing and is raised in increments until the patient is able to hear the presented sound (**threshold audiogram**). If the patient is unable to hear the sounds at normal levels, he or she has an hearing loss, which is quantitated in decibels (dB). In audiometry, all frequencies at the normal threshold of hearing are assigned the value of 0 dB (unlike the diagram on p. 363 B, green curve). Hearing loss can be caused by presbycusis ($\rightarrow$ p. 362), middle ear infection (impaired air conduction), and damage to the internal ear (impaired air and bone conduction) caused, for example, by prolonged exposure to excessive sound pressure (>90 dB, e.g. disco music, pneumatic drill etc.).

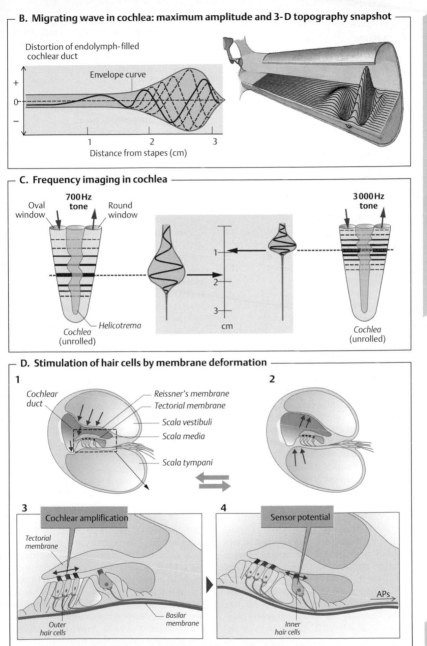

Plate 12.29 Conduction of Sound, Sound Sensors II

B. Migrating wave in cochlea: maximum amplitude and 3-D topography snapshot

Distortion of endolymph-filled cochlear duct

Envelope curve

+
0
−

Distance from stapes (cm)
1   2   3

C. Frequency imaging in cochlea

700 Hz tone

Oval window   Round window

Helicotrema
Cochlea (unrolled)

cm
1
2
3

3000 Hz tone

Cochlea (unrolled)

D. Stimulation of hair cells by membrane deformation

1

Cochlear duct
Reissner's membrane
Tectorial membrane
Scala vestibuli
Scala media
Scala tympani

2

3

Cochlear amplification

Tectorial membrane

Outer hair cells
Basilar membrane

4

Sensor potential

Inner hair cells

APs

Central Processing of Acoustic Information

Various qualities of sound must be coded for signal transmission in the acoustic pathway. These include the frequency, intensity and direction of sound waves as well as the distance of the sound source from the listener.

Frequency imaging. Tones of various frequencies are "imaged" along the cochlea, conducted in separate fibers of the auditory pathway and centrally identified. Assuming that a tone of 1000 Hz can just be distinguished from one of 1003 Hz (resembling true conditions), the frequency difference of 3 Hz corresponds to a relative *frequency differential threshold* of 0.003 (→ p. 352). This fine differential capacity is mainly due to frequency imaging in the cochlea, amplification by its outer hair cells (→ p. 366), and neuronal contrast along the auditory pathway (→ p. 313 D). This fine **tuning** ensures that a certain frequency has a particularly low threshold at its "imaging" site. Adjacent fibers are not recruited until higher sound pressures are encountered.

Intensity. Higher intensity levels result in higher action potential frequencies in afferent nerve fibers and recruitment of neighboring nerve fibers (→ **A**). The relative *intensity differential threshold* is 0.1 (→ p. 352), which is very crude compared to the frequency differential threshold. Hence, differences in loudness of sound are not perceived by the human ear until the intensity level changes by a factor of over 1.1, that is, until the sound pressure changes by a factor of over $\sqrt{1.1} = 1.05$.

Direction. *Binaural hearing* is needed to identify the direction of sound waves and is based on the following two effects. (1) Sound waves that strike the ear obliquely reach the averted ear later than the other, resulting in a *lag time*. The change in direction that a normal human subject can just barely detect (*direction threshold*) is roughly 3 degrees. This angle delays the arrival of the sound waves in the averted ear by about $3 \cdot 10^{-5}$ s (→ **B**, left). (2) Sound reaching the averted ear is also perceived as being *quieter*; differences as small as 1 dB can be distinguished. A lower sound pressure results in delayed firing of actions potentials, i.e., in increased *latency* (→ **B**, right). Thus, the impulses from the averted ear reach the CNS later (nucleus accessorius, → **D5**). Effects (1) and (2) are additive effects (→ **B**). The external ear helps to decide whether the sound is coming from front or back, above or below. Binaural hearing also helps to distinguish a certain voice against high background noise, e.g., at a party. Visibility of the speaker's mouth also facilitates comprehension.

Distance to the sound source can be determined because high frequencies are attenuated more strongly than low frequencies during sound wave conduction. The longer the sound wave travels, the lower the proportion of high frequencies when it reaches the listener. This helps, for instance, to determine whether a thunderstorm is nearby or far away.

Auditory pathway (→ **D**). The auditory nerve fibers with somata positioned in the *spiral ganglion* of the cochlea project from the cochlea (→ **D1**) to the anterolateral (→ **D2**), posteroventral and dorsal cochlear nuclei (→ **D3**). Afferents in these three nuclei exhibit *tonotopicity*, i.e., they are arranged according to tone frequency at different levels of complexity. In these areas, *lateral inhibition* (→ p. 313 D) enhances *contrast*, i.e., suppresses noise. Binaural comparison of intensity and transit time of sound waves (direction of sound) takes place at the next-higher station of the auditory pathway, i.e. in the *superior olive* (→ **D4**) and *accessory nucleus* (→ **D5**). The next stations are in the *nucleus of lateral lemniscus* (→ **D6**) and, after most fibers cross over to the opposite side, the *inferior quadrigeminal bodies* (→ **D7**). They synapse with numerous afferents and serve as a reflex station (e.g., muscles of the middle ear; → p. 366). Here, sensory information from the cochlear nuclei is compared with spatial information from the superior olive. Via connections to the superior quadrigeminal bodies (→ **D8**), they also ensure coordination of the auditory and visual space. By way of the thalamus (medial geniculate body, MGB; → **D9**), the afferents ultimately reach the *primary* **auditory cortex** (→ **D10**) and the surrounding secondary auditory areas (→ p. 311 E, areas 41 and 22). Analysis of complex sounds, short-term memory for comparison of tones, and tasks required for "eavesdropping" are some of their functions.

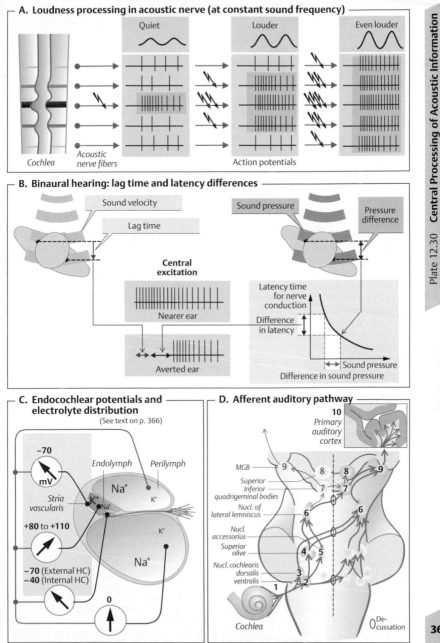

Plate 12.30   Central Processing of Acoustic Information

A. Loudness processing in acoustic nerve (at constant sound frequency)

Quiet   Louder   Even louder

Cochlea

Acoustic nerve fibers

Action potentials

B. Binaural hearing: lag time and latency differences

Sound velocity

Lag time

Sound pressure

Pressure difference

Central excitation

Nearer ear

Averted ear

Latency time for nerve conduction

Difference in latency

Sound pressure

Difference in sound pressure

C. Endocochlear potentials and electrolyte distribution

(See text on p. 366)

−70 mV

Endolymph   Perilymph

Stria vascularis

Na^+

K^+

K^+

Na^+

+80 to +110

−70 (External HC)
−40 (Internal HC)

Na^+

0

D. Afferent auditory pathway

10
Primary auditory cortex

MGB   9

8   8   9

Superior
Inferior
quadrigeminal bodies   7   7   7

Nucl. of
lateral lemniscus   6   6

Nucl.
accessorius
Superior
olive   4   5

Nucl. cochlearis
dorsalis   3
ventralis   1   2

Cochlea

De-
cussation

Voice and Speech

The human voice primarily functions as a **means of communication**, the performance of which is based on the human capacity of hearing (→ p. 363 B). As in wind instruments, the body contains a *wind space* (trachea, bronchi, etc.). Air is driven through the space between the *vocal cords* (rima glottidis) into the *air space* (*passages above the glottis*), which sets the vocal cords into vibration. The air space consists of the throat and oronasal cavities (→ **A**). The range of the human voice is so immense because of the large variety of muscles that function to modulate the intensity of the airstream (*loudness*), tension of the vocal cords, shape/width of the vocal cords (*fundamental tone*) and size/shape of the air space (*timbre, formants*) of each individual.

Joints and muscles of the **larynx** function to adjust the vocal cords and rima glottidis. A stream of air opens and closes the rima glottidis and sets off the rolling movement of the vocal cords (→ **B**). When a deep tone is produced, the fissure of the glottis remains closed longer than it opens (ratio of 5 : 1 at 100 Hz). This ratio drops to 1.4 : 1 in higher tones (400 Hz). The rima glottis remains open when whispering or singing falsetto (→ **C**, blue).

Motor signals originate in the motosensory cortex (→ p. 325 C/B, tongue/throat) and are conducted via the vagus nerve to the larynx. Sensory impulses responsible for voice production and the cough reflex are also conducted by the vagus nerve. Sensory fibers from the mucosa and muscle spindles of the larynx (→ p. 316) continuously transmit information on the position and tension of the vocal cords to the CNS. These reflexes and the close connection of the auditory pathway with bulbar and cortical motor speech centers are important for **fine adjustment of the voice**.

Vowels (→ **D**). Although their fundamental frequencies are similar (100–130 Hz), spoken vowels can be distinguished by their characteristic overtones (*formants*). Different formants are produced by modifying the shape of oral tract, i.e., mouth and lips (→ **D**). The three primary vowels [a:], [i:], [u:] make up the *vowel triangle*; [œ:], [ɔ:], [ø:], [y:], [æ:], and [ɛ] are intermediates (→ **D**).

The phonetic notation used here is that of the *International Phonetic Society*. The symbols mentioned here are as follows: [a:] as in glass; [i:] as in beat; [u:] as in food; [œ:] as in French *peur*; [ɔ:] as in bought; [ø:] as in French *peu* or in German *hören*; [y:] as in French *menu* or in German *trüb*; [æ:] as in bad; [ɛ:] as in head.

Consonants are described according to their site of articulation as *labial* (lips, teeth), e.g. P/B/W/F/M; *dental* (teeth, tongue), e.g. D/T/S/M; *lingual* (tongue, front of soft palate), e.g. L; *guttural* (back of tongue and soft palate), e.g. G/K. Consonants can be also defined according to their manner of articulation, e.g., *plosives* or *stop consonants* (P/B/T/D/K/G), *fricatives* (F/V/W/S/Ch) and *vibratives* (R).

The **frequency range** of the voice, including formants, is roughly 40–2000 Hz. Sibilants like /s/ and /z/ have higher-frequency fractions. Individuals suffering from presbyacusis or other forms of sensorineural hearing loss are often unable to hear sibilants, making it impossible for them to distinguish between words like "bad" and "bass." The **tonal range** (fundamental tone, → **C**) of the *spoken voice* is roughly one octave; that of the *singing voice* is roughly two octaves in untrained singers, and over three octaves in trained singers.

Language (see also p. 336). The main components of verbal communication are (a) auditory signal processing (→ p. 368), (b) central speech production and (c) execution of motor speech function. The centers for *speech comprehension* are mainly located in the posterior part of area 22, i.e., **Wernicke's area** (→ p. 311 E). Lesions of it result in a loss of language comprehension capacity (sensory aphasia). The patient will speak fluently yet often incomprehensibly, but does not notice it because of his/her disturbed comprehension capacity. The patient is also unable to understand complicated sentences or written words. The centers for *speech production* are mainly located in areas 44 and 45, i.e., **Broca's area** (→ p. 311 E). It controlls the primary speech centers of the sensorimotor cortex.

Lesions of this and other cortical centers (e.g., gyrus angularis) result in disorders of speech production (*motor aphasia*). The typical patient is either completely unable to speak or can only express himself in telegraphic style. Another form of aphasia is characterized by the forgetfulness of words (*anomic* or *amnestic aphasia*). Lesions of executive motor centers (corticobulbar tracts, cerebellum) cause various *speech disorders*. Auditory feedback is extremely important for speech. When a person goes deaf, speech deteriorates to an appreciable extent. Children born deaf do not learn to speak.

Plate 12.31 Voice and Speech

A. Larynx (cross-section)

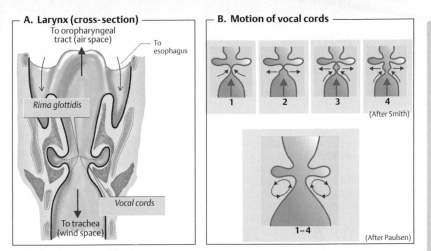

To oropharyngeal tract (air space)

To esophagus

Rima glottidis

Vocal cords

To trachea (wind space)

B. Motion of vocal cords

1 2 3 4

(After Smith)

1–4

(After Paulsen)

C. Vocal range and singing range

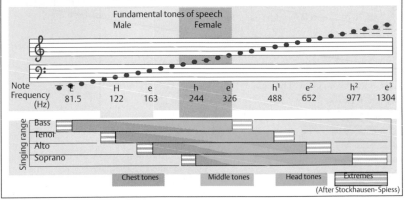

Fundamental tones of speech
Male   Female

Note	E	H	e	h	e¹	h¹	e²	h²	e³
Frequency (Hz)	81.5	122	163	244	326	488	652	977	1304

Singing range
Bass
Tenor
Alto
Soprano

Chest tones   Middle tones   Head tones   Extremes

(After Stockhausen-Spiess)

D. Vowel production

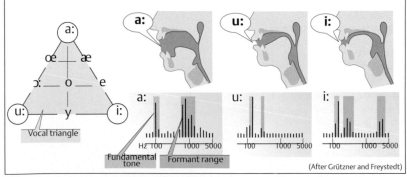

a:
œ — æ
ɔ: — o — e
u: y   i:

Vocal triangle

a: u: i:

a: u: i:

Hz 100   1000   5000
Fundamental tone   Formant range

(After Grützner and Freystedt)

Dimensions and Units

Physiology is the science of life processes and bodily functions. Since they are largely based on physical and chemical laws, the investigation, understanding, assessment, and manipulation of these functions is inseparably linked to the measurement of physical, chemical, and other **parameters**, such as blood pressure, hearing capacity, blood pH, and cardiac output. The **units** for measurement of these parameters are listed in this section. We have given preference to the *international system of SI units* (Système International d'Unités) for uniformity and ease of calculation. Non-SI units will be marked with an asterisk. *Conversion factors* for older units are also listed. Complicated or less common physiological units (e.g., wall tension, flow resistance, compliance) are generally explained in the book as they appear. However, some especially important terms that are often (not always correctly) used in physiology will be explained in the Appendix, e.g., concentration, activity, osmolality, osmotic pressure, oncotic pressure, and pH.

The seven base units of the SI system.

Unit	Symbol	Dimension
Meter	m	length
Kilogram	kg	mass
Second	s	time
Mole	mol	amount of substance
Ampere	A	electric current
Kelvin	K	temperature (absolute)
Candela	cd	luminous intensity

The base units are precisely defined autonomous units. All other units are derived by multiplying or dividing base units and are therefore referred to as **derived units**, e.g.:
— Area (length · length): $m \cdot m = m^2$
— Velocity (length/time): $m/s = m \cdot s^{-1}$.

If the new unit becomes too complicated, it is given a new name and a corresponding symbol, e.g., force = $m \cdot kg \cdot s^{-2} = N$ ($\rightarrow$ **Table 1**).

Fractions and Multiples of Units

Prefixes are used to denote decimal multiples and fractions of a unit since it is both tedious and confusing to write large numbers. We generally write 10 kg (kilograms) and 10 µg (micrograms) instead of 10 000 g and 0.00001 g, for example. The prefixes, which are usually varied in 1000-unit increments, and the corresponding symbols and conversion factors are listed in **Table 2**. Prefixes are used with base units and the units derived from them ($\rightarrow$ **Table 1**), e.g., 10^3 Pa = 1 kPa. Decimal increments are used in some cases (e.g., da, h, d, and c; $\rightarrow$ **Table 2**). **Time** is given in conventional nondecimal units, i.e., *seconds* (s), *minutes* (min), *hours* (h), and *days* (d).

Length, Area, Volume

The *meter* (m) is the SI unit of **length**. Other units of length have also been used.
Examples:

1 ångström (Å) = 10^{-10} m = 0.1 nm
1 micron (μ) = 10^{-6} m = 1 μm
1 millimicron (mμ) = 10^{-9} m = 1 nm

American and British units of length:
1 inch = 0.0254 m = 25.4 mm
1 foot = 0.3048 m
1 yard = 3 feet = 0.9144 m
1 (statute) mile = 1609.344 m ≈ 1.61 km
1 nautical mile = 1.853 km

The *square meter* (m^2) is the derived SI unit of **area**, and the *cubic meter* (m^3) is the corresponding unit of **volume**. When denoting the fractions or multiples of these units with prefixes (**Table 2**), please note that there are some peculiarities.
Examples:

1 m = 10^3 mm, but
1 mm^2 = 10^6 mm^2, and
1 m^3 = 10^9 mm^3

The *liter* (L or l)* is often used as a unit of *volume* for liquids and gases:

1 L = 10^{-3} m^3 = 1 dm^3
1 mL = 10^{-6} m^3 = 1 cm^3
1 µL = 10^{-9} m^3 = 1 mm^3.

Table 1 Derived units based on SI base units m, kg, s, cd, and A

Coulomb	C	electrical charge	$s \cdot A$
Farad	F	electrical capacitance	$C \cdot V^{-1} = m^{-2} \cdot kg^{-1} \cdot s^4 \cdot A^2$
Hertz	Hz	frequency	s^{-1}
Joule	J	heat, energy, work	$N \cdot m = m^2 \cdot kg \cdot s^{-2}$
Lumen	lm	light flux	$cd \cdot sr$
Lux	lx	light intensity	$lm \cdot m^{-2} = cd \cdot sr \cdot m^{-2}$
Newton	N	force	$m \cdot kg \cdot s^{-2}$
Ohm	Ω	electrical resistance	$V \cdot A^{-1} = m^2 \cdot kg \cdot s^{-3} \cdot A^{-2}$
Pascal	P	pressure	$N \cdot m^{-2} = m^{-1} \cdot kg \cdot s^{-2}$
Siemens	S	conductivity	$\Omega^{-1} = m^{-2} \cdot kg^{-1} \cdot s^3 \cdot A^2$
Steradian	sr	measure of solid angle[1]	$1 \, (m^2 \cdot m^{-2})$
Tesla	T	magnetic flux density	$Wb \cdot m^{-2} = kg \cdot s^{-2} \cdot A^{-1}$
Volt	V	electric potential	$W \cdot A^{-1} = m^2 \cdot kg \cdot s^{-3} \cdot A^{-1}$
Watt	W	electric power	$J \cdot s^{-1} = m^2 \cdot kg \cdot s^{-3}$
Weber	Wb	magnetic flux	$V \cdot s = m^2 \cdot kg \cdot s^{-2} \cdot A^{-1}$

[1] The solid angle of a sphere is defined as the angle subtended at the center of a sphere by an area (A) on its surface times the square of its radius (r^2). A steradian (sr) is the solid angle for which $r = 1$ m and $A = 1$ m^2, that is, $1 \, sr = 1 \, m^2/m^{-2}$.

Table 2 Prefixes for fractions and multiples of units of measure

Prefix	Symbol	Factor	Prefix	Symbol	Factor
deca-	da	10^1	deci-	d	10^{-1}
hecto-	h	10^2	centi-	c	10^{-2}
kilo-	k	10^3	milli-	m	10^{-3}
mega-	M	10^6	micro-	μ	10^{-6}
giga-	G	10^9	nano-	n	10^{-9}
tera-	T	10^{12}	pico-	p	10^{-12}
peta-	P	10^{15}	femto-	f	10^{-15}
exa-	E	10^{18}	atto-	a	10^{-18}

Conversion of American and British volume units into SI units:

 1 fluid ounce (USA) = 29.57 mL
 1 fluid ounce (UK) = 28.47 mL
 1 liquid gallon (USA) = 3.785 L
 1 liquid gallon (UK) = 4.546 L
 1 pint (USA) = 473.12 mL
 1 pint (UK) = 569.4 mL

Velocity, Frequency, Acceleration

Velocity is the distance traveled per unit time $(m \cdot s^{-1})$. This is an expression of **linear velocity**, whereas **"volume velocity"** is used to express the *volume flow per unit time*. The latter is expressed as $L \cdot s^{-1}$ or $m^3 \cdot s^{-1}$.

Frequency is used to describe how often a periodic event (pulse, breathing, etc.) occurs per unit time. The SI unit of frequency is s^{-1} or *hertz* (Hz). min^{-1} is also commonly used:

 $min^{-1} = 1/60\ Hz \approx 0.0167\ Hz.$

Acceleration, or *velocity change per unit time*, is expressed in $m \cdot s^{-1} \cdot s^{-1} = m \cdot s^{-2}$. Since *deceleration* is equivalent to negative acceleration, acceleration and deceleration can both be expressed in $m \cdot s^{-2}$.

Force and Pressure

Force equals mass times acceleration. **Weight** is a special case of force as weight equals mass times acceleration of gravity. Since the unit of mass is kg and that of acceleration $m \cdot s^{-2}$, force is expressed in $m \cdot kg \cdot s^{-2} = $ *newton* (N). The older units of force are converted into N as follows:

 $1\ dyn = 10^{-5}\ N = 10\ \mu N$
 $1\ pond = 9.8 \cdot 10^{-3}\ N = 9.8\ mN.$

Pressure equals *force per unit area*, so the SI unit of pressure is $N \cdot m^{-2} = $ *pascal* (Pa). However, the pressure of bodily fluids is usually measured in *mm Hg*. This unit and other units are converted into SI units as follows:

 $1\ mm\ H_2O \approx 9.8\ Pa$
 $1\ cm\ H_2O \approx 98\ Pa$
 $1\ mm\ Hg = 133.3\ Pa = 0.1333\ kPa$
 $1\ torr = 133.3\ Pa = 0.1333\ kPa$
 $1\ technical\ atmosphere\ (at) \approx 98.067\ kPa$
 $1\ physical\ atmosphere\ (atm) = 101.324\ kPa$
 $1\ dyne \cdot cm^{-2} = 0.1\ Pa$
 $1\ bar = 100\ kPa.$

Work, Energy, Heat, Power

Work equals *force times distance*, $N \cdot m = J$ (*joule*), or *pressure times volume*, $(N \cdot m^{-2}) \cdot m^3 = J$. **Energy** and **heat** are also expressed in J.

Other units of work, heat, and energy are converted into J as follows:

 $1\ erg = 10^{-7}\ J = 0.1\ \mu J$
 $1\ cal \approx 4.185\ J$
 $1\ kcal \approx 4185\ J = 4.185\ kJ$
 $1\ Ws = 1\ J$
 $1\ kWh = 3.6 \cdot 10^6\ J = 3.6\ MJ.$

Power equals *work per unit time* and is expressed in *watts* (W), where $W = J \cdot s^{-1}$. Heat flow is also expressed in W. Other units of power are converted into W as follows:

 $1\ erg \cdot s^{-1} = 10^{-7}\ W = 0.1\ \mu W$
 $1\ cal \cdot h^{-1} = 1.163 \cdot 10^{-3}\ W = 1.163\ mW$
 1 metric horse power (hp) = 735.5 W = 0.7355 kW.

Mass, Amount of Substance

The base unit of **mass** is the kilogram (kg), which is unusual insofar as the base unit bears the prefix "kilo". Moreover, 1000 kg is defined as a *metric ton** instead of as a megagram. **Weight** is the product of mass and gravity (see above), but weight scales are usually calibrated in units of mass (g, kg).

British and American units of mass are converted into SI units as follows.

 Avoirdupois weight:
 1 ounce (oz.) = 28.35 g
 1 pound (lb.) = 453.6 g
 Apothecary's and troy weight:
 1 ounce = 31.1 g
 1 pound = 373.2 g.

The mass of a molecule or an atom (**molecular** or **atomic mass**) is often expressed in *daltons* (Da)*. 1 Da = 1/12 the mass of a ^{12}C atom, equivalent to 1 kg/Avogadro's constant = 1 kg/ $(6.022 \cdot 10^{23})$:

 $1\ Da = 1.66 \cdot 10^{-27}\ kg$
 $1000\ Da = 1\ kDa.$

The **relative molecular mass (M_r)**, or *molecular "weight"*, is the molecular mass of a substance divided by 1/12 the mass of a ^{12}C atom. Since M_r is a ratio, it is a dimensionless unit.

The **amount of substance**, or *mole* (mol), is related to mass. One mole of substance contains as many elementary particles (atoms, molecules, ions) as 12 g of the nuclide of a ^{12}C atom = $6.022 \cdot 10^{23}$ particles. The *conversion factor between moles and mass* is therefore: 1 mole equals the mass of substance (in grams) corresponding to the relative molecular, ionic, or atomic mass of the substance. In other words, it expresses how much higher the mass of the atom, molecule, or ion is than 1/12 that of a ^{12}C atom.

Examples:
— Relative molecular mass of H_2O: 18
  → 1 mol H_2O = 18 g H_2O.
— Relative atomic mass of Na^+: 23
  → 1 mol Na^+ = 23 g Na^+.
— Relative molecular mass of $CaCl_2$:
  = $40 + (2 \cdot 35.5) = 111$
  → 1 mol $CaCl_2$ = 111 g $CaCl_2$.
($CaCl_2$ contains 2 mol Cl^- and 1 mol Ca^{2+}.)

The **equivalent mass** is calculated as moles divided by the valency of the ion in question and expressed in *equivalents* (Eq)*. The mole and equivalent values of *monovalent ions* are identical:
1 Eq Na^+ = 1/1 mol Na^+.
For bivalent ions, equivalent = $^1/_2$ mole:
1 Eq Ca^{2+} = $^1/_2$ mole Ca^{2+} or 1 mole Ca^{2+} = 2 Eq Ca^{2+}.

The **osmole (Osm)** is also derived from the mole (see below).

Electrical Units

Electrical **current** is the flow of charged particles, e.g., of electrons through a wire or of ions through a cell membrane. The number of particles moving per unit time is measured in *amperes* (A). Electrical current cannot occur unless there is an **electrical potential difference**, in short also called potential, voltage, or tension. Batteries and generators are used to create such potentials. Most electrical potentials in the body are generated by ionic flow (→ p. 32). The *volt* (V) is the SI unit of electrical potential (→ **Table 1**).

How much electrical current flows at a given potential depends on the amount of **electrical resistance**, as is described in *Ohm's law*

(voltage = current · resistance). The unit of electrical resistance is *ohm* (Ω) (→ **Table 1**). *Conductivity* is the reciprocal of resistance ($1/\Omega$) and is expressed in *siemens* (S), where S = Ω^{-1}. In physiology, resistance is related to the membrane surface area ($\Omega \cdot m^2$). The reciprocal of this defines the **membrane conductance** to a given ion: $\Omega^{-1} \cdot m^{-2} = S \cdot m^{-2}$ (→ p. 32).

Electrical **work** or **energy** is expressed in *joules* (J) or *watt seconds* (Ws), whereas electrical **power** is expressed in *watts* (W).

The electrical **capacitance** of a capacitor, e.g., a cell membrane, is the ratio of *charge* (C) to *potential* (V); it is expressed in *farads* (F) (→ **Table 1**).

Direct current (DC) always flows in one direction, whereas the direction of flow of **alternating current** (AC) constantly changes. The **frequency** of one cycle of change per unit time is expressed in *hertz* (Hz). Mains current is generally 60 Hz in the USA and 50 Hz in Europe.

Temperature

Kelvin (K) is the SI unit of temperature. The lowest possible temperature is 0 K, or *absolute zero*. The **Celsius** or **centigrade** scale is derived from the Kelvin scale. The temperature in degrees Celsius (°C) can easily be converted into K:
  °C = K − 273.15.
In the USA, temperatures are normally given in degrees Fahrenheit (°F). Conversions between Fahrenheit and Celsius are made as follows:
  °F = (9/5 · °C) + 32
  °C = (°F − 32) · 5/9.

Some important Kelvin, Celsius, and Fahrenheit temperature equivalents:

	K	°C	°F
Freezing point of water	+ 273	0°	+ 32°
Room temperature	293–298	20°–25°	68°–77°
Body core temperature	310	37°	98.6°
Fever	311–315	38°–42°	100°–108°
Boiling point of water (at sea level)	373	100°	212°

Concentrations, Fractions, Activity

The word *concentration* is used to describe many different relationships in physiology and medicine. Concentration of a substance X is often abbreviated as [X]. Some concentrations are listed below:

—*Mass concentration*, or the mass of a substance per unit volume (e.g., $g/L = kg/m^3$)

—*Molar concentration*, or the amount of a substance per unit volume (e.g., mol/L)

—*Molal concentration*, or the amount of substance per unit mass of solvent (e.g., mol/kg H_2O).

The SI unit of **mass concentration** is g/L (kg/m^3, mg/L, etc.). The conversion factors for older units are listed below:

$1 g/100 mL = 10 g/L$
$1 g\% = 10 g/L$
$1 \% (w/v) = 10 g/L$
$1 g‰ = 1 g/L$
$1 mg\% = 10 mg/L$
$1 mg/100 mL = 10 mg/L$
$1 \mu g\% = 10 \mu g/L$
$1 \gamma\% = 10 \mu g/L$.

Molarity is the **molar concentration**, which is expressed in mol/L (or mol/m^3, mmol/L, etc.). Conversion factors are listed below:

$1 M (molar) = 1 mol/L$
$1 N (normal) = (1/valency) \cdot mol/L$
$1 mM (mmolar) = 1 mmol/L$
$1 Eq/L = (1/valency) \cdot mol/L$.

In highly diluted solutions, the only difference between the molar and molal concentrations is that the equation "1 L H_2O = 1 kg H_2O" holds at only one particular temperature (4°C). Biological fluids are not highly diluted solutions. The volume of solute particles often makes up a significant fraction of the overall volume of the solution. One liter of plasma, for example, contains 70 mL of proteins and salts and only 0.93 L of water. In this case, there is a 7% difference between molarity and molality. Differences higher than 30% can occur in intracellular fluid. Although molarity is more commonly measured (volumetric measurement), molality plays a more important role in biophysical and biological processes and chemical reactions.

The **activity** (a) of a solution is a thermodynamic measure of its physicochemical efficacy. In physiology, the activity of ions is measured by ion-sensitive electrodes (e.g., for H^+, Na^+, K^+, Cl^-, or Ca^{2+}). The activity and molality of a solution are identical when the total **ionic strength** (μ) of the solution is very small, e.g., when the solution is an ideal solution. The ionic strength is dependent on the charge and concentration of all ions in the solution:

$$\mu \equiv 0.5 (z_1^2 \cdot c_1 + z_2^2 \cdot c_2 + \ldots + z_i^2 \cdot c_i) \qquad [13.1]$$

where z_i is the valency and c_i the molal concentration of a given ion "i", and 1, 2, etc. represent the different types of ions in the solution. Owing to the high ionic strength of biological fluids, the solute particles influence each other. Consequently, the activity (a) of a solution is always significantly lower than its molar concentration (c). Activity is calculated as $a = f \cdot c$, where f is the **activity coefficient**.

Example: At an ionic strength of 0.1 (as it is the case for a solution containing 100 mmol NaCl/kg H_2O), f = 0.76 for Na^+. The activity important in biophysical processes is therefore roughly 25% lower than the molality of the solution.

In solutions that contain weak electrolytes which do not completely dissociate, the molality and activity of free ions also depend on the *degree of electrolytic dissociation*.

Fractions ("fractional concentrations") are relative units:

— *Mass ratio*, i.e., mass fraction relative to total mass
— *Molar ratio*
— *Volume ratio*, i.e. volume fraction relative to total volume. The volume fraction (F) is commonly used in respiratory physiology.

Fractions are expressed in units of g/g, mol/mol, and L/L respectively, i.e. in "units" of 1, 10^{-3}, 10^{-6}, etc. The unabbreviated unit (e.g., g/g) should be used whenever possible because it identifies the type of fraction in question. The fractions %, ‰, ppm (parts per million), and ppb (parts per billion) are used for all types of fractions.

Conversion:
$1\% = 0.01$
$1‰ = 1 \cdot 10^{-3}$
$1 vol\% = 0.01 L/L$
$1 ppm = 1 \cdot 10^{-6}$
$1 ppb = 1 \cdot 10^{-9}$

Osmolality, Osmotic / Oncotic Pressure

Osmolarity (Osm/L), a unit derived from molarity, is the *concentration of all osmotically active particles* in a solution, regardless of which compounds or mixtures are involved.

However, measurements with osmometers as well as the biophysical application of osmotic concentration refer to the number of osmoles per unit volume of *solvent* as opposed to the total volume of the solution. This and the fact that volume is temperature-dependent are the reasons why **osmolality** (Osm/kg H_2O) is generally more suitable.

Ideal osmolality is derived from the molality of the substances in question. If, for example, 1 mmol (180 mg) of glucose is dissolved in 1 kg of water (1 L at 4°C), the molality equals 1 mmol/kg H_2O and the ideal osmolality equals 1 mOsm/kg H_2O. This relationship changes when electrolytes that dissociate are used, e.g., NaCl $\rightleftharpoons$ Na^+ + Cl^-. Both of these ions are osmotically active. When a substance that dissociates is dissolved in 1 kg of water, the ideal osmolality equals the molality times the number of dissociation products, e.g., 1 mmol NaCl/kg H_2O = 2 mOsm/kg H_2O.

Electrolytes weaker than NaCl do not dissociate completely. Therefore, their *degree of electrolytic dissociation* must be considered.

These rules apply only to ideal solutions, i.e., those that are extremely dilute. As mentioned above, bodily fluids are **nonideal (or real) solutions** because their real osmolality is lower than the ideal osmolality. The real osmolality is calculated by multiplying the ideal osmolality by the *osmotic coefficient* (g). The osmotic coefficient is concentration-dependent and amounts to, for example, approximately 0.926 for NaCl with an (ideal) osmolality of 300 mOsm/kg H_2O. The real osmolality of this NaCl solution thus amounts to $0.926 \cdot 300$ = 278 mOsm/kg H_2O.

Solutions with a real osmolality equal to that of plasma ($\approx$ 290 mOsm/kg H2 O) are said to be isosmolal. Those whose osmolality is higher or lower than that of plasma are hyperosmolal or hyposmolal.

Osmolality and Tonicity

Each osmotically active particle in solution (cf. real osmolality) exerts an **osmotic pressure** (π) as described by *van't Hoff's equation*:

$$\pi = R \cdot T \cdot c_{osm} \qquad [13.2]$$

where R is the universal gas constant (8.314 J $\cdot$ $K^{-1} \cdot Osm^{-1}$), T is the absolute temperature in K, and c_{osm} is the real osmolality in Osm $\cdot$ (m^3

$H_2O)^{-1}$ = mOsm $\cdot$ ($L H_2O)^{-1}$. If two solutions of different osmolality (Δc_{osm}) are separated by a water-permeable **selective membrane** , Δc_{osm} will exert an **osmotic pressure difference** ($\Delta\pi$) across the membrane in steady state if the membrane is less permeable to the solutes than to water. In this case, the *selectivity* of the membrane, or its relative impermeability to the solutes, is described by the **reflection coefficient** (σ), which is assigned a value between 1 (impermeable) and 0 (as permeable as water). The reflection coefficient of a *semipermeable membrane* is σ = 1. By combining **van't Hoff's** and **Staverman's equations**, the osmotic pressure difference ($\Delta\pi$) can be calculated as follows:

$$\Delta\pi = \sigma \cdot R \cdot T \cdot \Delta c_{osm}. \qquad [13.3]$$

Equation 13.3 shows that a solution with the same osmolality as plasma will exert the same osmotic pressure on a membrane in steady state (i.e., that the solution and plasma will be isotonic) only if σ = 1. In other words, the membrane must be strictly semipermeable.

Isotonicity, or equality of osmotic pressure, exists between plasma and the cytosol of red blood cells (and other cells of the body) in steady state. When the red cells are mixed in a urea solution with an osmolality of 290 mOsm/kg H_2O, isotonicity does not prevail after urea (σ < 1) starts to diffuse into the red cells. The interior of the red blood cells therefore becomes hypertonic, and water is drawn inside the cell due to osmosis ($\rightarrow$ p. 24). As a result, the erythrocytes continuously swell until they burst.

An **osmotic gradient** resulting in the subsequent flow of water therefore occurs in all parts of the body in which dissolved particles pass through water-permeable cell membranes or cell layers. This occurs, for example, when Na^+ and Cl^- pass through the epithelium of the small intestine or proximal renal tubule. The extent of this water flow or *volume flow* Jv ($m^3 \cdot s^{-1}$) is dependent on the *hydraulic conductivity* k ($m \cdot s^{-1} \cdot Pa^{-1}$) of the membrane (i.e., its permeability to water), the *area* A of passage (m^2), and the pressure difference, which, in this case, is equivalent to the osmotic pressure difference $\Delta\pi$ (Pa):

$$Jv = k \cdot A \cdot \Delta\pi \; [m^3 \cdot s^{-1}]. \qquad [13.4]$$

Since it is normally not possible to separately determine k and A of a biological membrane or cell layer, the product of the two (k · A) is often calculated as the *ultrafiltration coefficient* K_f (m³ s⁻¹ Pa⁻¹) (cf. p. 152).

The transport of osmotically active particles causes water flow. Inversely, flowing water drags dissolved particles along with it. This type of *solvent drag* (→ p. 24) is a form of convective transport.

Solvent drag does not occur if the cell wall is impermeable to the substance in question (σ = 1). Instead, the water will be retained on the side where the substance is located. In the case of the aforementioned epithelia, this means that the substances that cannot be reabsorbed from the tubule or intestinal lumen lead to osmotic diuresis (→ p. 172) and diarrhea respectively. The latter is the mechanism of action of saline laxatives (→ p. 262).

Oncotic Pressure / Colloid Osmotic Pressure

As all other particles dissolved in plasma, macromolecular proteins also exert an osmotic pressure referred to as *oncotic pressure* or *colloid osmotic pressure*. Considering its contribution of only 3.5 kPa (25 mm Hg) relative to the total osmotic pressure of the small molecular components of plasma, the oncotic pressure on a strictly semipermeable membrane could be defined as negligible. However, within the body, oncotic pressure is so important because the **endothelium** that lines the blood vessels allows small molecules to pass relatively easily ($\sigma \approx 0$). According to equation 13.3, their osmotic pressure difference $\Delta\pi$ at the endothelium is virtually zero. Consequently, only the oncotic pressure difference of proteins is effective, as the endothelium is either partly or completely impermeable to them, depending on the capillary segment in question. Because the protein reflection coefficient $\sigma \gg 0$ and the protein content of the plasma (ca. 75 g/L) are higher than that of the interstitium, these two factors *counteract filtration*, i.e., the blood pressure-driven outflow of plasma water from the endothelial lumen, making the endothelium an effective volume barrier between the plasma space and the interstitium.

If the blood pressure drives water out of the blood into the interstitium (filtration), the plasma protein concentration and thus the oncotic pressure difference π will rise (→ pp. 152, 208). This rise is much higher than equation 13.3 leads one to expect (→ **A**). The difference is attributable to specific biophysical properties of plasma proteins. If there is a pressure-dependent efflux or influx of water out of or into the bloodstream, these relatively high changes in oncotic pressure difference automatically exert a counterpressure that limits the flow of water.

pH, pK, Buffers

The **pH** indicates the hydrogen ion [H⁺] concentration of a solution. According to *Sörensen*, the pH is the negative common logarithm of the molal H⁺ concentration in mol/kg H₂O.

Examples:
1 mol/kg H₂O = 10^0 mol/kg H₂O = pH 0,
0.1 mol/kg H₂O = 10^{-1} mol/kg H₂O = pH 1,
and so on up to 10^{-14} mol/kg H₂O = pH 14.

Since glass electrodes are normally used to measure the pH, the **H⁺ activity** of the solution is actually being determined. Thus, the following rule applies:
pH = − log (f_H · [H⁺]),
where f_H is the activity coefficient of H⁺. Considering its ionic concentration (see above), the f_H of plasma is ≈ 0.8.

The logarithmic nature of pH must be considered when observing pH changes. For example, a rise in pH from 7.4 (40 nmol/kg H₂O) to pH 7.7 decreases the H⁺ activity by 20 nmol/kg H₂O, whereas an equivalent decrease (e.g., from pH 7.4 to pH 7.1) increases the H⁺ activity by 40 nmol/kg H₂O.

The **pK** is fundamentally similar to the pH. It is the negative common logarithm of the *dissociation constant* of an acid (K_a) or of a base (K_b):
$$pK_a = -\log K_a$$
$$pK_b = -\log K_b.$$
For an acid and its corresponding base, $pK_a + pK_b = 14$, so that the value of pK_a can be derived from that of pK_b and vice versa.

The *law of mass actions* applies when a weak acid (AH) dissociates:
$$AH \rightleftharpoons A^- + H^+ \qquad [13.5]$$

It states that the product of the molal concentration (indicated by square brackets) of the dissociation products divided by the concentration of the nondissociated substance remains constant:

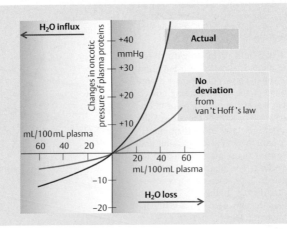

A. Physiological signification of deviations on oncotic pressure of plasma from van't Hoff's equation. A loss of water from plasma leads to a disproportionate rise in oncotic pressure, which counteracts the water loss. Conversely, the dilution of plasma due to the influx of water leads to a disproportionate drop in oncotic pressure, though less pronounced. Both of these are important mechanisms for maintaining a constant blood volume and preventing edema. (Adapted from Landis EM u. Pappenheimer JR. *Handbook of Physiology*. Section 2: Circulation, Vol. II. American Physiological Society: Washington D.C. 1963, S. 975.)

$$K_a = \frac{[A^-] \cdot [H^+] \cdot f_H}{[AH]} \qquad [13.6]$$

Converted into logarithmic form (and inserting H$^+$ activity for [H$^+$]), the equation is transformed into:

$$\log K_a = \log \frac{[A^-]}{[AH]} + \log ([H^+] \cdot f_H) \qquad [13.7]$$

or

$$-\log ([H^+] \cdot f_H) = -\log K_a + \log \frac{[A^-]}{[AH]} \qquad [13.8]$$

Based on the above definitions for pH and pK$_a$, it can also be converted into

$$\mathbf{pH = pK_a + \log \frac{[A^-]}{[AH]}} \qquad [13.9]$$

Because the concentration and not the activity of A$^-$ and AH is used here, pK$_a$ is concentration-dependent in nonideal solutions.

Equation 13.9 is the general form of the **Henderson–Hasselbalch equation** ($\rightarrow$ p. 138ff.), which describes the relationship between the pH of a solution and the concentration ratio of a dissociated to an undissociated form of a solute. If [A$^-$] = [AH], then the concentration ratio is 1/1 = 1, which corresponds to pH = pK$_a$ since the log of 1 = 0.

A weak acid (AH) and its dissociated salt (A$^-$) form a **buffer system** for H$^+$ and OH$^-$ ions:
Addition of H$^+$ yields A$^-$ + H$^+$ $\rightarrow$ AH
Addition of OH$^-$ yields AH + OH$^-$ $\rightarrow$ A$^-$ + H$_2$O.
The *buffering power* of a buffer system is greatest when [AH] = [A$^-$], i.e., when the pH of the solution equals the pK$_a$ of the buffer.

Example: Both [A$^-$] and [AH] = 10 mmol/L and pK$_a$ = 7.0. After addition of 2 mmol/L of H$^+$ ions, the [A$^-$]/[AH] ratio changes from 10/10 to 8/12 since 2 mmol/L of A$^-$ are consequently converted into 2 mmol/L of AH. Since the log of 8/12 $\approx$ -0.18, the pH decreases by 0.18 units to pH 6.82. If the initial $[A^-]/[AH]$ ratio had been 3/17, the pH would have dropped from an initial pH 6.25 (7 plus the log of 3/17 = 6.25) to pH 5.7 (7 + log of 1/19 = 5.7), i.e., by 0.55 pH units after addition of the same quantity of H^+ ions.

The titration of a buffer solution with H^+ (or OH^-) can be plotted to generate a **buffering curve** ($\rightarrow$ **B**). The steep part of the curve represents the range of the best buffering power. The pK_a value lies at the *turning point* in the middle of the steep portion of the curve. Substances that gain (or lose) more than one H^+ per molecule have more than one pK value and can therefore exert an optimal buffering action in several regions. Phosphoric acid (H_3PO_4) donates three H^+ ions, thereby successively forming $H_2PO_4^-$, HPO_4^{2-}, and PO_4^{3-}. The buffer pair $HPO_4^{2-}/H_2PO_4^-$ with a pK_a of 6.8 is important in human physiology ($\rightarrow$ p. 174ff.).

The absolute slope, $d[A^-]/d(pH)$, of a buffering curve (plot of pH *vs.* $[A^-]$) is a measure of **buffering capacity** ($mol \cdot L^{-1} \cdot [\Delta pH]^{-1}$; $\rightarrow$ p. 138).

Powers and Logarithms

Powers of ten are used to more easily and conveniently write numbers that are much larger or smaller than 1.

Examples:
$100 = 10 \cdot 10 = 10^2$
$1000 = 10 \cdot 10 \cdot 10 = 10^3$
$10\,000 = 10 \cdot 10 \cdot 10 \cdot 10 = 10^4$, etc.

In this case, the exponent denotes the amount of times ten is multiplied by itself. If the number is not an exact power of ten (e.g., 34 500), divide it by the next lowest decimal power (10 000) and use the quotient (3.45) as a multiplier to express the result as $3.45 \cdot 10^4$.

The number 10 can also be expressed exponentially (10^1). Numbers much smaller than 1 are annotated using negative exponents.

Examples:
$1 = 10 \div 10 = 10^0$
$0.1 = 10 \div 10 \div 10 = 10^{-1}$
$0.01 = 10 \div 10 \div 10 = 10^{-2}$, etc.

Similar to the large numbers above, numbers that are not exact powers of ten are expressed using multipliers, e.g.,
$0.04 = 4 \cdot 0.01 = 4 \cdot 10^{-2}$.

Note: When writing numbers smaller than 1, the (negative) exponent corresponds to the

13 Appendix

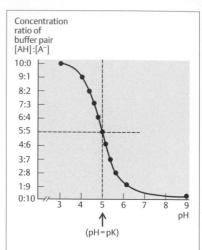

Concentration ratio of buffer pair [AH]:[A$^-$]

(pH = pK)

B. Buffering Curve. Graphic representation of the relationship between pH and the concentration ratio of buffer acid/buffer base [AH]/[A$^-$] as a function of pH. The numerical values are roughly equivalent to those of the buffer pair *acetic acid/acetate* (pK_a = 4.7). The buffering power of a buffer system is greatest when [AH] = [A$^-$], i.e., when the pH of the solution equals the pK_a of the buffer (broken lines).

position of the 1 after the decimal point; therefore, $0.001 = 10^{-3}$. When writing numbers greater than 10, the exponent corresponds to the number of decimal positions to the left of the decimal point minus 1; therefore, 1124.5 = $1.245 \cdot 10^3$.

Exponents can also be used to represent **units of measure**, e.g., m^3. As in the case of 10^3, the base element (meters) is multiplied by itself three times ($m \cdot m \cdot m$; $\rightarrow$ p. 372). Negative exponents are also used to express units of measure. As with $1/10 = 10^{-1}$, 1/s can be written as s^{-1}, mol/L as $mol \cdot L^{-1}$, etc.

There are specific rules for performing calculations with powers of ten. Addition and subtraction are possible only if the exponents are identical, e.g.,
$(2.5 \cdot 10^2) + (1.5 \cdot 10^2) = 4 \cdot 10^2$.

Unequal exponents, e.g., $(2 \cdot 10^3) + (3 \cdot 10^2)$, must first be equalized:
$(2 \cdot 10^3) + (0.3 \cdot 10^3) = 2.3 \cdot 10^3$.

The exponents of the multiplicands are added together when multiplying powers of 10, and the denominator is subtracted from the numerator when dividing powers of ten.

Examples:

$10^2 \cdot 10^3 = 10^{2+3} = 10^5$
$10^4 \div 10^2 = 10^{4-2} = 10^2$
$10^2 \div 10^4 = 10^{2-4} = 10^{-2}$

The usual mathematical rules apply to the multipliers of powers of ten, e.g.,

$(3 \cdot 10^2 \cdot (2 \cdot 10^3) = (2 \cdot 3) \cdot (10^{2+3}) = 6 \cdot 10^5$.

Logarithms. There are two kinds of logarithms: common and natural. Logarithmic calculations are performed using exponents alone. The **common** (**decimal**) **logarithm** (log or lg) is the power or exponent to which 10 must be raised to equal the number in question. The common logarithm of 100 (log 100) is 2, for example, because $10^2 = 100$. Decimal logarithms are commonly used in physiology, e.g., to define pH values (see above) and to plot the pressure of sound on a decibel scale ($\rightarrow$ p. 363).

Natural logarithms (ln) have a natural base of 2.71828…, also called *base e*. The common logarithm (log x) equals the natural logarithm of x (ln x) divided by the natural logarithm of 10 (ln 10), where ln 10 = 2.302585. The following rules apply when converting between natural and common logarithms:

log x = (ln x)/2.3
ln x = 2.3 · log x.

When performing mathematical operations with logarithms, the type of operation is reduced by one rank—multiplication becomes addition, potentiation becomes multiplication, and so on.

Examples:

log (a · b) = log a + log b
log (a/b) = log a - log b
log a^n = n · log a
log $\sqrt[n]{a}$ = (log a)/n

Special cases:

log 10 = ln e = 1
log 1 = ln 1 = 0
log 0 = ln 0 = $\pm \infty$

Graphic Representation of Data

Graphic plots of data are used to provide a clear and concise representation of measurements, e.g., body temperature over the time of day ($\rightarrow$ **C**). The axes on which the measurements (e.g., temperature and time) are plotted are called **coordinates**. The vertical axis is referred to as the *ordinate* (temperature) and the horizontal axis is the *abscissa* (time). It is customary to plot the first variable x (time) on the abscissa and the other dependent variable y (temperature) on the ordinate. The abscissa is therefore called the *x-axis* and the ordinate the *y-axis*. This method of graphically plotting data can be used to illustrate the connection between any two related dimensions imaginable, e.g., to describe the relationship between height and age, lung capacity and intrapulmonary pressure, etc. ($\rightarrow$ p. 117).

Plotting of data makes it easier to determine whether two variables correlate with each other. For example, the plot of height (ordinate) over age (abscissa) shows that the height increases during the growth years and reaches a plateau at the age of about 17 years. This means that height is related to age in the first phase of life, but is largely independent of age in the second phase. A correlation does not necessarily indicate a causal relationship. A decrease in the birth rate in Alsace-Lorraine, for example, correlated with a decrease in the number or nesting storks for a while.

When plotting variables of wide-ranging dimensions (e.g., 1 to 100 000) on a coordinate

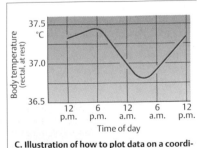

C. Illustration of how to plot data on a coordinate system. The plot in this example shows the relationship between body temperature (rectal, at rest) and time of day.

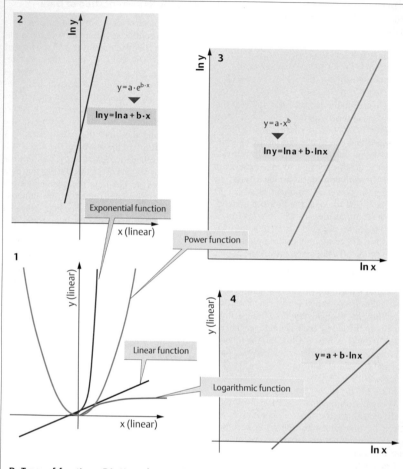

D. Types of functions. D1: Linear function (violet), exponential function (red), logarithmic function (blue), and power function (green) showing linear plotting of data on both axes. The three curves can be made into a straight line (linearized) by logarithmically plotting the data on the y-axis (**D2:** exponential function) or on the x-axis (**D4:** logarithmic function) or both (**D3:** power function).

system, it can be impossible to plot small values individually without having the axes become extremely long. This problem can be solved by plotting the data as powers of 10 or logarithms. For example, 1, 10, 100, and 1000 are written as 10^0, 10^1, 10^2, and 10^3 or as logarithms 0, 1, 2, and 3. This makes it possible to

obtain a relatively accurate graphic representation of very small numbers, and all the numbers fit on an axis of reasonable length (cf. sound curves on p. 363 B).

Correlations can be either linear or nonlinear. *Linear correlations* ($\rightarrow$ **D1**, violet line) obey the linear relationship

$$y = ax + b,$$

where a is the slope of the line and b is the point, or *intercept* (at x = 0), where it intersects the y-axis.

Many correlations are nonlinear. For simpler functions, graphic linearization can be achieved via a nonlinear (logarithmic) plot of the x and/or y values. This allows for the extrapolation of values beyond the range of measurement (see below) or for the generation of calibration curves from only two points. In addition, this method also permits the calculation of the "mean" correlation of scattered x-y pairs using *regression lines*.

An **exponential function** ($\rightarrow$ **D1**, red curve), such as

$$y = a \cdot e^{b \cdot x},$$

can be linearized by plotting $\ln y$ on the y-axis ($\rightarrow$ **D2**):

$$\ln y = \ln a + b \cdot x,$$

where b is the slope and $\ln a$ is the intercept.

A **logarithmic function** ($\rightarrow$ **D1**, blue curve), such as

$$y = a + b \cdot \ln x,$$

can be linearized by plotting $\ln x$ on the x-axis ($\rightarrow$ **D4**), where b is the slope and a is the intercept.

A **power function** ($\rightarrow$ **D1**, green curve), such as

$$y = a \cdot x^b,$$

can be graphically linearized by plotting $\ln y$ and $\ln x$ on the coordinate axes ($\rightarrow$ **D3**) because

$$\ln y = \ln a + b \cdot \ln x,$$

where b is the slope and $\ln a$ is the intercept.

Note: The condition x or y = 0 does not exist on logarithmic coordinates because $\ln 0 = \infty$. Nevertheless, $\ln a$ is still called the intercept in the equation when the logarithmic abscissa ($\rightarrow$ **D3,4**) is intercepted by the ordinate at $\ln x = 0$, i.e., x = 1.

Instead of plotting $\ln x$ and/or $\ln y$ on the x- and/or y-axis, they can be plotted on **logarithmic paper** on which the ordinate or abscissa (semi-log paper) or both coordinates (log-log paper) are plotted in logarithmic units. In such cases, a is no longer treated as the intersect because the position of a depends on site of intersection of the x-axis by the y-axis. All values > 0 are possible.

Other nonlinear functions can also be graphically linearized using an appropriate plotting method. Take, for example, the **Michaelis–Menten equation** ($\rightarrow$ **E1**), which applies to

many enzyme reactions and carrier-mediated transport processes:

$$J = J_{max} \cdot \frac{C}{K_M + C} \qquad [13.10]$$

where J is the actual rate of transport (e.g., in $mol \cdot m^{-2} \cdot s^{-1}$), J_{max} is the maximal transport rate, C ($mol \cdot m^{-3}$) is the actual concentration of the substance to be transported, and K_M is the concentration (half-saturation concentration) at $^1/_2 J_{max}$.

One of the three commonly used linear rearrangements of the Michaelis–Menten equation, the *Lineweaver–Burk plot*, states:

$$1/J = (K_M/J_{max}) \cdot (1/C) + 1/J_{max}, \qquad [13.11]$$

Consequently, a plot of $1/J$ on the y-axis and $1/C$ on the x-axis results in a straight line ($\rightarrow$ **E2**). While a plot of J over C ($\rightarrow$ **E1**) does not

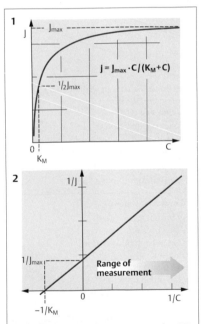

E. Two methods of representing the Michaelis–Menten equation: The data can be plotted as a curve of J over C (**E1**), or as $1/J$ over $1/C$ in linearized form (**E2**). In the latter case, J_{max} and K_M are determined by extrapolating the data outside the range of measurement.

permit accurate extrapolation of J_{max} (because an infinitely high concentration of C would be required), the linear rearrangement ($\rightarrow$ **E2**) makes it possible to generate a regression line that can be extrapolated to C = ∞ from the measured data. Since 1/C is equal to 1/∞ = 0,1/Jmax lies on the y-axis at x = 0 ($\rightarrow$ **E2**). The reciprocal of this value is J_{max}. Insertion of 1/J = 0 into equation 13.11 yields

$$0 = (K_M/J_{max}) \cdot (1/C) + 1/J_{max} \qquad [13.12]$$

or $1/K_M = -1/C$, so that K_M can be derived from the negative reciprocal of the x-axis intersect, which corresponds to 1/J = 0 ($\rightarrow$ **E2**).

The Greek Alphabet

α	A	alpha
β	B	beta
γ	Γ	gamma
δ	Δ	delta
ε	E	epsilon
ζ	Z	zeta
η	H	eta
ϑ, θ	Θ	theta
ι	I	iota
$\varkappa$	K	kappa
λ	Λ	lamda
μ	M	mu
ν	N	nu
ξ	Ξ	xi
o	O	omicron
π	Π	pi
ρ	P	rho
σ, ς	Σ	sigma
τ	T	tau
υ	Y	upsilon
ϕ	Φ	phi
χ	X	chi
ψ	Ψ	psi
ω	Ω	omega

Reference Values in Physiology

Total body and cells

Chemical composition of 1 kg fat-free body mass of an adult	720 g water, 210 g protein, 22.4 g Ca, 12 g P, 2.7 g K, 1.8 g Na, 1.8 g Cl, 0.47 g Mg
Distribution of water in adult (child) as percentage of body weight (cf. p. 168)	Intracellular: 40% (40%); interstitium: 15% (25%); plasma: 5% (5%)
Ion concentrations in ICF and ECF	See p. 93 C

Cardiovascular system

Weight of heart	250–350 g
Cardiac output at rest (maximal)	5–6 L/min (25 L/min); cf. p. 186
Resting pulse = sinus rhythm	60–75 min^{-1} or bpm
AV rhythm	40–55 min^{-1}
Ventricular rhythm	25–40 min^{-1}
Arterial blood pressure (Riva–Rocci)	120/80 mm Hg (16/10.7 kPa) systolic/diastolic
Pulmonary artery pressure	20/7 mm Hg (2.7/0.9 kPa) systolic/diastolic
Central venous pressure	3–6 mm Hg (0.4–0.8 kPa)
Portal venous pressure	3–6 mm Hg (0.4–0.8 kPa)
Ventricular volume at end of diastole/systole	120 mL/40 mL
Ejection fraction	0.67
Pressure pulse wave velocity	Aorta: 3–5 m/s; arteries: 5–10 m/s; veins: 1–2 m/s
Mean velocity of blood flow	Aorta: 0.18 m/s; capillaries: 0.0002–0.001 m/s; venae cavae: 0.06 m/s

Blood flow in organs at rest

(*See also pp. 187 A, 213 A*)	% of cardiac output	per gram of tissue
Heart	4%	0.8 mL/min
Brain	13%	0.5 mL/min
Kidneys	20%	4 mL/min
GI tract (drained by portal venous system)	16%	0.7 mL/min
Liver (blood supplied by hepatic artery)	8%	0.3 mL/min
Skeletal muscle	21%	0.04 mL/min
Skin and miscellaneous organs	18%	—

Lungs and gas transport

	Men	Women
Total lung capacity (TLC)	7 L	6.2 L
Vital capacity (VC); cf. p. 112	5.6 L	5 L
Tidal volume (V_T) at rest	0.6 L	0.5 L
Inspiratory reserve volume	3.2 L	2.9 L
Expiratory reserve volume	1.8 L	1.6 L
Residual volume	1.4 L	1.2 L
Max. breathing capacity in 30 breaths/min	110 L	100 L
Partial pressure of O_2	Air: 21.17 kPa	(159 mm Hg)
	Alveolar: 13.33 kPa	(100 mm Hg)
	Arterial: 12.66 kPa	(95 mm Hg)
	Venous 5.33 kPa	(40 mm Hg)
Partial pressure of CO_2	Air: 0.03 kPa	(0.23 mm Hg)
	Alveolar: 5.2 kPa	(39 mm Hg)
	Arterial: 5.3 kPa	(40 mm Hg)
	Venous: 6.1 kPa	(46 mm Hg)
Respiratory rate (at rest)	16 breaths/min	
Dead space volume	150 mL	
Oxygen capacity of blood	180–200 mL O_2/L blood = 8–9 mmol O_2/L blood	
Respiratory quotient	0.84 (0.7–1.0)	

Kidney and excretion

Renal plasma flow (RPF)	480–800 mL/min per 1.73 m² body surface area
Glomerular filtration rate (GFR)	80–140 mL/min per 1.73 m² body surface area
Filtration fraction (GFR/RPF)	0.19
Urinary output	0.7–1.8 L/day
Osmolality of urine	250–1000 mOsm/kg H_2O
Na^+ excretion	50–250 mmol/day
K^+ excretion	25–115 mmol/day
Glucose excretion	< 300 mg/day = 1.67 mmol/day
Nitrogen excretion	150–250 mg/kg/day
Protein excretion	10–200 mg/day
Urine pH	4.5–8.2
Titratable acidity	10–30 mmol/day
Urea excretion	10–20 g/day = 166–333 mmol/day
Uric acid excretion	300–800 mg/day = 1.78–6.53 mmol/day
Creatinine excretion	0.56–2.1 g/day = 4.95–18.6 µg/day

Nutrition and metabolism	Men	Women
Energy expenditure during various activities		
• Bed rest	6500 kJ/d	5400 kJ/d
	(1550 kcal/d)	(1300 kcal/d)
• Light office work	10 800 kJ/d	9600 kJ/d
	(2600 kcal/d)	(2300 kcal/d)
• Walking (4.9 km/h)	3.3 kW	2.7 kW
• Sports (dancing, horseback riding, swimming)	4.5–6.8 kW	3.6–5.4 kW
Functional protein minimum	1 g/kg body weight	
Vitamins, optimal daily intake	A: 10 000–50 000 IU; D: 400–600 IU	
(IU = international units)	E: 200–800 IU; K: 65–80 µg; B_1, B_2, B_5, B_6:	
	25–300 mg of each; B_{12}: 25–300 µg; folate:	
	0.4–1.2 mg; H: 25–300 µg; C: 500–5000 mg	
Electrolytes and trace elements,	Ca: 1–1.5 g; Cr: 200–600 µg; Cu: 0.5–2 mg;	
optimal daily intake	Fe: 15–30 mg; I: 50–300 µg; K^+: 0.8–1.5 g;	
	Mg: 500–750 mg; Mn: 15–30 mg;	
	Mo: 45–500 µg; Na^+: 2 g; P: 200–400 mg;	
	Se: 50–400 µg; Zn: 22–50 mg	

Nervous Systems, muscles		
Duration of an action potential	Nerve: 1–2 ms; skeletal muscle: 10 ms; myocardium: 200 ms	
Nerve conduction rate	See p. 49 C	

Blood and other bodily fluids	*(see also Tables 9.1, and 9.2 on pp. 186, 187)*	
Blood (in adults)	**Men:**	**Women:**
Blood volume (also refer to table on p. 88)	4500 mL	3600 mL
Hematocrit	0.40–0.54	0.37–0.47
Red cell count (RBC)	$4.5–5.9 \cdot 10^{12}$/L	$4.2–5.4 \cdot 10^{12}$/L
Hemoglobin (Hb) in whole blood	140–180 g/L	120–160 g/L
	(2.2–2.8 mmol/L)	(1.9–2.5 mmol/L)
Mean corpuscular volume (MCV)	80–100 fL	
Mean corpuscular Hb concentration (MCHC)	320–360 g/L	
Mean Hb concentration in single RBC (MCH)	27–32 pg	
Mean RBC diameter	7.2–7.8 µm	
Reticulocytes	0.4–2% (20–75 · 10^9/L)	
Leukocytes (also refer to table on p. 88)	3–11 · 10^9/L	
Platelets	170–360 · 10^9/L	180–400 · 10^9/L
Erythrocyte sedimentation rate (ESR)	< 10 mm in first hour	< 20 mm in first hour
Proteins		
Total	66–85 g/L serum	
Albumin	35–50 g/L serum	55–64 % of total
α_1-globulins	1.3–4 g/L serum	2.5–4 % of total
α_2-globulins	4–9 g/L serum	7–10 % of total
β-globulins	6–11 g/L serum	8–12 % of total
γ-globulins	13–17 g/L serum	12–20 % of total
Coagulation	*(See p. 102 for coagulation factors)*	
Thromboplastin time (Quick)	0.9–1.15 INR (international normalized ratio)	
Partial thromboplastin time	26–42 s	
Bleeding time	< 6 min	

Parameters of glucose metabolism

Glucose concentration in venous blood	3.9–5.5 mmol/L	(70–100 mg/dL)
Glucose concentration in capillary blood	4.4–6.1 mmol/L	(80–110 mg/dL)
Glucose concentration in plasma	4.2–6.4 mmol/L	(75–115 mg/dL)
Limit for diabetes mellitus in plasma	> 7.8 mmol/L	(> 140 mg/dL)
HBA_{1c} (glycosylated hemoglobin A)	3.2–5.2%	

Parameters of lipid metabolism

Triglycerides in serum	< 1.71 mmol/L	(< 150 mg/dL)
Total cholesterol in serum	< 5.2 mmol/L	(< 200 mg/dL)
HDL cholesterol in serum	> 1.04 mmol/L	(> 40 mg/dL)

Substances excreted in urine

Urea concentration in serum	3.3–8.3 mmol/L	(20–50 mg/dL)
Uric acid concentration in serum	150–390 µmol/L	(2.6–6.5 mg/dL)
Creatinine concentration in serum	36–106 µmol/L	(0.4–1.2 mg/dL)

Bilirubin

Total bilirubin in serum	3.4–17 µmol/L	(0.2–1 mg/dL)
Direct bilirubin in serum	0.8–5.1 µmol/L	(0.05–0.3 mg/dL)

Electrolytes and blood gases

Osmolality	280–300 mmol/kg H_2O	
Cations in serum:	Na^+	135–145 mmol/L
	K^+	3.5–5.5 mmol/L
	Ionized Ca^{2+}	1.0–1.3 mmol/L
	Ionized Mg^{2+}	0.5–0.7 mmol/L
Anions in serum:	Cl^-	95–108 mmol/L
	$H_2PO_4^- + HPO_4^{2-}$	0.8–1.5 mmol/L
pH		7.35–7.45
Standard bicarbonate		22–26 mmol/L
TotaL buffer bases		48 mmol/L
Oxygen saturation		96% arterial; 65–75% mixed venous
Partial pressure of O_2 at half saturation ($P_{0.5}$)		3.6 kPa (27 mmHg)

Cerebrospinal fluid

Pressure in relaxed horizontal position	1.4 kPa (10.5 mmHg)
Specific weight	1.006–1.008 g/L
Osmolality	290 mOsm/kg H_2O
Glucose concentration	45–70 mg/dL (2.5–3.9 mmol/L)
Protein concentration	0.15–0.45 g/L
IgG concentration	< 84 mg/dL
White cell count	< 5 WBC/µL

Important Equations in Physiology

1. Fick's law of diffusion for membrane transport (see also p. 20ff.)

$$J_{diff} = F \cdot D \cdot \frac{\Delta C}{\Delta x} \; [mol \cdot s^{-1}]$$

J_{diff} = net diffusion rate [$mol \cdot s^{-1}$];
A = area [m^2];
D = diffusion coefficient [$m^2 \cdot s^{-1}$];
ΔC = concentration difference [$mol \cdot m^{-3}$];
Δx = membrane thickness [m]

Alternative 1:

$$\frac{J_{diff}}{F} = P \cdot \Delta C \; [mol \cdot m^{-2} \cdot s^{-1}]$$

P = permeability coefficient [$m \cdot s^{-1}$];
J_{diff}, A an ΔC; see above

Alternative 2 (for gas diffusion)

$$\frac{\dot{V}_{diff}}{F} = K \cdot \frac{\Delta P}{\Delta x} \; [m \cdot s^{-1}]$$

$\dot{V}_{diff}$ = net diffusion rate [$m^3 \cdot s^{-1}$];
K = Krogh's diffusion coefficient [$m^2 \cdot s^{-1} \cdot Pa^{-1}$]
ΔP = partial pressure difference [Pa]

2. Van't Hoff–Stavermann equation (see also p. 377)

$$\Delta \pi = \sigma \cdot R \cdot T \cdot \Delta c_{osm} \; [Pa]$$

$\Delta \pi$ = osmotic pressure difference [Pa]
σ = reflection coefficient [dimensionless]
R = universal gas constant [$8.3144 J \cdot K^{-1} \cdot mol^{-1}$];
T = absolute temperature [K];
Δc_{osm} = concentration difference of osmotically active particles [$mol \cdot m^{-3}$].

3. Michaelis-Menten equation (see also pp. 28, 383ff.)

$$J_{sat} = J_{max} \cdot \frac{C}{K_M + C} \; [mol \cdot m^{-2} \cdot s^{-1}],$$

J_{sat} = substrate transport (turnover) [$mol \cdot m^{-2} \cdot s^{-1}$];
J_{max} = maximum substrate transport (turnover) [$mol \cdot m^{-2} \cdot s^{-1}$];
C = substrate concentration [$mol \cdot m^{-3}$];
K_M = Michaelis constant = substrate concentration at $^1/_2 J_{max}$ [$mol \cdot m^{-3}$].

4. Nernst equation (see also p. 32)

$$E_x = -61 \cdot z_x^{-1} \cdot \log \frac{[X]_i}{[X]_a} \; [mV]$$

E_x = equilibrium potential of ion X [mV];
z_x = valency of ion X;
$[X]_i$ = intracellular concentration of ion X [$mol \cdot m^{-3}$]
$[X_o]$ = extracellular concentration of ion X [$mol \cdot m^{-3}$].

5. Ohm's Law (see also pp. 32, 188)

a. *For ion transport at membrane*

$$I_x = g_x \cdot (E_m - E_x) \; [A \cdot m^{-2}]$$

I_x = ionic current of ion X per unit area of membrane [$A \cdot m^{-2}$];
g_x = conductance of membrane to ion X [$S \cdot m^{-2}$];
E_m = membrane potential [V];
E_x = equilibrium potential of ion X [V]

b. *For blood flow:*

$$\dot{Q} = \frac{\Delta P}{R} \; [L \cdot min^{-1}]$$

$\dot{Q}$ = flow rate (total circulation: cardiac output, CO) [$L \cdot min^{-1}$]
ΔP = mean blood pressure difference (systemic circulation: $\overline{P}_{aorta} - \overline{P}_{vena\;cava}$; lesser circulation: $\overline{P}_{pulmonary\;artery} - \overline{P}_{pulmonary\;vein}$) [mmHg]
R = flow resistance (systemic circulation: total peripheral resistance = TPR) [$mmHg \cdot min \cdot L^{-1}$].

6. Respiration-related equations (see also pp. 106, 120)

a. *Tidal volume (V_T):*

$$V_T = V_D + V_A \; [L]$$

b. *Respiratory volume per minute ($\dot{V}_E$ oder $\dot{V}_T$):*

$$\dot{V}_E = f \cdot V_T = (f \cdot V_D) + (f \cdot V_A) = \dot{V}_D + \dot{V}_A \; [L \cdot min^{-1}]$$

c. *O_2 consumption, CO_2 emission, and RQ (total body:)*

$$\dot{V}_{O_2} = \dot{V}_T \, (F_{IO_2} - F_{EO_2}) = CO \cdot avD_{O_2} \; [L \cdot min^{-1}]$$
$$\dot{V}_{CO_2} = \dot{V}_T \cdot F_{ECO_2} \; [L \cdot min^{-1}]$$
$$RQ = \frac{\dot{V}_{CO_2}}{\dot{V}_{O_2}}$$

V_D = dead space [L];
V_A = alveolar fraction of V_T [L];
f = respiration rate [min^{-1}];
$\dot{V}_D$ = dead space ventilation [L · min^{-1}];
$\dot{V}_A$ = alveolar ventilation [L · min^{-1}];
$\dot{V}_{O_2}$ = O_2 consumption [L · min^{-1}];
$\dot{V}_{CO_2}$ = CO_2 emission [L · min^{-1}];
F_{IO_2} = inspiratory O_2 fraction [L/L];
F_{EO_2} = exspiratory O_2 fraction [L/L];
F_{ECO_2} = exspiratory CO_2 fraction [L/L];
RQ = respiratory quotient (dimensionless)

d. *O_2 consumption and CO_2 emission (organ):*
$\dot{V}_{O_2} = \dot{Q} \cdot {}_{av}D_{O_2}$ [L · min^{-1}]
$\dot{V}_{CO_2} = \dot{Q} \cdot {}_{av}D_{CO_2}$ [L · min^{-1}]

$\dot{Q}$ = blood flow in organ [L · min^{-1}]
${}_{av}D_{O_2}$, ${}_{av}D_{CO_2}$ = arteriovenous
O_2 and CO_2 difference in total circulation
and organ circulation [L/L blood]

e. *Fick's principle:*
$$CO = \frac{\dot{V}_{O_2}}{{}_{av}D_{O_2}} \text{ [L · min}^{-1}]$$
CO = cardiac output [L · min^{-1}]>

f. *Gas partial pressure $\leftrightarrow$ gas concentration in liquids:*
$[X] = \alpha \cdot P_x$ [mmol/L plasma]

[X] = concentration of gas X [mmol · L^{-1}]
α = (Bunsen's) solubility coefficient
[mmol · L^{-1} · kPa^{-1}]
P_X = partial pressure of gas X [kPa]

g. *Bohr's formula* (see also p. 115)
$$V_D = V_T \frac{(F_{ACO_2} - F_{ECO_2})}{F_{ACO_2}}$$
V_D = dead space [L];
V_T = tidal volume [L];
F_{ACO_2} = alveolar CO_2 fraction
F_{ECO_2} = exspiratory CO_2 fraction [L/L]

h. *Alveolar gas equation* (see also p. 136)
$$P_{AO_2} = P_{IO_2} - \frac{P_{AO_2}}{RQ} \text{ [kPa]}$$
P_{AO_2} and P_{IO_2} = alveolar and inspiratory partial pressure of O_2 [kPa]
P_{ACO_2} = alveolar partial pressure of CO_2 [kPa]
RQ = respiratory quotient [dimensionless].

Henderson-Hasselbalch equation
(see also pp. 138ff., 379)
a. *General equation:*
$$pH = pK_a + \log \frac{[A^-]}{[AH]}$$

b. *for bicarbonate/CO_2 buffer (37 °C):*
$$pH = 6,1 + \frac{[HCO_3^-]}{\alpha \cdot P_{CO_2}}$$

pH = negative common logarithm of
H$^+$ activity
pK$_a$ = negative common logarithm of
dissociation constant of buffer acid in
denominator (AH or CO_2)
[A$^-$] and [HCO$_3^-$] = buffer base concentration; $\alpha \cdot P_{CO_2}$ = [CO_2]; see Equation 6f.

8. Equations for renal function
(see also p. 150ff.)

a. *Clearance of a freely filtrable substance X (C_X):*
$$C_X = \dot{V}_U \cdot \frac{U_X}{P_X} \text{ [L · min}^{-1}]$$

b. *Renal plasma flow*
$$RPF = \dot{V}_U \cdot \frac{U_{PAH}}{0,9 \cdot P_{PAH}} \text{ [L · min}^{-1}]$$

c. *Renal blood flow (RBF):*
$$RBF = \frac{RPF}{1 - HCT} \text{ [L · min}^{-1}]$$

d. *Glomerular filtration rate* (GFR):
$$GFR = \dot{V}_U \cdot \frac{U_{In}}{P_{In}} \text{ [L · min}^{-1}]$$

e. *Free water clearance* (C_{H_2O})
$$C_{H_2O} = \dot{V}_U \cdot \left(1 - \frac{U_{osm}}{P_{osm}}\right) \text{ [L · min}^{-1}]$$

f. *Filtration fraction*
$$FF = \frac{GFR}{RPF} \text{ [dimensionless]}$$

g. *Fractional excretion of substance X (FE_X):*
$$FE_X = \frac{C_X}{GFR} \text{ [dimensionless]};$$

h. *Fractional reabsorption of substance X (FR_X):*
$$FR_X = 1 - FE_X \text{ [dimensionslos]};$$

$\dot{V}_U$ = urinary excretion rate [L · min^{-1}]
U_X, U_{PAH}, U_{In} = urinary concentration of
substance X, para-aminohippuric acid, and
indicator (e.g., inulin, endogenous creatinine) [mol · L^{-1}] or [g · L^{-1}]

U_{osm} = osmolality of urine [$Osm \cdot L^{-1}$]
P_X, P_{PAH}, P_{In} = plasma concentration of substance X, para-aminohippuric acid, and indicator (e.g., inulin, endogenous creatinine) [$mol \cdot L^{-1}$] or [$g \cdot L^{-1}$]
P_{osm} = osmolality of plasma [$Osm \cdot L^{-1}$]
HCT = hematocrit [L of blood cells/L of blood]

9. Equations for filtration
(see also pp. 152, 208)

a. *Effective filtration pressure at capillaries* (P_{eff})
$$P_{eff} = P_{cap} - P_{int} - \pi_{cap} + \pi_{int} \ [mmHg]$$

b. *Effective filtration pressure at capillaries of renal glomerulus:*
$$P_{eff} = P_{cap} - P_{Bow} - \pi_{cap} \ [mmHg]$$

c. *Filtration rate ($\dot{Q}$ at glomerulus = GFR)*
$$\dot{Q} = P_{eff} \cdot F \cdot k \ [m^3 \cdot s^{-1}]$$

P_{cap} (P_{int}) = hydrostatic pressure in capillaries (interstitium) [mm Hg]
π_{cap} (π_{int}) = oncotic pressure in capillaries (interstitium) [mm Hg]
P_{eff} = mean effective filtration pressure [mm Hg]
A = area of filtration (m^2)
k = permeability to water (hydraulic conductance) [$m^3 \cdot s^{-1} \cdot mm \ Hg^{-1}$]

10. Law of Laplace
(see also pp. 118, 188, and 210)

a. *Elliptical hollow body (with radii r_1 and r_2)*
$$P_{tm} = T \left(\frac{1}{r_1} + \frac{1}{r_2} \right) \ [Pa];$$

b. *Elliptical hollow body, considering wall thickness:*
$$P_{tm} = S \cdot w \left(\frac{1}{r_1} + \frac{1}{r_2} \right) \ [Pa];$$

c. *Spherical hollow body ($r_1 = r_2 = r$):*
$$P_{tm} = 2 \frac{T}{r} \ [Pa] \ or \ P_{tm} = 2 \frac{S \cdot w}{r} \ [Pa];$$

d. *Cylindrical hollow body ($r_2 \to \infty$, therefore $1/r_2 = 0$):*
$$P_{tm} = \frac{T}{r} \ [Pa] \ or \ P_{tm} = \frac{S \cdot w}{r} \ [Pa]$$

P_{tm} = transmural pressure [Pa]
T = wall tension [$N \cdot m^{-1}$]
S = wall tension [$N \cdot m^{-2}$]
w = wall thickness [m]

11. Equations for cardiovascular function
(see also items 2, 5b, 6c, and 9 as well as p. 186ff.)

a. *Cardiac output (CO):*
$$CO = f \cdot SV \ [l \cdot min^{-1}]$$

b. *Hagen–Poiseuille equation*
$$R = \frac{8 \cdot l \cdot \eta}{\pi \cdot r^4} \ ;$$

f = heart rate [min^{-1}]
SV = stroke volume [L]
R = flow resistance in a tube [$Pa \cdot s \cdot m^{-3}$] of known length l [m] and inner radius r [m]
η = viscosity [$Pa \cdot s$]

Further Reading

Physiology

Berne RM, Levy MN, Koeppen BM, Stanton BA. *Physiology*. 4th ed. St Louis: Mosby; 1998.

Berne RM, Levy MN. *Principles of Physiology*. 3rd ed. St Louis: Mosby; 2000.

Greger R, Windhorst U. *Comprehensive Human Physiology*. Berlin: Springer; 1996.

Guyton AC, Hall JE. *Textbook of Medical Physiology*. 10th ed. Philadelphia: Saunders; 2000.

Vander AJ, Sherman JH, Luciano DS. *Human Physiology: The Mechanisms of Body Function*. 8h ed. New York: McGraw-Hill, 2000.

Cell Physiology

Alberts B, Bray D, Johnson A, et al. *Essential Cell Biology: An Introduction to the Molecular Biology of the Cell*. New York: Garland Publishers; 1998.

Byrne, JH, Schultz, SG. *An Introduction to Membrane Transport and Bioelectricity*. 2nd ed. New York: Raven Press; 1994.

Lodish H, Baltimore D, Berk A, Zipursky SL, Matsudaira P, Darnell J. *Molecular Cell Biology*. 3rd ed. New York: Scientific American Books; 1995.

Neurophysiology, Muscle

Bagshaw CR. Muscle Contraction. 2nd ed. London; New York: Chapman & Hall; 1993.

Bear MF, Connors BW, Paradiso MA. *Neuroscience—Exploring the Brain*. 2nd ed. Philadelphia: Lippincott Williams & Wilkins; 2001.

Kandel ER, Schwartz JH, Jessell TM. *Principles of Neural Science*. 4th ed. New York: McGraw-Hill; 2000.

Sensory Physiology

Doty RL. *Handbook of Olfaction and Gustation*. New York: Dekker; 1995

Kandel ER, Schwartz JH, Jessell TM. *Principles of Neural Science*. 4th ed. New York: McGraw-Hill; 2000.

Moore BCJ. *Hearing*. 2nd ed. London: Academic Press; 1995

Immunology

Goldsby RA, Kindt TJ, Osborne BA. *Kuby Immunology*, 4th ed. New York: W.H. Freeman; 2000

Respiration

Crystal RG, West JB, Barnes PJ. *The Lung: Scientific Foundations*. 2nd ed. Philadelphia: Lippincott Williams & Wilkins; 1997.

West JB. *Pulmonary Physiology and Pathophysiology: An Integrated, Case-based Approach*. Philadelphia: Lippincott Williams & Wilkins; 2001.

Exercise and Work Physiology

McArdle WD, Katch FI, Katch VL. *Exercise Physiology: Energy, Nutrition, and Human Performance*. 5th ed. Philadelphia: Lippincott Williams & Wilkins; 2001.

Wilmore JH, Costill DL. *Physiology of Sport and Exercise*. 2nd ed. Champaign, IL : Human Kinetics; 1999.

High Altitude Physiology

Ward MP, Milledge JS, West JB. *High Altitude Medicine and Physiology*. 3rd ed. London: Arnold; New York: co-published by Oxford University Press; 2000.

Heart and Circulation

Alexander RW, Schlant RC, Fuster V, eds. *Hurst's the Heart, Arteries and Veins*. 9th ed. New York: McGraw-Hill; 1998.

Katz AM. *Physiology of the Heart*. 3rd ed. Philadelphia: Lippincott Williams & Wilkins; 2001.

Renal, Electrolyte and Acid-Base Physiology

Halperin ML, Goldstein MB. *Fluid, Electrolyte, and Acid-Base Physiology: A Problem-based Approach*. 3rd. ed. Philadelphia: W.B. Saunders Co; 1999.

Schnermann JB, Sayegh SI. *Kidney Physiology*. Philadelphia: Lippincott-Raven; 1998.

Seldin DW, Giebisch G. *The Kidney: Physiology and Pathophysiology*. 3rd ed. Philadelphia: Lippincott Williams & Wilkins; 2000.

Valtin H, Schafer JA. *Renal Function: Mechanisms Preserving Fluid and Solute Balance in Health*. 3rd ed. Boston: Little, Brown, and Co; 1995.

Gastrointestinal Physiology

Arias IM, Boyer JL, Fausto N, Jakoby WB, Schachter D, Shafritz DA. *The Liver: Biology and Pathobiology*. 3rd ed. New York: Raven Press; 1994.

Johnson, LR. *Physiology of the Gastrointestinal Tract*. 3rd ed. New York: Raven Press; 1994.

Johnson LR, Gerwin TA. *Gastrointestinal Physiology*. 6th ed. St Louis: Mosby; 2001.

Endocrinology

Greenspan FS, Gardner DG. *Basic and Clinical Endocrinology*. 6th ed. Stamford, CT: Appleton & Lange; 2000.

Kacsoh B. *Endocrine Physiology*. Stamford, CT: Appleton & Lange; 2000.

Larsen PR, Williams RH, Wilson JD, Foster DW, Kronenberg HM. *Williams Textbook of Endocrinology*. 10th ed. Philadelphia: W.B. Saunders Co; 2002.

Reproduction and Sexual Physiology

Knobil E, Neill JD. *The Physiology of Reproduction*. 2nd ed. New York: Raven Press; 1994.

Masters WH, Johnson VE, Kolodny RC. *Human Sexuality*. 5th ed. New York: HarperCollins College Publishers; 1995.

Yen, SS, Jaffe RB. *Reproductive Endocrinology: Physiology, Pathophysiology, and Clinical Management*. 4th ed. Philadelphia: W.B. Saunders Co; 1999.

Animal Physiology

Land MF, Nilsson D-E. *Animal Eyes*. New York: Oxford University Press; 2002.

Randall DJ, Burggren WW, French K. *Eckert Animal Physiology: Mechanisms and Adaptations*. 5th ed. New York: W.H. Freeman and Co; 2001.

Schmidt-Nielsen K. *How Animals Work*. Cambridge, England: Cambridge University Press; 1972.

Schmidt-Nielsen K. *Animal Physiolog: Adaptation and Environment*. Cambridge, England; New York: Cambridge University Press; 1997.

Schmidt-Nielsen K, Davis KK. *The Camel's Nose: Memoirs of a Curious Scientist*. Washington, DC: Island Press; 1998.

Pathophysiology/Pathology/ Pharmacology

Bennett JC, Plum F. *Cecil Textbook of Medicine*. 20th ed. Philadelphia: W. B. Saunders Co; 1996.

Cotran RS, Kumar V, Collins T. *Robbins Pathologic Basis of Disease*. 6th ed. Philadelphia: W. B. Saunders Co; 1999

Hardman JG, Limbird L. *Goodman & Gilman's the Pharmacological Basis of Therapeutics*. 10th ed. New York: McGraw-Hill; 2001.

McCance KL, Huether SE. *Pathophysiology: The Biologic Basis for Disease in Adults & Children*. 4th ed. St Louis: Mosby; 2002.

McPhee S, Lingappa VR, Ganong W, Rosen A. *Pathophysiology of Disease: An Introduction to Clinical Medicine*. 3rd ed. Stamford, CT: Appleton & Lange; 1999.

Silbernagl S, Lang F. *Color Atlas of Pathophysiology*. New York: Thieme; 2000.

Dictionaries

Anderson DM. *Mosby's Medical, Nursing & Allied Health Dictionary*. 6th ed. St Louis: Mosby; 2002.

Dorland's Illustrated Medical Dictionary. 29th ed. Philadelphia: W.B. Saunders Co; 2000.

Smith AD, Datta SP, Smith G, Campbell PN, Bentley R, McKenzie HA. *Oxford Dictionary of Biochemistry and Molecular Biology*. Oxford, England; New York: Oxford University Press; 1997.

Stedman TL. *Stedman's Medical Dictionary*. 27th ed. Philadelphia: Lippincott Williams & Wilkins; 2000.

Venes D, Thomas CL. *Taber's Cyclopedic Medical Dictionary*. 19th ed. Philadelphia: F.A. Davis Co; 2001.

Index, Abbreviations

Index, Abbreviations

Index, Abbreviations

Index, Abbreviations

Index, Abbreviations

Index, Abbreviations

QM LIBRARY
(MILE END)